SCHAUM'S SOLVED PROBLEMS SERIES

3000 SOLVED PROBLEMS IN

PRECALCULUS

by

Philip Schmidt, Ph.D.
State University of New York
at New Paltz

McGRAW-HILL BOOK COMPANY
New York   St. Louis   San Francisco   Auckland   Bogotá   Caracas
Colorado Springs   Hamburg   Lisbon   London   Madrid   Mexico
Milan   Montreal   New Delhi   Oklahoma City   Panama   Paris
San Juan   São Paulo   Singapore   Sydney   Tokyo   Toronto

❚ Philip Schmidt, Ph.D., *Associate Professor of Secondary Education, SUNY at New Paltz, New York*

Dr. Schmidt has a B.S. from Brooklyn College (with a major in mathematics) and an M.A. in mathematics and a Ph.D. in mathematics education from Syracuse University. He was an associate professor at Berea College until 1985.

Other Contributors to This Volume

❚ C. B. Allendoerfer, Ph.D., *University of Washington*

❚ Frank Ayres, Jr., Ph.D., *Dickinson College*

❚ Raymond A. Barnett, Ph.D., *Merritt College*

❚ Gordon Fuller, Ph.D., *Texas Technical University*

❚ Thomas Koshy, Ph.D., *Framingham State College*

❚ Cletus O. Oakley, Ph.D., *Haverford College*

❚ Charles Rees, Ph.D., *University of New Orleans*

❚ Paul K. Rees, Ph.D., *Louisiana State University*

❚ Fred Safier, Ph.D., *City College of San Francisco*

❚ Fred Sparks, Ph.D., *Texas Technical University*

Project supervision was done by The Total Book.

Library of Congress Cataloging-in-Publication Data

Schmidt, Philip A.
   3000 solved problems in precalculus / by Philip Schmidt.
      p.   cm.—(Schaum's solved problems series)
   ISBN 0-07-055365-3
   1. Functions—Problems, exercises, etc. 2. Algebra—Graphic
methods—Problems, exercises, etc. I. Title. II. Title: Three
thousand solved problems in precalculus. III. Title: 3000 solved
problems in precalculus. IV. Title: Precalculus. V. Series.
QA331.3.S28 1988
515—dc19   88-18606
                                                                CIP

1 2 3 4 5 6 7 8 9 0 SHP/SHP 8 9 3 2 1 0 9 8

0-07-055365-3

CONTENTS

To the Student

There are many ways in which you can exploit this book. If you are currently enrolled in a precalculus course, use the table of contents to locate problems for the topic that you are currently studying. Notice that for the most important (and most common!) problem types, multiple examples are given. Read through several carefully, and then try to do the rest without looking at the solution given. Check yourself with the solutions given. Keep in mind that for any topic, you should do as many problems as possible. You may not be able to solve them all, for they range from very straightforward to moderate to difficult and very difficult. You will find that problem-solving techniques are scattered throughout the book: techniques for solving particular kinds of problems, as well as general mathematical techniques.

If you are using this book as a review (or study aid) prior to enrolling in a calculus course, be aware that it can also help you review (or learn, or relearn) topics in precalculus *while you're in* the calculus course.

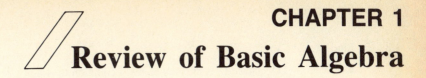

CHAPTER 1
Review of Basic Algebra

1.1 SETS AND THE NATURAL NUMBERS

1.1 Use the listing method to describe the set $\{x|x$ is a vowel in the word "consonant"$\}$.

▮ The word "consonant" contains the vowels "o" and "a." (*Recall*: We do not list the element "o" twice in the set even though it occurs twice in "consonant.") Thus, the set we are looking for is $\{o, a\}$.

1.2 Use set-builder notation to designate the set $\{1, 3, 5, 7, 9\}$.

▮ The numbers $1, 3, 5, 7, 9$ can easily be described as the odd integers that are greater than or equal to 1, and less than or equal to 9. In set-builder notation, the set is $\{x|1 \le x \le 9,$ and x is an odd integer$\}$. For Probs. 1.3, 1.4, and 1.5, let $A = \{1, 3, 5, 9, 14\}$ and $B = \emptyset$.

1.3 Find $A \cup B$.

▮ Remember that, by definition, $A \cup B = \{x|x \in A$ or $x \in B\}$. In this case, $1, 3, 5, 9$ and 14 belong to A, and B contains no elements. So, $A \cup B = \{1, 3, 5, 9, 14\}$.

1.4 Find $A \cap B$.

▮ $A \cap B = \{x|x \in A$ and $x \in B\}$. In this case, $1, 3, 5, 9$, and 14 belong to A, and there is nothing in B, so no element can belong to both sets. $A \cap B = \emptyset$.

1.5 Find $A - B$.

▮ $A - B = \{x|x \in A$ and $x \notin B\}$. Since, in this example, there are no elements in B, $A - B = \{x|x \in A\} = \{1, 3, 5, 9, 14\}$. (See Fig. 1.1.)

For Probs. 1.6, 1.7, and 1.8, let $A = \{1, 7, 11, 13\}$, $B = \{7, 14\}$, and $C = \{11, 18\}$.

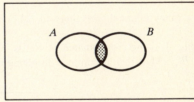

$A \cup B$ is shaded

$A \cap B$ is shaded

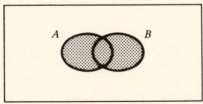

$A - B$ is shaded

Fig. 1.1

1.6 Find $(A - B) \cup C$.

▮ $A - B = \{x|x \in A$ and $x \notin B\} = \{1, 11, 13\}$ and $C = \{11, 18\}$. Thus, $(A - B) \cup C = \{1, 11, 13, 18\}$.

1.7 Find $B \times C$.

▮ Recall that $B \times C = \{(b, c)|b \in B$ and $c \in C\}$. In this example then, $B \times C = \{(7, 11), (7, 18), (14, 11), (14, 18)\}$.

1.8   Find $(A \cap C) \times C$.

    ❚ $A \cap C = \{11\}$ (since 11 is the only element common to A and C). $(A \cap C) \times C = \{11\} \times \{11, 18\} = \{(11, 11), (11, 18)\}$.

1.9   Let $A = \{1, 4\}$ and $B = \{1, 4, 6\}$.   Is $A \subseteq B$?   Is $A \subset B$?

    ❚   To determine whether $A \subseteq B$ ("A is a subset of B"), we need to find out whether every element of A is also an element of B.   In this case, A contains 1 and 4, and both of these are elements of B, so $A \subseteq B$. To determine whether $A \subset B$ ("A is a proper subset of B"), we need to find out if $A \subseteq B$ and $A \neq B$.
     In this example, we already have determined that $A \subseteq B$,   and since $A = \{1, 4\}$ and $B = \{1, 4, 6\}$, $A \neq B$.   So, $A \subset B$.

For Probs. 1.10 to 1.13, refer to Fig. 1.2.

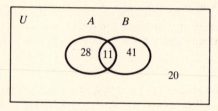

Fig. 1.2

1.10   How many elements are there in set B?

    ❚   We notice that set B contains 41 elements that are not in set A and shares 11 other elements with A. We conclude that B contains 52 elements.

1.11   How many elements are there in B'?

    ❚ B'(sometimes written as $\mathscr{C}B$) $= \{x \in U \,|\, x \notin B\}$,   where U is the universal set.   Here, U contains $28 + 11 + 41 + 20 = 100$ elements, and B contains 52 elements. Thus, B' contains 48 elements.

1.12   How many elements are there in $A \cap B$? $(A \cap B)'$?

    ❚   From the Venn diagram, we can easily see that $A \cap B$ contains 11 elements.   Thus, $(A \cap B)' = \mathscr{C}(A \cap B) = \{x \in U \,|\, x \notin A \cap B\}$ contains $100 - 11 = 89$ elements.

1.13   ❚   How many elements are there in $A \cap B'$?

    ❚   In Fig. 1.3, the crosshatched region is $A \cap B'$,   that is, the region containing those elements common to A and B'.   That region contains 28 elements (refer to Fig. 1.2).

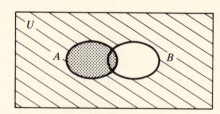

Fig. 1.3

1.14   How many subsets does $\{a, b\}$ contain? Exhibit these subsets.

    ❚   If set A contains n elements, where n is a whole number, then the number of subsets of $A = 2^n$.   In this case, $n = 2$,   so there are $2^2 = 4$ subsets.   They are $\{a, b\}, \emptyset, \{a\}, \{b\}$.

1.15   ❚   Suppose that A contains m elements and B contains n elements, where m and n are natural numbers. Find the number of subsets of $A \times B$.

    ❚ $A \times B$ will contain $m \cdot n$ elements (since each element of A must be paired with each element of B). So, $A \times B$ will contain 2^{mn} subsets. For Probs. 1.16 to 1.18, answer true or false, and explain your answer.

1.16   If $M \cap N = N$, then $N \subseteq M$.

    ▮ If $M \cap N = N$, then every element of N is also an element of $M \cap N$, which means it is an element of M. Thus, $N \subseteq M$. True.

1.17   The sets $\{ \ \}$, $\emptyset$, and $\{\emptyset\}$ all contain no elements.

    ▮ $\{ \ \}$ and $\emptyset$ both are empty, so they contain no elements. But $\{\emptyset\}$ contains the element $\emptyset$. So the answer is false.

1.18   If $M \cap N = \emptyset$, then at least one of M and N must be empty.

    ▮ Consider the set $M = \{1, 2\}$ and $N = \{e, f\}$. $M \cap N = \emptyset$ (they have no elements in common), but neither M nor N is empty. False.

1.19   Find a set containing only natural numbers that is equivalent to $\{e, f, \emptyset\}$.

    ▮ Two sets are equivalent if they are in a one-to-one correspondence with each other. The set $\{1, 2, 3\}$ contains only natural numbers, and since $\{1, 2, 3\}$, $\{e, f, \emptyset\}$ the sets are equivalent.

1.20   Suppose that m and n are natural numbers, and $m \geq n$. Must $m - n$ be a natural number? Why?

    ▮ No. Suppose $m = F$ and $n = F$; then m is natural, and $m \geq n$ ($F \geq F$). But $m - n = 0$, and 0 is *not* a natural number. (0 is a whole number; 0 is an integer, but the smallest natural number is 1.)

1.2   REAL NUMBERS AND THE NUMBER LINE

For Probs. 1.21 to 1.27, all letters represent real numbers. For each statement, tell which of the following properties or definitions are used.

Commutative	Identity	Division
Associative	Inverse	Zero
Distributive	Subtraction	Negative

1.21 $x + ym = x + my$

    ▮ Here, ym is being replaced by my. $ym = my$ is the *commutative* law for multiplication of real numbers.

1.22 $(2w + 8) + 3 = 2w + (8 + 3)$

    ▮ $(a + b) + c = a + (b + c)$ is the *associative* law for addition of real numbers. In this case, $a = 2w$, $b = 8$, $c = 3$.

1.23 $5(u + v) = 5u + 5v$

    ▮ $a(b + c) = ab + ac$ is the *distributive* law for the real numbers. Here, $a = 5$, $b = u$, $c = v$.

1.24 $8 - 12 = 8 + (-12)$

    ▮ If a and b are real numbers, then $a - b$ is defined as $a + (-b)$. Thus, $8 - 12 = 8 + (-12)$ is an application of the definition of subtraction.

1.25 $1 \cdot (-\frac{2}{3}) = -\frac{2}{3}$

    ▮ For any real number r, $1 \cdot r = r$. 1 is the *multiplicative identity*. In this case, r is simply replaced by $-\frac{2}{3}$.

1.26 $7 \div 9 = 7(\frac{1}{9})$

    ▮ If t and u are real, and $u \neq 0$, then $t \div u = t(1/u)$ by the definition of division. In this case, $t = 7$, $u = 9$ (and thus, $u \neq 0$), and we have a direct application of the definition of division.

1.27 $(^-u) + [-(-u)] = 0$

    ▮ For any real number a, $a + (-a) = 0$. $-a$ is the *additive identity* of a. Here, replace a by $-u$. $a + (-a) = 0 = (-u) + [-(-u)]$.

1.28   Let N = set of natural numbers
Z = set of integers
R = set of real numbers
Q = set of rational numbers
W = set of whole numbers
I = set of irrationals

Tell which of the above sets each of the following belongs to: (a) -3,   (b) $\sqrt{2}$,   (c) -1.414,   (d) 0.

▮   (a) -3 can be written as $-\frac{3}{1}$, and it is clearly an integer and also a real number.   Conclusion: $-3 \in Z, Q, R$.
(b) $\sqrt{2}$ is a real, irrational number (it is the square root of a non-perfect-square).   Conclusion: $\sqrt{2} \in R, I$.
(c) -1.414 is clearly a real number.   Also, it can be written as $-\frac{1414}{1000}$.   Conclusion: $-1.414 \in R, Q$.
(d) $0 = \frac{0}{1}$.   Conclusion: $0 \in R, W, Q, Z$.   (*Remark*: Remember 0 is *not* a natural number.)

For Probs. 1.29 to 1.33, simplify the given expression.   All letters represent real numbers.

1.29 $10 - (-7)$

▮ $10 - (-7) = 10 + 7 = 17$.

1.30 $[a + (-6)] + 6$

▮ $[a + (-6)] + 6 = a + [(-6) + 6] = a + 0 = a$.

1.31 $1 - (1 - b)$

▮ $1 - (1 - b) = 1 - 1 - (-b) = 1 - 1 + b = 0 + b = b$.

1.32 $2a - (b - 2a)$

▮ $2a - (b - 2a) = 2a - b - (-2a) = 2a - b + 2a = 2a + 2a - b = 4a - b$.

1.33 $(a - b)(-y)$

▮ $(a - b)(-y) = a(-y) - b(-y) = -ay + by$.

1.34   Convert the repeating decimal $0.04\overline{04}$ into a fraction.

▮   Let $x = 0.04\overline{04}$.   Then $100x = 4.04\overline{04}$,   and $100x - x = 4.\overline{04} - 0.\overline{04} = 4$.   So, $99x = 4$ and $x = \frac{4}{99}$.

1.35   *Prove*: $(a + b) + (-a) = b$,   for all real numbers a, b, justifying each step in your proof.

▮ $(a + b) + (-a) = (-a) + (a + b)$ (commutative law for addition) $= [(-a) + a] + b$ (associative law for addition) $= 0 + b$ (additive inverse) $= b$ (additive identity).

1.36   *Prove*: $a \cdot 0 = 0$ for every real number a.

▮ $a \cdot 0 = a(0 + 0)$ (additive identity) $= a \cdot 0 + a \cdot 0$ (distributive law).   So, $a \cdot 0 = a \cdot 0 + a \cdot 0$;   but then $a \cdot 0$ must be the additive identity since when you add it to $a \cdot 0$ you get $a \cdot 0$.   Conclusion: $a \cdot 0 = 0$.

1.37   Evaluate each of the following:

(a) $|-6|$, (b) $|1 - \sqrt{2}|$, (c) $7 - |-5|$, (d) $|-(-6)|$, (e) $|(-4) \cdot (6)|$.

▮   Recall that for any real number r, $|r| = \begin{cases} r, r \geq 0 \\ -r, r < 0 \end{cases}$.

(a)   Since $-6 < 0, |-6| = -(-6) = 6$.
(b)   Since $1 - \sqrt{2} < 0$, $|1 - \sqrt{2}| = -(1 - \sqrt{2}) = \sqrt{2} - 1$.
(c) $7 - |-5| = 7 - [-(-5)] = 7 - 5 = 2$.
(d) $|-(-6)| = |6| = 6$.
(e) $|(-4) \cdot 6| = |-24| = 24$ or $|(-4) \cdot 6| = |-4| \cdot |6| = 4 \cdot 6 = 24$.

1.38 Find the distance between -4 and 17 on the number line.

▎ If a and b are points on the number line, the distance between a and b, denoted $d(a, b)$ or $d(b, a)$, $= |a - b| = |b - a|$. In this case, $d(-4, 17) = |-4 - 17| = |-21| = 21$.

1.39 Tell whether each of the following is true or false, and why. All letters represent real numbers.

(a) $a \le a$. (b) If $a \le 3$ and $3 \le a$, then $a = 3$. (c) If $1 < a < 3$, then $a = 2$. (d) If $a < b$, then $-a < -b$. (e) If $a < b$, then $a^2 < b^2$.

▎ (a) $a \le a$ is true for all real numbers since "$\le$" means less than *or* equal to, and $a = a$ for all a.
(b) If $a \le 3$, then a is 3 or lies to the left of 3 on the number line. If $3 \le a$, then 3 is a or lies to the left of a on the number line. The only way this can happen is if $a = 3$. True.
(c) If $1 < a < 3$, then a could be 2, but it could also be 2.5 or $\pi - 1$, etc. False.
(d) Suppose $a < b$. Then $a - b < 0$. But if $a - b$ is negative, then $-(a - b)$ is positive. So $-(a - b) > 0$ or $-a + b > 0$, so $-a > -b$. So the given statement is false.
(e) This statement is false. Suppose $a = -2$ and $b = 1$. Then $a < b$. But $a^2 = 4$ and $b^2 = 1$, so $a^2 > b^2$.

1.40 *Prove:* $|x - y| = |y - x|$ for all real numbers x, y.

▎ Recall that for any real number a, $|a| = |-a|$. Thus, $|x - y| = |-(x - y)| = |-x + y| = |y - x|$.

1.3. INTEGRAL AND RATIONAL EXPONENTS

For Probs. 1.41 to 1.50, evaluate the given expression.

1.41 $2^9/2^4$

▎ $2^9/2^4 = 2^{9-4} = 2^5 = 32$.

1.42 $5^{19} \cdot 5^{-6}$

▎ $5^{19} \cdot 5^{-6} = 5^{19+(-6)} = 5^{19-6} = 5^{13}$.

1.43 $(2^6 \cdot 2^4)/(2^8 \cdot 2^9)$

▎ $(2^6 \cdot 2^4)/(2^8 \cdot 2^9) = 2^{10}/2^{17} = 2^{10-17} = 2^{-7} = 1/2^7 = 1/128$.

1.44 $[19,674(x + y)]^0$

▎ Recall that $p^0 = 1$ for any real number p. Thus, $[19,674(x + y)]^0 = 1$.

1.45 $\sqrt[5]{-32}$

▎ $\sqrt[5]{-32} = (-32)^{1/5} = [(-2)^5]^{1/5} = -2^{5(1/5)} = -2^1 = -2$.

1.46 $\sqrt[3]{\sqrt{\frac{1}{64}}}$

▎ $\sqrt[3]{\sqrt{\frac{1}{64}}} = \sqrt[3]{(\frac{1}{64})^{1/2}} = [(\frac{1}{64})^{1/2}]^{1/3} = (\frac{1}{64})^{1/6} = 1^{1/6}/64^{1/6} = 1/(2^6)^{1/6} = 1/2^1 = 1/2$.

1.47 $125^{-4/3}$

▎ $125^{-4/3} = (125^{1/3})^{-4} = 5^{-4} = 1/5^4 = 1/625$.

1.48 $(0.0001)^{3/4}$
$(0.0001)^{3/4} = [(0.0001)^{1/4}]^3 = (0.1)^3 = 0.001$.

1.49 $(\frac{8}{27})^{2/3}$

▎ $(\frac{8}{27})^{2/3} = [(\frac{8}{27})^{1/3}]^2 = (8^{1/3}/27^{1/3})^2 = (\frac{2}{3})^2 = \frac{4}{9}$.

1.50 $(2 + 2^{-1})/5 + (-8)^0 - 4^{3/2}$

▎ $(2 + 2^{-1})/5 + (-8)^0 - 4^{3/2} = (2 + \frac{1}{2})/5 - 1 - (4^{1/2})^3 = \frac{5}{2}/5 - 1 - 2^3 = \frac{5}{10} - 1 - 8 = \frac{1}{2} - 9 = -8\frac{1}{2}$.

For Probs. 1.51 to 1.63, simplify the given expression.

1.51 $x^2 \cdot x^{-9} \cdot x^{14}$

▮ $x^2 \cdot x^{-9} \cdot x^{14} = x^{2+(-9)+14} = x^7$.

1.52 $(xy^2)^{-6}$

▮ $(xy^2)^{-6} = x^6 \cdot (y^2)^{-6} = x^{-6} \cdot y^{-12} = x^{-6}/y^{12} = 1/x^6 y^{12}$.

1.53 $(2ab^2)^3(a^2c)^2$

▮ $(2ab^2)^3(a^2c)^2 = (2^3 a^3 b^6)(a^4 c^2) = (8a^3 b^6)(a^4 c^2) = 8a^7 b^8$.

1.54 $\dfrac{x^{-8} \cdot x^{-7}}{x^{-6}} \div \dfrac{x^{-5} \cdot x^{-4}}{x^{-3}}$

▮ $\dfrac{x^{-8} \cdot x^{-7}}{x^{-6}} \div \dfrac{x^{-5} \cdot x^{-4}}{x^{-3}} = \dfrac{x^{-15}}{x^{-6}} \cdot \dfrac{x^{-3}}{x^{-5} \cdot x^{-4}} = \dfrac{x^{-15}}{x^{-6}} \cdot \dfrac{x^{-3}}{x^{-9}} = x^{-9} \cdot x^6 = x^{-3} = \dfrac{1}{x^3}$.

1.55 $\dfrac{b^3 c^5}{a^7} \div \dfrac{b^2 c^7}{a^6}$

▮ $\dfrac{b^3 c^5}{a^7} \div \dfrac{b^2 c^7}{a^6} = \dfrac{b^3 c^5}{a^7} \cdot \dfrac{a^6}{b^2 c^7} = \dfrac{b^3 c^5}{b^2 c^7} \cdot \dfrac{a^6}{a^7} = bc^{-2} \cdot a^{-1} = \dfrac{b}{c^2 a}$.

1.56 $\left(\dfrac{2a^{-1}}{b^{-1}}\right)^2 \div \left(\dfrac{3a^{-2}}{b^{-3}}\right)^{-1}$

▮ $\left(\dfrac{2a^{-1}}{b^{-1}}\right)^2 \div \left(\dfrac{3a^{-2}}{b^{-3}}\right)^{-1} = \dfrac{4a^{-2}}{b^{-2}} \div \dfrac{3^{-1}a^2}{b^3} = \dfrac{4a^{-2}}{b^{-2}} \cdot \dfrac{b^3}{3^{-1}a^2} = \dfrac{4a^{-2}}{a^2} \cdot \dfrac{3b^3}{b^{-2}} = 12a^{-4}b^5 = \dfrac{12b^5}{a^4}$.

1.57 $(x^3/4x^{-6})^{-4/3}$

▮ $(x^3/4x^{-6})^{-4/3} = (x^9/4)^{-4/3} = (x^9)^{-4/3}/4^{-4/3} = x^{-12}/4^{-4/3} = 4^{4/3}/x^{12}$.

1.58 $\dfrac{(p^{2/3} q^{1/12})^2}{(p^{3/5})^{1/3}}$

▮ $\dfrac{(p^{2/3} q^{1/12})^2}{(p^{3/5})^{1/3}} = \dfrac{p^{4/3} q^{1/6}}{p^{1/5}} = p^{4/3-1/5} q^{1/6} = p^{17/15} q^{1/6}$.

1.59 $\dfrac{\sqrt{a} \cdot a^{-2/3}}{\sqrt[6]{a^5}} + \dfrac{a^{-5/6}}{\sqrt[3]{a^2} \cdot a^{-1/2}}$

▮ $\dfrac{\sqrt{a} \cdot a^{-2/3}}{\sqrt{a^5}} + \dfrac{a^{-5/6}}{\sqrt[3]{a^2} \cdot a^{-1/2}} = \dfrac{a^{1/2} a^{-2/3}}{a^{5/6}} + \dfrac{a^{-5/6}}{a^{2/3} \cdot a^{-1/2}} = \dfrac{a^{-1/6}}{a^{5/6}} + \dfrac{a^{-5/6}}{a^{1/6}} = \dfrac{a^{-1/6} + a^{4/6}(a^{-5/6})}{a^{5/6}} =$
$\dfrac{a^{-1/6} + a^{-1/6}}{a^{5/6}} = \dfrac{2a^{-1/6}}{a^{5/6}} = \dfrac{2}{a^{1/6}a^{5/6}} = \dfrac{2}{a^{6/6}} = \dfrac{2}{a}$.

1.60 $\left(\tfrac{27}{8} x^{-3} y^{1/2}\right)^{-4/3}$

▮ $\left(\tfrac{27}{8} x^{-3} y^{1/2}\right)^{-4/3} = \left(\tfrac{27}{8}\right)^{-4/3} \cdot (x^{-3})^{-4/3} \cdot (y^{1/2})^{-4/3} = [1/(\tfrac{27}{8})^{4/3}] \cdot x^{12/3} \cdot y^{-4/6} = 1/\tfrac{81}{16} \cdot x^4 \cdot y^{-2/3} =$
$\tfrac{16}{81} \cdot x^4 \cdot y^{-2/3} = 16x^4/81y^{2/3}$.

1.61 $\left(\dfrac{x^{-1/3} y^{1/2}}{x^{-1/4} y^{1/3}}\right)^6$

▮ $\left(\dfrac{x^{-1/3} y^{1/2}}{x^{-1/4} y^{1/3}}\right)^6 = \dfrac{(x^{-1/3})^6 (y^{1/2})^6}{(x^{-1/4})^6 (y^{1/3})^6} = \dfrac{x^{-2} y^3}{x^{-3/2} y^2} = x^{-1/2} y = \dfrac{y}{x^{1/2}}$.

1.62 $(9x^{1/3} y^{-1/2})^{3/2}(x^{-1/3} y^{1/4})$

▮ $(9x^{1/3} y^{-1/2})^{3/2}(x^{-1/3} y^{1/4}) = 9^{3/2}(x^{1/3})^{3/2}(y^{-1/2})^{3/2}(x^{-1/3} y^{1/4}) = 27x^{3/6} y^{-3/4} x^{-1/3} y^{1/4} = 27x^{1/6} y^{-2/4} =$
$27x^{1/6} y^{-1/2} = 27x^{4/6}/y^{1/2}$.

1.63 $(a^{m/3} b^{n/2})^{-6}$; where $m, n \geq 0$

▮ $(a^{m/3} b^{n/2})^{-6} = a^{-6m/3} b^{-6n/2} = a^{-2m} b^{-3n} = 1/a^{2m} b^{3n}$.

For Probs. 1.64 to 1.70, simplify and write in simplest radical form. Assume all letters and radicands represent positive real numbers.

1.64 $\sqrt{16m^4y^8}$

▎ $\sqrt{16m^4y^8} = \sqrt{16}\sqrt{m^4}\sqrt{y^8} = \sqrt{4^2}\sqrt{(m^2)^2}\sqrt{(y^4)^2} = 4m^2y^4.$

1.65 $\sqrt[5]{32a^{15}b^{10}}$

▎ (i) $\sqrt[5]{32a^{15}b^{10}} = \sqrt[5]{32}\sqrt[5]{a^{15}}\sqrt[5]{b^{10}} = \sqrt[5]{2^5}\sqrt[5]{(a^3)^5}\sqrt[5]{(b^2)^5} = 2a^3b^2.$

(ii) $\sqrt[5]{32a^{15}b^{10}} = 32^{1/5}(a^{15})^{1/5}(b^{10})^{1/5} = 2a^3b^2.$

1.66 $\sqrt{\sqrt[4]{5x}}$

▎ (i) *Recall* that $\sqrt[m]{\sqrt[n]{x}} = {}^{mn}\sqrt{x}$. So, $\sqrt{\sqrt[4]{5x}} = \sqrt[8]{5x}.$

(ii) $\sqrt{\sqrt[4]{5x}} = (\sqrt[4]{5x})^{1/2} = [(5x)^{1/4}]^{1/2} = (5x)^{1/8} = \sqrt[8]{5x}.$

1.67 $2a\sqrt[3]{8a^8b^{13}}$

▎ $2a\sqrt[3]{8a^8b^{13}} = 2a\sqrt[3]{2^3(a^2)^3a^2(b^4)^3b} = 2a\sqrt[3]{2^3}\sqrt[3]{(a^2)^3}a^2\sqrt[3]{(b^4)^3b} = 2a\cdot2\cdot a^2\sqrt[3]{a^2}\cdot b^4\sqrt[3]{b} = 4a^3b^4\sqrt[3]{a^2b}.$

1.68 $\sqrt[8]{3^6(u+v)^6}$

▎ $\sqrt[8]{3^6(u+v)^6} = \sqrt[8]{[3(u+v)]^6} = \sqrt[8]{\{[3(u+v)]^3\}^2} = \sqrt[4]{[3(u+v)]^3}.$

1.69 $\sqrt[3]{\dfrac{3y^5}{4x^4}}$

▎ $\sqrt[3]{\dfrac{3y^5}{4x^4}} = \sqrt[3]{\dfrac{3y^3y^2}{4x^3x}} = \dfrac{\sqrt[3]{3y^3y^2}}{\sqrt[3]{4x^3x}} = \dfrac{y\sqrt[3]{3y^2}}{x\sqrt[3]{4x}} = \dfrac{y\sqrt[3]{3y^2}}{x\sqrt[3]{4x}}\cdot\dfrac{\sqrt[3]{4^2x^2}}{\sqrt[3]{4^2x^2}}$ (rationalizing the denominator) $= \dfrac{y\sqrt[3]{48x^2y^2}}{x\sqrt[3]{4^3x^3}} =$

$\dfrac{y\sqrt[3]{48x^2y^2}}{x\cdot4\cdot x} = \dfrac{y\sqrt[3]{48x^2y^2}}{4x^2}.$

1.70 $\dfrac{2}{x^2-\sqrt{x^4+2x^2+1}}$

▎ $\dfrac{2}{x^2-\sqrt{x^4+2x^2+1}} = \dfrac{2}{x^2-\sqrt{(x^2+1)^2}} = \dfrac{2}{x^2-(x^2+1)} = \dfrac{2}{x^2-x^2-1} = -2.$

1.4 ALGEBRAIC EXPRESSIONS

For Probs. 1.71 to 1.84, perform the indicated operations and simplify.

1.71 $2(u-1)-(3u+2)-2(2u-3)$

▎ $2(u-1)-(3u+2)-2(2u-3) = 2u-2-3u-2-4u+6 = (2u-3u-4u)+(-2-2+6) = -5u+2.$

1.72 $(x^2-2x+3)+(4x^2-x+6)$

▎ $(x^2-2x+3)+(4x^2-x+6) = x^2-2x+3+4x^2-x+6 = (x^2+4x^2)+(-2x-x)+3+6 = 5x^2-3x+9.$

1.73 $(x^2-2x+3)-(-5x^3-7x+1)$

▎ $x^2-2x+3-(-5x^3-7x+1) = x^2-2x+3+5x^3+7x-1 = 5x^3+x^2+(7x-2x)+(3-1) = 5x^3+x^2+5x+2.$

1.74 $4\sqrt[3]{y}-\sqrt[3]{y}+2\sqrt{y}$

▎ $4\sqrt[3]{y}-\sqrt[3]{y}+\sqrt[2]{y} = (4\sqrt[3]{y}-\sqrt[3]{y})+\sqrt[2]{y} = 3\sqrt[3]{y}+\sqrt[2]{y}.$

1.75 $(x-2y)(x+3y)$

▎ $(x-2y)(x+3y) = x(x)+x(3y)+(-2y)(x)+(-2y)(3y) = x^2+3xy-2xy-6y^2 = x^2+xy-6y^2.$

1.76 $(\sqrt{s}+\sqrt{t})^2$

▎ $(\sqrt{s}+\sqrt{t})^2 = (\sqrt{s})^2+\sqrt{s}\sqrt{t}+(\sqrt{t})^2 = s+\sqrt{st}+t.$

1.77 $(s+t)^3$

▮ $(s+t)^3 = (s+t)^2(s+t) = (s^2 + 2st + t^2)(s+t) = s^2(s) + 2st(s) + t^2(s) + s^2(t) + 2st(t) + t^2(t) = s^3 + 3s^2t + 3st^2 + t^3$.

1.78 $(3-x)(x^2 - x - 1)$

▮ $(3-x)(x^2 - x - 1) = 3(x^2) + 3(-x) + 3(-1) + (-x)(x^2) + (-x)(-x) + (-x)(-1) = 3x^2 - 3x - 3 - x^3 + x^2 + x = -x^3 + 4x^2 - 2x - 3$.

1.79 $m - \{m - [m - (m-1)]\}$

▮ $m - \{m - [m - (m-1)]\} = m - \{m - [m - m + 1]\} = m - \{m - 1\} = m - m + 1 = 1$.

1.80 $(5\sqrt{x} + 2)(2\sqrt{x} - 3)$

▮ $(5\sqrt{x} + 2)(2\sqrt{x} - 3) = 5\sqrt{x}(2\sqrt{x}) + 5\sqrt{x}(-3) + 2(2\sqrt{x}) + 2(-3) = 10(\sqrt{x})^2 - 15\sqrt{x} + 4\sqrt{x} - 6 = 10x - 11\sqrt{x} - 6$.

1.81 $(x-y)^2 + 2x(1+y)$

▮ $(x-y)^2 + 2x(1+y) = (x^2 - 2xy + y^2) + (2x + 2xy) = x^2 - 2xy + 2xy + y^2 + 2x = x^2 + 2x + y^2$.

1.82 $3m^{3/4}(4m^{1/4} - 2m^8)$

▮ $3m^{3/4}(4m^{1/4} - 2m^8) = 3m^{3/4}4m^{1/4} + 3m^{3/4}(-2m^8) = 12m - 6m^{35/4}$.

1.83 $\dfrac{a^2}{6} - \dfrac{a}{5} + \dfrac{1}{15}$

▮ $\dfrac{a^2}{6} - \dfrac{a}{5} + \dfrac{1}{15} = \dfrac{5 \cdot a^2 + (-a) \cdot 6 + 2(1)}{30} = \dfrac{5a^2 - 6a + 2}{30}$.

1.84 $\dfrac{x^2 + 1}{x} + \dfrac{x-2}{x^2} + \dfrac{x-3}{2x^2}$

▮ $\dfrac{x^2 + 1}{x} + \dfrac{x-2}{x^2} + \dfrac{x-3}{2x^2} = \dfrac{2x(x^2 + 1) + 2(x-2) + 1(x-3)}{2x^2} = \dfrac{(2x^3 + 2x) + (2x - 4) + (x - 3)}{2x^2} = \dfrac{2x^3 + 5x - 7}{2x^2}$.

1.85 Evaluate the polynomial $2x^2 - 3x - 10$ when $x = -5$.

▮ Substituting -5 for x, $\quad 2x^2 - 3x - 10 = 2(-5)^2 - 3(-5) - 10 = 2(25) + 15 - 10 = 50 + 15 - 10 = 55$.

1.86 Evaluate the polynomial $p^2 + 2p + 8$ when $p = 2v$.

▮ Substituting $2v$ for p, we get $p^2 + 2p + 8 = (2v)^2 + 2(2v) + 8 = 4v^2 + 4v + 8$.

1.87 Evaluate the polynomial $y^2 - 3y - 2$ when $y = 1 - \sqrt{3}$.

▮ Substituting $1 - \sqrt{3}$ for y, we get $y^2 - 3y - 2 = (1 - \sqrt{3})^2 - 3(1 - \sqrt{3}) - 2 = (1 - 2\sqrt{3} + 3) - 3 + 3\sqrt{3} - 2 = -1 + (-2\sqrt{3} + 3\sqrt{3}) = \sqrt{3} - 1$.

1.88 Given that $x + 1/x = 5$, find the value of $x^2 + 1/x^2$.

▮ Recall that $(x + 1/x)^2 = x^2 + 2x \cdot (1/x) + (1/x)^2 = (x^2 + 1/x^2) + 2$. Thus, $(x + 1/x)^2 - 2 = x^2 + 1/x^2$. We know that $x + 1/x = 5$; thus, $5^2 - 2 = x^2 + 1/x^2$, and $x^2 + 1/x^2 = 25 - 2 = 23$.

1.89 The length of a rectangle is 8 meters (m) more than its width. (*a*) If x represents the width of the rectangle, write an algebraic expression in terms of x that represents the area. (*b*) Change the expression to a form without parentheses.

▮ (*a*) Consider Fig. 1.4. If x is the width and the length is 8 m more than the width, then the length must be $x + 8$. Recall that $A = 1 \cdot w$ for a rectangle. In this case, $A = x(x + 8)$.
(*b*) Without parentheses, $A = x^2 + 8x$.

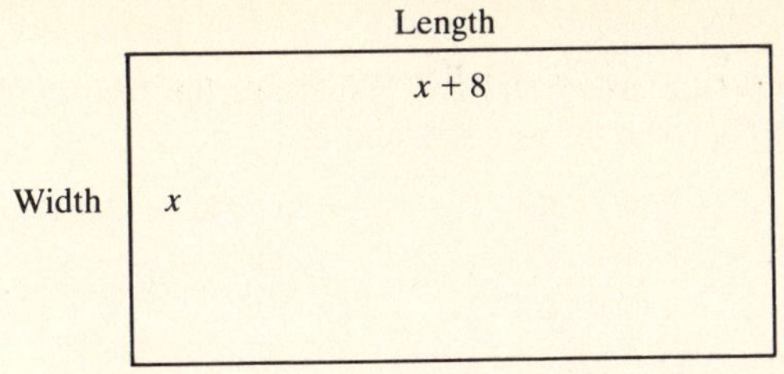

Length

x + 8

Width $\quad$ x

Fig. 1.4

1.90 Establish that the following two formulas are true: (**a**) $x^3 + y^3 = (x + y)(x^2 - xy + y^2)$, (**b**) $x^3 - y^3 = (x - y)(x^2 + xy + y^2)$.

$\blacksquare$ (**a**) $(x + y)(x^2 - xy + y^2) = x(x^2) + x(-xy) + x(y^2) + y(x^2) + y(-xy) + y(y^2) = x^3 - x^2y + xy^2 + yx^2 - xy^2 + y^3 = x^3 + y^3$.

(**b**) $(x - y)(x^2 + xy + y^2) = x(x^2) + x(xy) + x(y^2) - y(x^2) - y(xy) - y(y^2) = x^3 + x^2y + xy^2 - x^2y - xy^2 - y^3$

$= x^3 - y^3$. Thus, both formulas are established.

1.5 FACTORING AND FRACTIONAL EXPRESSIONS

For Probs. 1.91 to 1.117, factor the given polynomial completely (relative to the integers).

1.91 $x^2 + 5x + 4$

$\blacksquare$ We are looking for two binomials $(cx + d)$ and $(ex + f)$ such that $x^2 + 5x + 4 = (cx + d)(ex + f)$. Note that $(cx + d)(ex + f) = cex^2 + (de + cf)x + df$. Thus, $x^2 + 5x + 4 = cex^2 + (de + cf)x + df$. In order for this to take place, ce must be equal to the coefficient of x^2 (in this case, 1). df must be 4. We conclude that $de + cf = 5c = e = 1$, and either $f = 4, d = 1$ or $f = 1, d = 4$ or $f = 2, d = 2$. If $f = d = 2$, then $de + cf \neq 5$. Thus, $f = 4, d = 1$ or $f = 1, d = 4$. $x^2 + 5x + 4 = (x + 4)(x + 1)$. Note that $(x + 1)(x + 4)$ is also correct. We can check our answer by noting that $(x + 4)(x + 1) = x^2 + (4x + x) + 4 = x^2 + 5x + 4$.

1.92 $x^2 - x - 6$

$\blacksquare$ We repeat the procedure from Prob. 1.91: $x^2 - x - 6 = (cx + d)(ex + f) = cex^2 + (de + cf)x + df$. We find that $ce = 1$, $de + cf = -1$, and $df = -6$. So, $c = e = 1$, and either $d = -6, f = 1$ or $d = -1, f = 6$ or $d = -3, f = 2$ or $d = 3, f = -2$. Checking the quantity $de + ef$, we find that $c = -3, f = 2$ makes $de + cf = -1$. $x^2 - x - 6 = (x - 3)(x + 2)$. [Note that $(x + 2)(x - 3)$ is also correct.]

1.93 $2x^2 + 7x + 6$

$\blacksquare$ Again, we repeat the procedure given in Prob. 1.91. We want $ce = 2$, $de + cf = 7$, and $df = 6$. Then, $c = 2, e = 1$ (or the reverse is fine also), $de + cf = 7$, and $d = 6, f = 1$ or $d = 3, f = 2$ or $d = 1, f = -6$ or $d = 2, f = 3$. Note that since $ce \neq 1$, $d = 6, f = 1$ is different from $d = 1, f = 6$. Checking these choices, we find that if $d = 3, f = 2$, $de + cf = 7$. $2x^2 + 7x + 6 = (2x + 3)(x + 2)$.

1.94 $2x^2 - 2x - 4$

$\blacksquare$ (See Prob. 1.91 for a detailed outline of the method.) $2x^2 - 2x - 4 = (2x^{-4})(x + 1)$. Check by noting that $(2x - 4)(x + 1) = 2x^2 - 2x - 4$. Note that using the notation of Prob. 1.91, $ce = 2, df = -4$, and $cf + de = -2$. By trial and error, we conclude that $c = 2$, $e = 1$, $d = -4$, and $f = 1$.

1.95 $10x^2 - 14x - 12$

$\blacksquare$ Here, we will shortcut the full technique described in Prob. 1.91. The reader may still apply that technique and will find that it works. Here, we want factors whose product is $10x^2$. (Think! They are $10x$ and x, or $5x$ and $2x$.) We also want integral factors whose product is -12. (Think! They are ± 12 and ∓ 1, or ± 6 and ∓ 2, or ± 4 and ∓ 3.) By trial and error, looking for a middle term of $-14x$, we find $10x^2 - 14x - 12 = (5x + 3)(2x - 4)$. Note: $3(2x) + (-4)(5x) = -14x$.

1.96 $6x^2 - 16x + 8$

$\blacksquare$ $6x^2 - 16x + 8 = (2x - 4)(3x - 2)$. Notice that $(2x)(3x) = 6x^2$, $(-4)(-2) = +8$, and $(-4)(3x) + (-2)(2x) = -16x$.

1.97 $-x^2 - x + 12$

▮ Before we actually factor, ask yourself "What are the factors of 12 whose sum or difference is -1?" (The middle term is $-x$.) The answer is 4 and 3.

$$-x^2 - x + 12 = -x^2 + 3x - 4x + 12 = (-x + 3)(x + 4).$$

1.98 $2t^2 + 75t - 200$

▮ $2t^2 + 75t - 200 = (2t - 5)(t + 40).$

1.99 $2m^2 + 5m^n + 3n^2$

▮ Notice that the solution must be of the form $(2m + 3n)(m + n)$ or $(2m + n)(m + 3n)$. In the first case, the middle term is the correct one $(5mn)$. Thus, $2m^2 + 5mn + 3n^2 = (2m + 3n)(m + n)$.

1.100 $x^2 + x + 1$

▮ Try to find factors here. You will discover that they are difficult to find. Check the discriminant $b^2 - 4ac$ (where $ax^2 + bx + c$ is the form of the polynomial). In this case, $b^2 - 4ac$ is $1^2 - 4(1)(1)$, which is negative. A negative discriminant signals that the polynomial is not factorable.

1.101 $m^3 - 6m - 3$

▮ Again, as in Prob. 1.100, we check the discriminant since we are having trouble finding factors. In this case, the discriminant $b^2 - 4ac$ is $(-6)^2 - 4(1)(-3) = 36 + 12 = 48$, which is not a perfect square. If the discriminant is not a perfect square, then the polynomial is not factorable relative to the integers.

1.102 $4y^2 - 25$

▮ Recall that a polynomial of the form $r^2 - s^2$ is factored as follows: $r^2 - s^2 = (r + s)(r - s)$ (the "difference of two perfect squares" formula). See Fig. 1.5. In this case, we notice that $4y^2 = (2y)^2$ and $25 = 5^2$, and we conclude that $4y^2 - 25 = (2y + 5)(2y - 5)$. Check to see that $(2y + 5)(2y - 5) = 4y^2 - 25$ when you multiply.

(a) Difference of two squares:

$$r^2 - s^2 = (r + s)(r - s)$$

(b) Difference of two cubes:

$$r^3 - s^3 = (r - s)(r^2 + rs + s^2)$$

(c) Sum of two cubes:

$$(r^3 + s^3) = (r + s)(r^2 - rs + s^2)$$

Fig. 1.5

1.103 $27p^3 + 8q^3$

▮ This polynomial is of the form "sum of two cubes." $27p^3 + 8q^3 = (3p)^3 + (2q)^3$. Look at Fig. 1.5 for the correct formula. Then, $27p^3 + 8q^3 = (3p + 2q)[(3p)^2 - (3p)(2q) + (2q)^2] = (3p + 2q)(9p^2 - 6pq + 4q^2)$.

1.104 $t^6 - a^{12}$

▮ Refer to Fig. 1.5. $t^6 - a^{12} = (t^2)^3 - (a^4)^3$ is of the form "difference of two cubes." Using the proper formula, $t^6 - a^{12} = (t^2 - a^4)[(t^2)^2 + t^2a^4 + (a^4)^2] = (t^2 - a^4)(t^4 + t^2a^4 + a^8)$. But, $t^2 - a^4$ is of the form $(t)^2 - (a^2)^2$ which is the difference of two squares! $t^2 - a^4 = (t + a^2)(t - a^2)$. We conclude that $t^6 - a^{12} = (t + a^2)(t - a^2)(t^4 + t^2a^4 + a^8)$.

1.105 $t^2 + 25$

▮ Be careful! This is the *sum* of two squares. This is nonfactorable. Check the discriminant if you want proof. $b^2 - 4ac = 0^2 - 4(1)(25)$ which is negative.

1.106 $2x^4 - 24x^3 + 40x^2$

▮ Notice that each term contains a factor of $2x^2$. Factor out that x^2. $2x^4 - 24x^3 + 40x^2 = 2x^2(x^2 - 12x + 20)$. Now factor $(x^2 - 12x + 20)$. $(x^2 - 12x + 20) = (x - 10)(x - 2)$. Thus, $2x^4 - 24x^3 + 40x^2 = 2x^2(x - 10)(x - 2)$.

1.107 $x^2y + 7xy$

▮ Notice that each term contains a factor of xy. $x^2y + 7xy = xy(x + 7)$, and there is no further factoring possible.

1.108 $a^4 - b^4$

▮ $a^4 - b^4 = (a^2)^2 - (b^2)^2 = (a^2 - b^2)(a^2 + b^2) = (a + b)(a - b)(a^2 + b^2)$. Notice that we are applying the "difference of two squares" formula twice here.

1.109 $x^3 + (y - z)^3$

▮ This is a "sum of two cubes" problem. $x^3 + (y - z)^3 = [x + (y - z)][x^2 - x(y - z) + (y - z)^2] = (x + y - z)[x^2 - xy + xz + (y - z)^2]$.

1.110 $4x^2 - 9y^2 + 4x + 1$

▮ We will use a regrouping technique here. $4x^2 - 9y^2 + 4x + 1 = 4x^2 + 4x + 1 - 9y^2 = (2x + 1)(2x + 1) - 9y^2 = (2x + 1)^2 - 9y^2$ (difference of two squares) $= (2x + 1 - 3y)(2x + 1 + 3y)$.

1.111 $s^6 - t^6$

▮ $s^6 - t^6 = (s^2)^3 - (t^2)^3 = (s^2 - t^2)(s^4 + s^2t^2 + t^4) + (s + t)(s - t)(s^4 + s^2t^2 + t^4)$.

1.112 $s^{12} - t^{12}$

▮ $s^{12} - t^{12} = (s^6)^2 - (t^6)^2 = (s^6 + t^6)(s^6 - t^6)$
$= [(s^2)^3 + (t^2)^3] \quad [(s^2)^3 - (t^2)^3]$
sum of cubes difference of cubes
$= (s^2 + t^2)(s^4 - s^2t^2 + t^4)(s^2 - t^2)(s^2 + s^2t^2 + t^4)$
$= (s^2 + t^2)(s^4 - s^2t^2 + t^4)(s + t)(s - t)(s^2 + s^2t^2 + t^4)$

1.113 $y^2 - 2xy + x^2 = y - x$

▮ $y^2 - 2xy + x^2 - y + x = (y - x)(y - x) - y + x = (y - x)^2 - (y - x) = (y - x)(y - x - 1)$.

1.114 $27 + 8/t^3$

▮ Although t^3 is in the denominator here, the techniques we have employed thus far will still work. $27 + 8/t^3 = 3^3 + (2/t)^3 = (3 + 2/t)(9 - 6/t + 4/t^2)$.

1.115 $4a^4 + 8a^2b^2 + 9b^4$

▮ Direct attempts at factoring in this case do not seem to work. (You should try them!) Sometimes, the following trick works: Notice that $4a^4 + 12a^2b^2 + 9b^4$ is factorable: $4a^4 + 12a^2b^2 + 9b^4 = (2a^2 + 3b^2)(2a^2 + 3b^2)$. We then write $4a^2 + 8a^2b^2 + 9b^2 = (4a^2 + 12a^2b^2 + 9b^2) - 4a^2b^2$. Then, $4a^4 + 8a^2 + 9b^4 = (4a^2 + 12a^2b^2 + 9b^2) - 4a^2b^2 = (2a^2 + 3b^2)^2 - (2ab)^2$ (difference of two squares) $= (2a^2 + 3b^2 - 2ab)(2a^2 + 3b^2 + 2ab)$.

1.116 $x^4 - 13x^2y^2 + 4y^4$

▮ We again use the techniques described in Prob. 1.115.

$x^4 - 13x^2y^2 + 4y^4 = (x^4 + 4x^2y^2 + 4y^4) - 9x^2y^2$
$= (x^2 + 2y^2)^2 - 9x^2y^2$
$= (x^2 + 2y^2)^2 - (3xy)^2$
$= (x^2 + 2y^2 - 3xy)(x^2 + 2y^2 + 3xy)$
$= (x^2 - 3xy + 2y^2)(x^2 + 3xy + 2y^2)$
factorable factorable
$= (x - 2y)(x - y)(x + 2y)(x + y)$

1.117 $2y^{-2} - y^{-1} - 3$

▮ Don't be afraid of the negative exponents! $2y^{-2} - y^{-1} - 3 = (2y^{-1} - 3)(y^{-1} + 1)$. Remember that $(2y^{-1})(y^{-1}) = 2y^{-2}$.

For Probs. 1.118 to 1.142, perform the indicated operations and reduce to lowest terms. Notice that factoring is important in many of these problems.

1.118 $\dfrac{2(x^2 - 1)}{x + 1}$

▮ $\dfrac{2(x^2 - 1)}{x + 1} = \dfrac{2(x + 1)(x - 1)}{x + 1} = \dfrac{2\cancel{(x + 1)}(x - 1)}{\cancel{x + 1}} = 2(x - 1)$.

1.119 $\dfrac{2x^2 - 2}{x^3 - x^2}$

▮ $\dfrac{2x^2 - 2}{x^3 - x^2} = \dfrac{2(x^2 - 1)}{x^2(x - 1)} = \dfrac{2(x + 1)\cancel{(x - 1)}}{x^2\cancel{(x - 1)}} = \dfrac{2(x + 1)}{x^2}$.

1.120 $\dfrac{x^2 - 4}{x^2 - 4x - 12}$

▮ $\dfrac{x^2 - 4}{x^2 - 4x - 12} = \dfrac{\cancel{(x + 2)}(x - 2)}{(x - 6)\cancel{(x + 2)}} = \dfrac{x - 2}{x - 6}$.

1.121 $\dfrac{x^2 - 2x - 15}{x^2 - 3x - 10}$

▮ $\dfrac{x^2 - 2x - 15}{x^2 - 3x - 10} = \dfrac{\cancel{(x - 5)}(x + 3)}{\cancel{(x + 5)}(x + 2)} = \dfrac{x + 3}{x + 2}$.

1.122 $\dfrac{x^2 - 2x - 15}{x^3 - 3x^2 - 10x}$

▮ $\dfrac{x^2 - 2x - 15}{x^3 - 3x^2 - 10x} = \dfrac{x^2 - 2x - 15}{x(x^2 - 3x - 10)} = \dfrac{\cancel{(x - 5)}(x + 3)}{x\cancel{(x - 5)}(x + 2)} = \dfrac{x + 3}{x(x + 2)}$.

1.123 $\dfrac{d^5}{3a} \div \left(\dfrac{d^2}{6a^2} \cdot \dfrac{a}{4d^3} \right)$

$\dfrac{d^5}{3a} \div \left(\dfrac{d^2}{6a^2} \cdot \dfrac{a}{4d^3} \right) = \dfrac{d^5}{3a} \div \left(\dfrac{d^2}{4d^3} \cdot \dfrac{a}{6a^2} \right) = \dfrac{d^5}{3a} \div \left(\dfrac{1}{(4d)(6a)} \right) = \dfrac{d^5}{3a} \div \dfrac{1}{24ad} = \dfrac{d^5}{3a} \cdot \dfrac{24ad}{1} = 8d^6$.

1.124 $\dfrac{x^2}{12} + \dfrac{x}{18} - \dfrac{1}{30}$

▮ $\dfrac{x^2}{12} + \dfrac{x}{18} - \dfrac{1}{30} = \dfrac{x^2 \cdot 15 + x \cdot 10 - 1 \cdot 6}{180}$ (180 is the least common denominator) $= \dfrac{15x^2 + 10x - 6}{180}$.

1.125 $\dfrac{4m - 3}{18m^3} + \dfrac{3}{m} - \dfrac{2m - 1}{6m^2}$

▮ $\dfrac{4m - 3}{18m^3} + \dfrac{3}{m} - \dfrac{2m - 1}{6m^2} = \dfrac{(4m - 3) \cdot 1 + 3(18m^2) - (2m - 1)(3m)}{18m^3}$ ($18m^3$ is the least common denominator) $= \dfrac{4m - 3 + 54m^2 - (6m^2 - 3m)}{18m^3} = \dfrac{4m - 3 + 54m^2 - 6m^2 + 3m}{18m^3} = \dfrac{48m^2 + 7m - 3}{18m^3}$.

1.126 $\dfrac{t^2 - t - 6}{t^2 + t - 2} \cdot \dfrac{t^2 + 4t - 5}{t^2 + 6t + 5}$

▮ $\dfrac{t^2 - t - 6}{t^2 + t - 2} \cdot \dfrac{t^2 + 4t - 5}{t^2 + 6t + 5} = \dfrac{(t - 3)\cancel{(t + 2)}}{\cancel{(t + 2)}\cancel{(t - 1)}} \cdot \dfrac{\cancel{(t + 5)}\cancel{(t - 1)}}{\cancel{(t + 5)}(t + 1)} = \dfrac{t - 3}{t + 1}$.

1.127 $\dfrac{2x^2 + x - 1}{3x^2 + 2x - 1} \cdot \dfrac{3x^2 - 2x - 1}{2x^2 - 3x + 1}$

▮ $\dfrac{2x^2 + x - 1}{3x^2 + 2x - 1} \cdot \dfrac{3x^2 - 2x - 1}{2x^2 - 3x + 1} = \dfrac{\cancel{(2x - 1)}(x + 1)}{(3x - 1)\cancel{(x + 1)}} \cdot \dfrac{(3x + 1)\cancel{(x - 1)}}{\cancel{(2x - 1)}\cancel{(x - 1)}} = \dfrac{3x + 1}{3x - 1}$.

1.128 $\dfrac{2a^2 - 5a + 3}{a^2 + a - 2} \div \dfrac{3a^2 - 8a - 3}{a^2 - a - 6}$

▮ $\dfrac{2a^2 - 5a + 3}{a^2 + a - 2} \div \dfrac{3a^2 - 8a - 3}{a^2 - a - 6} = \dfrac{2a^2 - 5a + 3}{a^2 + a - 2} \cdot \dfrac{a^2 - a - 6}{3a^2 - 8a - 3} = \dfrac{(2a + 1)\cancel{(a - 3)}}{\cancel{(a + 2)}(a - 1)} \cdot \dfrac{(a - 3)\cancel{(a + 2)}}{(3a + 1)\cancel{(a - 3)}} =$

$\dfrac{(2a + 1)(a - 3)}{(a - 1)(3a + 1)}$.

1.129 $\dfrac{y^2 - y - 6}{y^2 - 2y + 1} \cdot \dfrac{y^2 + 3y - 4}{9y - y^3}$

▮ $\dfrac{y^2 - y - 6}{y^2 - 2y + 1} \cdot \dfrac{y^2 + 3y - 4}{9y - y^3} = \dfrac{(y - 3)(y + 2)}{(y - 1)(y - 1)} \cdot \dfrac{(y + 4)(y - 1)}{y(3 - y)(3 + y)} = \dfrac{-\cancel{(3 - y)}(y + 2)}{\cancel{(y - 1)}(y - 1)} \cdot \dfrac{(y + 4)\cancel{(y - 1)}}{y\cancel{(3 - y)}(3 + y)}$. Notice that we change $(y - 3)$ to $-(3 - y)$ so that we can cancel the $3 - y$ in the denominator. The above expression $= \dfrac{-(y + 2)(y + 4)}{y(y - 1)(3 - y)}$.

1.130 $\dfrac{2 - x}{2x + x^2} \cdot \dfrac{x^2 + 4x + 4}{x^2 - 4}$

▮ $\dfrac{2 - x}{2x + x^2} \cdot \dfrac{x^2 + 4x + 4}{x^2 - 4} = \dfrac{2 - x}{x(2 + x)} \cdot \dfrac{(x + 2)(x + 2)}{(x + 2)(x - 2)} = \dfrac{-\cancel{(x - 2)}}{x\cancel{(2 + x)}} \cdot \dfrac{\cancel{(x + 2)}\cancel{(x + 2)}}{\cancel{(x + 2)}\cancel{(x - 2)}} = \dfrac{-1}{x}$. [Notice the $2 - x = -(x - 2)$ trick here. Also, remember that $2 + x$ and $x + 2$ are equal, and so they cancel!]

1.131 $\dfrac{5}{x^2 - 4} + \dfrac{7}{x - 2}$

▮ $\dfrac{5}{x^2 - 4} + \dfrac{7}{x - 2} = \dfrac{5}{(x + 2)(x - 2)} + \dfrac{7}{x - 2} = \dfrac{5(1) + 7(x + 2)}{(x + 2)(x - 2)} = \dfrac{5 + 7x + 14}{(x + 2)(x - 2)}$ [$(x + 2)(x - 2)$ is the least common denominator] $= \dfrac{7x + 19}{x^2 - 4}$.

1.132 $\dfrac{3x}{2x - 3} - \dfrac{7x}{2x + 1}$

▮ $\dfrac{3x}{2x - 3} - \dfrac{7x}{2x + 1} = \dfrac{3x(2x + 1) - 7x(2x - 3)}{(2x - 3)(2x + 1)} = \dfrac{6x^2 + 3x - (14x^2 - 51x)}{(2x - 3)(2x + 1)} = \dfrac{-8x^2 + 54x}{(2x - 3)(2x + 1)} =$

$\dfrac{2x(27 - 4x)}{(2x - 3)(2x + 1)}$.

1.133 $\dfrac{d + 5}{3d - 1} - \dfrac{2d - 1}{3d + 1}$

▮ $\dfrac{d + 5}{3d - 1} - \dfrac{2d - 1}{3d + 1} = \dfrac{(d + 5)(3d + 1) - (2d - 1)(3d - 1)}{(3d - 1)(3d + 1)} = \dfrac{3d^2 + 16d + 5 - (6d^2 - 5d + 1)}{(3d - 1)(3d + 1)} =$

$\dfrac{-3d^2 + 21d + 4}{(3d - 1)(3d + 1)}$.

1.134 $\dfrac{2x}{x^2 - y^2} + \dfrac{1}{x + y} - \dfrac{1}{x - y}$

▮ Noting that $x^2 - y^2 = (x + y)(x - y)$, we have $\dfrac{2x}{x^2 - y^2} + \dfrac{1}{x + y} - \dfrac{1}{x - y} = \dfrac{2x(1) + 1(x - y) - 1(x + y)}{(x^2 - y^2)} =$

$\dfrac{2x + x - y - x - y}{x^2 - y^2} = \dfrac{2x - 2y}{x^2 - y^2} = \dfrac{2\cancel{(x - y)}}{(x + y)\cancel{(x - y)}} = \dfrac{2}{x + y}$.

1.135 $\dfrac{1}{x-y} - \dfrac{2}{x+y} + \dfrac{3}{(2x-1)(x-y)}$

▮ $\dfrac{1}{x-y} - \dfrac{2}{x+y} + \dfrac{3}{(2x-1)(x-y)}$

$= \dfrac{1(x+y)(2x-1) - 2(x-y)(2x-1) + 3(x+y)}{(x-y)(x+y)(2x-1)}$

$= \dfrac{1(2x^2+2xy-x-y) - 2(2x^2-2xy-x+y) + (3x+3y)}{(x-y)(x+y)(2x-1)}$

$= \dfrac{2x^2+2xy-x-y-4x^2+4xy+2x-2y+3x+3y}{(x-y)(x+y)(2x-1)}$

$= \dfrac{-2x^2+6xy+4x}{(x-y)(x+y)(2x-1)}$

1.136 $\dfrac{1}{x-1} + \dfrac{2x}{(x-1)^2} - \dfrac{4-x}{(x-1)^3}$

▮ $\dfrac{1}{x-1} + \dfrac{2x}{(x-1)^2} - \dfrac{4-x}{(x-1)^3} = \dfrac{1(x-1)^2 + 2x(x-1) - 1(4-x)}{(x-1)^3} = \dfrac{(x^2-2x+1) + (2x^2-2x) - (4-x)}{(x-1)^3}$

$= \dfrac{3x^2-3x-3}{(x-1)^3} = \dfrac{3(x^2-x-1)}{(x-1)^3}$.

1.137 $\dfrac{2+1/t}{2-1/t}$

▮ Notice that this expression involves a complex fraction. There are two basic ways of solving problems like this one. (i) Get rid of the complex fraction first, and then simplify. $\dfrac{2+1/t}{2-1/t} = \dfrac{2+1/t}{2-1/t} \cdot \dfrac{t}{t}$ (t is the least common denominator of $1/t$ and $1/t$) $= \dfrac{2t+1}{2t-1}$.

(ii) Simplify the numerator and denominator first. $\dfrac{2+1/t}{2-1/t} = \dfrac{(2t+1)/t}{(2t-1)/t} = \dfrac{2t+1}{t} \div \dfrac{2t-1}{t} = \dfrac{2t+1}{t} \cdot \dfrac{t}{2t-1} = \dfrac{2t+1}{2t-1}$.

1.138 $\dfrac{x+y}{x^{-1}-y^{-1}}$

▮ $\dfrac{x+y}{x^{-1}-y^{-1}} = \dfrac{x+y}{1/x-1/y} = \dfrac{x+y}{1/x-1/y} \cdot \dfrac{xy}{xy}$ (where xy is the least common denominator of $1/x$ and $1/y$) $= \dfrac{xy(x+y)}{xy/x - xy/y} = \dfrac{xy(x+y)}{y-x}$

1.139 $\dfrac{x+y}{x^{-1}+y^{-1}}$

▮ $\dfrac{x+y}{x^{-1}+y^{-1}} = \dfrac{x+y}{1/x+1/y} = \dfrac{x+y}{1/x+1/y} \cdot \dfrac{xy}{xy} = \dfrac{xy(x+y)}{y+x} = xy$.

1.140 $\dfrac{x/y - 2 + y/x}{x/y - y/x}$

▮ $\dfrac{x/y - 2 + y/x}{x/y - y/x} = \dfrac{x/y - 2 + y/x}{x/y - y/x} \cdot \dfrac{xy}{xy} = \dfrac{x^2 - 2xy + y^2}{x^2 - y^2} = \dfrac{(x-y)(x-y)}{(x+y)(x-y)} = \dfrac{x-y}{x+y}$.

1.141 $\dfrac{s^2/(s-t) - s}{t^2/(s-t) + t}$

▮ $\dfrac{s^2/(s-t) - s}{t^2/(s-t) + t} = \dfrac{s^2/(s-t) - s}{t^2/(s-t) + t} \cdot \dfrac{s-t}{s-t} = \dfrac{s^2 - s(s-t)}{t^2 + y(s-t)} = \dfrac{s^2 - s^2 + st}{t^2 + ts - t^2} = \dfrac{st}{ts} = 1$.

1.142 $1 + \dfrac{1}{1 + 1/(1 + 1/x)}$

■ $\dfrac{1}{1+1/x} = \dfrac{1}{1+1/x} \cdot \dfrac{x}{x} = \dfrac{x}{x+1}$. Thus, since $1+\dfrac{1}{x} = \dfrac{1}{\dfrac{x}{x+1}}$, $1+\dfrac{1}{1+\dfrac{1}{1+1/x}} = 1+\dfrac{1}{1+\dfrac{x}{x+1}} =$

$1+\dfrac{1}{1+\dfrac{x}{x+1}} \cdot \dfrac{x+1}{x+1} = 1+\dfrac{x+1}{1(x+1)+x(1)} = 1+\dfrac{x+1}{2x+1} = \dfrac{1(2x+1)+(x+1)}{2x+1} = \dfrac{3x+2}{2x+1}$.

For Probs. 1.143 to 1.146, rationalize the denominators, and reduce each fraction to lowest terms.

1.143 $\dfrac{\sqrt{x}}{1+\sqrt{x}}$

■ $\dfrac{\sqrt{x}}{1+\sqrt{x}} = \dfrac{\sqrt{x}}{1+\sqrt{x}} \cdot \dfrac{1-\sqrt{x}}{1-\sqrt{x}} = \dfrac{\sqrt{x}-x}{1-x}$. ($1-\sqrt{x}$ is the conjugate of $1+\sqrt{x}$. When we multiply $1+\sqrt{x}$ by $1-\sqrt{x}$, we get $1-x$, and the radical in the denominator is gone.)

1.144 $\dfrac{s\sqrt{3}}{\sqrt{3}-1}$

■ $\dfrac{s\sqrt{3}}{\sqrt{3}-1} = \dfrac{s\sqrt{3}}{\sqrt{3}-1} \cdot \dfrac{\sqrt{3}+1}{\sqrt{3}+1} = \dfrac{s+s\sqrt{3}}{3-1} = \dfrac{s+s\sqrt{3}}{2}$.

1.145 $\dfrac{1-\sqrt{x+1}}{1+\sqrt{x+1}}$

■ $\dfrac{1-\sqrt{x+1}}{1+\sqrt{x+1}} = \dfrac{1-\sqrt{x+1}}{1+\sqrt{x+1}} \cdot \dfrac{1-\sqrt{x+1}}{1-\sqrt{x+1}}$ (Remember! $1-\sqrt{x+1}$ is the conjugate of $1+\sqrt{x+1}$. It is always the case that the conjugate of $\sqrt{m}+\sqrt{n}$ is $\sqrt{m}-\sqrt{n}$; the conjugate of $\sqrt{m}-\sqrt{n}$ is $\sqrt{m}+\sqrt{n}$.) $= \dfrac{1-2(x+1)-(x+1)}{1-(x+1)} = \dfrac{1-2\sqrt{x+1}-x-1}{1-x-1} = \dfrac{-2\sqrt{x+1}-x}{-x}$.

1.146 $\dfrac{x^2}{3-\sqrt{x+3}}$

■ $\dfrac{x^2}{3-\sqrt{x+3}} = \dfrac{x^2}{3-\sqrt{x+3}} \cdot \dfrac{3+\sqrt{x+3}}{3+\sqrt{x+3}} = \dfrac{x^2(3+\sqrt{x+3})}{3-(x+3)} = \dfrac{x^2(3+\sqrt{x+3})}{-x} = -x(3+\sqrt{x+3})$.

For Probs. 1.147 and 1.148, rationalize the numerators.

1.147 $\dfrac{\sqrt{2+h}+\sqrt{2}}{h}$

■ $\dfrac{\sqrt{2+h}+\sqrt{2}}{h} = \dfrac{\sqrt{2+h}+\sqrt{2}}{h} \cdot \dfrac{\sqrt{2+h}-\sqrt{2}}{\sqrt{2+h}-\sqrt{2}} = \dfrac{2+h-2}{h(\sqrt{2+h}-\sqrt{2})} = \dfrac{h}{h(\sqrt{2+h}-\sqrt{2})} = \dfrac{1}{\sqrt{2+h}-\sqrt{2}}$.

(Notice that we used the same procedure used for rationalizing the denominator!)

1.148 $\dfrac{2\sqrt{17}-3x}{x+y}$

■ $\dfrac{2\sqrt{17}-3x}{x+y} = \dfrac{2\sqrt{17}-3x}{x+y} \cdot \dfrac{2\sqrt{17}+3x}{2\sqrt{17}+3x} = \dfrac{4\cdot17+9x^2}{(x+y)(2\sqrt{17}+3x)} = \dfrac{68+9x^2}{(x+y)(2\sqrt{17}+3x)}$.

For Probs. 1.149 and 1.150, combine into single terms.

1.149 $\sqrt{\dfrac{5a}{8}} - \sqrt{\dfrac{2a}{5}}$

■ $\sqrt{\dfrac{5a}{8}} - \sqrt{\dfrac{2a}{5}} = \dfrac{\sqrt{5a}}{\sqrt{8}} - \dfrac{\sqrt{2a}}{\sqrt{5}} = \dfrac{\sqrt{5a}}{2\sqrt{2}} - \dfrac{\sqrt{2a}}{\sqrt{5}} = \dfrac{\sqrt{5}\cdot\sqrt{5a}-2\sqrt{2}\cdot\sqrt{2a}}{2\sqrt{10}} = \dfrac{5\sqrt{a}-4\sqrt{a}}{2\sqrt{10}} = \dfrac{\sqrt{a}}{2\sqrt{10}} =$

$\dfrac{\sqrt{a}}{2\sqrt{10}} \cdot \dfrac{\sqrt{10}}{\sqrt{10}} = \dfrac{\sqrt{10a}}{20}$.

1.150 $\sqrt{\dfrac{1}{3}} - \dfrac{2}{\sqrt{3}} + \sqrt{12}$

■ $\sqrt{\dfrac{1}{3}} - \dfrac{2}{\sqrt{3}} + \sqrt{12} = \dfrac{1}{\sqrt{3}} - \dfrac{2}{\sqrt{3}} + \sqrt{12} = \dfrac{1-2+\sqrt{12}\cdot\sqrt{3}}{\sqrt{3}} = \dfrac{-1+\sqrt{36}}{\sqrt{3}} = \dfrac{-1+6}{\sqrt{3}} = \dfrac{5}{\sqrt{3}} = \dfrac{5}{\sqrt{3}} \cdot \dfrac{\sqrt{3}}{\sqrt{3}} = \dfrac{5\sqrt{3}}{3}$.

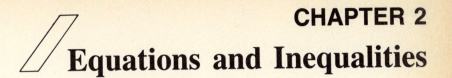

CHAPTER 2
Equations and Inequalities

2.1 LINEAR EQUATIONS

For Probs. 2.1 to 2.18, solve the given equation, if possible.   In each case, the replacement set for the variable is the set of real numbers.

2.1 $7x + 3 = 19x + 5$

$$
\begin{aligned}
7x + 3 &= 19x + 5 \\
-3 & \qquad -3 \quad \text{(Add } -3 \text{ to both sides.)} \\
\hline
7x &= 19x + 2 \\
-19x & \quad -19x \quad \text{(Add } -19x \text{ to both sides.)} \\
\hline
-\tfrac{1}{12} \cdot (-12x) &= 2 \qquad \cdot(-\tfrac{1}{12}) \text{ (Multiply both sides by } -\tfrac{1}{12}.) \\
x &= -\tfrac{2}{12} = -\tfrac{1}{6}
\end{aligned}
$$

Note that we added -3 to both sides so that constant terms could only be found on one side of the equation [in this case, the right-hand side (RHS)].   We then add $-19x$ to both sides to isolate the terms involving a variable.   We then multiplied by $-\tfrac{1}{12}x$ to transform $-12x$ into x.

2.2 $4x - 9 = 17 - 3x$

$$
\begin{aligned}
4x - 9 &= 17 - 3x \\
+9 & \quad +9 \qquad \text{(Add 9.)} \\
\hline
4x &= 26 - 3x \\
3x & \qquad 3x \quad \text{(Add } 3x.) \\
\hline
\tfrac{1}{7} \cdot 7x &= 26 \cdot \tfrac{1}{7} \qquad \text{(Multiply by } \tfrac{1}{7}.) \\
x &= \tfrac{26}{7}
\end{aligned}
$$

2.3 $9 + 14t = 17t - 11$

$$
\begin{aligned}
9 + 14t &= 17t - 11 \\
-14t &= -14t \\
\hline
9 &= 3t - 11 \\
+11 & \qquad +11 \\
\hline
\tfrac{1}{3} \cdot 20 &= 3t \cdot \tfrac{1}{3} \\
\tfrac{20}{3} &= t
\end{aligned}
$$

Note that in this example, we first isolated the variable and then added the constant.   It could have been done in reverse order.   See Prob. 2.4.

2.4 $9 + 14t = 17t - 11$

$$
\begin{aligned}
9 + 14t &= 17t - 11 \\
-9 & \qquad\qquad -9 \\
\hline
14t &= 17t - 20 \\
-17t & \quad -17t \\
\hline
-\tfrac{1}{3} \cdot (-3t) &= -20 \cdot (-\tfrac{1}{3}) \\
t &= \tfrac{20}{3}
\end{aligned}
$$

2.5 $5(2t - 6) = 4(3t - 1)$

First we eliminate the parentheses, $10t - 30 = 12t - 4$.   Don't forget the distributive law!   Now we proceed as we did in Probs. 2.1 to 2.4.

$$
\begin{aligned}
10t - 30 &= 12t - 4 \\
+30 & +30 \\
\hline
10t &= 12t - 26 \\
-12t & -12t \\
\hline
-2t &= -26 \\
t &= 13 \qquad \text{(Multiply both sides by } -\tfrac{1}{2}.\text{)}
\end{aligned}
$$

2.6 $4(3x - 6) + 2(3x - 5) = 14x$

❚ First, remove parentheses, $12x - 24 + 6x - 10 = 14x$.   Next, collect like terms on the left-hand side (LHS). $(12x + 6x) - 24 - 10 = 14x$.

$$
\begin{aligned}
18x - 34 &= 14x \\
-18x & -18x \\
\hline
-34 &= -4x \\
\tfrac{34}{4} &= x \\
x &= \tfrac{17}{2}
\end{aligned}
$$

2.7 $-2(4x + 2) = -3 - 4x - (x - 2)$

❚ Proceed as in Prob. 2.6.   Do *not* attempt to solve until you have simplified the LHS and RHS.
$-8x - 4 = -3 - 4x - x + 2$.

$$
\begin{aligned}
-8x - 4 &= -1 - 5x \\
+1 & +1 \\
\hline
-8x - 3 &= -5x \\
8x & 8x \\
\hline
-3 &= 3x \\
x &= -1
\end{aligned}
$$

2.8 $(x - 1)(2x + 1) = (x + 1)(2x - 1)$

❚ Remove parentheses:

$$
\begin{aligned}
2x^2 - x - 1 &= 2x^2 + x - 1 \\
-2x^2 & -2x^2 \\
\hline
-x - 1 &= x - 1 \\
+1 & +1 \\
\hline
-x &= x \\
x & x \\
\hline
0 &= 2x \\
x &= 0
\end{aligned}
$$

2.9 $(3x - 1)(4x + 3) = (2x + 3)(6x + 10)$

❚ Proceed as in Prob. 2.8.

$$
\begin{aligned}
\cancel{12x^2} + 5x - 3 &= \cancel{12x^2} + 38x + 30 \\
-5x - 30 & -5x - 30 \\
\hline
-33 &= 33x \\
-1 &= x
\end{aligned}
$$

2.10 $x/3 - 2 = \tfrac{1}{10} - x$

❚ Don't let the fractions bother you.   Proceed in exactly the same way as in earlier examples.

$$
\begin{aligned}
x/3 - 2 &= \tfrac{1}{10} - x \\
+2 & +2 \\
\hline
x/3 &= \tfrac{21}{10} - x \qquad (\tfrac{21}{10} = 2 + \tfrac{1}{10}) \\
+x &\phantom{=\tfrac{21}{10}} +x \\
\hline
4x/3 &= \tfrac{21}{10} \\
x &= \tfrac{21}{10} \cdot \tfrac{3}{4} = \tfrac{63}{40} \qquad \text{(Multiply both sides by } \tfrac{3}{4}.\text{)}
\end{aligned}
$$

There is an alternate way to handle this kind of problem.   See Prob. 2.11.

2.11 $x/3 - 2 = \frac{1}{10} - x$

▮ Combine fractions first on the LHS and RHS: $x/3 - \frac{2}{1} = \frac{1}{10} - x/1$,   or $x - \frac{6}{3} = 1 - 10x/10$.   Cross multiply, $10(x - 6) = 3(1 - 10x)$,   and proceed as we previously did.

$$
\begin{array}{rcl}
10x - 60 = & 3 - 30x \\
+ 30x & + 30x \\
\hline
40x - 60 = & 3 \\
+ 60 & +60 \\
\hline
40x = & 63 \\
x = & \frac{63}{40}
\end{array}
$$

2.12 $\dfrac{3}{2x - 1} + 4 = \dfrac{6x}{2x - 1}$

▮ $\dfrac{3}{2x - 1} + 4 = \dfrac{6x}{2x - 1}$ is similar to the equation in Prob. 2.13.   Combining fractions on the LHS, we get $\dfrac{3 + 4(2x - 1)}{2x - 1} = \dfrac{6x}{2x - 1}$ or $\dfrac{-1 + 8x}{2x - 1} = \dfrac{6x}{2x - 1}$.   Multiply both sides by $2x - 1$. $-1 + 8x = 6x$, $2x = 1$, or $x = \frac{1}{2}$.

2.13 $\dfrac{2}{x + 3} = \dfrac{4}{x + 4}$

▮ Cross multiply: $2(x + 4) = 4(x + 3)$, $2x + 8 = 4x + 12$, $-4 = 2x$,   or $-2 = x$.   See Prob. 2.14.

2.14 $\dfrac{2x}{x + 3} = \dfrac{4}{x + 4} + 2$

▮ $\dfrac{2x}{x + 3} = \dfrac{4 + 2(x + 4)}{x + 4}$.   (Do you see where this term came from?) $\dfrac{2x}{x + 3} = \dfrac{2x + 12}{x + 4}$.   Cross multiply: $2x(x + 4) = (2x + 12)(x + 3)$, $2x^2 + 8x = 2x^2 + 18x + 36$, $-36 = 10x$,   or $-3.6 = x$.

2.15 $4x + 9 = 4x + 11$

▮ Subtract $4x$ from both sides (i.e., add $-4x$ to both sides).   Then $9 = 11$.   Since $9 \neq 11$,   there is no solution.

2.16 $4t + 9 = -2(-2t - \frac{9}{2})$

▮ $4t + 9 = 4t + 9$ (using the distributive law).   Since $4t + 9 = 4t + 9$,   for any value of t, *any* real number is a solution.   There are infinitely many solutions.

2.17 $\dfrac{5}{x - 3} = \dfrac{33 - x}{x^2 - 6x + 9}$

▮ $\dfrac{5}{x - 3} = \dfrac{33 - x}{(x - 3)^2}$:   Cross multiply: $5(x - 3)^2 = (x - 3)(33 - x)$.   Divide both sides by $x - 3$.   This is legal since $x \neq 3$ (why?). $5(x - 3) = 33 - x$, $5x - 15 = 33 - x$, $6x = 48$,   or $x = 8$.

2.18 $\dfrac{n - 5}{6n - 6} = \dfrac{1}{9} - \dfrac{n - 3}{4n - 4}$

▮ $\dfrac{n - 5}{6(n - 1)} = \dfrac{4(n - 1) - 9(n - 3)}{9(4)(n - 1)}$,   or $\dfrac{n - 5}{6(n - 1)} = \dfrac{9(3n - 1)}{9(4)(n - 1)}$.   Multiply both sides by $n - 1$, $\dfrac{n - 5}{6} = \dfrac{3n - 1}{4}$,   and continue. $4(n - 5) = 6(3n - 1)$, $4n - 20 = 18n - 6$, $-14 = 14n$,   or $-1 = n$.

2.19   Check the result in Prob. 2.1.

▮ In Prob. 2.1, we obtained $x = -\frac{1}{6}$ as the solution for $7x + 3 = 19x + 5$. $7(-\frac{1}{6}) + 3 \overset{?}{=} 19(-\frac{1}{6}) + 5$, $-\frac{7}{6} + 3 \overset{?}{=} -\frac{19}{6} + 5$, $\dfrac{-7 + 19}{6} \overset{?}{=} \dfrac{-19 + 30}{6}$, $\frac{11}{6} = \frac{11}{6}$.   Thus, $x = -\frac{1}{6}$ is a correct solution.   This technique works for all equations.

2.20   Check the result in Prob. 2.17.

▮ $\dfrac{5}{x - 3} = \dfrac{33 - x}{x^2 - 6x + 9}$ where $x = 8$. $\dfrac{5}{8 - 3} \overset{?}{=} \dfrac{33 - 8}{8^2 - 6(8) + 9}$, $1 = \frac{25}{25}$.

2.21 Solve $p + q = rs$ for q.

▮ $$p + q = rs$$
$$\underline{-p \qquad\qquad -p}$$
$$q = rs - p$$

Notice that we treat r, s, and p as constants since we are solving for the *unknown* q.

2.22 Check the solution in Prob. 2.21.

▮ $p + q = rs$ where $q = rs - p$. $p + (rs - p) \overset{?}{=} rs$, $p + rs - p \overset{?}{=} rs$, $rs = rs$. The solution is correct.

2.23 Solve $p + q = rs$ for s, where $r \neq 0$.

▮ Multiply $p + q = rs$ on both sides by $1/r$. $(1/r)(p + q) = rs(1/r)$, or $(p + q)/r = s$ $(r \neq 0)$.

2.24 Solve $y = \dfrac{2x - 3}{3x + 5}$ for x.

▮ $y(3x + 5) = 2x - 3$, $3yx + 5y = 2x - 3$, $3yx - 2x = -3 - 5y$, $x(3y - 2) = -3 - 5y$, or $x = \dfrac{-3 - 5y}{3y - 2}$.
You should check this solution.

2.25 Let m, n be real numbers with m larger than n. Then there exists a positive real p such that $m = n + p$. Find the fallacy in the following argument:

$$m = n + p$$
$$(m - n)m = (m - n)(n + p)$$
$$m^2 - mn = mn + mp - n^2 - np$$
$$m^2 - mn - mp = mn - n^2 - np$$
$$m(m - n - p) = n(m - n - p)$$
$$m = n.$$

▮ To get to the last step from the one before it, we divided by $m - n - p$. But if $m = n + p$ (see step 1), then $m - n - p = 0$, and division by 0 is illegal.

For Probs. 2.26 to 2.30, find k if the given number is a solution of the given equation.

2.26 12, $2x + 5 = 3x + k$

▮ If $x = 12$ is a solution, then $2(12) + 5$ must equal $3(12) + k$. Then $24 + 5 = 36 + k$, $29 = 36 + k$, or $-7 = k$.

2.27 2, $x^2 + kx + 2 = 0$

▮ As we did in Prob. 2.26, we substitute 2 for x. Then $2^2 + k(2) + 2 = 0$, $4 + 2k + 2 = 0$, $2k = -6$, or $k = -3$.

2.28 $-\frac{1}{2}$, $(2x + 3)(4x + 5) = (4x + 1)(2x + k)$

▮ $[2(-\frac{1}{2}) + 3][4(-\frac{1}{2}) + 5] = [4(-\frac{1}{2}) + 1][2(-\frac{1}{2}) + k]$. Therefore, $2(3) = (-1)(-1 + k)$, $6 = 1 - k$, or $k = -5$. See Prob. 2.29.

2.29 Consider Prob. 2.28 again, but this time simplify before substituting $x = -\frac{1}{2}$.

▮ $8x^2 + 22x + 15 = 8x^2 + 2x + 4xk + k$. Then $8(-\frac{1}{2})^2 + 22(-\frac{1}{2}) + 15 = 8(-\frac{1}{2})^2 + 2(-\frac{1}{2}) + 4(-\frac{1}{2})k + k$. After doing the arithmetic, we find $k = -5$.

2.30 $-\frac{1}{2}$, $x^2(3kx - 6k - 1) + 3x + k = 0$

▮ $(-\frac{1}{2})^2[3k(-\frac{1}{2}) - 6k - 1] + 3(-\frac{1}{2}) + k = 0$, $\frac{1}{4}(-\frac{3}{2}k - 6k - 1) - \frac{3}{2} + k = 0$, $-\frac{3}{8}k - \frac{3}{2}k - \frac{1}{4} - \frac{3}{2} + k = 0$, $-\frac{7}{8}k = \frac{7}{4}$, $-28k = 56$, or $k = -2$.

2.31 The sum of two consecutive even integers is 10. Find the integers.

▮ Let $x =$ one of the integers; then $x + 2$ must be the other. (Consecutive even integers have a difference of 2.) Thus, $x + (x + 2) = 10$, $2x + 2 = 10$, $2x = 8$, or $x = 4$. If $x = 4$, then $x + 2 = 6$. The integers are 4 and 6. *Check*: 4 and 6 are consecutive evens. $4 + 6 = 10$. Thus, their sum is 10.

2.32 Find three consecutive even integers such that the sum of the first and third is twice the second.

▮ Let $x =$ the first, $x + 2 =$ the second, $x + 4 =$ the third. Then, $x + (x + 4)$ (the sum of the 1st and 3rd) $= 2(x + 2)$ (twice the second), $2x + 4 = 2x + 4,$ or $0 = 0.$ Thus, any three consecutive integers is a solution, such as 10, 12, 14.

2.33 Find three consecutive integers such that the sum of the first and twice the second is 30 more than the third.

▮ Let $x,$ $x + 1,$ $x + 2$ be the integers. Then, $x + 2(x + 1) = (x + 2) + 30,$ $3x + 2 = x + 32,$ $2x = 30,$ or $x = 15;$ thus, the integers are $15(x),$ $16(x + 1),$ and $17(x + 2).$ *Check:* 15, 16, 17 are consecutive. $15 + 2(16) = 47,$ and $30 + 17 = 47.$

2.34 The perimeter of a rectangle is 30 m, and its length is twice its width. Find the length and width of the rectangle.

▮ Let $x =$ the width. Then, $2x =$ the length (i.e., twice the width). Then, $2x + 2x + x + x = 30,$ $6x = 30,$ $x = 5$ m (width), and $2x = 10$ m (length). *Check:* $10 = 2(5).$ Also, $10 + 10 + 5 + 5 = 30.$

2.35 A rectangle 24 m long has the same area as a square that is 12 m on a side. Find the dimensions of the rectangle.

▮ If the side of the square is 12 m, then the area of the square $= S^2 = 144$ m$^2.$ The area of the rectangle $A, = 1w = 144.$ It is known that $l = 24.$ Thus, $24w = 144w = 6$ m. *Check:* $6m \cdot 24m = 144m^2 =$ area of square.

2.36 Train A leaves station Q at the same time train B leaves station $R.$ A travels at a rate of 30 mi/h directly toward $R;$ B travels at 45 mi/h directly toward $Q.$ Find how many miles A must travel before the trains meet, if the stations are 60 mi apart. (See Fig. 2.1.)

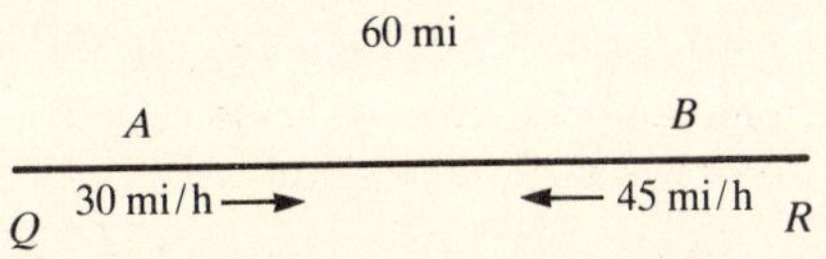

Fig. 2.1

▮ Let $x =$ the distance A travels until the trains meet. Then $60 - x =$ the distance B travels. Put that information in Fig. 2.2. Also, insert the respective rates of the trains. Note that if $d = rt,$ then $t = d/r.$ Fill in these times. Notice that the two times are equal, since both trains must travel until they meet. So, $x/30 = (60 - x)/45,$ $45x = 1800 - 30x,$ $75x = 1800,$ $x = 24 =$ the distance A travels. See Prob. 2.37.

	$d =$	r $\times$	t
A	x	30	$\dfrac{x}{30}$
B	$60 - x$	45	$\dfrac{60 - x}{45}$

Fig. 2.2

2.37 Using Prob. 2.36, find how long B travels until the trains meet.

▮ The time B travels is $(60 - x)/45.$ We found $x = 24$ in Prob. 2.36. $(60 - 24)/45 = \frac{36}{45} = \frac{4}{5}$ h $=$ 48 min.

2.38 The sale price on a camera after a 20 percent discount is $72. What was the price before the discount?

▮ Let $x =$ price before discount. Then, $x - 20\%x =$ price after discount. $x - 2x = 72,$ $8x = 72,$ or $x = 72 \times \frac{10}{8} = \$90.$

2.39 Suppose John can paint a particular room in 6 h. At what minimum rate must his helper be able to paint the room alone if they must together complete the job in $3\frac{3}{7}$ h?

▮ Let $x =$ the time in hours it would take John's helper to paint the room. If it takes x hours to complete, then $1/x$ of the job could be completed in 1 h. Also, if John can complete the job in 6 h, he will complete $\frac{1}{6}$ of it in 1 h. Then, in 1 h $\frac{1}{6} + 1/x = 1/\frac{24}{7}$ $(\frac{24}{7} = 3\frac{3}{7})$, $\frac{1}{6} + 1/x = \frac{7}{24}$, $(x+6)/6x = \frac{7}{24}$, $24x + 144 = 42x$, $144 = 18x$, or $x = 8$ h. His helper must be able to paint the room working alone in 8 h. *Check*: $\frac{1}{6} + \frac{1}{8} = \frac{7}{24} = 1/\frac{24}{7}$.

2.40 Barbara is twice as old as Mary, and Dick is three times as old as Barbara. Their average age is 36. How old is Barbara?

▮ Let $x =$ Mary's age. Then $2x =$ Barbara's age, and $6x =$ Dick's age. $\dfrac{6x + 2x + x}{3} = 36$ (average of ages), $9x = 108$, so $x = 12$ years old, $2x = 24$ years old $=$ Barbara's age, and $6x = 72$ years old $=$ Dick's age.

2.41 Janet has nickels and dimes in her pocket. Their total value is $1, and there are twice as many dimes as nickels. How many nickels does she have?

▮ Let $x =$ number of nickels, $2x =$ number of dimes. Then, $5(x) + 10(2x) = 100$ $(100¢ = \$1)$, $5x + 20x = 100$, $25x = 100$, or $x = 4 =$ number of nickels. *Check*: $2x = 8 =$ number of dimes. $4(5) + 8(10) = 100¢ = \$1$.

2.2 NONLINEAR EQUATIONS

For Probs. 2.42 to 2.71, solve the given equation. Leave any answer containing a radical in simplest radical form.

2.42 $x^2 = 4$

▮ If $x^2 = 4$, then x is that number or numbers which, when squared, is 4. Thus, $x = \pm 2$. Notice that, to get this result, we found the square roots ($+$ and $-$) of the constant term 4 isolated on one side of the equation. Notice that there is no linear term in this equation. *Check*: $2^2 = 4$; $(-2)^2 = 4$.

2.43 $2t^2 = 18$

▮ $2t^2 = 18$, $t^2 = 9$, or $t = \pm 3$. *Check*: $2(3)^2 = 18$; $2(-3)^2 = 18$.

2.44 $p^2 = 14$

▮ $p^2 = 14$, or $p = \pm\sqrt{14}$. *Check*: $(\sqrt{14})^2 = 14$; $(-\sqrt{14})^2 = 14$.

2.45 $2t^2 = 80$

▮ $2t^2 = 80$, $t^2 = 40$, or $t = \pm\sqrt{40} = \pm\sqrt{4 \cdot 10} = \pm\sqrt{4}\sqrt{10} = \pm 2\sqrt{10}$. *Check*: $2(2\sqrt{10})^2 = 2(4 \cdot 10) = 80$; $2(-2\sqrt{10})^2 = 2(4 \cdot 10) = 80$.

2.46 $x^2 - 8 = 0$

▮ $x^2 - 8 = 0$, $x^2 = 8$, or $x = \pm\sqrt{8} = \pm\sqrt{4}\sqrt{2} = \pm 2\sqrt{2}$.

2.47 $t^2 + 5 = 20$

▮ $t^2 + 5 = 0$, or $t^2 = -5$. But no real number squared is negative. Thus, the solutions must be complex numbers. If, $t^2 = 5$, then $t = \pm\sqrt{-5} = \pm i\sqrt{5}$. (See Sec. 2.6 for a review of complex numbers. Also, see Chap. 10 for a complete treatment of complex numbers.)

2.48 $4x^2 - 7 = 0$

▮ $4x^2 - 7 = 0$, $4x^2 = 7$, $x^2 = \dfrac{7}{4}$, $x = \pm\sqrt{\dfrac{7}{4}} = \pm\dfrac{\sqrt{7}}{\sqrt{4}} = \pm\dfrac{\sqrt{7}}{2}$.

2.49 $(n+5)^2 = 9$

▮ $(n+5)^2 = 9$, or $n + 5 = \pm 3$. *Possibility 1*: $n + 5 = 3$, or $n = -2$. *Possibility 2*: $n + 5 = -3$, or $n = 8$. So the answer is $n = -2, -8$.

2.50 $x^2 + 3x + 2 = 0$

▮ Factoring the LHS: $(x+2)(x+1) = 0$ ($x+2$ or $x+1$ or both must be 0). If $x + 2 = 0$, then $x = -2$. If $x + 1 = 0$, then $x = -1$. *Check*: $(-2)^2 + 3(-2) + 2 = 4 - 6 + 2 = 0$, or $0 = 0$. Also, $(-1)^2 + 3(-1) + 2 = 0$, $1 - 3 + 2 = 0$, or $0 = 0$.

2.51 $t^2 + 12t + 35 = 0$

$\blacksquare$ Factoring, we get $(t+5)(t+7) = 0$. If $t+5 = 0$, then $t = -5$. If $t+7 = 0$, then $t = -7$.

2.52 $4x^2 - 4x - 3 = 0$

$\blacksquare$ $(2x-3)(2x+1) = 0$. If $2x-3 = 0$, then $x = \frac{3}{2}$. If $2x+1 = 0$, then $x = -\frac{1}{2}$.

2.53 $2t^2 - 2t = 12$

$\blacksquare$ $2t^2 - 2t = 12$, $2t^2 - 2t - 12 = 0$, or $(t-3)(2t+4) = 0$. If $(t-3) = 0$, then $t = 3$. If $2t = -4$, $t = -2$. Note that we would have divided through by 2 initially. See Prob. 2.13.

2.54 $2t^2 - 2t = 12$

$\blacksquare$ $2t^2 - 2t = 12$, $t^2 - t = 6$, $t^2 - t - 6 = 0$, or $(t-3)(t+2) = 0$. If $t-3 = 0$, then $t = 3$. If $t+2 = 0$, then $t = -2$.

2.55 $10x^2 - 52x + 64 = 0$

$\blacksquare$ $(2x-4)(5x-16) = 0$. If $2x-4 = 0$, then $x = 2$. If $5x-16 = 0$, then $x = \frac{16}{5}$.

2.56 $p^4 - 2p^2 + 1 = 0$

$\blacksquare$ $(p^2-1)(p^2-1) = 0$, $p^2 = 1$, or $p = \pm 1$.

2.57 $x^3 + 3x^2 + 2x = 0$

$\blacksquare$ $x(x^2 + 3x + 2) = 0$, $x(x+2)(x+1) = 0$, or $x = 0, -2, -1$.

2.58 $x^2 + x - 1 = 0$

$\blacksquare$ This does not appear factorable. Using the quadratic formula $x = \dfrac{-b \pm \sqrt{b^2 - 4ac}}{2a}$ $(a \neq 0)$, we have $a = 1$, $b = 1$, $c = -1$, and $x = \dfrac{-1 \pm \sqrt{1^2 - 4(1)(-1)}}{2(1)} = \dfrac{-1 \pm \sqrt{5}}{2}$. Thus, $x = \dfrac{-1 + \sqrt{5}}{2}$ or $x = \dfrac{-1 - \sqrt{5}}{2}$.

2.59 $2t^2 + t - 4 = 0$

$\blacksquare$ Again, factoring does not appear to work. We have $a = 2$, $b = 1$, $c = -4$; $t = \dfrac{-1 \pm \sqrt{1^2 - 4(1)(-4)}}{2(2)} = \dfrac{-1 \pm \sqrt{17}}{4}$.

2.60 $2x^2 + 4x + 1 = 0$

$\blacksquare$ Here, $a = 2$, $b = 4$, $c = 1$. Thus, $x = \dfrac{-4 \pm \sqrt{16 - 8}}{4} = \dfrac{-4 \pm \sqrt{8}}{4} = \dfrac{-4 \pm 2\sqrt{2}}{4} = \dfrac{-2 \pm \sqrt{2}}{2}$.

2.61 $2x^2 + 8x - 42 = 0$

$\blacksquare$ The left-hand side here is factorable into $(2x-6)(x+7)$. If, however, we had not noticed that, we could have solved the equation as follows: $x = \dfrac{-8 \pm \sqrt{64 - 4(2)(-42)}}{4} = \dfrac{-8 \pm \sqrt{400}}{4} = \dfrac{-8 \pm 20}{4}$. So, $x = \dfrac{-8+20}{4} = 3$ or $x = \dfrac{-8-20}{4} = -7$.

2.62 $x^2 + 4x + 1 = 0$

$\blacksquare$ Here we will illustrate the method of "completing the square." $x^2 + 4x + \textcircled{4}$ is a perfect square, since $x^2 + 4x + \textcircled{4} = (x+2)^2$. To obtain the circled 4 above, we find the coefficient of the x term (here it is 4), divide it by 2, and square it: $(4/2)^2 = 2^2 = 4$. $x^2 + 4x + 1 \neq x^2 + 4x + 4$, but $x^2 + 4x + 1 = (x^2 + 4x + 4) - 3$. So, the equation $x^2 + 4x + 1 = 0$ is rewritten as $(x^2 + 4x + 4) - 3 = 0$, $(x+2)^2 - 3 = 0$, $(x+2)^2 = 3$, $x+2 = \pm\sqrt{3}$, $x = \pm\sqrt{3} - 2$, or $x = \sqrt{3} - 2, -\sqrt{3} - 2$.

2.63 $t^2 - 6t + 2 = 0$

▮ $t^2 - 6t + 2 = (t^2 - 6t + 9) - 7$. Thus, $t^2 - 6t + 2 = 0$ is rewritten as $(t^2 - 6t + 9) - 7 = 0$ or $t^2 - 6t + 9 = 7$, $(t-3)^2 = 7$, $t - 3 = \pm\sqrt{7}$, or $t = 3 \pm \sqrt{7}$.

2.64 $2x^2 + 3x - 9 = 0$

▮ $2(x^2 + \frac{3}{2}x - \frac{9}{2}) = 0$ or $x^2 + \frac{3}{2}x - \frac{9}{2} = 0$. Computing the square we get: $(x^2 + \frac{3}{2}x + \frac{9}{4}) - \frac{9}{2} - \frac{9}{4} = 0$, $(x + \frac{3}{2})^2 = \frac{27}{4}$, $x + \frac{3}{2} = \pm\sqrt{\frac{27}{4}}$, $x + \frac{3}{2} = \pm\sqrt{27}/2$, $x = (\pm\sqrt{27}/2)/2 - \frac{3}{2}$, $x = \sqrt{27}/2 - \frac{3}{2} = (\sqrt{27} - 3)/2$, or $x = -\sqrt{27}/2 - \frac{3}{2} = -(\sqrt{27} - 3)/2$.

2.65 $-x^2 + 3x - 2 = 0$

▮ By factoring we get $(-x + 1)(x - 2) = 0$, or $x = 1, 2$. Or we can rewrite the equation as $x^2 - 3x + 2 = 0$ and factor. See Prob. 2.66.

2.66 $-x^2 + 3x - 2 = 0$

▮ Using the quadratic formula: $x = \dfrac{-3 \pm \sqrt{9 - 4(-1)(-2)}}{2} = \dfrac{-3 \pm \sqrt{1}}{-2} = 1$ or 2. See Prob. 2.67.

2.67 $-x^2 + 3x - 2 = 0$

▮ Rewrite the equation as $x^2 - 3x + 2 = 0$ and complete the square. $(x^2 - 3x + \frac{9}{4}) + 2 - \frac{9}{4} = 0$, $(x - \frac{3}{2})^2 = \frac{1}{4}$, $x - \frac{3}{2} = \pm\sqrt{\frac{1}{4}} = \pm\frac{1}{2}$, or $x = 1, 2$.

2.68 $(x^2 - 3)(x^2 - 4) = 0$

▮ Either $(x^2 - 3)$ or $(x^2 - 4)$ (or both) $= 0$. So, $x^2 - 3 = 0$, $x = \pm\sqrt{3}$, or $x^2 - 4 = 0$, $x = \pm 2$. Thus, $x = 3, -3, 2$, or -2.

2.69 $\left(\dfrac{2x + 1}{3x + 1}\right)^2 - 9 = 0$

▮ $\left(\dfrac{2x + 1}{3x + 1}\right)^2 = 9$, $\dfrac{2x + 1}{3x + 1} = \pm\sqrt{9} = \pm 3$. If $\dfrac{2x + 1}{3x + 1} = 3$, then $2x + 1 = 9x + 3$, $-7x = 2$, or $x = -\frac{2}{7}$.
If $\dfrac{2x + 1}{3x + 1} = -3$, then $2x + 1 = -9x - 3$, $11x = -4$, or $x = -\frac{4}{11}$.

2.70 $(2x + 3)^2 + 4(2x + 3) + 3 = 0$

▮ Let $u = 2x + 3$. Then, $u^2 + 4u + 3 = 0$, $(u + 1)(u + 3) = 0$, $u = -1, -3$. Thus, if $2x + 3 = -1$, then $x = -2$. If $2x + 3 = -3$, then $x = -3$.

2.71 $8y^{-2} - 6y^{-1} + 1 = 0$

▮ Let $u = 1/y = y^{-1}$. Then $u^2 = (y^{-1})^2 = y^{-2}$, $8u^2 + 6u + 1 = 0$, $(4u + 1)(2u + 1) = 0$, or $u = -\frac{1}{4}, -\frac{1}{2}$. Then, $1/y = -\frac{1}{4}$ or $-\frac{1}{2}$, or $y = -4, -2$.

For Probs. 2.72 and 2.73, find the discriminant.

2.72 $Z^2 + 5Z - 6 = 0$

▮ The discriminant of a quadratic equation $ax^2 + bx + c = 0$ is $b^2 - 4ac$. In this case, $b^2 - 4ac = 5^2 - 4(1)(-6) = 25 + 24 = 49$.

2.73 $S^2 + 5S = 3$

▮ $a = 1$, $b = 5$, $c = -3$. $b^2 - 4ac = 25 - 4(1)(-3) = 25 + 12 = 37$.

For Probs. 2.74 to 2.76, determine whether the equation has real roots or not.

2.74 $x^2 + 11x + 11 = 0$

▮ Recall that when the discriminant is positive or zero, the equation has real roots. If the discriminant is zero, it has one real root; otherwise, it has two. If the discriminant is negative, the equation has imaginary roots. Here, $b^2 - 4ac = 11^2 - 4(1)(11) > 0$. Conclusion: Two real roots.

2.75 $x^2 - 3x + \frac{9}{4} = 0$

$\blacksquare$ $b^2 - 4ac = (-3)^2 - 4(1)(\frac{9}{4}) = 9 - 9 = 0$. Conclusion: One real root.

2.76 $2x^2 + x + 1 = 0$

$\blacksquare$ $b^2 - 4ac = 1^2 - 4(2)(1) < 0$. Conclusion: No real roots.

For Probs. 2.77 to 2.81, reduce the given equation to a quadratic and solve.

2.77 $\sqrt{x + 2} = x$

$\blacksquare$ Square both sides. $x + 2 = x^2$, $x^2 - x - 2 = 0$, $(x - 2)(x + 1) = 0$, or $x = 2, -1$. *Check*: When $x = 2$, $\sqrt{x + 2} = \sqrt{2 + 2} = \sqrt{4} = 2$. This checks. When $x = -1$, $\sqrt{x + 2} = \sqrt{-1 + 2} = \sqrt{1} = 1 \neq -1$. This does *not* check. Conclusion: $x = 2$. (*Note*: When you square an equation, you may pick up extraneous roots. It is imperative that you check all solutions.)

2.78 $3 + \sqrt{2x - 1} = 0$

$\blacksquare$ $3 + 2x - 1 = 0$, $\sqrt{2x - 1} = -3$, $2x - 1 = 9$, $2x = 10$, or $x = 5$. *Check*: $3 + \sqrt{10 - 1} = 3 + 3 = 6 \neq 0$. This equation has no solution. Notice that it did not reduce to a quadratic.

2.79 $\sqrt{3w - 2} - \sqrt{w} = 2$

$\blacksquare$ $\sqrt{3w - 2} = 2 + \sqrt{w}$, $3w - 2 = (2 + \sqrt{w})^2$ (squaring), $3w - 2 = 4 + 4\sqrt{w} + w$, $4\sqrt{w} = 6 - 2w$, $16w = (6 - 2w)^2$ (squaring), $16w = 36 - 24w + 4w^2$, $4w^2 - 40w + 36 = 0$, $w^2 - 10w + 9 = 0$, $(w - 9)(w - 1) = 0$, or $w = 9, 1$. *Check*: $\sqrt{3w - 2} - \sqrt{w} \stackrel{?}{=} 2$. If $w = 9$, $\sqrt{27 - 2} - \sqrt{9} = 5 - 3 = 2$. This checks. If $w = 1$, $\sqrt{3 - 2} - \sqrt{1} = \sqrt{1} - \sqrt{1} = 0$. This does *not* check. Conclusion: $w = 9$.

2.80 $y - 6 + \sqrt{y} = 0$

$\blacksquare$ $\sqrt{y} = 6 - y$, $y = (6 - y)^2 = 36 - 12y + y^2$, $y^2 - 13y + 36 = 0$, $(y - 9)(y - 4) = 0$, or $y = 9, 4$. *Check*: $y = 9$ is extraneous. $y = 4$ is the solution. See Prob. 2.81.

2.81 $y - 6 + \sqrt{y} = 0$

$\blacksquare$ Let $u = \sqrt{y}$, then $u^2 = y$, so $u^2 - 6 + u = 0$, $u^2 + u - 6 = 0$, $(u + 3)(u - 2) = 0$, or $u = 3, 2$. $u = -3$ is extraneous, but $u = \sqrt{y}$, so $\sqrt{y} = 2, 4$.

2.82 Find two consecutive positive integers whose product is 210.

$\blacksquare$ Let $x =$ the first number. Then $x + 1 =$ the second number. If their product is 210, then $x(x + 1) = 210$, $x^2 + x = 210$, $x^2 + x - 210 = 0$, $(x - 14)(x + 15) = 0$, $x = 14, -15$. $x = -15$ is extraneous (it is nonpositive). Thus, $x = 14$, $x + 1 = 15$ (check this!).

2.83 Find the length of a side of an equilateral triangle if the triangle's area is the same as the area of a square with a side of 5 m.

$\blacksquare$ If the square has side $x = 5$ m, then its area $= 25$ m^2. Let $S =$ the length of the side of the triangle. Then the area of the triangle is $S^2\sqrt{3}/4$. Thus, $S^2\sqrt{3}/4 = 25$, $S^2\sqrt{3} = 100$, $S^2 = 100/\sqrt{3} = 100\sqrt{3}/3$, $S = \pm\sqrt{100\sqrt{3}/3}$. The negative result is extraneous (length can never be negative). So, $S = \sqrt{\dfrac{100\sqrt{3}}{3}} = \sqrt{\dfrac{100}{3}}\sqrt{\sqrt{3}} = \dfrac{10}{\sqrt{3}}\sqrt[4]{3} = \dfrac{10}{3^{1/2}} \cdot 3^{1/4} = \dfrac{10}{3^{1/4}} = 10 \cdot 3^{-1/4}$ m.

2.84 A number x has the property that the sum of x and twice its reciprocal is three times the number. Find x.

$\blacksquare$ Let $x =$ the number and $1/x =$ the reciprocal. Then $x + 2(1/x) = 3x$, $x + 2/x = 3x$, $\dfrac{x^2 + 2}{x} = 3x$, $x^2 + 2 = 3x^2$, $2x^2 = 2$, $x^2 = 1$, or $x = \pm 1$. $x = 1$ and $x = -1$ are both solutions. *Check*: $1 + 2(1/1) = 3 = 3(1)$, $-1 + 2(1/-1) = -3 = 3(-1)$.

2.85 If P dollars are invested at r percent compounded annually, at the end of 2 years the amount will be $A = P(1 + r)^2$. At what interest rate will \$1000 increase to \$1400 in 2 years?

$\blacksquare$ $A = P(1 + r)^2$, where $A = 1440$ and $P = 1000$. $1440 = 1000(1 + r)^2$, $(1 + r)^2 = 1.44$, $1 + r = \pm\sqrt{1.44}$ (reject the negative!), $1 + r = \sqrt{1.44} = 1.2$, or $r = 0.2$. The rate is 20 percent.

2.86 Solve for h: $h^2 + q^2 = 5x$

▮ $h^2 + q^2 = 5x$, $h^2 = 5x = q^2$, $h = \pm\sqrt{5x - q^2}$.

2.87 Solve for I: $P = EI - RI^2$

▮ $P = EI - RI^2$, $-RI^2 + EI - P = 0$, or $RI^2 - EI + P = 0$. Letting $a = R$, $b = -E$, $c = P$,
$I = \dfrac{E \pm \sqrt{E^2 - 4RP}}{2R}$.

2.88 Show that if $ax^2 + bx + c = 0$ has one real root, then $b = \pm\sqrt{4ac}$.

▮ If $ax^2 + bx + c = 0$ has one real root, then $b^2 - 4ac = 0$. $b^2 = 4ac$, or $b = \pm\sqrt{4ac}$.

2.89 Show that if $ax^2 + bx + a = 0$ has one real root, then $b = 2a$ or $b = -2a$.

▮ If $ax^2 + bx + a = 0$ has one real root, then $b^2 - 4(a)(a) = 0$. Then, $b^2 - 4a^2 = 0$, $b^2 = 4a^2$, $b^2/a^2 = 4$ or $(b/a)^2 = 4$, so, $b/a = 2$ or $-2a$. Thus, $b = 2a$ or $-2a$.

2.90 Find all values of R so that $x^2 + (R + 3)x + 4R = 0$ has one real root.

▮ The discriminant $(R + 3)^2 - 4(1)(4R) = 0$. Thus, $R^2 + 6R + 9 - 16R = 0$, $R^2 - 10R + 9 = 0$, $(R - 9)(R - 1) = 0$, or $R = 9, 1$.

2.91 The height (in feet) of a ball thrown vertically upwards above the ground in t s is given by $h = 128t - 16t^2$. In how many seconds will the ball be 192 ft high?

▮ If the height is 192 ft, then $192 = 128t - 16t^2$, $16t^2 - 128t + 192 = 0$, $t^2 - 8t + 12 = 0$, or $(t - 6)(t - 2) = 0$; after 2 s and after 6 s. *Note*: The ball goes upward, stops, and turns downward. That is why it reaches 192 ft twice. It first reaches 192 ft after 2 s.

2.3 LINEAR INEQUALITIES

Probs. 2.92 to 2.96, rewrite the given sentence using inequality symbols and the equal sign if necessary.

2.92 Five times t is more than three times y.

▮ $5t > 3y$

2.93 Five times t is three more than three times y.

▮ $5t = 3 + 3y$

2.94 Five times t is more than three more than three times y.

▮ $5t > 3 + 3y$

2.95 Five times t is less than or equal to three less than three times negative y.

▮ $5t \le 3(-y) - 3$ or $5t \le -3y - 3$

2.96 $3y$ less than four times z is nonpositive.

▮ $4z - 3y \le 0$

For Probs. 2.97 to 2.100, a and b are real numbers such that $a < b$. How are the given pairs of numbers related?

2.97 $a - 4$ and $b - 4$.

▮ If $a < b$, then $a - 4 < b - 4$

2.98 $-3a$ and $-3b$

▮ If $a < b$, then $3a < 3b$ and $-3a > -3b$ (the inequality sign reverses).

2.99 $2-a$ and $2-b$.

▮ If $a<b$,   then $-a>-b$,   and $-a+2>-b+2$,   or $2-a>2-b$.

2.100 $-a/4$ and $-b/4$.

▮ If $a<b$,   then $-a>-b$,   and $-a\cdot\frac{1}{4}>-b\cdot\frac{1}{4}$,   or $-a/4>-b/4$.

For Probs. 2.101 to 2.103, tell whether the given number is a solution of the given inequality.

2.101 $-x+5\geq 0$;   3

▮ Substituting 3 for x, $-x+5=-3+5=2$ and $2\geq 0$;   3 does satisfy the inequality.

2.102 $4/x+3\geq 1/x$; $\frac{1}{2}$

▮ When $x=\frac{1}{2}$, $4/x+3=4/\frac{1}{2}+3=11$,   and $1/x=2$,   so $11\geq 2$; $\frac{1}{2}$ does satisfy the inequality.

2.103 $x^{-1}+1<x^{-2}-2$;   1

▮ When $x=1$, $x^{-1}+1=1^{-1}+1=2$ and $x^{-2}-2=1^{-2}-2=-1$,   so $2>-1$;   1 does not satisfy the inequality.

For Probs. 2.104 to 2.122, solve the given inequality (if possible).

2.104 $5t>1+2t$

▮ $5t>1+2t$, $3t>1$,   or $t>\frac{1}{3}$.

2.105 $14s\leq -3s-4$

▮ $17s\leq -4$,   or $s\leq -\frac{4}{17}$.

2.106 $-3s>4s+2$

▮ $-7s>2$,   or $s<-\frac{2}{7}$.   (Notice the sign reversal!)

2.107 $4x+9\geq -7x-16$

▮ $4x+9\geq -7x-16$, $11x\geq -25$,   or $x\geq -\frac{25}{11}$.

2.108 $-3(4t-8)\leq 6(t+5)$

▮ $-3(4t-8)\leq 6(t+5)$, $-12t+24\leq 6t+30$, $-18t\leq 6$, $t\geq -\frac{6}{18}$,   or $t\geq -\frac{1}{3}$.   See Prob. 2.109.

2.109 $-3(4t-8)\leq 6(t+5)$

▮ $-12t+24\leq 6t+30$, $-6\leq 18t$, $-\frac{6}{18}\leq t$, $-\frac{1}{3}\leq t$,   or $t\geq -\frac{1}{3}$.

Notice the difference between the methods in Probs. 2.108 and 2.109.   It makes little difference whether the unknown is isolated on the RHS or LHS of the equation.

2.110 $\frac{1}{6}-\frac{2}{3}z\geq z/12+8$

▮ $\frac{1}{6}-\frac{2}{3}z\geq z/12+8$, $\frac{6}{12}-4z/12\geq z/12+\frac{96}{12}$, $-\frac{90}{12}\geq \frac{5z}{12}$, $-90\geq 5z$, $-\frac{90}{5}\geq z$, $z\leq -\frac{90}{5}$,   or $z\leq -18$.

2.111 $9(x-3)-(2x+4)\leq 4x+\frac{3}{2}$

▮ $-x+3-2x-4\leq 2x+\frac{3}{2}$, $-3x-1\leq 2x+\frac{3}{2}$, $-5x\leq \frac{5}{2}$,   or $x\leq -\frac{1}{2}$.

2.112 $\dfrac{x-2}{6}+\dfrac{2x+3}{3}\geq 0$

▮ $\dfrac{x-2}{6}+\dfrac{2(2x+3)}{6}\geq 0$, $(x-2)+2(2x+3)\geq 0$ (here we multiply both sides by 6, and 0)$\cdot(6=0)$, $x-2+4x+6\geq 0$, $5x+4\geq 0$, $5x\geq -4$,   or $x\geq -\frac{4}{5}$.

2.113 $\dfrac{5x}{x+5} \le 2 - \dfrac{25}{x+5}$

▮ Multiply both sides by $x+5$. Then, $5x \le 2(x+5)-25$, $5x \le 2x+10-25$, $3x \le -15$, or $x \le -5$. But $x \ne 5$, since $x+5$ appears in the denominator. Thus, $x < 5$ is the solution. Notice that we can see that $\dfrac{5x}{x+5} = 2 - \dfrac{25}{x+5}$ has *no* solution.

2.114 $5x - 3 < \frac{1}{3}(15x - 9)$

▮ $5x - 3 < 5x - 3$ or $0 < 0$. But $0 = 0$. Conclusion: No solution.

2.115 $2x/5 - \frac{1}{2}(x-3) \le 2x/3 - \frac{3}{10}(x+2)$

▮ $2x/5 - x/2 + \frac{3}{2} \le 2x/3 - 3x/10 - \frac{6}{10}$, $\dfrac{12x - 15x + 45}{30} \le \dfrac{20x - 9x - 18}{30}$, $-3x + 45 \le 11x - 18$, $-14x \le -63$, or $x \ge \frac{63}{14}$.

2.116 $2x - a < -3x + p$

▮ $-5x < p + a$, or $x > \dfrac{p+a}{-5}$.

2.117 $-tx + q \le sx - r$. Solve for x.

▮ $-tx - sx \le -r - q$, or $x(-t-s) \le -r - q$. *Case 1:* If $-t-s > 0$, then $x \le \dfrac{-r-q}{-t-s}$. *Case 2:* If $-t-s < 0$, then $x \le \dfrac{-r-q}{-t-s}$. *Case 3:* If $-t-s = 0$, then $0 \le -r-q$, or $r \le -q$. Then, if $r \le -q$, any real x satisfies the equation. If $r > -q$, there is no solution.

2.118 $-6 < x + 2 < 10$

▮ $-6 < x + 2 < 10$, or $-8 < x < 8$ (adding -2 to all three parts of the inequality).

2.119 $-6 \le 2x - 3 \le -1$

▮ $-3 \le 2x \le 2$, or $-\frac{3}{2} \le x \le 1$. (See Prob. 3.120.)

2.120 $-6 \le 2x - 3 < -1$ (See Prob. 3.119.) Then $-\frac{3}{2} \le x < 1$.

2.121 $-10 < 16 - 2x < 10$

▮ $-26 < -2x < -6$. Dividing by -2, $13 > x > 3$. Notice the reversal of both inequality signs.

2.122 $5 < \dfrac{x-3}{4} \le 10$

▮ $20 < x - 3 \le 40$, or $23 < x \le 43$.

For Probs. 2.123 to 2.127, an inequality is given. It is one of those in the above group. Graph the solution, and indicate the solution using interval notation.

2.123 $14s \le -3s - 4$ (2.105)

▮

$(-\infty, -\frac{4}{17}]$

2.124 $4x + 9 \ge -7x - 16$ (2.107)

▮

$[-\frac{25}{11}, \infty)$

2.125 $\dfrac{5x}{x+5} \le 2 - \dfrac{25}{x+5}$ (2.113)

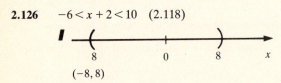

$(\infty, -5)$

2.126 $-6 < x + 2 < 10$ (2.118)

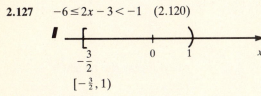

$(-8, 8)$

2.127 $-6 \le 2x - 3 < -1$ (2.120)

[figure: number line with bracket at $-\frac{3}{2}$ and parenthesis at 1]

$-\dfrac{3}{2}$ 0 1 x

$[-\frac{3}{2}, 1)$

For Probs. 2.128 to 2.132, $a, b < 0$ and $a > b$. Tell whether each statement is true or false. If it is false, give a counterexample.

2.128 $a^2 > b^2$

▮ False; For example, $-5 > -6$, but $25 < 36$.

2.129 $ab < b^2$

▮ True; If $a < b$, then when we multiply by b on both sides, we reverse the inequality since $b < 0$.

2.130 $a(a-b) > b(a-b)$

▮ True; $a > b$ implies $a - b > 0$. Multiplying $a > b$ on both sides by $a - b$ maintains the inequality sign.

2.131 $ab(a-b) > 0$

▮ True; $a > b$ implies $a - b > 0$, which means $a(a-b) < 0$ and $ab(a-b) > 0$ (note that the signs are reversed).

2.132 $\dfrac{(a-b)^3}{b} > 0$

▮ False; $a - b > 0$, thus $(a-b)^3 > 0$ and $(a-b)^3/b < 0$. For example, let $a = -1$ and $b - 2$.

2.133 If both a and b are negative and b/a is greater than 1, then is $a - b$ positive or negative?

▮ $b/a > 1$. Then $b < a$. (Careful! The sign reverses since $a < 0$.) So $0 < a - b$ and $a - b > 0$.

2.134 Assume that $m > n > 0$; then $mn > n^2$, $mn - m^2 > n^2 - m^2$, $m(n-m) > (n+m)(n-m)$, $m > n + m$, or $0 > n$. But we assumed that $n > 0$. Find the error.

▮ To get from $n(n-m) > (n+m)(n-m)$, we divided by $n - m$. But $n - m < 0$; thus, we needed to reverse the inequality sign. However, notice that we did not.

2.135 *Prove*: For any real numbers x, y, and z, if $x < y$ then $x + z < y + z$.

▮ If $x < y$, then $x - y < 0$. But $x - y = (x-y) = (z-z)$ which means $(x-y) + (z-z) < 0$, $(x+z) - (y+z) < 0$, or $x + z < y + z$.

2.136 *Prove*: If $a < b$ and $b < c$ (where a, b, and c are real numbers), then $a < c$.

▮ If $a < b$, then $a - b < 0$; if $b < c$, then $b - c < 0$. Then, $(a-b) + (b-c) < 0$ and $(a-c) + (b-b) < 0$ or $a - c < 0$. Therefore, $a < c$.

2.137   The area of a square does not exceed 25 cm². Find the possible integral values of a side of the square.

▮ $A = S^2 \leq 25 \text{ cm}^2$. If $S = 1 \text{ cm}$, then $S^2 = 1 \text{ cm}^2$. If $S = 2 \text{ cm}$, then $S^2 = 4 \text{ cm}^2$. If $S = 5 \text{ cm}$, then $S^2 = 25 \text{ cm}^2$. Conclusion: $1 \text{ cm} \leq 3 \leq 5 \text{ cm}$.

For Probs. 2.138 to 2.140 use the following: Linda bought some pizzas at $5.40 a piece. The total bill was more than $12, but did not exceed $23.

2.138   Let $x = $ the number of pizzas bought. Use the above information to write an inequality in x.

▮ $x = $ number of pizzas bought. $5.40 = $ price of 1 pizza. Then $5.40x = $ total amount spent. We know then that $5.40x \leq 23$.

2.139   Solve the inequality in Prob. 2.138.

▮ $5.4x < 23$ or $x < 23/5.4$.

2.140   Find how many pizzas Linda may have purchased.

▮ $x < 23/5.4$ where $x = $ number of pizzas bought. $23/5.4 = 4.25 \ldots$ which means she bought 1, 2, 3, or 4. She could not have purchased 5 or more.

For Probs. 2.141 and 2.142, let a and b be positive integers whose sum is less than 12.

2.141   If we know that a is either 1 or 2, how many possible choices are there for b?

▮ $a + b < 12$ $(a, b > 0)$. If $a = 1$, then $1 + b < 12$ or $b < 11$, so b can be $1, 2, \ldots, 10$ (10 choices). If $a = 2$, then $2 + b < 12$ or $b < 10$, so b can be $1, 2, \ldots, 9$ (9 choices). Total number of choices $= 19$.

2.142   How many possibilities are there altogether for a and b?

▮
$a = 1$ 10 choices for b.
$a = 2$ 9 choices for b.
$a = 3$ 8 choices for b.
$a = 4$ 7 choices for b.
$a = 5$ 6 choices for b.
$a = 6$ 5 choices for b.
$a = 7$ 4 choices for b.
$a = 8$ 3 choices for b.
$a = 9$ 2 choices for b.
$a = 10$ 1 choice for b.
   Total: 55 possibilities.

2.4 ABSOLUTE VALUE

For Probs. 2.143 to 2.192, solve each equation or inequality, if possible.

2.143 $|x| = 9$

▮ Recall that $|x|$ has two common equivalent definitions.

$$|x| = \begin{cases} x, & x \geq 0 \\ -x, & x < 0 \end{cases} \quad \text{or} \quad |x| = \sqrt{x^2}$$

If $|x| = 9$, then $x = 9$ or $x = -9$. *Check*: $|9| = 9$, $|-9| = 9$. Note that this check works using either of the definitions above: $\sqrt{9^2} = 9$ and $\sqrt{(-9)^2} = 9$.

2.144 $|x| > 9$

▮ Recall that $|x - a| < b$ can be interpreted as all x whose distance from a is less than b.

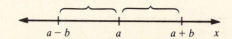

In Fig. 2.3, notice that the interval $(a - b, a + b)$ satisfies that criterion. In our problem, $a = 0$, $b = 9$. Thus $|x| < 9$ has the solution $-9 < x < 9$ $(-9 = 0 - 9, 9 = 0 + 9)$. More generally: Let $a > 0$; then

> $|x| = a$ has the solution $x = \pm a$
>
> $|x| < a$ has the solution $-a < x < a$
>
> $|x| > a$ has the solution $x > a$, $x < -a$
>
> $|cx| + d < a$ can be written as $-a < cx + d < a$
>
> $|cx| + d > a$ can be written as $cx + d > a$, $cx + d < -a$

Fig. 2.3

2.145 $|x| \geq 14$

▮ $|x| \geq 4$ has solutions $x \geq 4$ and $x \leq -4$. (See Fig. 2.3.) The graph of this solution set is

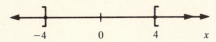

2.146 $|t| \geq 0$

▮ Then $t \geq 0$ or $t \leq 0$ which implies that t can be *any* real number. Note that this makes sense. $|t|$ is never negative!

2.147 $|x| < -5$

▮ This inequality has no solution: $|x| \geq 0$ for all x.

2.148 $|x - 4| = 6$

▮ Then $x - 4 = 6$ or $x - 4 = -6$. Then $x = 10$ or $x = -2$.

2.149 $|-t + 4| = 2$

▮ $-t + 4 = 2$ or $-t + 4 = -2$. Thus, $t = 2$ or $t = 6$.

2.150 $|2t| = 8$

▮ $2t = 8$ or $2t = -8$. Thus, $t = 4$ or $t = -4$.

2.151 $|-3x| = 15$

▮ (i) $-3x = 15$ or $-3x = -15$. Thus, $x = -5$ or $x = 5$. (ii) $|-3x| = |-3||x| = 3|x|$. $3|x| = 15$, $|x| = 5$, so $x = 5$ or $x = -5$.

2.152 $|2x + 1| = 4$

▮ $2x + 1 = 4$ or $2x + 1 = -4$. Thus, $x = \frac{3}{2}$ or $x = -\frac{5}{2}$.

2.153 $|2s - 6| = 10$

▮ $2s - 6 = 10$ or $2s - 6 = -10$, $2s = 16$ or $2s = -4$, so $s = 8$ or $s = -2$.

2.154 $|-2t + 4| = -2$

▮ $-2t + 4 = -2$ or $-2t + 4 = 2$, $-2t = -6$ or $-2t = -2$, so $t = 3$ or $t = 1$.

2.155 $|2x| < 1$ (See Fig. 4.1.)

▮ Then $-1 < 2x < 1$ or $-\frac{1}{2} < x < \frac{1}{2}$. Using interval notation, $(-\frac{1}{2}, \frac{1}{2})$ is the solution set.

2.156 $|2s| \leq 2$

▮ $-2 \leq 2s \leq 2$ or $-1 \leq s \leq 1$. The graph of the solution set is

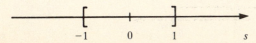

2.157 $|2t| > 1$ (See Fig. 4.1.)

❚ $2t > 1$ or $2t < -1$. Then $t > \frac{1}{2}$ or $t < -\frac{1}{2}$.

2.158 $|x - 1| < 2$

❚ $-2 < x - 1 < 2$ or $-1 < x < 3$. In interval form, $(-1, 3)$.

2.159 $|t - z| \le 4$

❚ $-4 \le t - 2 \le 4$, or $-2 \le t \le 6$.

2.160 $|t - 2| \ge 5$

❚ $t - 2 \ge 5$ or $t - 2 \le -5$. Thus, $t \ge 7$ or $t \le -3$.

2.161 $|2x + 1| < 7$

❚ $-7 < 2x + 1 < 7$, $-8 < 2x < 6$, or $-4 < x < 3$.

2.162 $|2x - 5| < 10$

❚ $-10 < 2x - 5 < 10$, $-5 < 2x < 15$, or $-\frac{5}{2} < x < \frac{15}{2}$.

2.163 $|-2x + 2| < 2$

❚ $-2 < -2x + 2 < 2$, $-4 < -2x < 0$, or $2 > x > 0$. (Careful! The signs reversed because of division by -2.)

2.164 $|-3t - 2| < 6$

❚ $-6 < -3t - 2 < 6$, $-4 < -3t < 8$, or $\frac{4}{3} > t > -\frac{8}{3}$.

2.165 $|2x + 1| > 6$

❚ $2x + 1 > 6$ or $2x + 1 < -6$, $2x > 5$ or $2x < -7$, so $x > \frac{5}{2}$ or $x < -\frac{7}{2}$.

2.166 $|2s - 2| > 8$

❚ $2s - 2 > 8$ or $2s - 2 < -8$, $2s > 10$ or $2s < -6$, so $s > 5$ or $s < -3$.

2.167 $|2x - 4| \ge 10$

❚ $2x - 4 \ge 10$ or $2x - 4 \le -10$, $2x \ge 14$ or $2x \le -6$, so $x \ge 7$ or $x \le -3$.

2.168 $|-x + 11| \ge 5$

❚ $-x + 1 \ge 5$ or $-x + 1 \le -5$, $-x \ge 4$ or $-x \le -6$, so $x \le -4$ or $x \ge 6$.

2.169 $|-s - 1| \ge 7$

❚ $-s - 1 \ge 7$ or $-s - 1 \le -7$, $-s \ge 8$ or $-s \le -6$, so $s \le -8$ or $s \ge 6$.

2.170 $|-2x + 4| > 6$

❚ $-2x + 4 > 6$ or $-2x + 4 < -6$, $-2x > 2$ or $-2x < -10$, so $x < -1$ or $x > 5$.

2.171 $|-3x - 2| \ge 10$

❚ $-3x - 2 \ge 10$ or $-3x - 2 \le -10$, $-3x \ge 12$ or $-3x \le -8$, so $x \le -4$ or $x \ge \frac{8}{3}$.

2.172 $|1 - x| < 5$

❚ $-5 < 1 - x < 5$, $-6 < -x < 4$, or $6 > x > -4$. (Careful!)

2.173 $|2 - t| \le 6$

❚ $-6 \le 2 - t \le 6$, $-8 \le -t \le 4$, or $8 \ge t \ge -4$.

2.174 $|-3-s| \le 5$

▮ $-5 \le -3-s \le 5$, $-2 \le -s \le 8$, or $2 \ge s \ge -8$.

2.175 $|1-2x| \le 8$

▮ $-8 \le 1-2x \le 8$, $-9 \le -2x \le 7$, or $\frac{9}{2} \ge x \ge -\frac{7}{2}$.

2.176 $|2-3x| < 6$

▮ $-6 < 2-3x < 6$, $-8 < -3x < 4$, or $\frac{8}{3} > x > -\frac{4}{3}$.

2.177 $|-3-4x| \le 12$

▮ $-12 \le -3-4x \le 12$, $-9 \le -4x \le 15$, or $\frac{9}{4} \ge x \ge -\frac{15}{4}$.

2.178 $|1-t| > 11$

▮ $1-t > 11$ or $1-t < -11$, $-t > 10$ or $-t < -12$, so $t < -10$ or $t > 12$.

2.179 $|2-s| > 10$

▮ $2-s > 10$ or $2-s < -10$, $-s > 8$ or $-s < -12$, so $s < -8$ or $s > 12$.

2.180 $|2-3x| \ge 13$

▮ $2-3x \ge 13$ or $2-3x \le -13$, $-3x \ge 11$ or $-3x \le 15$, so $x \le -\frac{11}{3}$ or $x \ge 5$.

2.181 $|-1-2x| > 2$

▮ $-1-2x > 2$ or $-1-2x < -2$, $-2x > 3$ or $-2x < -1$, $x < -\frac{3}{2}$ or $x > \frac{1}{2}$.

2.182 $\left|\dfrac{x-1}{2}\right| < 3$

▮ (i) $-3 < \dfrac{x-1}{2} < 3$, $-3 < x/2 - \frac{1}{2} < 3$, $-\frac{5}{2} < x/2 < \frac{7}{2}$, $-5 < x < 7$.

(ii) $|x-1| < 6$ ($|2| = 2$, and cross multiply). Then $-6 < x-1 < 6$, or $-5 < x < 7$.

2.183 $\left|\dfrac{2x-3}{4}\right| \le 4$

▮ $|2x-3| \le 16$, $-16 \le 2x-3 \le 16$, $-13 \le 2x \le 19$, or $-\frac{13}{2} \le x \le \frac{19}{2}$.

2.184 $\left|\frac{1}{2}(1-2x)\right| < 10$

▮ $|1-2x| < 20$, $-20 < 1-2x < 20$, $-21 \le -2x \le 19$, or $\frac{21}{2} \ge x \ge -\frac{10}{2}$.

2.185 $|x+1| = x+1$

▮ Recall that $|p| = p$ when $p \ge 0$. Thus, $|x+1| = x+1$, when $x+1 \ge 0$, or $x \ge -1$.

2.186 $|2s-3| = 3-2s$

▮ If $|2s-3| = 3-2s$, then $|2s-3| = -(2s-3)$. But $|p| = -p$ only when $p \le 0$. Thus, $2s-3 \le 0$, $2s \le 3$, so $s \le \frac{3}{2}$.

2.187 $|x+2| < x+2$

▮ No solution. $|x+2| \ge x+2$ for all x. Note that $|x+2| < x+2$ would be rewritten as $-(x+2) < x+2 < x+2$ which is impossible.

2.188 $|t^2| = 9$

▮ Then $t^2 = 9$ or $t^2 = -9$, but $t^2 \ne 9$. So, $t^2 = 9$, $t = \pm 3$.

2.189 $|x^2 - 2| = 7$

▮ Then $x^2 - 2 = \pm 7$. If $x^2 - 2 = 7$, then $x^2 = 9$, $x = \pm 3$. If $x^2 - 2 = -7$, then $x^2 = -5$ which is impossible.

2.190 $|2t - 1| < -3$

▮ No solution. $|2t - 1| \geq 0$ for all t.

2.191 $|-\frac{5}{2}(|x - 16|)| \geq 0$

▮ x can be any real number, since $|a| \geq 0$ for all a.

2.192 Solve $|2x - p| > s$ for x.

▮ $2x - p > s$ or $2x - p < -s$, $2x > p + s$ or $2x < -s + p$, so $x > \dfrac{p + s}{2}$ or $2x < \dfrac{-s + p}{2}$.

2.5 NONLINEAR INEQUALITIES

For Probs. 2.193 to 2.205, solve the given inequality. Graph the solution set for each even-numbered exercise, and express each odd-numbered solution set in interval form.

2.193 $x^2 - 2x + 1 > 0$

▮ Factoring the LHS, we get $(x - 1)(x - 1) > 0$, which means that (1) Both factors on the LHS are > 0, or (2) both factors on the LHS are < 0. Thus, (1) $x - 1 > 0$, $x > 1$, or (2) $x - 1 < 0$, $x < 1$. ($x - 1$ is the only factor!) Conclusion: $x > 1$ or $x < 1$. In interval form: $(-\infty, 1)$ or $(1, \infty)$. Using set notation: $(-\infty, 1) \cup (1, \infty)$.

2.194 $x^2 - 3x + 2 > 0$

▮ $(x - 2)(x - 1) > 0$. (1) $x - 2 > 0$, $x - 1 > 0$; $x > 2$, $x > 1$, or (2) $x - 2 < 0$, $x - 1 < 0$; $x < 2$, $x < 1$. Simplifying: (1) If $x > 2$ and $x > 1$, then $x > 2$, since $x > 2$ guarantees us that $x > 1$. (2) If $x < 2$ and $x < 1$, then $x < 1$, since $x < 1$ guarantees that $x < 2$. Conclusion: $x > 2$ or $x < 1$.

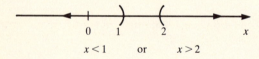

$$x < 1 \qquad \text{or} \qquad x > 2$$

2.195 $x^2 - 3x + 2 \geq 0$

▮ See Prob. 5.194 above. All $>$ signs will be replaced by $\geq$ signs. Conclusion: $x \geq 2$ or $x \leq 1$, or in interval form, $[2, \infty) \cup (-\infty, 1]$.

2.196 $x^2 - 3x + 2 < 0$

▮ If $(x - 2)(x - 1) < 0$, then (1) $x - 2 > 0$, $x - 1 < 0$, or (2) $x - 2 < 0$, $x - 1 > 0$. In (1) $x > 2$ and $x < 1$, which is impossible. In (2) $x < 2$ and $x > 1$. Conclusion: $x < 2$ and $x > 1$, or $1 < x < 2$.

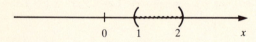

2.197 $x^2 - 3x + 2 \leq 0$

▮ See Prob. 5.196 above. Replacing $<$ by $\leq$, we get $x \leq 2$ and $x \geq 1$. In interval form, $[1, 2]$.

2.198 $x^2 + x - 6 \leq 0$

▮ $(x + 3)(x - 2) \leq 0$. Then (1) $x + 3 \leq 0$ and $x - 2 \geq 0$, or (2) $x + 3 \geq 0$ and $x - 2 \leq 0$. Thus, $x \leq -3$ and $x \geq 2$, which is impossible, or $x \geq -3$ and $x \leq 2$. Conclusion: $x \geq -3$ and $x \leq 2$, i.e., $-3 \leq x \leq 2$.

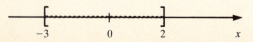

2.199 $x^2 + x - 6 > 0$

▮ $(x + 3)(x - 2) > 0$. Then (1) $x + 3 > 0$, $x - 2 > 0$, or (2) $x + 3 < 0$, $x - 2 < 0$. Thus, $x > -3$ and $x > 2$, which means $x > 2$, or $x < -3$ and $x < 2$, which means $x < -3$. Conclusion: $x > 2$ or $x < -3$, or, in interval form, $(\infty, -3) \cup (2, \infty)$.

2.200 $2x^2 - 5x - 3 \le 0$

▮ $(2x + 1)(x - 3) \le 0$. Then (1) $2x + 1 \le 0$, $x - 3 \ge 0$, or (2) $2x + 1 \ge 0$, $x - 3 \le 0$. Thus, $x \le -\frac{1}{2}$ and $x \ge 3$ (which is impossible); or $x \ge -\frac{1}{2}$ and $x \le 3$. Conclusion: $x \ge -\frac{1}{2}$ and $x \le 3$.

2.201 $3x^2 + x > 0$

▮ $x(3x + 1) > 0$. Then (1) $x > 0$ and $3x + 1 > 0$, or (2) $x < 0$ and $3x + 1 < 0$. Then, $x > 0$ and $x > -\frac{1}{3}$, which means $x > 0$, or $x < 0$ and $x < -\frac{1}{3}$, which means $x < -\frac{1}{3}$. Conclusion: $x > 0$ or $x < -\frac{1}{3}$, or, in interval form, $(0, \infty) \cup (-\infty, -\frac{1}{3})$.

2.202 $3x^2 < -2x + 1$

▮ If $3x^2 < -2x + 1$, then $3x^2 + 2x - 1 < 0$. Thus, $(3x - 1)(x + 1) < 0$ which means (1) $3x - 1 < 0$, $x + 1 > 0$ or (2) $3x - 1 > 0$, $x + 1 < 0$. Then $x < \frac{1}{3}$ and $x > -1$, or $x > \frac{1}{3}$ and $x < -1$ which is impossible. Conclusion: $x < \frac{1}{3}$ and $x > -1$.

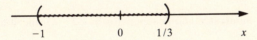

2.203 $S^2 \le 1$

▮ $S^2 \le 1$ means $S^2 - 1 \le 0$. Then $(S - 1)(S + 1) \le 0$, which means (1) $S - 1 \le 0$, $S + 1 \ge 0$, or (2) $S - 1 \ge 0$, $S + 1 \le 0$. Then $S \le 1$ and $S \ge -1$, or $S \ge 1$ and $S \le -1$ which is impossible. Conclusion: $S \le 1$ and $S \ge -1$ or $[-1, 1]$. See Prob. 2.204.

2.204 $S^2 \le 1$

▮ Here is another method. If $S^2 \le 1$, then $\sqrt{S^2} \le \sqrt{1}$. Recall that $\sqrt{S^2} = |S|$. Then $|S| \le 1$ or $-1 < S < 1$. Notice the result is the same as in Prob. 2.203.

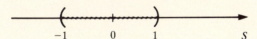

2.205 $x^3 - x > 0$

▮ $x(x^2 - 1) > 0$. Thus, (1) $x > 0$, $x^2 - 1 > 0$, or (2) $x < 0$, $x^2 - 1 < 0$. If (1) $x > 0$, $(x - 1)(x + 1) > 0$, then $x - 1 > 0$, $x + 1 > 0$ which means $x > 1$, $x > -1$, which means $x > 1$; or $x - 1 < 0$, $x + 1 < 0$ which means $x < 1$, $x < -1$ which means $x < -1$. Thus, $x > 0$ and $(x > 1$ or $x < -1)$, or $x > 0$ and $x > 1$, or $x > 1$. If (2) $x < 0$, $x - 1 < 0$, then $x + 1 > 0$ which means $x < 1$ and $x > -1$; or $x - 1 > 0$, $x + 1 < 0$ which means $x > 1$ and $x < -1$ which is impossible. Thus, $x < 0$ and $(x < 1$ and $x > -1)$, which means $x > -1$ and $x < 0$. Conclusion: $x > 1$ or $(x > -1$ and $x < 0)$, or, in interval form, $(1, \infty) \cup (-1, 0)$.

2.6   REVIEW OF COMPLEX NUMBERS

For Probs. 2.206 to 2.215, perform the indicated operations.   Write the answer in the form $a + bi$.

2.206 $(1 + i) + (3 - 2i)$

▮ Recall that $(a + bi) \pm (c + di) = (a + c) + (b + d)i$. Then $(1 + i) + (3 - 2i) = 4 + (-1)i = 4 - i$.

2.207 $(-6 + 4i) + (2 - i)$

▮ $(-6 + 4i) + (2 - i) = -4 + 3i$.

2.208 $(2-i)-(3-4i)$

▮ $(2-i)-(3-4i)=(2-3)+(-1+4)i=-1+3i.$

2.209 $(-3-i)-(-2-3i)$

▮ $-3-i-(-2-3i)=(-3+2)+(-1+3)i=-1+2i.$

2.210 $(2-i)+(4i-3)$

▮ $(2-i)+(4i-3)=(2-i)+(-3+4i)=-1+3i.$

2.211 $3i+(4-2i)$

▮ $3i+(4-2i)=(0+3i)+(4-2i)=4+i.$

2.212 $(2+i)(3+2i)$

▮ Recall that $(a+bi)(c+di)=(ac-bd)+(bc+ad)i.$ Then $(2+i)(3+2i)=2(3)-1(2)+(3+4)i=4+7i.$

2.213 $(3-i)(i-6)$

▮ $(3-i)(i-6)=(3-i)(-6+i)=(3)(-6)-(-1)(1)+(6+3)i=19+9i.$

2.214 $\dfrac{2}{3+i}$

▮ We simplify this expression by multiplying the numerator and denominator by $3-i,$ the conjugate of $3+i.$ $\dfrac{2}{3+i}=\dfrac{2}{3+i}\cdot\dfrac{3-i}{3-i}=\dfrac{6-2i}{9+1}=\dfrac{6-2i}{10}=\dfrac{3}{5}-\dfrac{i}{5}$.

2.215 $\dfrac{2+i}{1-i}$

▮ $\dfrac{2+i}{1-i}=\dfrac{2+i}{1-i}\cdot\dfrac{1+i}{1+i}=\dfrac{1+3i}{1+1}=\dfrac{1+3i}{2}=\dfrac{1}{2}+\dfrac{3}{2}\,i.$

For Probs. 2.216 to 2.221, find the indicated power of $i.$

2.216 i^3

▮ $i^3=i^2i=(-1)i=-i.$

2.217 i^4

▮ $i^4=(i^2)(i^2)=(-1)(-1)=1.$

2.218 i^7

▮ $i^7=i^4i^3=1(-i)=-i.$

2.219 i^8

▮ $i^8=i^4i^4=1\cdot1=1.$

2.220 i^{15}

▮ $i^{15}=(i^4)^3i^3=1^3\cdot(-i)=-i.$

2.221 i^{30}

▮ $i^{30}=(i^4)^7i^2=1^7(-1)=1(-1)=-1.$

For Probs. 2.222 to 2.225, perform the indicated operations and write all answers in the form $a+bi.$

2.222 $(4-\sqrt{-16})+(2+\sqrt{-25})$

▮ $\sqrt{-16}=4i;\ \sqrt{-25}=5i.$ (*Recall*: $\sqrt{-16}=\sqrt{-1\cdot16}=\sqrt{-1}\sqrt{16}=i\cdot4=4i.$) $(4-\sqrt{-16})+(2+\sqrt{-25})=(4-4i)+(2+5i)=6+i.$

2.223 $(2 + i) - (4 + \sqrt{-49})$

$\quad\blacksquare\quad$ $(2 + i) - (4 + \sqrt{-49}) = (2 + i) - (4 + 7i) = -2 - 6i.$

2.224 $(2 + \sqrt{-9})(1 + \sqrt{-4})$

$\quad\blacksquare\quad$ $(2 + \sqrt{-9})(1 + \sqrt{-4}) = (2 + 3i)(1 + 2i) = -4 + 7i.$

2.225 $\dfrac{1}{i}$

$\quad\blacksquare\quad$ $\dfrac{1}{i} = \dfrac{1}{0 + i} = \dfrac{1}{0 + i} \cdot \dfrac{0 - i}{0 - i} = \dfrac{-i}{-i^2} = \dfrac{-i}{1} = -i.$

2.226 Prove that $i^{4k} = 1$ for all natural numbers k.

$\quad\blacksquare\quad$ $i^{4k} = (i^4)^k = 1^k = 1.$

2.227 Solve for x and y: $\quad 3 - 2i = 4xi + 2y$.

$\quad\blacksquare\quad$ Two complex numbers $a + bi$ and $c + di$ are equal if and only if $a = c$ and $b = d$. $\quad 3 - 2i = 4xi + 2y$ is rewritten as $\quad 3 - 2i = 2y + 4xi$; then $\quad 2y = 3$ and $\quad 4x = -2$, or $\quad y = \frac{3}{2}$ and $\quad x = -\frac{1}{2}$.

For Probs. 2.228 to 2.231, solve the given equation.

2.228 $x^2 - x + 1 = 0$

$\quad\blacksquare\quad$ $x = \dfrac{-b \pm \sqrt{b^2 - 4ac}}{2a} = \dfrac{1 \pm \sqrt{1^2 - 4(1)(1)}}{2} = \dfrac{1 \pm \sqrt{-3}}{2} = \dfrac{1 \pm i\sqrt{3}}{2}.$

2.229 $2x^2 - 2x + 3 = 0$

$\quad\blacksquare\quad$ $x = \dfrac{2 \pm \sqrt{4 - 24}}{4} = \dfrac{2 \pm \sqrt{-20}}{4} = \dfrac{2 \pm 2\sqrt{-5}}{4} = \dfrac{1 \pm i\sqrt{5}}{2}.$

2.230 $x^2 + 5 = 0$

$\quad\blacksquare\quad$ $x^2 = -5$ or $x = \pm\sqrt{-5} = \pm i\sqrt{5}.$

2.231 $x^3 + 8 = 0$

$\quad\blacksquare\quad$ $x^3 = -8$ or $x = \sqrt[3]{-8} = -2.$ (Compare this with Prob. 2.230 above.)

2.232 Prove that 0 is the additive identity for the complex numbers.

$\quad\blacksquare\quad$ Write 0 as $0 + 0i$. Then $(a + bi) + 0 = (a + bi) + (0 + 0i) = (a + 0) + (b + 0)i = a + bi$. Thus, $0 + 0i = 0 = $ additive identity.

2.233 Find the additive inverse of $c + di$, and prove that it is the additive inverse.

$\quad\blacksquare\quad$ We claim $-c - di$ is the additive inverse. $c + di + (-c - di) = c + (-c) + [d + (-d)]i = 0 + 0i = $ the additive identity. Thus, $-c - di$ is the additive inverse for $c + di$.

2.234 Prove that 1 is the multiplicative identity for the complex numbers.

$\quad\blacksquare\quad$ $1 = 1 + 0i$. $(a + bi)(1) = (a + bi)(1 + 0i) = a(1) - b(0) + (b + 0)i = a + bi$. Therefore, $1 = 1 + 0i$ is the multiplicative identity.

2.235 Find x and y such that $(3 + 2i)(x + yi) = 1$.

$\quad\blacksquare\quad$ $(3 + 2i)(x + yi) = (3x - 2y) + (2x + 3y)i$. Then $(3 + 2i)(x + yi) = (3x - 2y) + (2x + 3y)i = 1 = 1 + 0i$. Thus, $3x - 2y = 1$ and $2x + 3y = 0$. Solving these simultaneously (see Chapter 5 for the background on this if you need it),

$$
\begin{array}{r}
6x - 4y = 2 \\
6x + 9y = 0 \\
\hline
-13y = 2 \\
y = -\tfrac{2}{13}
\end{array}
$$

$6x - 4(-\tfrac{2}{13}) = 2, \quad 6x + \tfrac{8}{13} = 2 = \tfrac{26}{13}, \quad 6x = \tfrac{18}{13}, \quad$ or $\quad x = \tfrac{18}{78} = \tfrac{9}{39}.$

CHAPTER 3

Graphs, Relations, and Functions

3.1 THE CARTESIAN COORDINATE SYSTEM

3.1 Plot each of the following points in the cartesian coordinate system: $A(0, 1)$, $B(-1, 3)$, $C(-1, -4)$, $D(1, 3)$.

▮ See Fig. 3.1.

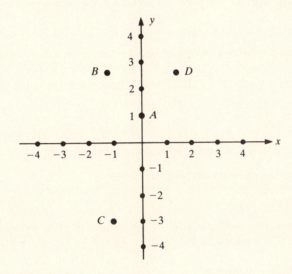

Fig. 3.1

For Probs. 3.2 to 3.5, find the distance between the given points.

3.2 $A(0, 1)$ and $B(1, 0)$.

▮ The distance between two points A and B in the plane $= d(A, B) = AB = \sqrt{(x_2 - x_1)^2 + (y_2 - y_1)^2}$, where A has coordinates (x_1, y_1) and B has coordinates (x_2, y_2). Then, in this case, $d(A, B) = \sqrt{(1-0)^2 + (0-1)^2} = \sqrt{1+1} = \sqrt{2}$.

3.3 $A(0, 0)$ and $B(3, 4)$.

▮ $AB = \sqrt{(3-0)^2 + (4-0)^2} = \sqrt{9+16} = \sqrt{25} = \sqrt{5}$. Notice that since $(x_1 - x_2)^2 = (x_2 - x_1)^2$, the order of subtraction is unimportant.

3.4 $A(1, 3)$ and $B(-1, 4)$.

▮ $d(A, B) = \sqrt{[1-(-1)]^2 + (4-3)^2} = \sqrt{2^2 + 1^2} = \sqrt{5}$.

3.5 $P(1, 3)$ and $Q(1, 5)$

▮ $PQ = \sqrt{(1-1)^2 + (3-5)^2} = \sqrt{0+4} = 2$. Notice that since P and Q have the same abscissa, we can also get the result by finding $|5-3| = |3-5| = 2$. This would also be the case if the ordinates were identical.

For Probs. 3.6 to 3.11, find the midpoint of the segment joining the two given points.

3.6 $(1, 2)$ and $(3, 1)$.

▮ The midpoint M of the segment joining (x_1, y_1) and (x_2, y_2) has coordinates $\left(\dfrac{x_1 + x_2}{2}, \dfrac{y_1 + y_2}{2}\right)$. Thus, in this case M has coordinates $\left(\dfrac{1+3}{2}, \dfrac{2+1}{2}\right) = (2, \tfrac{3}{2})$.

3.7 $(-3, 1)$ and $(3, 1)$.

▮ The midpoint has coordinates $\left(\dfrac{-3+3}{2}, \dfrac{1+1}{2}\right) = (0, 1)$.

3.8 $(-3, -5)$ and $(-4, -6)$.

▮ M has coordinates $\left(\dfrac{-3-4}{2}, \dfrac{-5-6}{2}\right) = \left(\dfrac{-7}{2}, \dfrac{-11}{2}\right)$.

3.9 $(\sqrt{2}, 1)$ and $(2, \sqrt{3})$.

▮ M is the point with coordinates $\left(\dfrac{\sqrt{2}+2}{2}, \dfrac{1+\sqrt{3}}{2}\right)$.

3.10 $(p, 2q)$ and $(q, 2p)$.

▮ M has coordinates $\left(\dfrac{p+q}{2}, \dfrac{2p+2q}{2}\right) = \left(\dfrac{p+q}{2}, p+q\right)$.

For Probs. 3.11 and 3.12, tell whether the points all lie on a circle with given center P.

3.11 $P(0, 0)$; $(0, 1)$, $(-1, 0)$, $(\sqrt{2}/2, \sqrt{2}/2)$.

▮ We need to find the distance from each given point to $(0, 0)$. If these distances are all the same, the points all lie on a circle with P as the center. $d((0, 1, P) = \sqrt{1^2 + 0^2} = 1$. $d((-1, 0), P) = \sqrt{(-1)^2 + 0^2} = 1$. $d((\sqrt{2}/2, \sqrt{2}/2), P) = \sqrt{(\sqrt{2}/2)^2 + (\sqrt{2}/2)^2} = 1$. Thus, the three points do lie on the circle.

3.12 $P(1, 2)$; $(4, 7)$, $(-2, -2)$, $(6, -1)$.

▮ $d((4, 7), P) = \sqrt{3^2 + 5^2}$. $d((-2, -3), P) = \sqrt{3^2 + 5^2}$. $d((6, -1), P) = \sqrt{5^2 + 3^2}$. Since the three distances are the same, the points do lie on the same circle with center P.

For Probs. 3.13 to 3.17, find the equation of the given circle.

3.13   The center is at $(0, 0)$,   radius $= 4$.

▮ The equation of a circle with center (h, k) and radius r is $(x - h)^2 + (y - k)^2 = r^2$. In this case, $h = 0$, $k = 0$, $r = 4$. The equation of the circle is $(x - 0)^2 + (y - 0)^2 = 4^2$, or $x^2 + y^2 = 16$.

3.14   The center is at $(1, 2)$,   radius $= 6$.

▮ $h = 1$, $k = 2$, $r = 6$. The equation of the circle is $(x - 1)^2 + (y - 2)^2 = 6^2$, or $(x - 1)^2 + (y - 2)^2 = 36$.

3.15   The center is at $(-2, -3)$, radius $= 1$.

▮ $h = -2$, $k = -3$, $r = 1$. The equation of the circle is $[x - (-2)]^2 + [y - (-3)]^2 = 1^2$, or $(x + 2)^2 + (y + 3)^2 = 1$.

3.16   The center is at $(-2, 3)$,   radius $= \sqrt{2}$.

▮ $h = -2$, $k = 3$, $r = \sqrt{2}$. $(x + 2)^2 + (y - 3)^2 = 2$.

3.17   The center is at $A(0, 0)$ and the point $B(1, 1)$ lies on the circle.

▮ Since $(1, 1)$ lies on the circle, $d(A, B) =$ radius of the circle. $d(A, B) = \sqrt{1^2 + 1^2} = \sqrt{2}$. Then, $h = 0$, $k = 0$, $r = \sqrt{2}$. The equation is $x^2 + y^2 = 2$.

For Probs. 3.18 to 3.21, find the center and radius of the given circle.

3.18 $x^2 + (y - 1)^2 = 4$

▮ Since $(x - h)^2 + (y - k)^2 = r^2$ is the equation of the circle, in this case $h = 0$, $k = 1$, $r = 2$. The center is at $(0, 1)$ and the radius is 2.

3.19 $(x + 1)^2 + (y + 2)^2 = 3$

▮ $h = -1$, $k = -2$, $r = \sqrt{3}$. The center is at $(-1, -2)$ and the radius is $\sqrt{3}$.

3.20 $x^2 + y^2 - 4y = 0$

▮ We must put this equation in the form $(x - h)^2 + (y - k)^2 = r^2$. Complete the square on $y^2 - 4y$, making sure you add the required 4 to both sides: $x^2 + (y^2 - 4y + 4) = 0 + 4$. $x^2 + (y - 2)^2 = 4$. The center is at $(0, 2)$ and the radius is 2.

3.21 $x^2 + y^2 + 2x - 2y = -1$

▮ Regrouping terms, we obtain $(x^2 + 2x) + (y^2 - 2y) = -1$. Completing the square, $(x^2 + 2x + 1) + (y^2 - 2y + 1) = -1 + (1 + 1)$, or $(x + 1)^2 + (y - 1)^2 = 1$. The center is at $(-1, 1)$ and the radius is 1.

3.22 Graph the equation in Prob. 3.21.

▮ See Fig. 3.2.

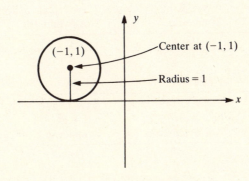

Fig. 3.2

For Probs. 3.23 to 3.30, decide whether the graph of the equation exhibits symmetry with respect to the x axis or origin. Do not graph the equation to determine whether the symmetry exists.

3.23 $y = x + 1$

▮ The tests for symmetry are the following:
(1) Replace x with $-x$. If the equation *does not* change, the graph is symmetric with respect to the y axis.
(2) Replace y with $-y$. If the equation *does not* change, the graph is symmetric with respect to the x axis.
(3) Replace x with $-x$ and y with $-y$. If the equation *does not* change, the graph is symmetric with respect to the origin.

(1) In $y = x + 1$, replace x with $-x$: then $y = -x + 1$. This is a change. The graph is not symmetric with respect to the y axis. (2) Replace y with $-y$. Then $-y = x + 1$, or $y = -x - 1$. Not symmetric with respect to the x axis. (3) Replace x with $-x$ and y with $-y$. Then $-y = -x + 1$, or $y = x - 1$. Not symmetric with respect to the origin.

3.24 $y = x$

x axis	y axis	Origin
$-y = x$	$y = -x$	$-y = -x$
$y = x$		$y = x$
Not Symmetric	Not Symmetric	Symmetric

3.25 $x^2 + y^2 = 1$

x axis	y axis	Origin
$x^2 + (-y)^2 = 1$	$(-x)^2 + y^2 = 1$	$(-x)^2 + (-y)^2 = 1$
$x^2 + y^2 = 1$	$x^2 + y^2 = 1$	$x^2 + y^2 = 1$
Symmetric	Symmetric	Symmetric

3.26 $x^2 + (y - 1)^2 = 1$

x axis	y axis	Origin
$x^2 + (-y - 1)^2 = 1$	$(-x)^2 + (y - 1)^2 = 1$	$(-x)^2 + (-y - 1)^2 = 1$
$x^2 + (y + 1)^2 = 1$	$x^2 + (y - 1)^2 = 1$	$x^2 + (y + 1)^2 = 1$
Not Symmetric	Symmetric	Not Symmetric

3.27 $y = x^2 + 1$

    ▌ Since $(-x)^2 = x^2$, we see y axis symmetry. It does not exhibit x axis or origin symmetry.

3.28 $y = |2x|$

▌

x axis	y axis	Origin
$-y = |2x|$	$y = |2(-x)|$	$-y = |2(-x)|$
or $y = -|2x|$	or $y = |-2x|$	$-y = |2x|$
	$y = |2x|$	$y = -|2x|$
Not Symmetric	Symmetric	Not Symmetric

3.29 $y = x^3$

▌

x axis	y axis	Origin
$-y = x^3$	$y = (-x)^3$	$-y = (-x)^3$
$y = x^3$	or $y = -x^3$	$y = x^3$
Not Symmetric	Not Symmetric	Symmetric

3.30 $|x| + 1 = y$

▌

x axis	y axis	Origin
$-y = |x| + 1$	$y = |-x| + 1$	$-y = |-x| + 1$
	or $y = |x| + 1$	$y = -|x| + 1$
Not Symmetric	Symmetric	Not Symmetric

3.31   Graph $y = x + 1$.

    ▌ See Fig. 3.3. See Prob. 3.23 above. Notice the lack of symmetry first uncovered in Prob. 3.23.

x	0	1	2
y	1	2	3

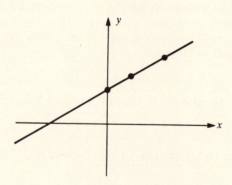

Fig. 3.3

3.32   Graph $x^2 + (y - 1)^2 = 1$.

    ▌ See Fig. 3.4. See Prob. 3.26 above. Here we have a circle with center $(0, 1)$ and radius 1. Notice y axis symmetry.

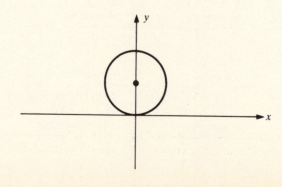

Fig. 3.4

3.33 Graph $y = |x| + 1$.

▎ See Fig. 3.5. See Prob. 3.30 above. Notice that we only calculated y for positive x values since we already knew the graph had y axis symmetry.

x	0	1	2
y	1	2	3

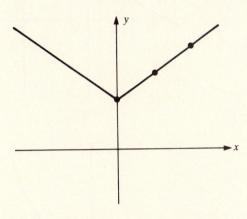

Fig. 3.5

3.34 Show that $(5, 2\sqrt{3})$, $(2, -\sqrt{3})$, and $(8, -\sqrt{3})$ form an equilateral triangle.

▎ Let A be the point $(5, 2\sqrt{3})$, let B be the point $(2, -\sqrt{3})$, and let C be the point $(8, -\sqrt{3})$. Then, $d(A, B) = \sqrt{(5-2)^2 + (3\sqrt{3})^2} = \sqrt{9 + 27} = 6$, $d(B, C) = \sqrt{(8-2)^2 + (0)^2} = \sqrt{36} = 6$, and $d(A, C) = \sqrt{3^2 + (3\sqrt{3})^2} = 6$. Since the three lengths are equal, it is an equilateral triangle.

For Probs. 3.35 and 3.36 determine whether the given points are the vertices of a right triangle.

3.35 $(-3, 2)$, $(1, -2)$, $(8, 5)$

▎ Call the three points A, B, and C, respectively. Then, $AB = \sqrt{(-4)^2 + (4)^2} = \sqrt{32}$, $BC = \sqrt{(-7)^2 + (-7)^2} = \sqrt{98}$, $AC = \sqrt{(-11)^2 + (-3)^2} = \sqrt{130}$. Then $AB^2 + BC^2 = 32 + 98 = 130 = AC$. Since these lengths satisfy the pythagorean theorem, the three points are the vertices of a right triangle.

3.36 $(-4, -1)$, $(0, 7)$, $(6, -6)$

▎ Calling the points A, B, and C, respectively, $AB^2 = 16 + 64 = 80$, $BC^2 = 36 + 169 = 205$, $AC^2 = 100 + 25 = 125$. Since $80 + 125 = 205$, we do have a right triangle.

3.2 RELATIONS AND FUNCTIONS

For Probs. 3.37 to 3.46 tell whether the relation shown is a function. In all cases, x is the independent variable.

3.37

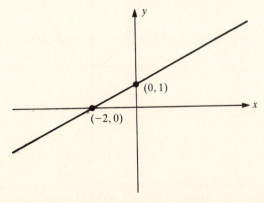

Fig. 3.6

▎ Fig. 3.6 is a function. No two ordered pairs have the same abscissa. Notice (Fig. 3.7) that for any line drawn at x_0 perpendicular to the x axis on the x axis, it intersects the function only once.

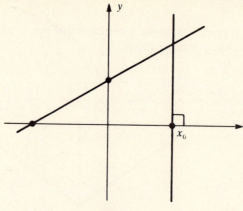

Fig. 3.7

3.38

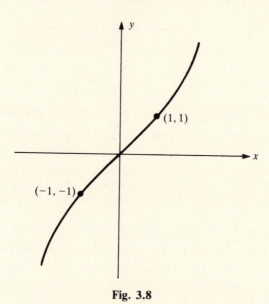

Fig. 3.8

▎ Fig. 3.8 is a function. All ordered pairs have different abscissas. Notice that this graph reminds us of $y = x^3$.

3.39

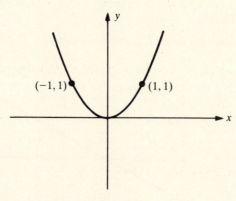

Fig. 3.9

▎ Be careful here! Fig. 3.9 exhibits a function. For each x value, there is only one y value; thus, no two ordered pairs have the same abscissa. However, unlike Probs. 3.37 and 3.38, two ordered pairs share an ordinate. For example, $(1, 1)$ and $(-1, 1)$.

3.40

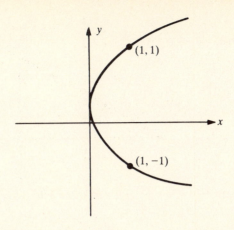

Fig. 3.10

▮ Fig. 3.10 is not a function.   Since $(1, 1)$ and $(1, -1)$ have the same first element, it violates the definition of a function.

3.41

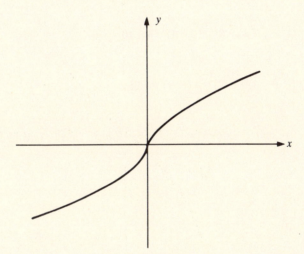

Fig. 3.11

▮ Fig. 3.11 is a function.   Draw a perpendicular to the x axis at any point on the x axis, and note that it intersects the function only once.

3.42 $f(x) = x - 9$

▮ f is a function.   For every x, $x - 9$ is only one value. It is impossible to get two values of f from one x.

3.43 $f(x) = x^2 - 9$

▮ f is a function.   If we plug in a value for x, we get only one value for f.

3.44 $y = |x|$

▮ This equation does define a function.   For each x, $|x|$ is only one value.

3.45 $x = |y|$

▮ Remember that x is the independent variable.   But for $x = 1$ (for example), if $1 = |y|$,   then $y = \pm 1$.   Thus, $(1, 1)$ and $(1, -1)$ are both in this relation.   The relation is not a function.

3.46 $x = y^2$

▮ This is not a function.   Two ordered pairs share an x value.   For example, $(4, 2)$ and $(4, -2)$.

For Probs. 3.47 to 3.56, find the indicated value if $f(x) = x - 9$, $g(x) = x^2 - 9$, $h(x) = |x|$,   and $F(x) = x^3 - x + 4$.

3.47 $f(0)$

▮ $f(x) = x - 9$, so $f(0) = 0 - 9 = -9$.

3.48 $g(3)$

▮ $g(x) = x^2 - 9$, so $g(3) = 3^2 - 9 = 0$.

3.49 $h(-7)$

▮ $h(x) = |x|$, so $h(-7) = |7| = 7$.

3.50 $g(a + b)$

▮ $g(x) = x^2 - 9$, so $g(a + b) = (a + b)^2 - 9 = a^2 + 2ab + b^2 - 9$.

3.51 $F(2) - g(3)$

▮ $F(2) - g(3) = (2^3 - 2 + 4) - (3^2 - 9) = 10 - 0 = 10$.

3.52 $F(1) + F(2)$

▮ $F(1) + F(2) = (1^2 - 1 + 4) - (2^3 - 2 + 4) = 4 - 10 = -6$.

3.53 $F(0) \cdot f(0)$

▮ $F(0) \cdot f(0) = (0^3 - 0 + 4)(0 - 9) = 4 \cdot (-9) = -36$.

3.54 $3g(a)[-2h(-b)]$

▮ $3g(a) = 3(a^2 - 9)$ and $-2h(-b) = -2|-b| = -2|b|$. Thus, $3g(a)[-2h(-b)] = -6|b|(a^2 - 9)$.

3.55 $\dfrac{f(0)}{F(0)}$

▮ $\dfrac{f(0)}{F(0)} = \dfrac{0 - 9}{0^3 - 0 + 4} = -\dfrac{9}{4}$.

3.56 $\dfrac{g(1 + h) - g(1)}{h}$

▮ $\dfrac{g(1 + h) - g(1)}{h} = \dfrac{(1 + h)^2 - 9 - (1^2 - 9)}{h} = \dfrac{(1 + h)^2 - 1^2}{h} = \dfrac{1 + 2h + h^2 - 1}{h} = \dfrac{h^2 + 2h}{h} = h + 2$.

For Probs. 3.57 to 3.66, find the domain of the given relation. In all cases, x is the independent variable.

3.57 $3x + 2 = y$

▮ For any real value x, $3x + 2 = y$ is real. Since any real number can replace x, the domain is the set of all real numbers $\mathcal{R}$).

3.58 $y = x^2$

▮ Since any real number squared yields a real number, y is defined for any real x. The domain $= \mathcal{R}$.

3.59 $y = |x| - 1$

▮ For any real x, $|x| - 1$ is real. The domain $= \mathcal{R}$.

3.60 $y = \dfrac{3}{2 - x}$

▮ We need to be certain that $2 - x \neq 0$. The domain is the set of all reals except for $x = 2$; i.e. $(-\infty, 2) \cup (2, \infty)$ or $\mathcal{R} - \{2\}$.

3.61 $y = \sqrt{3 - x^2}$

▮ We must ensure that $3 - x^2 \geq 0$. Then $3 \geq x^2$ or $x^2 \leq 3$. Then, $-\sqrt{3} \leq x \leq \sqrt{3}$ is the domain.

3.62 $x^2 + y^2 = 4$

▮ Since x is the independent variable, we need to solve for y to examine the domain. If $x^2 + y^2 = 4$, then $y^2 = 4 - x^2$ or $y = \pm\sqrt{4 - x^2}$. Then $4 - x^2 \geq 0$ if and only if $4 \geq x^2$ or $x^2 \leq 4$; i.e., $-2 \leq x \leq 2$ is the domain of this relation. Note that this relation is not a function.

3.63 $y = \sqrt{\dfrac{2}{x - 2}}$

▮ We need to be cautious here. We must be certain that $x - 2 \neq 0$ and $\dfrac{2}{x - 2} \geq 0$. But $\dfrac{2}{x - 2} \geq 0$ means that $x - 2 \geq 0$, since the numerator is ≥ 0 always. Thus, $x - 2 \neq 0$ and $x - 2 \geq 0$. Thus, $x - 2 > 0$, or $x > 2$ is the domain.

3.64 $x^2 + y^2 = 0$

▮ Since $x^2 \geq 0$ and $y^2 \geq 0$ for all real x and y, the only way $x^2 + y^2 = 0$ is for $x = 0$ and $y = 0$. Thus, the domain is $\{0\}$.

3.65 $\{(x, y) | x \in \mathcal{R}, \ y \in \mathcal{R}, \ y = x\}$

▮ The domain is $\mathcal{R}$. For any real x, y is that value and is therefore always defined.

3.66 $\{(1, 2), (2, -4), (-4, 2)\}$

▮ The domain is the set of x values; in this case, the domain $= \{1, 2, -4\}$.

For Probs. 3.67 to 3.76 find the range of the given function.

3.67 $f(x) = 2x - 3$

▮ Any real y can be expressed as $2x - 3$ for some real x. For example, if $-4 = 2x - 3$, $2x = -1$, or $x = -\frac{1}{2}$. The range $= \mathcal{R}$.

3.68 $f(x) = x^2$

▮ For every $x \in \mathcal{R}$, $x^2 \geq 0$. The range is the nonnegative reals, or $\mathcal{R}^+ \cup \{0\}$ or $[0, \infty)$.

3.69 $f(x) = x^4 - 2$

▮ For any real x, $x^4 - 2 \geq -2$, since $x^4 \geq 0$. The range is $[-2, \infty)$.

3.70 $f(x) = |x| + 3$

▮ $|x| \geq 0$; thus, $|x| + 3 \geq 3$. Range $= [3, \infty)$.

3.71 $g(x) = 2 - |x|$

▮ $|x| \geq 0$; thus, $-|x| \leq 0$ and $2 - |x| \leq 2$. Range $= (-\infty, 2]$.

3.72 $h(x) = 4$

▮ Since 4 is the function's value for any x, the range is $\{4\}$.

3.73 $f(x) = \sqrt{2 - x}$

▮ If $x = 2$, $f(x) = 0$. If $x < 2$, $f(x) > 0$. (Note that x cannot be > 2). The range is $[0, \infty)$.

3.74 $f(x) = 1/x$

▮ $x \neq 0$. For $x > 0$, $0 < 1/x < 1$. For $x < 0$, $-1 < 1/x < 0$. The range is $(-1, 0) \cup (0, 1)$.

3.75 $\{(1, 0), (0, 1), (2, a)\}$

▮ The range is the set of ordinates, which in this case is $\{0, 1, a\}$.

3.76 $\{(x, y) | x \in \mathcal{R}, y \in \mathcal{R} | y = x^3\}$

▮ For every real x, y is defined. Also, for every real y, y is the cube of some real x. Range $= \mathcal{R}$.

For Probs. 3.77 to 3.82 tell whether the function is one-to-one.

3.77 $f(x) = 2x + 3$

▮ A function is one-to-one if whenever $x_1 \neq x_2$, $f(x_1) \neq f(x_2)$. In this case, if $x_1 \neq x_2$, $2x_1 + 3 \neq 2x_2 + 3$, so the function is one-to-one.

3.78 $f(x) = |x| + 1$

▮ $|x| + 1$ is the same value for a and $-a$. Thus, f is not one-to-one.

3.79 $g(x) = \sqrt{1 - x}$

▮ $\sqrt{1 - x}$ is not the same value for two different x choices. This function is one-to-one.

3.80 $f(x) = \{(1, 2), (2, 1)\}$

▮ This is one-to-one. No two ordered pairs have the same second element.

3.81 $h(x) = \{(1, 3), (2, 4), (3, 1), (3, 4)\}$

▮ This function is not one-to-one. Notice that $(2, 4)$ and $(3, 4)$ have the same second element.

3.82 Give an example of two functions that are unequal, but whose domains and range are the same.

▮ Let $f(x) = x$ and $g(x) = x^2$. Then $f \neq g$, but the domain of $f =$ domain of g and the range of $f =$ range of g.

3.3 THE GRAPH OF A FUNCTION

For Probs. 3.83 to 3.86, refer to Fig. 3.12 below which is the graph of $y = f(x)$.

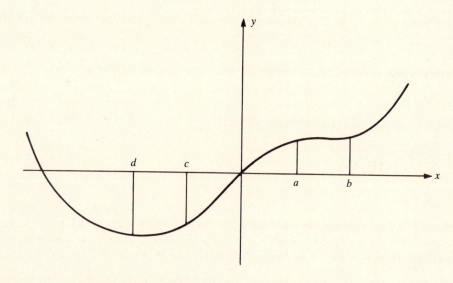

Fig. 3.12

3.83 Is the relation $y = f(x)$ a function?

▮ Yes; for every x value there is exactly one y value.

3.84 Find all intervals over which $y = f(x)$ is constant.

▮ Notice the horizontal segments. They occur on $[d, c]$ and $[a, b]$. These are the only intervals over which $y = f(x)$ is constant.

3.85   Find all intervals over which $y = f(x)$ is increasing and all intervals over which it is decreasing.

▮ Approaching d from the left, we see that the y values are decreasing.   From c to a, they are increasing, and from b on (infinitely far!), f is increasing.   Decreasing: $(-\infty, d)$.   Increasing: $(c, a), (b, \infty)$.

3.86   Find all intervals over which $y = f(x)$ is nonincreasing and those over which it is nondecreasing.

▮ Nonincreasing means decreasing or constant; nondecreasing means increasing or constant.   Refer to Probs. 3.84 and 3.85 above.   Nonincreasing: $(-\infty, c]$.   Nondecreasing: $[d, \infty)$.

For Probs. 3.87 and 3.88, answer true or false, and explain your answer.

3.87   If $y = f(x)$ is an increasing function, then it is a nondecreasing function.

▮ True.   Since nondecreasing means increasing or constant, increasing implies nondecreasing.

3.88   If $y = g(x)$ is a nonincreasing function, then $y = g(x)$ is a decreasing function.

▮ False. $y = 3$ is nonincreasing (it's constant), but it is not decreasing.

For Probs. 3.89 to 3.109, graph the given function.

3.89 $f(x) = 3$.

▮ See Fig. 3.13.   Notice that this is of the form $f(x) = a$,   which is a constant function.   The vertical axis is the y or $f(x)$ axis.

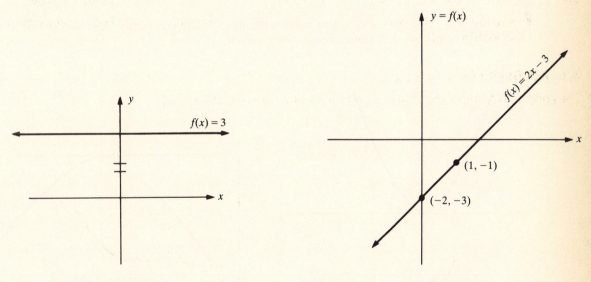

Fig. 3.13   Fig. 3.14

3.90 $f(x) = 2x - 3$

▮ See Fig. 3.14.   This function is of the form $f(x) = ax + b$,   which is the form for a linear function. Since we know the graph is a straight line, we find two points.   Find a third point as well to make certain you did not make an error.   The third point must lie on the line.

x	0	1
y	-3	-1

3.91 $x = 4$

▮ See Fig. 3.15.   Notice that this is not a function of x, the independent variable.   It is in the form $x = a$ for a vertical line.

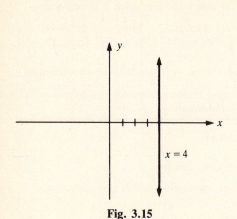

Fig. 3.15

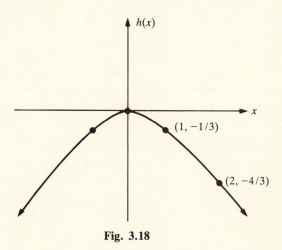

Fig. 3.16

3.92 $f(x) = x^2$

▮ See Fig. 3.16. Notice that, since $f(-x) = x^2 = f(x)$, this function's graph is symmetric about the y axis. Also, $f(x) \geq 0$ for all x. This graph is a parabola. See Chap. 14 for more about the conic sections in general.

x	-2	-1	0	1	2
y	4	1	0	1	4

3.93 $g(x) = 2x^2$

▮ See Fig. 3.17. See Prob. 3.92. This function has the property that each y value is double what the y value is for $f(x)$ in Prob. 3.92 (for corresponding x values).

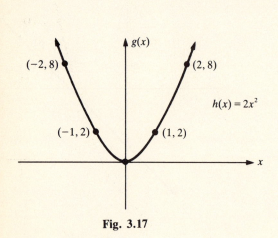

Fig. 3.17

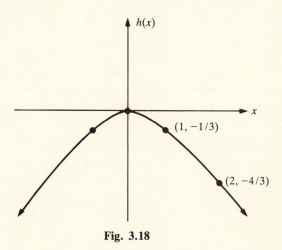

Fig. 3.18

3.94 $h(x) = -\frac{1}{3}x^2$

▮ See Fig. 3.18. Again we look at $f(x) = x^2$ in Prob. 3.92 above. Notice that for corresponding x values we multiply the y values by $-\frac{1}{3}$. Thus $(0, 0)$ remains $(0, 0)$; $(1, 1)$ becomes $(1, -\frac{1}{3})$, etc.

3.95 $k(x) = x^2 - 1$

▮ See Fig. 3.19. Take the ordered pairs in $f(x) = x^2$ and subtract 1 from each y value [since $k(x) = x^2 - 1$]. $(0, 0)$ becomes $(0, -1)$; $(1, 1)$ becomes $(1, 0)$, etc. Note that we could have used y-axis symmetry and found several ordered pairs to plot the graph directly instead of using Prob. 3.92.

3.96 $f(x) = x^3$

▮ See Fig. 3.20. This function is symmetric about the origin, since $y = x^3$ and $-y = (-x)^3$.

x	0	1	-1	2	-2
y	0	1	-1	8	-8

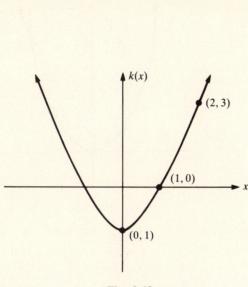

Fig. 3.19

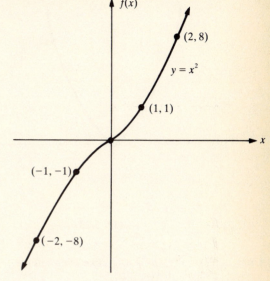

Fig. 3.20

3.97 $g(x) = x^3 - 1$

▮ See Fig. 3.21.   See Prob. 3.96 above.   We subtract 1 from each y value.   You can also graph g directly.

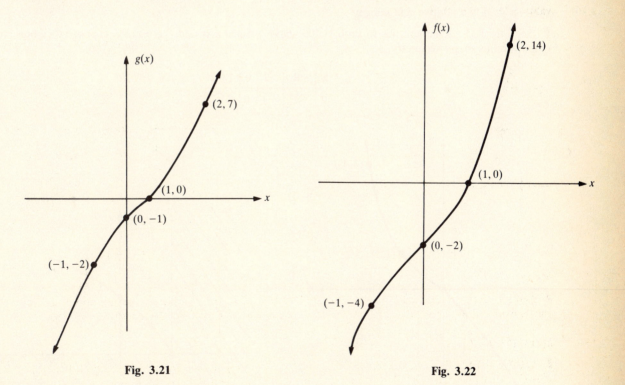

Fig. 3.21   **Fig. 3.22**

3.98 $h(x) = 2(x^3 - 1)$

▮ See Fig. 3.22.   See Prob. 3.97 above.   We double each y value.

3.99 $f(x) = x^4$

▮ See Fig. 3.23.   This graph will be symmetric about the y axis since $f(-x) = f(x)$.   Notice that it is just a very steep version of $y = x^2$.

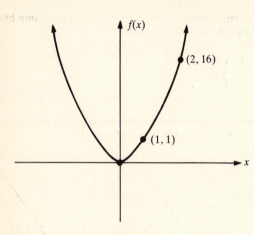

Fig. 3.23

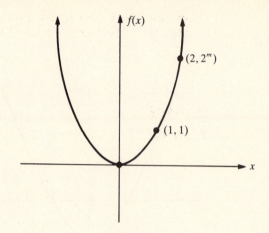

Fig. 3.24

3.100 $f(x) = x^m$, m is a positive even integer.

▌   See Fig. 3.24.   If m is a positive even integer, then $f(-x) = (-x)^m = x^m$;   thus, the graph is symmetric about the y axis.

x	-2	-1	0	1	2
y	2^m	1	0	1	2^m

3.101 $g(x) = x^n$, n is a positive odd integer.

▌   See Fig. 3.25.   Compare this to Prob. 3.100 above.   Since n is odd, $g(-x) = -g(x) = -x^n$;   thus, the graph exhibits origin symmetry.

x	-2	-1	0	1	2
y	-2^n	-1	0	1	2^n

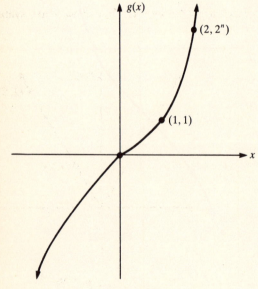

Fig. 3.25

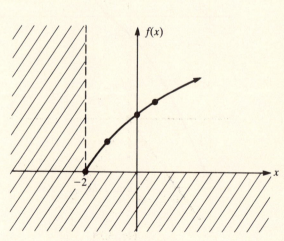

Fig. 3.26

3.102 $f(x) = \sqrt{x + 2}$

▌   Fig. 3.26 exhibits none of the three symmetries.   Notice that $f(x) \geq 0$ for all x, and that $x + 2 \geq 0$

or $x \geq -2$. No part of the graph can exist in the shaded areas. Notice that the graph is one-half of a parabola. The "negative branch" is "missing" because of the square root.

x	-2	-1	0	2
$f(x)$	0	1	$\sqrt{2}$	2

3.103 $g(x) = 3\sqrt{x+2} - 1$

▮ See Fig. 3.27. Using Prob. 3.102 above, we triple each y value and then subtract 1.

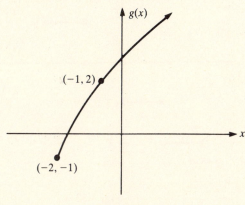

Fig. 3.27

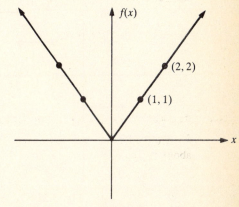

Fig. 3.28

3.104 $f(x) = |x|$

▮ Fig. 3.28 is symmetric about the y axis. Also, $f(x) \geq 0$ for all x.

x	-1	0	1	2	-2
$f(x)$	1	0	1	2	2

3.105 $g(x) = -2|x|$

▮ See Fig. 3.29. Using Prob. 3.104 above, we multiply each y value by -2. Notice the y axis symmetry.

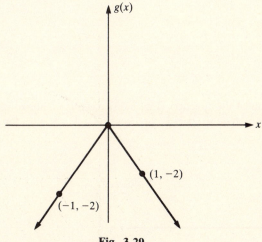

Fig. 3.29

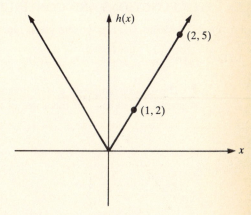

Fig. 3.30

3.106 $h(x) = 2|x| + 1$

▮ See Fig. 3.30. Again using Prob. 3.104, we find the following graph. Note that you could have found ordered pairs directly.

3.107 $f(x) = [x]$

⬤   See Fig. 3.31. f here is the greatest-integer function: For every x, $f(x)$ is the "greatest integer that is not greater than x."   Then $[0] = 0$, $[1] = 1$, $[-1] = -1$, $[2] = 2$, $[-2] = -2$,   etc. However, $[\frac{1}{2}] = 0$, $[\frac{7}{8}] = 0$, $[4.2] = 4$, $[-2.5] = -3$.   The graph then looks as follows:

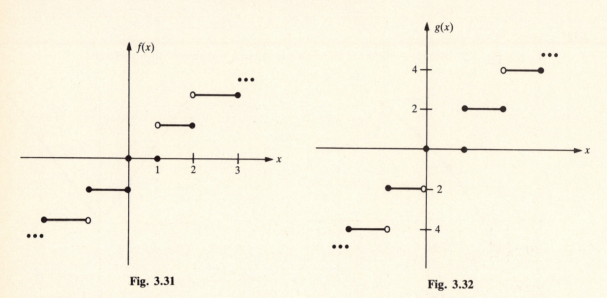

Fig. 3.31   Fig. 3.32

3.108 $g(x) = 2[x]$

⬤   See Fig. 3.32.   We will do this in two ways.   See Prob. 3.109 for the second way where we will use Prob. 3.107.   Here we will plot points directly.

x	-3	-2.5	-1	0	1	1.5	1.7	3.5
$g[x] = 2[x]$	-6	-6	-2	0	2	2	2	2

3.109 $g(x) = 2[x]$

⬤   Here, we simply double each y value in Prob. 3.107 and obtain the same graph as Fig. 3.32. $[0] = 0$, $[1] = 2$, $[-1] = -2$, $[2] = 4$, $[-2] = -4$, $[\frac{1}{2}] = 0$, $[\frac{7}{8}] = 0$, $[4.2] = 8$,   etc.

3.4   STEP FUNCTIONS AND CONTINUITY

For Probs. 3.110 to 3.121 graph the given function.

3.110 $f(x) = \begin{cases} 1, & x \geq 1 \\ 0, & x < 1 \end{cases}$

⬤   This is an example of a step function.   The graph of the function (Fig. 3.33) consists of horizontal segments (or steps).   In this case, in the xy-coordinate system, $y = 1$ at 1 and $x > 1$; $y = 0$ for all other x.

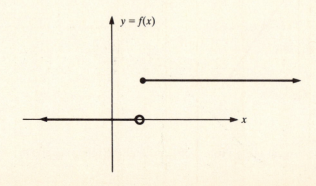

Fig. 3.33

3.111 $g(x) = \begin{cases} 1, & x > 1 \\ -3, & x \le 1 \end{cases}$

▮ See Fig. 3.34.

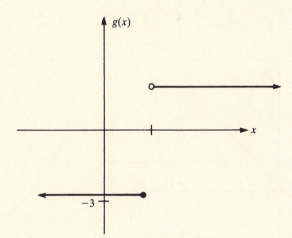

Fig. 3.34

3.112 $h(x) = \begin{cases} 1, & x > 1 \\ 2, & -2 < x \le 1 \\ -1, & x \le -2 \end{cases}$

▮ See Fig. 3.35.

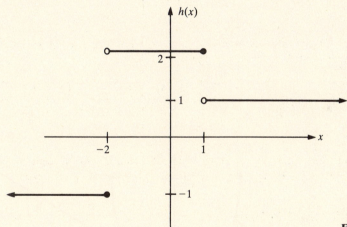

Fig. 3.35

3.113 $f(x) = \begin{cases} 1, & x \ge 1 \\ x, & x < 1 \end{cases}$

▮ See Fig. 3.36.

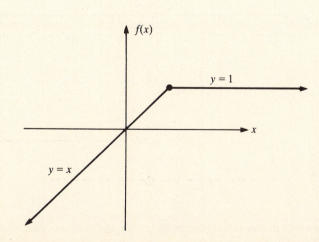

Fig. 3.36

3.114 $g(x) = \begin{cases} 1, & x \geq 2 \\ x, & x < 2 \end{cases}$

   ▮   See Fig. 3.37

Compare this $g(x)$ to $f(x)$ in Prob. 3.113.   Later in this section we will look at these again to discuss continuity.

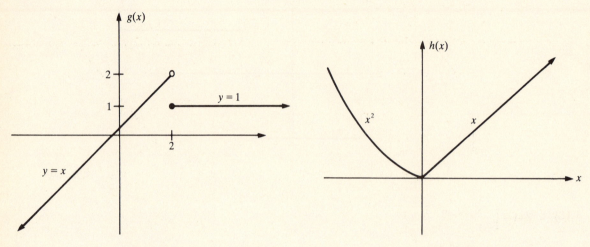

Fig. 3.37   **Fig. 3.38**

3.115 $h(x) = \begin{cases} x, & x > 0 \\ x^2, & x \leq 0 \end{cases}$

   ▮   See Fig. 3.38.

3.116 $f(x) = \begin{cases} x, & x > 2 \\ x^2, & x \leq 2 \end{cases}$

   ▮   See Fig. 3.39.

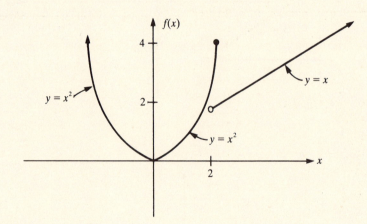

Fig. 3.39

3.117 $g(x) = \begin{cases} x, & x > 2 \\ x^3, & x \leq 2 \end{cases}$

   ▮   See Fig. 3.40.

3.118 $g(x) = \begin{cases} x + 1, & x > 2 \\ x - 1, & x \leq 2 \end{cases}$

   ▮   See Fig. 3.41.

Notice that the two rays in this graph are parallel.   We will look at other examples like this in the section on linear functions.

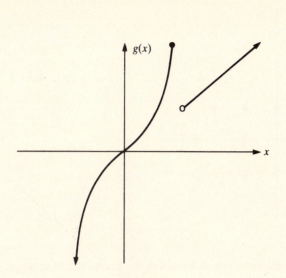

Fig. 3.40

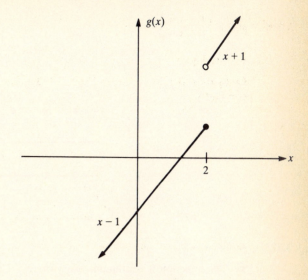

Fig. 3.41

3.119 $h(x) = \begin{cases} x - 2, x > 2 \\ -x + 1, x \le 2 \end{cases}$

▮ See Fig. 3.42.

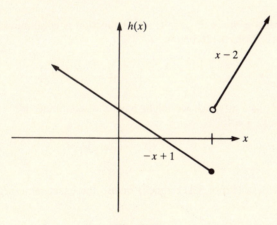

Fig. 3.42

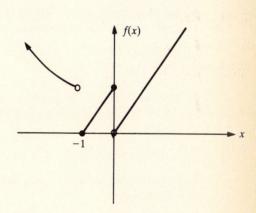

Fig. 3.43

3.120 $f(x) = \begin{cases} x, x > 0 \\ x + 1, -1 < x \le 0 \\ x^2, x \le -1 \end{cases}$

▮ See Fig. 3.43

3.121 $g(x) = \begin{cases} 1, & x \text{ is an integer} \\ -1, & \text{otherwise} \end{cases}$

▮ See Fig. 3.44.

For Probs. 3.122 to 3.131, tell whether the given function is continuous for every x in its domain. If it is not, tell where it is discontinuous. Some of these are from Probs. 3.110 to 3.121 above.

3.122 $f(x) = x^2$

▮ This function is continuous everywhere. There are no breaks in the graph of this function. Until we study continuity more formally (see Chap. 15), we will use this simple criterion.

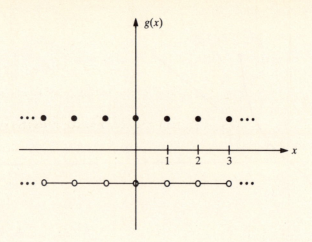

Fig. 3.44

3.123 $f(x) = x^3 - 1$

▮   This function is continuous everywhere.   Clearly there are no breaks in the graph of this function.

3.124 $g(x) = ax^2 + bx + c;$ $a, b, c \in \mathcal{R}; b \neq 0.$

▮   In general, all quadratic and linear functions are continuous everywhere.   This is a quadratic function (or linear if $a = 0$).   It is continuous everywhere.

3.125 $f(x) = \begin{cases} 1, x \geq 1 \\ 0, x < 1 \end{cases}$

▮   See Prob. 3.110 above.   Fig. 3.33 clearly indicates a break at $x = 1$.   This function is continuous for all x except $x = 1$;   it is discontinuous at $x = 1$.   See Prob. 3.126 below.

3.126 $f(x) = \begin{cases} 1, x \geq 1 \\ 0, x < 1 \end{cases}$

▮   Without looking at the graph, we note that the place at which we might have a discontinuity is $x = 1$. Since $1 \neq 0$,   there is clearly going to be a jump at $x = 1$.

3.127 $g(x) = \begin{cases} 1, x > 1 \\ -3, x \leq 1 \end{cases}$

▮   Discontinuous at $x = 1$.   You can look at Fig. 3.34 (Prob. 3.111 above) or use the exact same argument as in Prob. 3.126.

3.128 $h(x) = \begin{cases} 1, x \geq 1 \\ x, x < 1 \end{cases}$

▮   Looking at Fig. 3.37 (Prob. 3.114 above), we see that $h(x)$ is continuous everywhere.   See Prob. 3.130.

3.129 $h(x) = \begin{cases} 1, x \geq 1 \\ x, x < 1 \end{cases}$

▮   Without looking at the graph, we note that the only candidate for discontinuity is $x = 1$.   Since $1 = x$ when $x = 1$,   there clearly will not be a jump on the graph. $h(x)$ is continuous everywhere.

3.130 $g(x) = \begin{cases} 1, x \geq 2 \\ x, x < 2 \end{cases}$

▮   See Fig. 3.37 (Prob. 3.114 above).   Clearly the function is discontinuous when $x = 2$.   See Prob. 3.131 below.

3.131 $g(x) = \begin{cases} 1, x \geq 2 \\ x, x < 2 \end{cases}$

▮   Without looking at the graph, we investigate $g(x)$ when $x = 2$ (the only possible discontinuity for g). Since $1 \neq x$ when $x = 2$,   there will be a jump in the graph. g is discontinuous when $x = 2$.

3.5 LINEAR FUNCTIONS

For Probs. 3.132 to 3.137 find the slope and y intercept (if they exist).

3.132 $y = 3x + 1$

▮ Since when $y = mx + b$, $m = $ slope and $b = y$ intercept, in this case m (slope) $= 3$ and b (y intercept) $= 1$.

3.133 $2y = 3x + 1$

▮ Solve for y; $y = \frac{3}{2}x + \frac{1}{2}$. $m = \frac{3}{2}$ (slope); $b = \frac{1}{2}$ (y intercept).

3.134 $-3y = x + 6$

▮ If $-3y = x + 6$, then $y = -x/3 - 2$. Slope $= -\frac{1}{3}$; y intercept $= -2$.

3.135 $x + y + 4 = 0$

▮ $y = -x - 4$. Slope $= -1$; y intercept $= -4$.

3.136 $y = 3$

▮ Then $y = mx + b$ where $m = 0$, $b = 3$. Note that this is a horizontal line.

3.137 $x = -2$

▮ This is a vertical line; it has no slope and no y intercept.

For Probs. 3.138 to 3.144, graph the linear functions given. Use the slope and y intercept graphing technique.

3.138 $y = 3x + 2$

▮ See Fig. 3.45. After plotting the y intercept, make use of the fact that the slope is 3. That means that $(y_2 - y_1)/(x_2 - x_1) = 3$ for any (x_1, y_1), (x_2, y_2) on the line. Thus, if x changes by 1, then y changes by 3. In the graph we went from $(0, 2)$ to $(1, 5)$. Therefore, $y_2 - y_1 = 5 - 2$ is a change of 3, and $x_2 - x_1 = 1 - 0$ is a change of 1.

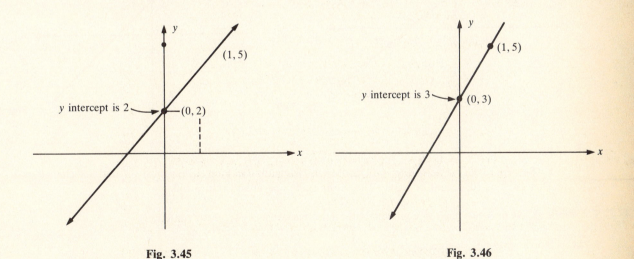

Fig. 3.45   Fig. 3.46

3.139 $y = 2x + 3$

▮ See Fig. 3.46. To get $(1, 5)$ we note that $m = 2$. Thus, for a change of 1 in x, we have a change of 2 for y.

3.140 $3y = 3x + 6$

▮ See Fig. 3.47. Then $y = x + 2$. We get the point $(1, 3)$ from $(0, 2)$ using the fact that $m = 1$.

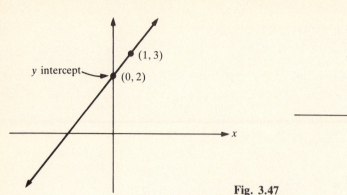

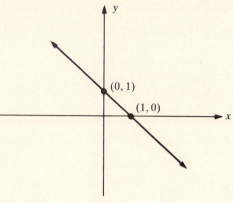

Fig. 3.47

Fig. 3.48

3.141 $y = -x + 1$

▮ See Fig. 3.48. Notice that $b = 1$ and $m = -1$; then for each change of 1 in y, the x change is -1. Notice that in the graph we went from $(0, 1)$ to $(1, 0)$. Therefore, $y_2 - y_1 = 0 - 1$ is a change of -1, and $x_2 - x_1 = 1 - 0$ is a change of 1.

3.142 $y = -2x + 1$

▮ See Fig. 3.49. If x changes by 1, then y changes by -2. See Prob. 3.143.

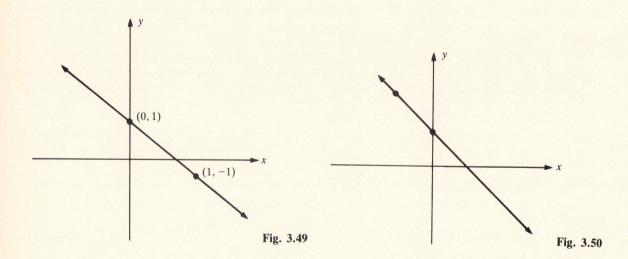

Fig. 3.49

Fig. 3.50

3.143 $y = -2x + 1$

▮ See Fig. 3.50. Alternately, we can use $(0, 1)$ together with an x change of -1 and a y change of 2. That gives us $(0, 1)$ and $(-1, 3)$. This is the same line as in Prob. 3.142.

For Probs. 3.144 to 3.161, find an equation of the line that satisfies the given conditions.

3.144 Slope $(m) = 1$, y intercept $(b) = 2$.

▮ $y = mx + b$ is slope-intercept form. $y = 1x + 2 = x + 2$. $y = x + 2$.

3.145 Slope 2, y intercept $= 3$.

▮ $y = mx + b = 2x + 3$. $y = 2x + 3$.

3.146 Slope -1, y intercept $= \frac{1}{2}$.

▮ $y = mx + b = -x + \frac{1}{2}$. $y = -x + \frac{1}{2}$.

3.147 Slope $= 0$, y intercept $= 3$.

▮ $y = 0x + 3 = 3$. $y = 3$.

3.148 No slope, the x intercept is 7.

▮ "No slope" signals a vertical line. This one is $x = 7$.

3.149 No slope, passes through the origin.

▮ Vertical line, x intercept is 0. This is $x = 0$ (the y axis!).

3.150 The line is the x axis.

▮ This is a horizontal line with y intercept $= 0$. Then $y = 0$.

3.151 Contains the points $(0, 0)$ and $(1, 1)$.

Since $m = \dfrac{y_2 - y_1}{x_2 - x_1} = \dfrac{1 - 0}{1 - 0} = 1$, $y - y_1 = m(x - x_1)$ or $y - 0 = 1(x - 0)$ or $y = x$. [Note; we could have used $y - y_2 = m(x - x_2)$ and obtained the same result.]

3.152 Contains the points $(1, -1)$ and $(2, 3)$.

▮ $m = \dfrac{3 - (-1)}{2 - 1} = \dfrac{4}{1} = 4$. Thus, $y - 3 = 4(x - 2)$. $y = 4x - 5$.

3.153 Contains the point $(2, 1)$ and is parallel to $y = x - 3$.

▮ If the line is parallel to $y = x - 3$, then its slope $= m = 1$. Thus, $y - y_1 = m(x - x_1)$, or $y - 1 = 1(x - 2)$. $y = x - 1$.

3.154 Contains the point $(3, 1)$ and is perpendicular to $2x + y = 3$.

▮ $2x + y = 3$ is the equivalent of a line with slope -2. The slope of our line here must be the negative reciprocal of -2, or $\frac{1}{2}$. Thus, $y - 1 = \frac{1}{2}(x - 3)$, or $y = x/2 - \frac{1}{2}$.

3.155 Parallel to $x + 2y = 12$, and has y intercept $= -5$.

▮ $x + 2y = 12$, so $2y = -x + 12$, or $y = -x/2 + 6$. Our line has $m = -\frac{1}{2}$. Then $b = -5$, and $y = -\frac{1}{2}x - 5$.

3.156 Perpendicular to $x + 2y = 12$, y intercept $= -5$.

▮ See Prob. 3.155. Then $m = 2$, $b = -5$. $y = 2x - 5$.

3.157 Vertical line containing $(7, -6)$.

▮ The line must be of the form $x = a$. Since $x = 7$ is on the line, the line is $x = 7$.

3.158 Vertical line containing $(7, -6)$.

▮ The line must be of the form $x = a$. Since $x = 7$ is on the line, the line is $x = 7$.

3.159 Horizontal line containing the y intercept of $x - y = 3$.

▮ The y intercept of $x - y = 3$ is -3. Thus, our line has equation $y = -3$.

3.160 x intercept is 3; y intercept is 2.

▮ The line contains $(3, 0)$ and $(0, 2)$. Then $m = (2 - 0)/(0 - 3) = -\frac{2}{3}$ and $y - 2 = -\frac{2}{3}(x - 0)$, or $y = -\frac{2}{3}x + 2$.

3.161 x intercept is 4, y intercept is 6.

▮ $m = (6 - 0)/(0 - 4) = -\frac{3}{2}$ [points $(4, 0)$ and $(0, 6)$ are on the line]. Then, $y = -\frac{3}{2}x + 6$.

For Probs. 3.162 to 3.164, evaluate the given line or function at the given value.

3.162 $f(x) = 3x - 5$.   Find $f(0)$.

▮ $f(0) = 3(0) - 5 = -5$.

3.163 $f(t) = 2t + 1$.   Find $f(-3)$.

▮ $f(-3) = 2(-3) + 1$.

3.164 $f(t) = t - 5$.   Find $f(a + b)$.

▮ $f(a + b) = (a + b) - 5 = a + b - 5$.

3.165   Find the point of intersection of $2x + 1 = y$ and $x - y = 3$.

▮   Since $y = 2x + 1$ and $y = x - 3$,   they will meet when $2x + 1 = x - 3$.   Then $x = -4$;   thus, since $y = x - 3$, $y = -4 - 3 = -7$.   The intersection point is $(-4, 7)$.

3.166   Find all points of intersection of $y = x^2$ and $y = x$.

▮   Since $x^2 = x$ when the graphs intersect, $x^2 - x = 0$, $x(x - 1) = 0$, $x = 0$, $x = 1$.   When $x = 0$, $y = 0$.   When $x = 1$, $y = 1$.   Thus, $(0, 0)$ and $(1, 1)$ are the intersection points.

For Probs. 3.167 to 3.171, use the quadrilateral with vertices $A(0, 2)$, $B(4, -1)$, $C(1, -5)$,   and $D(-3, -2)$.

3.167   Show $\overline{AB} \| \overline{DC}$.

▮ m for $\overline{AB}$ is $\dfrac{2 - (-1)}{0 - 4} = -\dfrac{3}{4}$,   and m for $\overline{CD}$ is $-\dfrac{5 - (-2)}{1 - (-3)} = -\dfrac{3}{4}$.   So they are parallel since their slopes are equal and they are not the same line.

3.168   Show that $\overline{DA} \| \overline{CB}$.

▮ m for $\overline{DA} = -\dfrac{2 - 2}{-3 - 0} = -\dfrac{4}{-3} = \dfrac{4}{3}$ and m for $\overline{CB} = -\dfrac{1 - (-5)}{4 - 1} = \dfrac{4}{3}$.   The slopes are the same, and they are parallel.

3.169   Show that $\overline{AB} \perp \overline{BC}$.

▮ m for $\overline{AB} = -\frac{3}{4}$ and m for $\overline{BC} = \frac{4}{3}$.   Since $-\frac{3}{4} = -1/\frac{4}{3}$,   the lines are perpendicular.

3.170   Show that $\overline{AD} \perp \overline{DC}$.

▮ m for $\overline{AD} = \frac{4}{3}$ and m for $\overline{DC} = -\frac{3}{4}$.   Since $\frac{4}{3} = -1/-\frac{3}{4}$,   they are perpendicular.

3.171   Find the perpendicular bisector of $\overline{AD}$.

▮ $\overline{AD}$ has slope $\frac{4}{3}$, and passes through $(0, 2)$.   Its equation is then $y - 2 = \frac{4}{3}(x - 0)$ or $y = \frac{4}{3}x + 2$. Then the perpendicular bisector has slope $-\frac{3}{4}$.   The midpoint of $\overline{AD}$ is $\left(\dfrac{0 - 3}{2}, \dfrac{2 - 2}{2}\right) = (-\frac{3}{2}, 0)$.   The perpendicular bisector has $m = -\frac{3}{4}$,   and passes through $(-\frac{3}{2}, 0)$.   Its equation is $y = -\frac{3}{4}(x + \frac{3}{2})$.

3.172   If $f(-x) = 3$, $f(3) = 5$,   and f is linear, find $f(x)$.

▮   The line contains $(-1, 3)$ and $(3, 5)$.   Thus, $m = \dfrac{5 - 3}{3 - (-1)} = \dfrac{1}{2}$ and $y - 5 = \frac{1}{2}(x - 3)$ or $y = x/2 + \frac{7}{2}$.

3.173   Determine whether $y = 3/x + 1$ is a linear function.

▮   It is not.   Notice that equations of the form $xy = a$ are nonlinear unless $a = 0$.   If $a = 0$,   we find the graph is two lines (see Prob. 3.172).   Any function otherwise containing an xy term is nonlinear.

3.174   Discuss the graph of $xy = 0$.

▮   If $xy = 0$,   then $x = 0$ or $y = 0$.   Thus, the graph is the x axis and the y axis.

3.6 THE ALGEBRA OF FUNCTIONS

For Probs. 3.175 to 3.184, find $(f+g)(x)$ and $(f-g)(x)$, and their respective domains.

3.175 $f(x) = x$, $g(x) = 2x$

▮ $(f+g)(x) = x + 2x = 3x$; $(f-g)(x) = x - 2x = -x$. Domain of both $= \mathcal{R}$ (all reals).

3.176 $f(x) = x^2$, $g(x) = x - 1$

▮ $(f+g)(x) = x^2 + (x-1) = x^2 + x - 1$; $(f-g)(x) = x^2 - (x-1) = x^2 - x + 1$. Domain of both $= \mathcal{R}$.

3.177 $f(x) = x - 2$, $g(x) = 1/x$

▮ $(f+g)(x) = x - 2 + 1/x$; $(f-g)(x) = x - 2 - 1/x$. Domain of both $= (-\infty, 0) \cup (0, \infty)$ (x can't be zero).

3.178 $f(x) = \dfrac{1}{x-1}$, $g(x) = \sqrt{x}$

▮ $(f+g)(x) = \dfrac{1}{x-1} + \sqrt{x}$; $(f-g)(x) = \dfrac{1}{x-1} - \sqrt{x}$. Domain of both $= \{x \in \mathcal{R} \mid x \geq 0, x \neq 1\}$.

3.179 $f(x) = \sqrt{x+1}$, $g(x) = \sqrt{x-1}$

▮ $(f+g)(x) = \sqrt{x+1} + \sqrt{x-1}$; $(f-g)(x) = \sqrt{x+1} - \sqrt{x-1}$. Domain of both $= \{x \in \mathcal{R} \mid x \geq 1\}$. Note that if $x < 1$, $\sqrt{x-1}$ is not defined.

3.180 $f(x) = |x|$, $g(x) = [x]$

$(f+g)(x) = |x| + [x]$; $(f-g)(x) = |x| - [x]$. Domain for both is $\mathcal{R}$.

3.181 $f(x) = x^2$, $g(x) = \dfrac{1}{1-x^2}$

▮ $(f+g)(x) = x^2 + \dfrac{1}{1-x^2}$; $(f-g)(x) = x^2 - \dfrac{1}{1-x^2}$. Note that we need $1 - x^2 \neq 0$. Then, $x^2 \neq 1$ or $x \neq 1, -1$. Domain (for both) $= \{x \in \mathcal{R} \mid x \neq \pm 1\}$.

3.182 $f(x) = x^3 + 1$, $g(x) = 0$

▮ $(f+g)(x) = x^3 + 1 + 0 = x^3 + 1$; $(f-g)(x) = x^3 + 1 - 0 = x^3 + 1$. Note that if $g(x) = 0$, then $f(x) + g(x) = 0$ for all $f(x)$. $g(x) = 0$ is the zero function behaving for functions the way 0 does for real numbers.

3.183 $f(x) = \sqrt[4]{x-9}$, $g(x) = x$

▮ $(f+g)(x) = \sqrt[4]{x-9} + x$; $(f-g)(x) = \sqrt[4]{x-9} - x$. Domain $= \{x \in \mathcal{R} \mid x \geq 9\}$.

3.184 $f(x) = x + \sqrt{x}$, $g(x) = f(x) + x^2$

▮ $(f+g)(x) = \underbrace{(x + \sqrt{x})}_{f(x)} + \underbrace{(x + \sqrt{x} + x^2}_{\substack{f(x) + x^2 \\ g(x)}} = 2x + 2\sqrt{x} + x^2$

$(f-g)(x) = (x + \sqrt{x}) - (x + \sqrt{x} + x^2) = -x^2$.
Domain for $f + g = [0, \infty)$. Domain for $f - g = \mathcal{R}$.

For Probs. 3.185 to 3.189, find $fg(x)$ and $(f/g)(x)$ and their respective domains.

3.185 $f(x) = 3x$, $g(x) = 2x$

▮ $fg(x) = f(x) \cdot g(x) = 3x \cdot 2x = 6x^2$; $(f/g)(x) = f(x)/g(x) = 3x/2x = \frac{3}{2}$, $(2x \neq 0)$. Domain $fg = \mathcal{R}$. Domain $f/g = (-\infty, 0) \cup (0, \infty)$. Notice that $g(x) = 2x$, and $g(x)$ is the denominator of f/g.

3.186 $f(x) = 1 + x$, $g(x) = x$

▮ $fg(x) = (1+x)x = x^2 + x$; $\dfrac{f}{g}(x) = \dfrac{1+x}{x}$ $(x \neq 0)$. Domain $fg = \mathcal{R}$. Domain $f/g = \mathcal{R} \setminus \{0\}$.

3.187 $f(x) = \sqrt{x+1}$, $g(x) = x - 3$

▮ $fg(x) = (x-3)\sqrt{x+1}$; $\dfrac{f}{g}(x) = \dfrac{\sqrt{x+1}}{x-3}$, $(x \neq 3)$. Domain $fg = \{x \in \mathscr{R} \,|\, x \geq 1\}$. Domain $f/g = \{x \in \mathscr{R} \,|\, x \geq -1$ and $x \neq 3\}$.

3.188 $f(x) = \sqrt{x-2}$, $g(x) = \sqrt{x-2}$

▮ $fg(x) = \sqrt{x-2}\sqrt{x-2} = (\sqrt{x-2})^2 = x - 2$; $\dfrac{f}{g}(x) = \dfrac{\sqrt{x-2}}{\sqrt{x-2}} = 1 (x \neq 2)$. Domain $fg = \{x \in \mathscr{R} \,|\, x \geq 2\}$. Domain $f/g = \{x \in \mathscr{R} \,|\, x > 2\}$.

3.189 $f(x) = x^3$, $g(x) = 1$

▮ $fg(x) = x^3 \cdot 1 = x^3$; $(f/g)(x) = x^3/1 = x^3$. Domain $fg = $ domain $f/g = \mathscr{R}$. Notice that $g(x) = 1$ functions as a multiplicative identity for function multiplication.

For Probs. 3.190 to 3.199, find $f \circ g(x)$.

3.190 $f(x) = x$, $g(x) = 2x + 1$

▮ $f \circ g(x) = f(g(x)) = f(2x + 1) = 2x + 1$.

3.191 $f(x) = x + 3$, $g(x) = x^2$

▮ $f \circ g(x) = f(x^2) = x^2 + 3$.

3.192 $f(x) = x^2$, $g(x) = 2x + 5$

▮ $f \circ g(x) = f(2x + 5) = (2x + 5)^2 = 4x^2 + 20x + 25$.

3.193 $f(x) = 1$, $g(x) = 3x^3$

▮ $f \circ g(x) = f(3x^3) = 1$. (Do you see this? $f(x) = 1$ for all x!)

3.194 $f(x) = x^3 + 4$, $g(x) = 3$

▮ $f \circ g(x) = f(3) = 3^3 + 4 = 31$.

3.195 $f(x) = a + x^2$, $g(x) = x$

▮ $f \circ g(x) = f(x) = a + (x)^2 = a + x$.

3.196 $f(x) = x^2 + x$, $g(x) = x^2 + x$

▮ $f \circ g(x) = f(x^2 + x) = (x^2 + x)^2 + (x^2 + x) = (x^2 + x)(x^2 + x + 1)$.

3.197 $f(x) = \dfrac{1}{1+x}$, $g(x) = \dfrac{1}{2+x}$

▮ $f \circ g(x) = f\left(\dfrac{1}{2+x}\right) = \dfrac{1}{1 + 1/(2+x)} = \dfrac{2+x}{3+x}$.

3.198 $f(x) = \frac{1}{2}$, $g(x) = 2$

▮ $f \circ g(x) = f(2) = \frac{1}{2}$.

3.199 $f(x) = x/2$, $g(x) = 2x$

▮ $f \circ g(x) = f(2x) = 2x/2 = x$. Look at the results in Probs. 3.198 and 3.199. Don't be tricked in the future.

For Probs. 3.200 to 3.206, find $g \circ f(x)$. These are functions from Probs. 3.190 through 3.199 above. Check the results to notice that $f \circ g(x)$ and $g \circ f(x)$ are not always the same function.

3.200 $f(x) = x$, $g(x) = 2x + 1$

▮ $g \circ f(x) = g(x) = 2x + 1$. [See Prob. 3.190; $f \circ g(x) = g \circ f(x)$ in this case.]

3.201 $f(x) = x + 3, \quad g(x) = x^2$

▮ $g \circ f(x) = g(x+3) = (x+3)^3.$ [See Prob. 3.191; $f \circ g(x) \neq g \circ f(x)$ in this case.]

3.202 $f(x) = x^2, \quad g(x) = 2x + 5$

▮ $g \circ f(x) = g(x^2) = 2(x^2) + 5 = 2x^2 + 5.$ See Prob. 3.192.

3.203 $f(x) = 1, \quad g(x) = 3x^3$

▮ $g \circ f(x) = g(1) = 3(1^3) = 3.$ See Prob. 3.193.

3.204 $f(x) = a + x^2, \quad g(x) = x$

▮ $g \circ f(x) = g(a + x^2) = a + x^2.$ See Prob. 3.195.

3.205 $f(x) = \dfrac{1}{1 + x}, \quad g(x) = \dfrac{1}{3 + 2x}$

▮ $g \circ f(x) = g\left(\dfrac{1}{1+x}\right) = \dfrac{1}{2 + 1/(1+x)} = \dfrac{1+x}{3 + 2x}.$ See Prob. 3.197.

3.206 $f(x) = x/2, \quad g(x) = 2x$

▮ $g \circ f(x) = g(x/2) = 2(x/2) = x.$ See Prob. 3.199, and notice that $f \circ g(x) = g \circ f(x) = x.$

For Probs. 3.207 to 3.211, let $f(x) = 2x - 4.$

3.207 Find domain and range of f and f^{-1}.

▮ First note that f is one-to-one, so we are assured of the existence of f^{-1}. Also, the domain of f = range of f^{-1} and the range of f = domain of f^{-1}. Since the domain f = range $f = \mathcal{R}$, the domain f^{-1} = range $f^{-1} = \mathcal{R}$.

3.208 Find $f^{-1}(x)$.

▮ $y = 2x - 4.$ Reverse the x and y: $x = 2y - 4,$ $2y = x + 4,$ or $y = x/2 + 2,$ $f^{-1}(x) = x/2 + 2.$

3.209 Find $f^{-1} \circ f(x)$.

▮ $f^{-1} \circ f(x) = f^{-1}(2x - 4) = \dfrac{2x - 4}{2} + 2 = x - 2 + 2 = x.$ $f^{-1} \circ f(x) = x.$

3.210 Find $f \circ f^{-1}(x)$.

▮ $f \circ f^{-1}(x) = f(x/2 + 2) = 2(x/2 + 2) - 4 = x + 4 - 4 = x.$ Thus, $f \circ f^{-1}(x) = x.$

3.211 Graph $f(x)$ and $f^{-1}(x)$ in the same coordinate system. Show that these two graphs are symmetric about $y = x.$

▮ See Fig. 3.51.

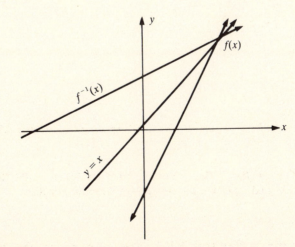

Fig. 3.51

For Probs. 3.212 to 3.216, let $f(x) = x^2 + 1, \quad x \geq 0.$

3.212 Find the domain and range of f^{-1}.

▮ Notice first that $f(x)$ here is one-to-one. Note that if we did not add the restriction $x \geq 0$, f would not be one-to-one, and we would not have an inverse function. The domain of f^{-1} = range of $f = [1, \infty)$ since $x^2 + 1 \geq 1$ for all x. The range of f^{-1} = domain of $f = [0, \infty)$ since we are given that $x \geq 0$.

3.213 Find $f^{-1}(x)$.

▮ $y = x^2 + 1$; $x \geq 0$. Thus if $x = y^2 + 1$, $y \geq 0$, then $y^2 = x - 1$, or $y = \sqrt{x-1} = f^{-1}(x)$.

3.214 Find $f^{-1} \circ f(x)$.

▮ $f^{-1} \circ f(x) = f^{-1}(x^2 + 1) = \sqrt{(x^2 + 1) - 1} = \sqrt{x^2} = |x|$. But $x \geq 0$, so $|x| = x$.

3.215 Find $f \circ f^{-1}(x)$.

▮ $f \circ f^{-1}(x) = f(\sqrt{x-1}) = (\sqrt{x-1})^2 + 1 = x - 1 + 1 = x$.

3.216 Graph $f(x)$, $f^{-1}(x)$, and $y = x$ on the same axis system.

▮ See Fig. 3.52.

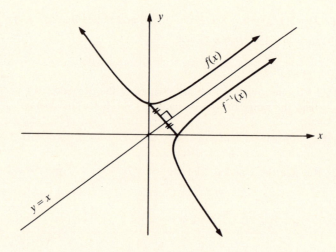

Fig. 3.52

For Probs. 3.217 to 3.227, let D be the set of reals other than 0 and 1. Consider $f_1, f_2, \ldots, f_6$ defined on D as follows:

$$f_1(x) = x \qquad f_2(x) = 1/x \qquad f_3(x) = 1 - x$$

$$f_4(x) = \frac{x}{x-1} \qquad f_5(x) = \frac{1}{1-x} \qquad f_6(x) = \frac{x-1}{x}$$

Find the following functions.

3.217 $f_2 \circ f_3(x)$

▮ $f_2 \circ f_3(x) = f_2(1 - x) = \dfrac{1}{1-x} = f_5(x)$.

3.218 $f_3 \circ f_5(x)$

▮ $f_3 \circ f_5(x) = f_3\left(\dfrac{1}{1-x}\right) = 1 - \dfrac{1}{1-x} = \dfrac{-x}{1-x} = \dfrac{x}{x-1} = f_4(x)$.

3.219 $f_1 \circ f_2(x)$

▮ $f_1 \circ f_2(x) = f_1(1/x) = 1/x = f_2(x)$.

3.220 $f_2 \circ f_1(x)$

▮ $f_2 \circ f_1(x) = f_2(x) = 1/x = f_2(x)$.

3.221 $f_4 \circ f_4(x)$

▮ $f_4 \circ f_4(x) = f_4\left(\dfrac{x}{x-1}\right) = \dfrac{x/(x-1)}{x/(x-1)-1} \dfrac{x-1}{x-1} = \dfrac{x}{x-(x-1)} = \dfrac{x}{x-x+1} = x = f_1(x).$ Notice that f_4 is its own inverse.

3.222 $f_3 \circ f_3(x)$

▮ $f_3 \circ f_3(x) = f_3(1-x) = 1 - (1-x) = 1 - 1 + x = x = f_1(x).$

3.223 $f_5 \circ f_6(x)$

▮ $f_5 \circ f_6(x) = f_5\left(\dfrac{x-1}{x}\right) = \dfrac{1}{1-(x-1)/x} = \dfrac{x}{x-(x-1)} = \dfrac{x}{x-x+1} = x = f_1(x).$

3.224 $f_3 \circ (f_4 \circ f_5)(x)$

▮ $f_4 \circ f_5(x) = f_4\left(\dfrac{1}{1-x}\right) = \dfrac{1/(1-x)}{1/(1-x)-1} = \dfrac{1}{x}.$ Then $f_3 \circ (f_4 \circ f_5)(x) = f_3 \circ f_2(x) = 1 = \dfrac{1}{x} = \dfrac{x-1}{x} = f_6(x).$

3.225 $f_4^{-1}(x)$

▮   See Prob. 3.221.   Since $f_4 \circ f_4(x) = x,$ $f_4^{-1}(x) = f_4(x).$ See Prob. 3.226 for an alternate method.

3.226 $f_3^{-1}(x).$

▮ $f_3(x) = 1 - x.$ If $y = 1 - x,$ then when $x = 1 - y,$ $y = 1 - x = f^{-1}(x).$ Notice that $f_3(x) = f_3^{-1}(x).$ See Prob. 3.222.

3.227 $f_6^{-1}(x)$

▮ $y = \dfrac{x-1}{x}.$ If $x = \dfrac{y-1}{y}$ then $x = 1 - \dfrac{1}{y}.$ $\dfrac{1}{y} = 1 - x$ or $y = \dfrac{1}{1-x} = f_6^{-1}(x) = f_5(x).$

CHAPTER 4
Polynomial and Rational Functions

4.1 POLYNOMIALS

For each of Probs. 4.1 to 4.10, write the given equation in standard form.

4.1 $4x^2 + 2x^3 - 6 + 5x = 0$

▮ In standard form, polynomial equations are written from the highest power of x to the lowest. In this case, the standard form is $2x^3 + 4x^2 + 5x - 6 = 0$.

4.2 $-3x^3 + 6x - 4x^2 + 2 = 0$

▮ $3x^3 + 4x^2 - 6x - 2 = 0$.

4.3 $2x^5 + x^3 + 4 = 0$

▮ $2x^5 + 0x^4 + x^3 + 0x^2 + 0x + 4 = 0$.

4.4 $x^3 + \frac{1}{2}x^2 - x + 2 = 0$

▮ $x^3 + \frac{1}{2}x^2 - x + 2 = 0$ is rewritten as $2x^3 + x^2 - 2x + 4 = 0$ since in standard form all coefficients are integers.

4.5 $4x^4 + 6x^3 - 8x^2 + 12x - 10 = 0$

▮ $2x^4 + 3x^3 - 4x^2 + 6x - 5 = 0$, since in standard form $(a_0, \ldots, a_n) = 1$ for all coefficients $a_0, a_1, \ldots, a_n$.

4.6 $2x - 4x^4 = 0$

▮ $-4x^4 + 0x^3 + 0x^2 + 2x + 0 = 0$, and then notice that $gcd(-4, 2) \neq 1$. So the *answer* is $-2x^4 + 0x^3 + 0x^2 + x + 0 = 0$.

4.7 $\frac{x^2}{3} + \frac{x^4}{12} + 3 = 0$

▮ Multiplying by 12, $x^4 + 4x^2 + 36 = 0$. Inserting zeros, $x^4 + 0x^3 + 4x^2 + 0x + 36 = 0$.

4.8 $\frac{x^5}{2} - 1 = 0$

▮ $x^5 + 0x^4 + 0x^3 + 0x - 2 = 0$.

4.9 $2x^4 - 6x^3 + 2 = 0$

▮ $x^4 - 3x^2 + 1 = 0$. Insert zeros: $x^4 - 3x^2 + 0x + 1 = 0$.

4.10 $(x + 2)^2 + 5 = 0$

▮ $(x^2 + 4x + 4) + 5 = 0$, or $x^2 + 4x + 9 = 0$.

For each of Probs. 4.11 to 4.15, rewrite the equation in the standard form $y - k = a(x - h)^2$.

4.11 $y = x^2 + 1$

▮ $y = x^2 + 1$ is written as $y - 1 = x^2$, or $y - 1 = 1(x - 0)^2$.

4.12 $y = -2x^2 + 3$

▮ $y - 3 = -2x^2$, or $y - 3 = -2(x - 0)^2$.

4.13 $y = 4 - (x-1)^2$

$y - 4 = -(x-1)^2$, or $y - 4 = -1(x-1)^2$.

4.14 $y = x^2 - 4x + 6$

$y = (x^2 - 4x + 4) + 2$ (by completing the square); $y = (x-2)^2 + 2$, or $y - 2 = 1(x-2)^2$.

4.15 $y = (1-x)(2-x)$

$y = 2 - 3x + x^2 = (x^2 - 3x + \frac{9}{4}) + (-\frac{1}{4})$ (since $\frac{9}{4} - \frac{1}{4} = 2$); $y = (x - \frac{3}{2})^2 - \frac{1}{4}$, or $y + \frac{1}{4} = 1(x - \frac{3}{2})^2$.

For Probs. 4.16 to 4.20, find the vertex of the given parabola.

4.16 $y + 1 = 2(x-3)^2$

If $y - k = a(x-h)^2$, then (h, k) is the vertex of the given parabola. In this case, $k = -1$ and $h = 3$. $(3, -1)$ is the vertex.

4.17 $y = -2x^2 + 3$

Then $y - 3 = -2(x-0)^2$. Thus, $k = 3$ and $h = 0$. The vertex is at $(0, 3)$.

4.18 $y = x^2 - 4x + 6$

$y = (x^2 - 4x + 2) + 4 = (x-2)^2 + 4$; $y - 4 = (x-2)^2$. The vertex is at $(2, 4)$.

4.19 $y = x(x+2)$

$y = x^2 + 2x = (x^2 + 2x + 1) - 1$ (since $1 - 1 = 0$); $y + 1 = (x+1)^2$. The vertex is at $(-1, -1)$.

4.20 $y = 2x^2 - 8x + 5$

$y - 5 = 2(x^2 - 4x) = 2(x^2 - 4x + 4) - 8$; $y + 3 = 2(x-2)^2$. The vertex is at $(2, -3)$.

For Probs. 4.21 to 4.24, identify the given parabola as opening upward or downward.

4.21 $y = x^2 - 5x + 6$

If $y = ax^2 + bx + c$, then the parabola opens upward if $a > 0$, and downward if $a < 0$. In this case, $a = 1 > 0$, and the parabola opens upward.

4.22 $y = 2 - x^2$

If $y = 2 - x^2$, then $y = -x^2 + 2$, and $a = -1 < 0$; the parabola opens downward.

4.23 $y = x(1-x)$

If $y = x - x^2$, then $y = -x^2 + x$, and $a = -1 < 0$, and the parabola opens downward.

4.24 $y = (2-x)(3-x)$

$y = 6 - 5x + x^2$; $a = 1$ and the parabola opens upward.

For Probs. 4.25 to 4.29, find the maximum (or minimum) value of the quadratic function.

4.25 $f(x) = 2(x-3)^2 - 1$

If $y = 2(x-3)^2 - 1$, then $y + 1 = 2(x-3)^2$. The vertex is at $(3, -1)$, and the parabola is opening upward (since $a > 0$, where a is the coefficient of x^2). Thus, the function has a minimum value of -1.

4.26 $f(x) = 2x^2 + 3$

If $y = -2x^2 + 3$, then $y - 3 = -2(x-0)^2$. The vertex is at $(0, 3)$, and since $a < 0$, the maximum value is 3.

4.27 $f(x) = 4 - (x-1)^2$

$\blacksquare$ If $y = 4 - (x-1)^2$, then $y - 4 = -(x-1)^2$. Thus, f has a maximum value of 4.

4.28 $f(x) = x^2 - 4x + 6$

$\blacksquare$ $y = (x-2)^2 + 2$ (since $4 + 2 = 6$). Thus, $a > 0$ (f has a minimum) and the minimum value is 2.

4.29 $f(x) = (x-1)(x+3)$

$\blacksquare$ $y = x^2 + 2x - 3$; $a > 0$, so f has a minimum. $y = (x+1)^2 - 4$ (since $1 - 4 = -3$), so $y + 4 = (x+1)^2$, and the minimum value is -4.

For Probs. 4.30 to 4.35, tell whether the given function is a polynomial function.

4.30 $y = x^2$

$\blacksquare$ This is a polynomial function, since $y = a_0 x^n + a_1 x^{n-1} + \cdots + a_{n-1}x + a_n$ where n is a positive integer or zero, and all a_i are complex numbers.

4.31 $y = 1/x^2$

$\blacksquare$ This is not a polynomial function since $y = x^{-2}$, and -2 is not a positive integer or zero.

4.32 $y = (x+2)^x$

$\blacksquare$ This is not a polynomial function. Note that x is the independent variable and thus may not be a positive integer or zero.

4.33 $y = 2x + 4 - 2i$

$\blacksquare$ This is a polynomial function. If $y = 2x + (4 - 2i)$, then all conditions of the definition in Prob. 4.30 above are satisfied. Note that the constant term $a_n = 4 - 2i$.

4.34 $y = 2x^2 + \sqrt{3}x - i$

$\blacksquare$ This is a polynomial function. The coefficients 2 and $\sqrt{3}$ are complex as is the constant term $-i$.

4.35 $y = ix^3 - \sqrt{4x} - 2$

$\blacksquare$ This is not a polynomial function since $x^{1/2}$ appears, and $\frac{1}{2}$ is nonintegral.

For Probs. 4.36 to 4.39, determine whether the graph of the polynomial will cross or touch the x axis at each zero.

4.36 $f(x) = x^2(x-1)$

$\blacksquare$ The zeros here are 0 with a multiplicity of 2, and 1 with a multiplicity of 1. Thus, the graph touches at 0, and the graph crosses at 1. (Recall that the graph of a polynomial crosses at a zero of odd multiplicity and touches at a zero of even multiplicity.)

4.37 $f(x) = x(x-1)^2(x+2)$

$\blacksquare$ The zeros here are 0 with a multiplicity of 1 (crosses), 1 with a multiplicity of 2 (touches), and -2 with a multiplicity of 1 (crosses).

4.38 $f(x) = (x^2 - 1)(x+2)^3$

$\blacksquare$ The zeros here are 1 with multiplicity of 1 (crosses), -1 with a multiplicity of 1 (crosses), and -2 with a multiplicity of 3 (crosses). (Do you see where 1 and -1 come from? If $x^2 - 1 = 0$, then $x = \pm 1$.)

4.39 $f(x) = (x+1)^3(x+2)^4(x+3)^5$

$\blacksquare$ Touches at -2; crosses at -1 and -3.

For Probs. 4.40 to 4.42, find the remainder when the division is performed by actually performing the division.

4.40 $(2x^2 - 4x + 8) \div (x - 2)$

$$\begin{array}{r} 2x \\ x-2\overline{)2x^2 - 4x + 8} \\ \underline{2x^2 - 4x} \\ +8 \end{array}$$

Remainder $= 8$.

4.41 $(3x^2 + x - 7) \div (x - 1)$

$$\begin{array}{r} 3x + 4 \\ x-1\overline{)3x^2 + x - 7} \\ \underline{3x^2 - 3x} \\ 4x - 7 \\ \underline{4x - 4} \\ -3 \end{array}$$

Remainder $= -3$.

4.42 $(3x^3 + 4x^2 - 7x + 5) \div (x + 1)$

$$\begin{array}{r} 3x^2 + x - 8 \\ x+1\overline{)3x^3 + 4x^2 - 7x + 5} \\ \underline{3x^3 + 3x^2} \\ x^2 - 7x \\ \underline{x^2 + x} \\ -8x + 5 \\ \underline{-8x - 8} \\ 13 \end{array}$$

Remainder $= 13$.

For Probs. 4.43 and 4.44, find the remainder using the remainder theorem.

4.43 $(x^3 - 5x^2 - 3x + 15) \div (x + 2)$

▮ The remainder theorem tells us that the remainder upon division by $x - r$ will be $f(r)$. In this case, $r = -2$, so $f(-2) = (-2)^3 - 5(-2)^2 - 3(-2) + 15 = -7$.

4.44 $(2x^4 + 6x^3 + 3x - 4) \div (x + 3)$

▮ $f(r) = f(-3) = 2(-3)^4 + 6(-3)^3 + 3(-3) - 4 = -13$.

For Probs. 4.45 to 4.52, show that the second expression is a factor of the first by means of the factor theorem.

4.45 $2x^3 - 3x^2 + 2x - 8, \quad x - 2$

▮ By the factor theorem, we must show that r is a root of $f(x) = 0$ to show that $x - r$ is a factor of $f(x)$. In this case, $f(x) = 2 \cdot 2^3 - 3 \cdot 2^2 + 2 \cdot 2 - 8 = 0$. Thus, $x - 2$ is a factor.

4.46 $3x^3 + 5x^2 - 6x + 18, \quad x + 3$

▮ $f(x) = 3x^3 + 5x^2 - 6x + 18$; $r = -3$. Then, $f(-3) = 3 \cdot (-3)^3 + 5 \cdot (-3)^2 - 6 \cdot (-3) + 18 = 0$. Thus, $x + 3$ is a factor.

4.47 $2x^3 - 3x^2 + x - 6, \quad x - 2$

▮ $f(x) = 2x^3 - 3x^2 + x - 6$; $r = 2$. Then $f(2) = 2 \cdot 2^3 - 3 \cdot 2^2 + 2 - 6 = 0$. Thus, $x - 2$ is a factor.

4.48 $2x^3 - 7x^2 - 3x - 4, \quad x - 4$

▮ $f(x) = 2x^3 - 7x^2 - 3x - 4$; $r = 4$. Then $f(4) = 2 \cdot 4^3 - 7 \cdot 4^2 - 3 \cdot 4 - 4 = 0$. Thus, $x - 4$ is a factor.

4.49 $3x^4 - x^3 - 3x^2 + \frac{1}{3}, \quad x - \frac{1}{3}$

▮ $f(x) = 3x^4 - x^3 - 3x^2 + \frac{1}{3}$; $r = \frac{1}{3}$. Then $f(\frac{1}{3}) = 3 \cdot (\frac{1}{3})^4 - (\frac{1}{3})^3 - 3(\frac{1}{3})^2 + \frac{1}{3} = 0$. Thus, $x - \frac{1}{3}$ is a factor.

4.50 $x^4 - 81, \quad x + 3$

▮ $f(-3) = -3^4 - 81 = 0$. Thus, $x + 3$ is a factor. See Prob. 4.51.

4.51 $x^9 + a^9$, $x + a$

▮ $f(x) = x^9 + a^9$; $r = -a$. Then $f(-a) = (-a)^9 + a^9 = -a^9 + a^9 = 0$. Thus, $x + a$ is a factor. See Prob. 4.52.

4.52 $x^n + a^n$, $x + a$, where n is an odd positive integer.

▮ $f(x) = x^n + a^n$; $r = -a$. Then $f(-a) = (-a)^n + a^n = -a^n + a^n = 0$ (since n is an odd positive integer). Thus, $x + a$ is a factor.

For Probs. 4.53 to 4.56, write equations with integral coefficients having the given numbers and no others as roots.

4.53 $1, 2, -3$

▮ By the factor theorem, $x - 1$, $x - 2$, $x + 3$ must be factors. Then $f(x) = (x - 1)(x - 2)(x + 3) = (x^2 - 3x + 2)(x + 3) = x^3 - 7x + 6 = 0$.

4.54 $-1, -2, -3$

▮ By the factor theorem, $x + 1$, $x + 2$, $x + 3$ must be factors. Then $f(x) = (x + 1)(x + 2)(x + 3) = (x^2 + 3x + 2)(x + 3) = x^3 + 6x^2 + 11x + 6 = 0$.

4.55 $-4, 3, 0$

▮ $(x + 4)(x - 3)(x) = 0$; $x^3 + x^2 - 12x = 0$.

4.56 $\sqrt{3}, -\sqrt{3}, 2$

▮ $(x - \sqrt{3})(x + \sqrt{3})(x - 2) = 0$; $(x^2 - 3)(x - 2) = 0$; $x^3 - 2x^2 - 3x + 6 = 0$.

4.2 GRAPHING POLYNOMIAL FUNCTIONS

For Probs. 4.57 to 4.83, sketch the graph of the given polynomial function.

4.57 $f(x) = x^3$

▮ See Fig. 4.1. If $f(x) = x^3 = 0$, then $x = 0$. The graph crosses the x axis at $x = 0$ since the multiplicity is odd. When $x = 0$, $y = 0$, so the graph crosses the y axis when $x = 0$. Since, $f(-x) = -f(x)$, the graph is symmetric about the origin. As x grows larger, so does y; as x grows smaller, so does y. Notice the steps we have gone through here. We will use them throughout this section.

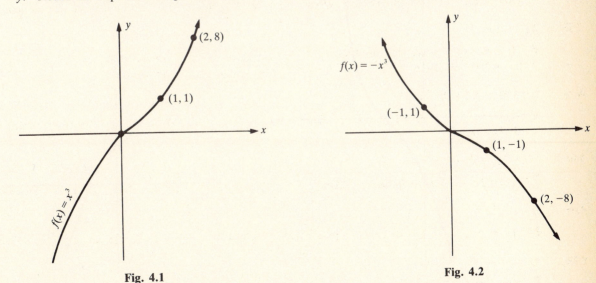

Fig. 4.1   Fig. 4.2

4.58 $f(x) = -x^3$

▮ See Fig. 4.2. If $f(x) = -x^3 = 0$, then $x = 0$. The graph crosses the x axis at $(0, 0)$. If $x = 0$, then $f(0) = 0$. If $f(-x) = -f(x)$, then f is symmetric about the origin; $f(1) = -1$, $f(-1) = 1$, $f(2) = -8$, $f(-2) = 8$.

4.59 $f(x) = 3x^3$

▮ See Fig. 4.3. $f(0) = 0$, and $f(-x) = -f(x)$, so the graph is symmetric about the origin. $f(1) = 3$, $f(-1) = -3$, $f(2) = 24$.

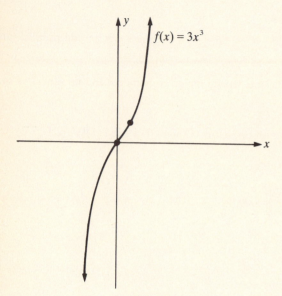

Fig. 4.3

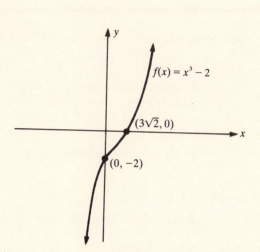

Fig. 4.4

4.60 $f(x) = x^3 - 2$

▮ See Fig. 4.4. $f(x) = 0$ when $x^3 - 2 = 0$. So, $x^3 = 2$, and $x = \sqrt[3]{2} \approx 1.26$. Note that the multiplicity is odd. Also, the graph will show no symmetry since all three symmetry tests fail: $f(-x) \neq f(x)$ and $f(-x) \neq -f(x)$. Substituting $-x$ for x, $-y$ for y, or both does not produce equivalent equations.

4.61 $f(x) = 2(x^3 - 2)$

▮ See Fig. 4.5 and Prob. 4.60. We can sketch this graph by observing that all y values in Prob. 4.60 are doubled.

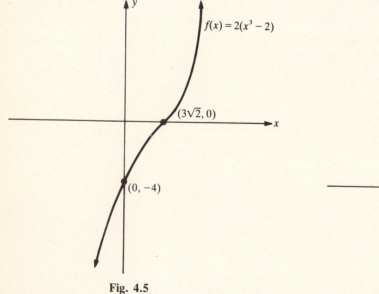

Fig. 4.5

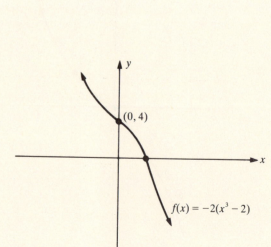

Fig. 4.6

4.62 $f(x) = -2(x^3 - 2)$

▮ See Fig. 4.6 and Probs. 4.60 and 4.61.

4.63 $f(x) = (x - 2)^3$

┃ See Fig. 4.7. $f(0) = -8$, and when $(x - 2)^3 = 0$, $x = 2$.

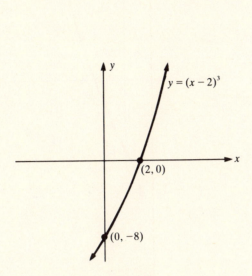

Fig. 4.7

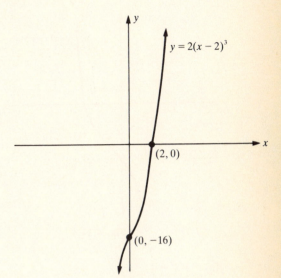

Fig. 4.8

4.64 $f(x) = 2(x - 2)^2$

┃ See Fig. 4.8 and Prob. 4.63.

4.65 $f(x) = -2(x - 2)^3$

┃ See Fig. 4.9 and Prob. 4.64.

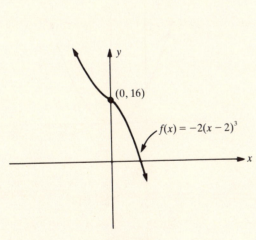

Fig. 4.9

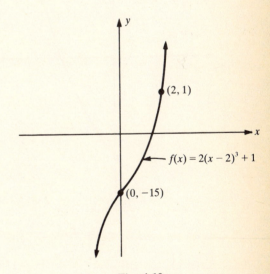

Fig. 4.10

4.66 $f(x) = 2(x - 2)^3 + 1$

┃ See Fig. 4.10 and Prob. 4.64.

4.67 $f(x) = (x^3 - 4)$

┃ See Fig. 4.11. When $x^3 - 4 = 0$, $x = \sqrt[3]{4} \approx 1.587$. $f(0) = -4$.

4.68 $f(x) = 3(x^3 - 4)$

┃ See Fig. 4.12 and Prob. 4.67. Triple each y value.

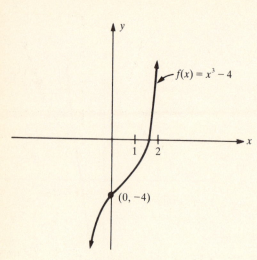

Fig. 4.11

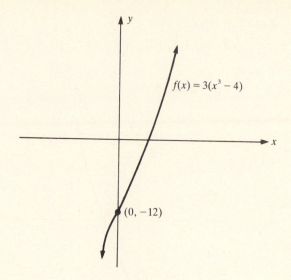

Fig. 4.12

4.69 $f(x) = 3(x^3 - 4) + 2$

▮ See Fig. 4.13 and Probs. 4.67 and 4.68.   Add 2 to each y value in Prob. 4.68.

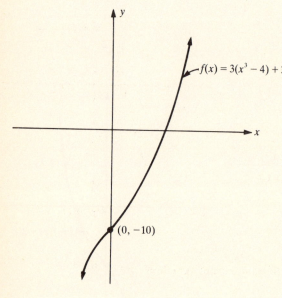

Fig. 4.13

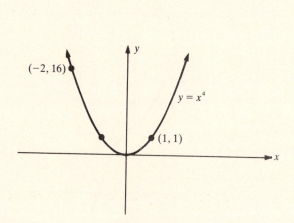

Fig. 4.14

4.70 $f(x) = x^4$

▮ See Fig. 4.14. If $y = x^4$, then $y = (-x)^4 = x^4$; the graph is symmetric about the y axis. $f(0) = 0$, $f(1) = 1$, $f(-1) = 1$, $f(2) = 16$, $f(-2) = 16$.

4.71 $f(x) = 3x^4$

▮ See Fig. 4.15 and Prob. 4.70.   This graph will have all ordinate values in Prob. 4.70 multiplied by 3.

4.72 $f(x) = 3x^4 - 2$

▮ See Fig. 4.16 and Prob. 4.71.   Subtract 2 from each ordinate in the graph of $y = 3x^4$.

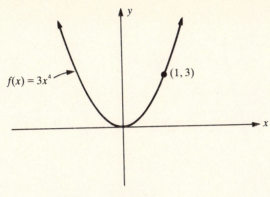

Fig. 4.15

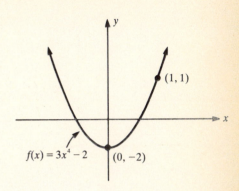

Fig. 4.16

4.73 $f(x) = x^5 + 1$

▮ See Fig. 4.17. This graph passes none of our symmetry tests. $f(0) = 1$ and if $x^5 + 1 = 0$, then $x^5 = -1$ and $x = -1$. Notice that as x increases, y increases, just as was the case with $y = x^3$.

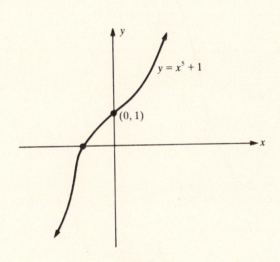

Fig. 4.17

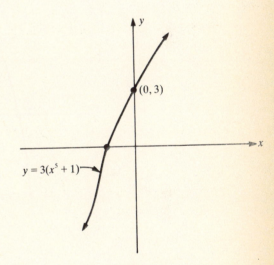

Fig. 4.18

4.74 $f(x) = 3(x^5 + 1)$

▮ See Fig. 4.18 and Prob. 4.73. We triple each ordinate in the graph of $y = x^5 + 1$.

4.75 $f(x) = 3(x^5 + 1) - 6$

See Fig. 4.19 and Prob. 4.74. We subtract 6 from each ordinate in the graph of $y = 3(x^5 + 1)$.

4.76 $y = (x - 3)(x^2 - 1)$

▮ See Fig. 4.20. If $(x - 3)(x^2 - 1) = 0$, then $x = 3, \pm 1$. Each of these zeros has a multiplicity of one, so the graph will cross the x axis at these points. Also, the three symmetry tests fail. Investigating the behavior of this function, we find that (a) it is negative for $x < -1$, (b) it is positive for $-1 < x < 1$, (c) it is negative for $1 < x < 3$, (d) it is positive for $x > 3$.

4.77 $y = 2(x - 3)(x^2 - 1)$

▮ See Fig. 4.21 and Prob. 4.76.

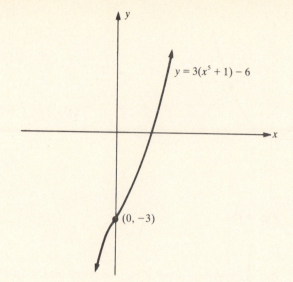

$$y = 3(x^5 + 1) - 6$$

$(0, -3)$

Fig. 4.19

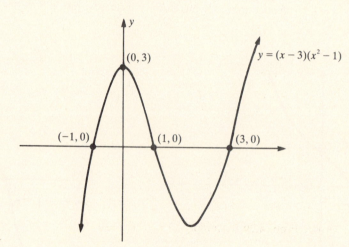

$(0, 3)$

$y = (x - 3)(x^2 - 1)$

$(-1, 0)$ $(1, 0)$ $(3, 0)$

Fig. 4.20

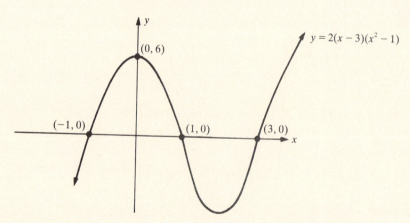

$(0, 6)$

$y = 2(x - 3)(x^2 - 1)$

$(-1, 0)$ $(1, 0)$ $(3, 0)$

Fig. 4.21

4.78 $y = 2(x - 3)(x^2 - 1) + 1$

▮ See Fig. 4.22 and Prob. 4.77.   Add 1 to each y value in the graph of $y = 2(x - 3)(x^2 - 1)$.

4.79 $y = (x^2 - 1)(x^2 - 9)$

▮ See Fig. 4.23.   This graph has 4 zeros: $x = \pm 1$, $x = \pm 3$,   all with odd multiplicity.   Also, replacing x by $-x$ does not change the equation, so the graph exhibits y-axis symmetry. $f(0) = 9$. $f > 0$ for $x < -3$, $x > 3$, $-1 < x < 1$;   and $f < 0$ for $-3 < x < -1$, $1 < x < 3$.

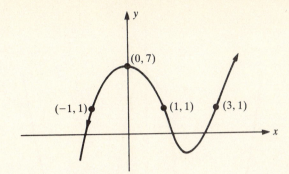

Fig. 4.22

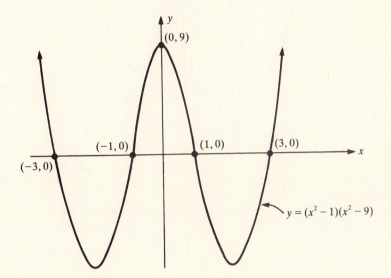

$y = (x^2 - 1)(x^2 - 9)$

Fig. 4.23

4.80 $y = 2(x^2 - 1)(x^2 - 9)$

▮ See Fig. 4.24 and Prob. 4.79.   Multiply all y values in the graph of $y = (x^2 - 1)(x^2 - 9)$ by 2.

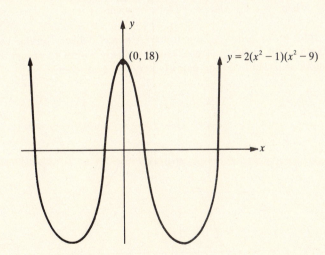

$y = 2(x^2 - 1)(x^2 - 9)$

Fig. 4.24

4.81 $y = 2(x^2 - 1)(x^2 - 9) + 1$

▮ See Fig. 4.25 and Prob. 4.80.   Add 1 to each y value in the graph of $y = 2(x^2 - 1)(x^2 - 9)$.

4.82 $f(x) = x^2 - 6x + 5$

▮ See Fig. 4.26.   Since this is in nonfactored form, we first factor the equation's RHS: $x^2 - 6x + 5 = (x - 1)(x - 5)$.   Clearly, 1 and 5 are zeros.   Also, $f(0) = 5$,   and the function is negative between 1 and 5.

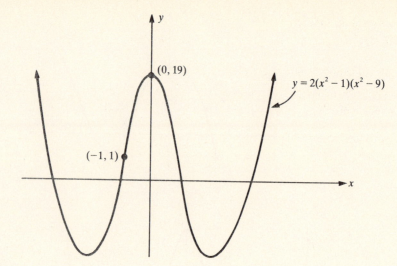

Fig. 4.25

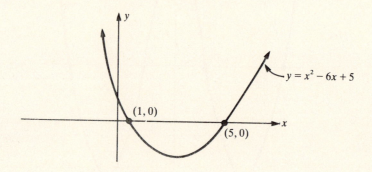

Fig. 4.26

4.83 $y = -x^2 + 2x + 8$

▮ See Fig. 4.27.   Factoring, we find: $-x^2 + 2x + 8 = -(x^2 - 2x - 8) = -(x^2 - 2x + 1 - 9) = -(x^2 - 2x + 1) + 9 = -(x - 1)^2 + 9$. $f(0) = 8$,   and if $f(x) = 0$, $-(x - 1)^2 + 9 = 0$, $-(x - 1)^2 = -9$, $(x - 1)^2 = 9$.   Then $x - 1 = 3$ when $x = 4$,   and $x - 1 = -3$ when $x = -2$.

For Probs. 4.84 to 4.87, refer to the following situation: A parcel delivery service will deliver only packages with length plus girth not exceeding 108 inches.   A packaging company wishes to design a box with a square base that will have a minimum volume and will meet the delivery service's restrictions.

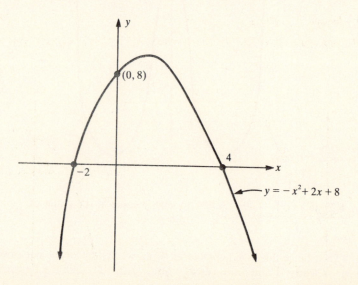

Fig. 4.27

4.84 Write the volume of the box $V(x)$ in terms of x.

▮ Volume $= l \cdot w \cdot h = x \cdot x \cdot (108 - x - x - x - x) = x^2(108 - 4x)$.

4.85 What is the domain of V in Prob. 4.84?

▮ $108 - 4x$ must be nonnegative in order that $V \geq 0$. Thus, $0 \leq x \leq 27$. If $x = 0$ or $x = 27$, the volume of the box is 0.

4.86 Graph V for the domain in Prob. 4.85.

▮ See Fig. 4.28.

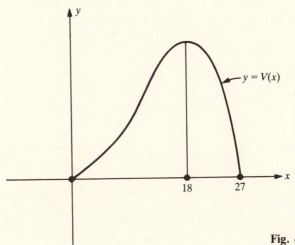

Fig. 4.28

4.87 From Fig. 4.28 estimate (to the nearest inch) the dimensions of the box with maximum volume? What is this maximum volume?

▮ To the nearest inch, the maximum occurs when $x \approx 18$. $V(18) = 18 \cdot 18 \cdot 36 = 11,664$ square inches $\approx$ maximum volume.

For Probs. 4.88 to 4.96, refer to the graphs in Figs. 4.29 to 4.34.

4.88 Which of the figures represents a polynomial function which is always increasing?

▮ Fig. 4.34. Note that if $a > b$, then $f(a) > f(b)$ for all a, b. None of the other graphs have this property. The graph of Fig. 4.34 is a polynomial of degree 1.

4.89 Which of the figures are symmetric about the y axis?

▮ Figs. 4.29 and 4.30. For both of these we see that if (x, y) is on the graph, so is $(-x, y)$.

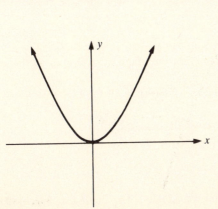

Fig. 4.29

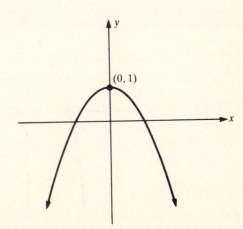

$(0, 1)$

Fig. 4.30

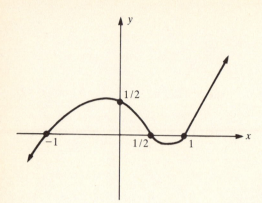

Fig. 4.31

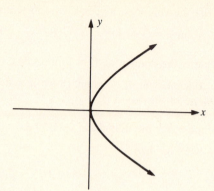

Fig. 4.32

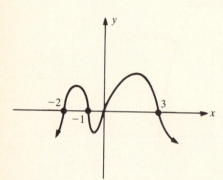

Fig. 4.33

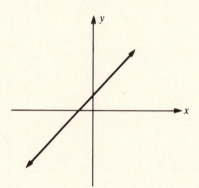

Fig. 4.34

4.90 Which of the figures exhibits x axis symmetry?

▮ Only Fig. 4.32.   Notice that if (x, y) is on the graph, so is $(x, -y)$.

4.91 Which figure has 4 zeros?

▮ Fig. 4.33.   Notice that the graph crosses the x axis four times.   Each zero must be odd for this to happen.

4.92 Which figures exhibit zeros with even multiplicity?

▮ Only Fig. 4.29.   The graph touches the x axis at 0 without crossing it.

4.93 Which figure does not represent $y = f(x)$?

▮ In Fig. 4.32, the graph is not a function of x.   If you "plug in" an x, you get two y values.

4.94 In which figure are there two intervals in which the function is positive?

▮ In Fig. 4.31, $f > 0$ when x is in the intervals $(-1, \frac{1}{2})$ and $(1, \infty)$.   In Fig. 4.33, $f > 0$ when x is in the intervals $(-2, -1)$ and $(0, 3)$.

4.95 Which figure is symmetric about $(0, 0)$?

▮ None; there is no figure in the collection where, for all (x, y) on the graph, $(-x, -y)$ is on the graph.

4.96 Which figures must represent a function of the form $P(x) = ax^n$,   where n is an even positive integer, and $a > 0$?

▮ The graph in Fig. 4.29.   If n were odd, the graph would cross the x axis at $(0, 0)$.

4.3 SYNTHETIC DIVISION

For Probs. 4.97 to 4.134, use synthetic division to find the quotient and remainder.   The method is fully described in Prob. 4.97.

4.97 $(x^3 + x^2 + x + 3) \div (x - 1)$

$$
\begin{array}{cccc}
1 & 1 & 1 & 3 \; \lfloor 1 \\
 & 1 & 2 & 3 \\
\hline
1 & +2 & +3 & +6
\end{array}
$$

$$\underbrace{}_{\text{quotient}} \qquad \underbrace{}_{\text{remainder}}$$

The quotient is $1x^2 + 2x + 3 = x^2 + 2x + 3$ and the remainder is 6.

Outline of methods: (a) Write down the coefficients of the dividend from left to right in decreasing order of the powers of x.   Insert 0 for any missing terms.

$$1 \quad 1 \quad 1 \quad 3$$

(b) To the right of this, write the a of the divisor $x - a$.

$$1 \quad 1 \quad 1 \quad 3 \; \lfloor 1$$

(c) On the *third* line (skip one line) rewrite the leading coefficient.

$$\begin{array}{cccc} 1 & 1 & 1 & 3 \; \lfloor 1 \\ \hline \\ 1 \end{array}$$

(d) Multiply this leading coefficient by r, and write it on the second line in the second column as shown.   Then add the terms in column two and write the sum on the third line as shown.

$$\begin{array}{cccc} 1 & 1 & 1 & 3 \; \lfloor 1 \\ & 1 \\ \hline 1 & 2 \end{array}$$

(e) Multiply the sum obtained in (d) by r, and repeat the procedure in (d) for each column.

$$
\begin{array}{cccc}
1 & 1 & 1 & 3 \; \lfloor 1 \\
1 & 2 & 3 & 3 \\
\hline
1 & 2 & 3 & 6
\end{array}
$$

$$\underbrace{\text{Coefficients of quotient}} \qquad \text{Remainder}$$

The quotient is $(1)x^2 + (2)x + 3 = x^2 + 2x + 3$,   and the remainder is 6.

4.98 $(x^3 - x - 1) \div (x - 1)$

$$
\begin{array}{ccccc}
1 & 0 & -1 & -1 & \lfloor 1 \\
 & 1 & 1 & 0 \\
\hline
1 & 1 & 0 & -1
\end{array}
$$

The quotient is $x^2 + x$,   and the remainder is -1.

4.99 $(x^3 + x^2 + x - 1) \div (x - 1)$

$$
\begin{array}{ccccc}
1 & 1 & 1 & -1 & \lfloor -1 \\
 & -1 & 0 & -1 \\
\hline
1 & 0 & 1 & -2
\end{array}
$$

The quotient is $x^2 + 1$,   and the remainder is -2.

4.100 $(x^3 + x^2 + x + 1) \div (x - 1)$

$$
\begin{array}{ccccc}
1 & 1 & 1 & 1 & \lfloor 1 \\
 & 1 & 2 & 3 \\
\hline
1 & 2 & 3 & 4
\end{array}
$$

The quotient is $x^2 + 2x + 3$,   and the remainder is 4.

4.101 $(2x^3 + x^2 + x + 1) \div (x + 1)$

$$
\begin{array}{ccccc}
2 & 1 & 1 & 1 & \lfloor -1 \\
 & -2 & 1 & -2 \\
\hline
2 & -1 & 2 & -1
\end{array}
$$

The quotient is $2x^2 - x + 2$,   and the remainder is -1.

4.102 $(2x^3 + 2x^2 + x + 1) \div (x + 1)$

$$
\begin{array}{rrrr|r}
2 & 2 & 1 & 1 & \underline{-1} \\
  & -2 & 0 & -1 & \\
\hline
2 & 0 & 1 & 0 &
\end{array}
$$

The quotient is $2x^2 + 1$,   and the remainder is 0.   (What does that mean?)

4.103 $(2x^3 + 2x^2 + 2x + 1) \div (x + 1)$

$$
\begin{array}{rrrr|r}
2 & 2 & 2 & 1 & \underline{-1} \\
  & -2 & 0 & -2 & \\
\hline
2 & 0 & 2 & -1 &
\end{array}
$$

The quotient is $2x^2 + 2$,   and the remainder is -1.

4.104 $(2x^3 + 2x^2 + 2x + 2) \div (x + 1)$

$$
\begin{array}{rrrr|r}
2 & 2 & 2 & 2 & \underline{-1} \\
  & -2 & 0 & -2 & \\
\hline
2 & 0 & 2 & 0 &
\end{array}
$$

The quotient is $2x^2 + 2$,   and the remainder is 0.

4.105 $(2x^3 + 2x^2 + 2x + 2) \div (x + 2)$

$$
\begin{array}{rrrr|r}
2 & 2 & 2 & 2 & \underline{-2} \\
  & -4 & 4 & -12 & \\
\hline
2 & -2 & 6 & -10 &
\end{array}
$$

The quotient is $2x^2 - 2x + 6$,   and the remainder is -10.

4.106 $(2x^3 + 2x^2 + 2x + 2) \div (x + 3)$

$$
\begin{array}{rrrr|r}
2 & 2 & 2 & 2 & \underline{-3} \\
  & -6 & 12 & -42 & \\
\hline
2 & -4 & 14 & -40 &
\end{array}
$$

The quotient is $2x^2 - 4x + 14$,   and the remainder is -40.

4.107 $(2x^3 + 2x^2 + 2x + 2) \div (x - 3)$

$$
\begin{array}{rrrr|r}
2 & 2 & 2 & 2 & \underline{3} \\
  & 6 & 24 & 78 & \\
\hline
2 & 8 & 26 & 80 &
\end{array}
$$

The quotient is $2x^2 + 8x + 26$,   and the remainder is 80.

4.108 $(5x^3 - 2x^2 - 2x - 2) \div (x + 1)$

$$
\begin{array}{rrrr|r}
5 & -2 & -2 & -2 & \underline{-1} \\
  & -5 & 7 & -5 & \\
\hline
5 & -7 & 5 & -7 &
\end{array}
$$

The quotient is $5x^2 - 7x + 5$,   and the remainder is -7.

4.109 $(5x^3 - 2x^2 - 2x - 2) \div (x - 1)$

$$
\begin{array}{rrrr|r}
5 & -2 & -2 & -2 & \underline{1} \\
  & 5 & 3 & 1 & \\
\hline
5 & 3 & 1 & -1 &
\end{array}
$$

The quotient is $5x^2 = 3x + 1$,   and the remainder is -1.

4.110 $(5x^3 - 2x^2 - 2x - 2) \div (x + 3)$

$$
\begin{array}{rrrr|r}
5 & -2 & -2 & -2 & \underline{-3} \\
  & -15 & 51 & -147 & \\
\hline
5 & -17 & 49 & -149 &
\end{array}
$$

The quotient is $5x^2 - 17x + 49$,   and the remainder is -149.

4.111 $(5x^3 - 2x^2 - 2x - 2) \div (x + 2)$

∎

$$
\begin{array}{rrrr|r}
5 & -2 & -2 & -2 & \underline{-2} \\
  & -10 & 24 & -44 & \\
\hline
4 & -12 & 22 & -46 &
\end{array}
$$

The quotient is $5x^2 - 12x + 22$, and the remainder is -46.

4.112 $(5x^3 - 2x^2 - 2x - 2) \div (x - 2)$

∎

$$
\begin{array}{rrrr|r}
5 & -2 & -2 & -2 & \underline{2} \\
  & 10 & 16 & 28 & \\
\hline
5 & 8 & 14 & 26 &
\end{array}
$$

The quotient is $5x^2 + 8x + 14$, and the remainder is 26.

4.113 $(5x^3 - 3x^2 - 2x - 2) \div (x + 2)$

∎

$$
\begin{array}{rrrr|r}
5 & -3 & -2 & -2 & \underline{-2} \\
  & -10 & 26 & -48 & \\
\hline
5 & -13 & 24 & -50 &
\end{array}
$$

The quotient is $5x^2 - 13x + 24$, and the remainder is -50.

4.114 $(x^4 + x^3 + x^2 + x + 1) \div (x - 1)$

∎

$$
\begin{array}{rrrrr|r}
1 & 1 & 1 & 1 & 1 & \underline{1} \\
  & 1 & 2 & 3 & 4 & \\
\hline
1 & 2 & 3 & 4 & 5 &
\end{array}
$$

The quotient is $x^3 + 2x^2 + 3x + 4$, and the remainder is 5.

4.115 $(x^4 + x^3 + x^2 + x + 1) \div (x + 1)$

∎

$$
\begin{array}{rrrrr|r}
1 & 1 & 1 & 1 & 1 & \underline{-1} \\
  & -1 & 0 & -1 & 0 & \\
\hline
1 & 0 & 1 & 0 & 1 &
\end{array}
$$

The quotient is $x^3 + x$, and the remainder is 1.

4.116 $(x^4 + 3x^2 + 2x + 2) \div (x - 1)$

∎

$$
\begin{array}{rrrrr|r}
1 & 0 & 3 & 2 & 2 & \underline{1} \\
  & 1 & 1 & 4 & 6 & \\
\hline
1 & 1 & 4 & 6 & 8 &
\end{array}
$$

The quotient is $x^3 + x^2 + 4x + 6$, and the remainder is 8.

4.117 $(x^4 + 3x^2 + 2x + 2) \div (x + 1)$

∎

$$
\begin{array}{rrrrr|r}
1 & 0 & 3 & 2 & 2 & \underline{-1} \\
  & -1 & 1 & -4 & 2 & \\
\hline
1 & -1 & 4 & -2 & 4 &
\end{array}
$$

The quotient is $x^3 - x^2 + 4x - 2$, and the remainder is 4.

4.118 $(x^4 + 2x + 2) \div (x - 1)$

∎

$$
\begin{array}{rrrrr|r}
1 & 0 & 0 & 2 & 2 & \underline{1} \\
  & 1 & 1 & 1 & 3 & \\
\hline
1 & 1 & 1 & 3 & 5 &
\end{array}
$$

The quotient is $x^3 + x^2 + x + 3$, and the remainder is 5.

4.119 $(x^4 + x + 2) \div (x - 3)$

∎

$$
\begin{array}{rrrrr|r}
1 & 0 & 0 & 1 & 2 & \underline{3} \\
  & 3 & 9 & 27 & 84 & \\
\hline
1 & 3 & 9 & 28 & 86 &
\end{array}
$$

The quotient is $x^3 + 3x^2 + 9x + 28$, and the remainder is 86.

4.120 $(2x^5 + x^4 + x^3 + x^2 + 1) \div (x - 1)$

$$
\begin{array}{rrrrrr|l}
2 & 1 & 1 & 1 & 1 & 1 & \underline{1} \\
  & 2 & 3 & 4 & 5 & 6 & \\
\hline
2 & 3 & 4 & 5 & 6 & 7 &
\end{array}
$$

The quotient is $2x^4 + 3x^3 + 4x^2 + 5x + 6$,   and the remainder is 7.

4.121 $(2x^5 + x^4 + x^3 + x^2 + x + 1) \div (x - 2)$

$$
\begin{array}{rrrrrr|l}
2 & 1 & 1 & 1 & 1 & 1 & \underline{2} \\
  & 4 & 10 & 22 & 46 & 94 & \\
\hline
2 & 5 & 11 & 23 & 47 & 95 &
\end{array}
$$

The quotient is $2x^4 + 5x^3 + 11x^2 + 23x + 47$,   and the remainder is 95.

4.122 $(2x^5 + x^4 + x^3 + x^2 + x + 3) \div (x + 4)$

$$
\begin{array}{rrrrrr|l}
2 & 1 & 1 & 1 & 1 & 3 & \underline{-4} \\
  & -8 & 28 & -116 & 460 & -1844 & \\
\hline
2 & -7 & 29 & -115 & 461 & -1841 &
\end{array}
$$

The quotient is $2x^4 - 7x^3 + 29x^2 - 115x + 461$,   and the remainder is -1841.

4.123 $(x^3 - 1) \div (x - 1)$

$$
\begin{array}{rrrr|l}
1 & 0 & 0 & -1 & \underline{1} \\
  & 1 & 1 & 1 & \\
\hline
1 & 1 & 1 & 0 &
\end{array}
$$

The quotient is $x^2 + x + 1$,   and the remainder is 0.   (What does that mean?)

4.124 $(x^3 - 1) \div (x + 1)$

$$
\begin{array}{rrrr|l}
1 & 0 & 0 & -1 & \underline{1} \\
  & -1 & 1 & -1 & \\
\hline
1 & -1 & 1 & -2 &
\end{array}
$$

The quotient is $x^2 - x + 1$,   and the remainder is -2.

4.125 $(x^3 + 1) \div (x + 2)$

$$
\begin{array}{rrrr|l}
1 & 0 & 0 & -1 & \underline{-2} \\
  & -2 & 4 & -8 & \\
\hline
1 & -2 & 4 & -9 &
\end{array}
$$

The quotient is $x^2 - 2x + 4$,   and the remainder is -9.

4.126 $(x^4 + 1) \div (x + 1)$

$$
\begin{array}{rrrrr|l}
1 & 0 & 0 & 0 & 1 & \underline{-1} \\
  & -1 & 1 & -1 & 1 & \\
\hline
1 & -1 & 1 & -1 & 2 &
\end{array}
$$

The quotient is $x^3 - x^2 + x - 1$,   and the remainder is 2.

4.127 $(x^4 + 1) \div (x - 1)$

$$
\begin{array}{rrrrr|l}
1 & 0 & 0 & 0 & 1 & \underline{1} \\
  & 1 & 1 & 1 & 1 & \\
\hline
1 & 1 & 1 & 1 & 2 &
\end{array}
$$

The quotient is $x^3 + x^2 + x + 1$,   and the remainder is 2.

4.128 $(x^4 - 1) \div (x + 1)$

$$
\begin{array}{rrrrr|l}
1 & 0 & 0 & 0 & -1 & \underline{-1} \\
  & -1 & 1 & -1 & 1 & \\
\hline
1 & -1 & 1 & -1 & 0 &
\end{array}
$$

The quotient is $x^3 - x^2 + x - 1$,   and the remainder is 0.

4.129 $(x^4 - 1) \div (x - 1)$

$$
\begin{array}{rrrrr|l}
1 & 0 & 0 & 0 & -1 & \underline{1} \\
  & 1 & 1 & 1 & 1 & \\
\hline
1 & 1 & 1 & 1 & 0 &
\end{array}
$$

The quotient is $x^3 + x^2 + x + 1$,   and the remainder is 0.

4.130 $(x^5 + 32) \div (x + 2)$

$$
\begin{array}{rrrrrr|l}
1 & 0 & 0 & 0 & 0 & 32 & \underline{-2} \\
  & -2 & 4 & -8 & 16 & -32 & \\
\hline
1 & -2 & 4 & -8 & 16 & 0 &
\end{array}
$$

The quotient is $x^4 - 2x^3 + 4x^2 - 8x + 16$,   and the remainder is 0.

4.131 $(x^5 - 32) \div (x - 2)$

$$
\begin{array}{rrrrrr|l}
1 & 0 & 0 & 0 & 0 & -32 & \underline{2} \\
  & 2 & 4 & 8 & 16 & 32 & \\
\hline
1 & 2 & 4 & 8 & 16 & 0 &
\end{array}
$$

The quotient is $x^4 + 2x^3 + 4x^2 + 8x + 16$,   and the remainder is 0.

4.132 $(2x^4 - 5x - x - 4) \div (x - \frac{1}{2})$

$$
\begin{array}{rrrrr|l}
2 & -5 & 0 & -1 & -4 & \underline{\frac{1}{2}} \\
  & 1 & -2 & -1 & -1 & \\
\hline
2 & -4 & -2 & -2 & -5 &
\end{array}
$$

The quotient is $2x^3 - 4x^2 - 2x - 2$,   and the remainder is -5.

4.133 $(4x^4 + 3x^2 - 4x + 3) \div (x - \frac{1}{2})$

$$
\begin{array}{rrrrr|l}
4 & 0 & 3 & -4 & 3 & \underline{\frac{1}{2}} \\
  & 2 & 1 & 2 & -1 & \\
\hline
4 & 2 & 4 & -2 & 2 &
\end{array}
$$

The quotient is $4x^3 + 2x^2 + 4x - 2$,   and the remainder is 2.

4.134 $(8x^3 + 2x^2 + 5x - 3) \div (x + \frac{3}{4})$

$$
\begin{array}{rrrr|l}
8 & 2 & 5 & -3 & \underline{-\frac{3}{4}} \\
  & -6 & 3 & -6 & \\
\hline
8 & -4 & 8 & -9 &
\end{array}
$$

The quotient is $8x^2 - 4x + 8$,   and the remainder is -9.

For Probs. 4.135 to 4.138, evaluate by synthetic division.

4.135 $f(-2)$ if $f(x) = x^3 - 7x^2 + 12x - 3$.

▮ We will divide by $x + 2$;   the remainder will be $f(-2)$.

$$
\begin{array}{rrrr|l}
1 & -7 & 12 & -3 & \underline{-2} \\
  & -2 & 18 & -60 & \\
\hline
1 & -9 & 30 & -63 &
\end{array}
$$

The remainder is $-63 = f(-2)$.

4.136 $f(3)$ if $f(x) = x^3 - 7x^2 + 12x - 3$

▮ Divide by $x - 3$.

$$
\begin{array}{rrrr|l}
1 & -7 & 12 & -3 & \underline{3} \\
  & 3 & -12 & 0 & \\
\hline
1 & -4 & 0 & -3 &
\end{array}
$$

The remainder is $-3 = f(3)$.

4.137 $f(5)$ if $f(x) = x^3 - 4x^2 - 20x + 55$

⬛
$$
\begin{array}{rrrr|r}
1 & -4 & -20 & 55 & \underline{5} \\
  & 5 & 5 & -75 & \\
\hline
1 & 1 & -15 & -20 &
\end{array}
$$

The remainder is $-20 = f(5)$.

4.138 $f(\frac{1}{2})$ if $f(x) = x^6 + 1$

⬛
$$
\begin{array}{rrrrrrr|r}
1 & 0 & 0 & 0 & 0 & 0 & 1 & \underline{\frac{1}{2}} \\
  & \frac{1}{2} & \frac{1}{4} & \frac{1}{8} & \frac{1}{16} & \frac{1}{32} & \frac{1}{64} & \\
\hline
1 & \frac{1}{2} & \frac{1}{4} & \frac{1}{8} & \frac{1}{16} & \frac{1}{32} & \frac{65}{64} &
\end{array}
$$

The remainder is $f(\frac{1}{2}) = \frac{65}{64}$.

For Probs. 4.139 to 4.141, use synthetic division to show that the first polynomial is a factor of the second.

4.139 $x - 3$, $x^3 - 18x + 27$

⬛ We will divide and show that we obtain a zero remainder.
$$
\begin{array}{rrrr|r}
1 & 0 & -18 & 27 & \underline{3} \\
  & 3 & 9 & -27 & \\
\hline
1 & 3 & -9 & 0 &
\end{array}
$$

The remainder is 0, so $x - 3$ is a factor.

4.140 $x + 1$, $x^4 + 2x^3 + 2x^2 + 2x + 1$

⬛
$$
\begin{array}{rrrrr|r}
1 & 2 & 2 & 2 & 1 & \underline{-1} \\
  & -1 & -1 & -1 & -1 & \\
\hline
1 & 1 & 1 & 1 & 0 &
\end{array}
$$

The remainder is 0, so $x + 1$ is a factor.

4.141 $x + \frac{1}{2}$, $x^4 = 3x^2/4 + x/2 + \frac{3}{8}$

⬛
$$
\begin{array}{rrrrr|r}
1 & 0 & -\frac{3}{4} & \frac{1}{2} & \frac{3}{8} & \underline{-\frac{1}{2}} \\
  & -\frac{1}{2} & \frac{1}{4} & \frac{1}{4} & -\frac{3}{8} & \\
\hline
1 & -\frac{1}{2} & -\frac{1}{2} & \frac{3}{4} & 0 &
\end{array}
$$

The remainder is 0, so $x + \frac{1}{2}$ is a factor.

4.142 For what values of k does $kx^2 + x - 4$ yield the same remainder when divided by $x + 1$ and $x - 1$?

⬛
$$
\begin{array}{rrr|r}
k & 1 & -4 & \underline{1} \\
  & k & k+1 & \\
\hline
k & k+1 & k-3 &
\end{array}
\qquad
\begin{array}{rrr|r}
k & 1 & -4 & \underline{-1} \\
  & -k & k-1 & \\
\hline
k & -k+1 & k-5 &
\end{array}
$$

We want the remainders to be equal, so $k - 3 = k - 5$, but $-3 \neq -5$. Thus, no such values of k exist.

4.143 For what values of k is $x^2 + kx - 2$ divisible by $x = k$?

⬛
$$
\begin{array}{rrr|r}
1 & k & -2 & \underline{k} \\
  & k & 2k^2 & \\
\hline
1 & 2k & 2k^2 - 2 &
\end{array}
$$

We want the remainder to be 0, so $0 = 2k - 2 = k^2 - 1$. Thus, $k = \pm 1$.

4.144 If the polynomial $P(x) = Ax^4 - Ax^2 + x - 1$ is such that $P(3) = 0$, what is $P(2)$?

⬛
$$
\begin{array}{rrrrr|r}
A & 0 & -A & 1 & -1 & \underline{3} \\
  & 3A & 9A & 24A & 72A + 3 & \\
\hline
A & 3A & 8A & 24A + 1 & 72A + 2 &
\end{array}
$$

If $P(3) = 0$, then $72A + 2 = 0$ and $72A = -2$. Thus, $A = -\frac{1}{36}$. Then $P(x) = -\frac{1}{36}x^4 + \frac{1}{36}x^2 + x - 1$; $P(2)$ is the remainder after division by $x - 2$.

$$
\begin{array}{ccccc|c}
-\frac{1}{36} & 0 & \frac{1}{36} & 1 & -1 & \underline{2} \\
 & -\frac{1}{18} & -\frac{1}{9} & -\frac{1}{6} & \frac{10}{6} & \\
\hline
-\frac{1}{36} & -\frac{1}{18} & -\frac{1}{12} & \frac{5}{6} & \frac{2}{3} &
\end{array}
$$

Thus, $P(2) = \frac{2}{3}$.

4.145 If the polynomial $P(x) = 3x^3 - kx^2 + 1$ is such that $P(1) = 2$, what is $P(3)$?

$$
\begin{array}{ccccc|c}
3 & -k & 0 & 1 & \underline{1} \\
 & 3 & 3-k & 3-k & \\
\hline
3 & 3-k & 3-k & 4-k &
\end{array}
$$

If $P(1) = 2$, then $4 - k = 2$ and $k = 2$.

$$
\begin{array}{cccc|c}
3 & -2 & 0 & 1 & \underline{3} \\
 & 9 & 21 & 63 & \\
\hline
3 & 7 & 21 & 64 &
\end{array}
$$

Thus, $P(3)$ is the remainder and is equal to 64.

4.146 When $x^2 + 4x - 4$ is divided by $x + r$, the remainder is -8. Find r.

$$
\begin{array}{ccc|c}
1 & 4 & -4 & \underline{-r} \\
 & -r & -4r + r^2 & \\
\hline
1 & 4-r & r^2 - 4r - 4 &
\end{array}
$$

The remainder is -8, so $r^2 - 4r - 4 = -8$. Then, $r^2 - 4r + 4 = 0$, $(r-2)^2 = 0$, or $r = 2$.

4.147 When $6x^2 + x - 9$ is divided by $x - r$, the remainder is -9. Find r.

$$
\begin{array}{ccc|c}
6 & 1 & -9 & \underline{r} \\
 & 6r & 6r^2 + r & \\
\hline
6 & 6r+1 & 6r^2 + r - 9 &
\end{array}
$$

Then $6r^2 + r - 9 = -9$, $6r^2 + r = 0$, $r(6r+1) = 0$, and $r = 0$ or $r = -\frac{1}{6}$.

4.148 Prove that $x^n + a^n$ is divisible by $x + a$ when n is an odd positive integer.

$$
\begin{array}{cccccc|c}
1 & 0 & 0 & \cdots & 0 & a^n & \underline{-a} \\
 & -a & a^2 & \cdots & a^{n-1} & -a^n & \\
\hline
1 & -a & a^2 & \cdots & a^{n-1} & 0 &
\end{array}
$$

The remainder is 0, so $x + a$ divides $x^n + a^n$.

4.4 THE FUNDAMENTAL THEOREM OF ALGEBRA

4.149 State the fundamental theorem of algebra.

$\blacksquare$ Every polynomial $P(x)$ of degree $d \geq 1$, with complex coefficients, has at least one complex zero. (Remember! Real numbers are also complex.)

For Probs. 4.150 to 4.159, find a polynomial equation of lowest degree which has the roots listed.

4.150 $0, 1$

$\blacksquare$ $(x - 0)(x - 1) = x(x - 1) = 0$, or $x^2 - x = 0$.

4.151 $3, 4$

$\blacksquare$ $(x - 3)(x - 4) = 0$, or $x^2 - 7x + 12 = 0$.

4.152 $0, 1, 2$

$\blacksquare$ $x(x - 1)(x - 2) = x(x^2 - 3x + 2) = 0$, or $x^3 - 3x^2 + 2x = 0$.

4.153 $1, 2, -3$

▮ $(x-1)(x-2)(x+3) = (x-1)(x^2+x-6) = 0$,   or $x^3 - 7x + 6 = 0$.

4.154 $1, i$

▮ $(x-1)(x-i) = x^2 - x - ix + i = 0$,   or $x^2 + (-1-i)x + i = 0$.

4.155 i

▮ $x - i = 0$;   remember, first-degree polynomials *are* polynomials.

4.156 $2i, i$

▮ $(x-2i)(x-i) = x^2 - 2ix - ix + 2i^2 = 0$,   or $x^2 - 3ix - 2 = 0$.

4.157 $-1, i$

▮ $(x+1)(x-i) = x^2 + x - ix - i = 0$,   or $x^2 + (1-i)x - i = 0$.

4.158 $1, -1, i$

▮ $(x-1)(x+1)(x-i) = (x-1)(x^2+x-ix-i) = x^3 + (1-i)x^2 - (1-i)x - ix - i = x^3 + (-i)x^2 + (-1)x - i = 0$,   or $x^3 - ix^2 - x - i = 0$.

4.159 $1+i, -1-i$

▮ $[x-(1+i)][x-(-1-i)] = (x-1-i)(x+1+i) = 0$,   or $x^2 - 2i = 0$.

For Probs. 4.160 to 4.163, find a polynomial of degree 3 having the given zeros.

4.160 $1, 2$

▮   For 1 and 2 to be the only zeros, $x-1$ and $x-2$ must be the only linear factors.   Thus, $(x-1)^2(x-2)$ is one possibility.   In polynomial form, $(x-1)^2(x-2) = x^3 - 4x^2 + 5x - 2$.

4.161 $1, -1$

▮   Then $x-1$ and $x+1$ are the only linear factors. $(x-1)^2(x+1)$ is one possibility.   (What is the other?)   In polynomial form, $(x-1)^2(x+1) = x^3 - x^2 - x + 1$.

4.162 $1+\sqrt{3}, \ 1-\sqrt{3}, \ -1-\sqrt{3}$

▮   Then $x-(1+\sqrt{3}), \ x-(1-\sqrt{3}), \ x-(-1-\sqrt{3})$ are the three factors. $[x-(1+\sqrt{3})]$ $[x-(1-\sqrt{3})][x-(-1-\sqrt{3})] = x^3 - (1-\sqrt{3})x^2 - (4+2\sqrt{3})x - (2\sqrt{3}+2)$ in polynomial form.

4.163 a, b

▮ $(x-a)(x-b) = x^2 - ax - bx + ab = x^2 + (-a-b)x + ab$ in polynomial form.

For Probs. 4.164 to 4.166, find a polynomial $P(x)$ of lowest degree, with leading coefficient 1 (monic) that has the indicated set of zeros.   You may leave the answer in factored form, but indicate the degree of the polynomial.

4.164 $2, 3, 4$ (multiplicity of 2)

▮ $x-2, \ x-3, \ (x-4)^2$ are the factors. $P(x) = (x-2)(x-3)(x-4)^2$ where the degree of $P(x) = 4$.

4.165 $1, -1$ (multiplicity of 3), 0

▮ $P(x) = (x-1)(x+1)^3 x$ where degree $P(x) = 5$.

4.166 $(2-3i), \ (2+3i), \ -4$ (multiplicity = 2)

▮ $P(x) = (x+4)^2[x-(2-3i)][x-(2+3i)]$ where the degree of $P(x) = 4$.

For Probs. 4.167 to 4.169, find the zeros of the given polynomial.

4.167 $x^2(x-2)(x+3)=0$

▮ If $(x-a)^m$ is a factor of $P(x)$, then a is a zero. In this case $x-0$, $x-2$, $x+3$ are factors, so $0, 2, -3$ are zeros.

4.168 $(x+i)^2(x-1)^3(x+4)=0$

▮ $x=-i$, 1, and -4 are all zeros.

4.169 $(2x-1)(2x+5)(x-3)=0$

▮ If $2x-1=0$, $P(x)=0$. If $2x+5=0$, $P(x)=0$. Thus, the zeros are $x=\frac{1}{2}, -\frac{5}{2}, 3$.

For Probs. 4.170 to 4.173, find the remaining zeros of each given polynomial, using the given zero(s).

4.170 x^3-3x^2+x+1, 1

▮ If 1 is a zero, then $x-1$ is a factor. Using synthetic division:

$$
\begin{array}{rrrr|r}
1 & -3 & 1 & 1 & \underline{1} \\
  & 1 & -2 & -1 & \\
\hline
1 & -2 & -1 & 0 &
\end{array}
$$

x^2-2x-1 is a factor; but if $x^2-2x-1=0$, then $x=\dfrac{-b\pm\sqrt{b^2-4ac}}{2a}=1\pm\sqrt{2}$. Thus, the remaining zeros are $1+\sqrt{2}$ and $1-\sqrt{2}$.

4.171 x^3-3x^2+2, $1+\sqrt{3}$

▮ Then $1-\sqrt{3}$ must be a zero, so $[x-(1+\sqrt{3})]$ is a factor of $P(x)$ as well. $[x-(1+\sqrt{3})][x-(1-\sqrt{3})]=x^2-2x-2$. Since $(x^3-3x^2+2)\div(x^2-2x-2)=x-1$, 1 and $1-\sqrt{3}$ are the other zeros.

4.172 x^4+5x^2+4, i

▮ If i is a zero, then $-i$ is a zero as well. $(x-i)(x+i)=x^2-i^2=x^2+1$. We divide x^4+5x^2+4 by x^2+1 and obtain x^2+2. When $x^2+2=0$, $x^2=-4$ or $x=\pm 2i$. Thus, other zeros are $-i$ and $\pm 2i$.

4.173 $x^4-x^3+6x^2-26x+20$, $-1-3i$

▮ If $-1-3i$ is a zero, then $-1+3i$ is as well. $[x-(-1+3i)][x-(-1-3i)]=x^2+2x+(1-9i^2)=x^2+2x+10$. We divide $x^4-x^3+6x^2-26x+20$ by x^2-3x+2 and obtain $x^2-3x+2=(x-2)(x-1)$. Thus, the other zeros are $2, 1$, $-1+3i$.

For Probs. 4.174 to 4.176, tell how many roots the given polynomial has.

4.174 $x^6-3x+2=0$

▮ This is a sixth-degree equation. It *must* have six roots.

4.175 $x^{10}-1=0$

▮ This polynomial equation is of degree 10. It has 10 roots. By the way, these are the 10 tenth roots of unity. See Chap. 10.

4.176 $(2i)x^2-\sqrt{2}=0$

▮ This is a second-degree equation. It has two roots. Do not let the imaginary coefficient fool you!

For Probs. 4.177 to 4.179, consider the given polynomial and tell what are the possible combinations of real and complex zeros?

4.177 $P(x)=2x^3-3x^2+x-5$

▮ Every polynomial of odd degree with real coefficients has at least 1 real zero. Thus, the possibilities are 1 real, 2 complex; or 3 real.

4.178 $P(x) = x^4 + 2x + 1$

▮ For every nonreal complex zero, its conjugate must be a zero as well. Thus, the possibilities are 4 complex; 2 complex, 2 real; or 4 real.

4.179 $P(x) = x^7 + x^4 - x - 1$

▮ This $P(x)$ is of odd degree, so it must have at least 1 real zero. The possibilities are 1 real, 6 complex; 3 real, 4 complex; 5 real, 2 complex; or 7 real.

For Probs. 4.180 to 4.182, consider the equation $x^3 - 8 = 0$.

4.180 What are the possible combinations of real and complex roots?

▮ $P(x) = x^3 - 8$ is of odd degree with real coefficients. Thus, it could have 1 real zero, 2 complex zeros; or 3 real zeros. Thus, $x^3 - 8 = 0$ could have 1 real root, 2 complex roots; or 3 real roots.

4.181 How many cube roots of 8 are there?

▮ If $x^3 - 8 = 0$, then $x^3 = 8$, or $x = \sqrt[3]{8}$. Thus, there must be three cube roots of 8. See Prob. 4.180.

4.182 Find the cube roots of 8.

▮ $\sqrt[3]{8} = 2 = $ a real cube root of 8. Using synthetic division,

$$\begin{array}{r} 1 \quad 0 \quad 0 \quad -8 \; \underline{|2} \\ \quad 2 \quad 4 \quad 8 \\ \hline 1 \quad 2 \quad 4 \quad 0 \end{array}$$

$(x^3 - 8) \div (x - 2) = x^2 + 2x + 4$. If $x^2 + 2x + 4 = 0$, then $x = \dfrac{-2 \pm \sqrt{4 - 4(1)(4)}}{2(1)} = \dfrac{-2 \pm \sqrt{4 - 16}}{2} = \dfrac{-2 \pm \sqrt{-12}}{2} = \dfrac{-2 \pm 2i\sqrt{3}}{2} = -1 \pm i\sqrt{3}$. Thus, $x = -1 + i\sqrt{3}$, or $x = -1 - i\sqrt{3}$. The three cube roots of 8 are 2, $-1 + i\sqrt{3}$, and $-1 - i\sqrt{3}$.

4.5 FACTORS AND ZEROS

For Probs. 4.183 to 4.188, find all possible candidates for rational zeros of the given polynomial. See Prob. 4.183 for a description of the method.

4.183 $2x^2 - 5x + 2$

▮ If p/q (a rational number in lowest terms) is a zero of $P(x)$, then $p|a_0$ and $q|a_n$, where a_0 is the constant term. In this case, any zero p/q must be such that $p|2$, (p divides the constant term) and $q|2$ (q divides the coefficient of x^2). Thus, $p = \pm 1, \pm 2$; $q = \pm 1, \pm 2$; and $p/q = \pm 1, \pm 2, \pm \frac{1}{2}$.

4.184 $3x^2 - 7$

▮ If p/q is a zero, then $p|7$ and $q|3$. Thus, $p = \pm 1, \pm 7$; $q = \pm 1, \pm 3$; and $p/q = \pm 1, \pm \frac{1}{3}, \pm 7, \pm \frac{7}{3}$.

4.185 $x^2 + 2x + 4$

▮ If p/q is a zero, then $p|4$ and $q|1$. Thus, $p = \pm 1, \pm 2, \pm 4$; $q = \pm 1$; and $p/q = \pm 1, \pm 2, \pm 4$.

4.186 $3x^3 + x^2 + 2x + 6$

▮ If p/q is a zero, then $p|6$ and $q|3$. Thus, $p = \pm 1, \pm 2, \pm 3, \pm 6$; $q = \pm 1, \pm 3$; and $p/q = \pm 1, \pm \frac{1}{3}, \pm 2, \pm \frac{2}{3}, \pm 3, \pm 6$.

4.187 $x^5 - 4$

▮ If p/q is a zero, then $p|4$ and $q|1$. Thus, $p = \pm 1, \pm 2, \pm 4$; $q = \pm 1$; and $p/q = \pm 1, \pm 2, \pm 4$.

4.188 $3u^4 + u^2 - 15$

▮ If p/q is a zero, then $p|15$ and $q|3$. Thus, $p = \pm 1, \pm 3, \pm 5, \pm 15$; $q = \pm 1, \pm 3$; and $p/q = \pm 1, \pm \frac{1}{3}, \pm 3, \pm 5, \pm \frac{5}{3}, \pm 15$.

For Probs. 4.189 to 4.194, find the rational zeros of the given polynomials.

4.189 $2x^2 - 5x + 2$

▮ Using the method in Probs. 4.183 to 4.188 above, the *possible* rational zeros are $\pm \frac{1}{2}, \pm 2, \pm 1$. Using synthetic division

$$
\begin{array}{rrr|r}
2 & -5 & 2 & \underline{1} \\
  & 2 & -3 & \\
\hline
2 & -3 & -1 &
\end{array}
$$

we find that 1 is not a zero (remainder $\neq 0$). Continuing with synthetic division for each possible rational zero, we find $\frac{1}{2}$ and 2 are the rational zeros.

4.190 $x^3 + 3x^2 + x - 2$

▮ The possible zeros are $\pm 2, \pm 1$.

$$
\begin{array}{rrrr|r}
1 & 3 & 1 & -2 & \underline{-2} \\
  & -2 & -2 & 2 & \\
\hline
1 & 1 & -1 & 0 &
\end{array}
$$

Thus, -2 is a zero. Checking with synthetic division, we find that 2 and ± 1 do *not* yield a zero remainder. Thus, -2 is the only rational zero.

4.191 $t^3 + 1$

▮ The only possibilities are ± 1.

$$
\begin{array}{rrrr|r}
1 & 0 & 0 & 1 & \underline{-1} \\
  & -1 & 1 & -1 & \\
\hline
1 & -1 & 1 & 0 &
\end{array}
$$

-1 is the only rational root. Check to see that division by 1 does *not* yield a 0 remainder.

4.192 $x^4 - x^3 - 5x^2 + 3x + 2$

▮ The possible rational zeros are $\pm 2, \pm 1$.

$$
\begin{array}{rrrrr|r}
1 & -1 & -5 & 3 & 2 & \underline{1} \\
  & 1 & 0 & -5 & -2 & \\
\hline
1 & 0 & -5 & -2 & 0 &
\end{array}
$$

1 is a rational root. Checking, you will find that -2 is the other rational root.

4.193 $2x^3 - x^2 - 5x - 2$

▮ Possible rational zeros are $+\frac{1}{2}, \pm 2, \pm 1$. Using synthetic division, we find $-\frac{1}{2}, \frac{1}{2}, 1, 2$ to be rational zeros.

$$
\begin{array}{rrrr|r}
2 & -1 & -5 & -2 & \underline{-\frac{1}{2}} \\
  & -1 & -1 & 2 & \\
\hline
2 & -2 & -4 & 0 &
\end{array}
$$

Check the other roots using synthetic division.

4.194 $32u^5 - 1$

▮ The possibilities are $\pm 1, \pm \frac{1}{2}, \pm \frac{1}{4}, \pm \frac{1}{8}, \pm \frac{1}{16}$. Using synthetic division, or noting (without it) that $32(\frac{1}{2})^5 - 1 = 0$, we find $\frac{1}{2}$ to be the only rational root.

For Probs. 4.195 to 4.197, find all roots for the given polynomial equation.

4.195 $x^3 + 3x^2 + x - 2$

▮ The possible rational roots are $\pm 1, \pm 2$. Using synthetic division, we find that -2 is a root.

$$\begin{array}{rrrr|r} 1 & 3 & 1 & -2 & \underline{-2} \\ & -2 & -2 & 2 & \\ \hline 1 & 1 & -1 & 0 & \end{array}$$

Thus, the remainder is 0, and $(x^3 + 3x^2 + x - 2) \div (x + 2) = x^2 + x - 1$. If $x^2 + x - 1 = 0$, then, using the quadratic formula, $x = \dfrac{-1 \pm \sqrt{5}}{2}$. The three roots are -2, $\dfrac{-1 \pm \sqrt{5}}{2}$.

4.196 $t^3 + 1$

▮ The possible rational roots are ± 1.

$$\begin{array}{rrrr|r} 1 & 0 & 0 & 0 & \underline{-1} \\ & -1 & 1 & -1 & \\ \hline 1 & -1 & 1 & 0 & \end{array}$$

Then -1 is a root. If $x^2 - x + 1 = 10$, $x = \dfrac{1 \pm \sqrt{3}i}{2}$. The 3 roots are -1, $\dfrac{1 \pm \sqrt{3}i}{2}$.

4.197 $x^4 - x^3 - 5x^2 + 3x + 2$

▮ The possible rational roots are $\pm 1, \pm 2$. Using synthetic division, we find $-2, 1$ to be roots, since $(x^4 - x^3 - 5x^2 + 3x + 2) \div (x + 2) = x^3 - 3x^2 + x + 1$, and $(x^3 - 3x^2 + x + 1) \div (x - 1) = x^2 - 2x - 1$.

$$\begin{array}{rrrrr|r} 1 & -1 & -5 & 3 & 2 & \underline{-2} \\ & -2 & 6 & -2 & -2 & \\ \hline 1 & -3 & 1 & 1 & 0 & \end{array} \qquad \begin{array}{rrrr|r} 1 & -3 & 1 & 1 & \underline{1} \\ & 1 & -2 & -1 & \\ \hline 1 & -2 & -1 & 0 & \end{array}$$

If $x^2 - 2x - 1 = 0$, then, $x = 1 \pm \sqrt{2}$. The roots are $1 \pm \sqrt{2}, -2, 1$.

For Probs. 4.198 to 4.202, use Descartes' rule of signs to discuss the number of positive and negative zeros of each given polynomial.

4.198 $P(x) = 2x^2 + x - 4$

▮ $P(x) = 2x^2 + x - 4$ which has *one* variation of sign ($+$ to $-$). $P(-x) = 2x^2 - x - 4$ which has *one* variation of sign ($+$ to -1). Descartes' rule of signs tells us that there will be 1 positive zero [one variation in sign in $P(x)$] and 1 negative zero [one variation in sign in $P(-x)$]. Note that the other possibility by Descartes' rule does not apply here. It is not possible to subtract an even integer from 1 to get a number of zeros. See Prob. 4.199.

4.199 $M(x) = 7x^2 + 2x + 4$

▮ $M(x) = 7x^2 + 2x + 4$ which has zero variations. $M(-x) = 7x^2 - 2x + 4$ which has two variations. Thus, there are 0 positive zeros and *either* 2 or 0 negative zeros. (Here, we find that since there are possibly 2 negative zeros, there are also possibly $2 - 2 = 0$ negative zeros.)

4.200 $P(x) = x^3 + 4x^2 + x - 1$

▮ $P(x) = x^3 + 4x^2 + x - 1$ which has one variation. $P(-x) = -x^3 + 4x^2 - x - 1$ which has two variations.

4.201 $P(x) = x^4 + x^2 + 1$

▮ $P(x) = P(-x) = x^4 + x^2 + 1$. There are zero variations, which means no positive or negative zeros.

4.202 $S(x) = x^5 + x^4 + x^3 - x^2 + 1$

▮ $S(x)$ has two variations. $S(-x) = -x^5 + x^4 - x^3 - x^2 + 1$ which has three variations. Thus, $S(x)$ has 2 or 0 positive and 3 or 1 negative zeros.

For Probs. 4.203 to 4.208, find the smallest positive integer and largest negative integer that are, respectively, upper and lower bounds for the real zeros of each given polynomial.

4.203 $P(x) = x^2 - 2x + 3$

▮ We will use synthetic division and divide by $1, 2, 3, \ldots,$ until the quotient row is nonnegative. Then, we use $-1, -2, \ldots,$ until it alternates in sign. In this case, we find:

$$
\begin{array}{rrr|l}
1 & -2 & 3 & \underline{1} \\
  & 1 & -1 & \\
\hline
1 & -1 & 2 & \text{(not nonnegative)}
\end{array}
\qquad
\begin{array}{rrr|l}
1 & -2 & 3 & \underline{2} \\
  & 2 & 0 & \\
\hline
1 & 0 & 3 & \text{(nonnegative)}
\end{array}
$$

Thus, 2 is an upper bound.

$$
\begin{array}{rrr|l}
1 & -2 & 3 & \underline{-1} \\
  & -1 & 3 & \\
\hline
1 & -3 & 6 & \text{(alternates)}
\end{array}
$$

Thus, -1 is a lower bound.

4.204 $Q(x) = x^2 - 3x - 2$

▮

$$
\begin{array}{rrr|l}
1 & -3 & -2 & \underline{1} \\
  & 1 & -2 & \\
\hline
1 & -2 & 0 &
\end{array}
\qquad
\begin{array}{rrr|l}
1 & -3 & -2 & \underline{2} \\
  & 2 & -2 & \\
\hline
1 & -1 & 0 &
\end{array}
$$

$$
\begin{array}{rrr|l}
1 & -3 & -2 & \underline{3} \\
  & 3 & 0 & \\
\hline
1 & 0 & -2 &
\end{array}
\qquad
\begin{array}{rrr|l}
1 & -3 & -2 & \underline{4} \\
  & 4 & 4 & \\
\hline
1 & 1 & 2 & \text{(nonnegative)}
\end{array}
$$

Thus, 4 is an upper bound.

$$
\begin{array}{rrr|l}
1 & -3 & -2 & \underline{1} \\
  & -1 & 4 & \\
\hline
1 & -4 & 2 & \text{(alternates)}
\end{array}
$$

Thus, -1 is a lower bound.

4.205 $M(k) = x^3 - 3x + 5$

▮

$$
\begin{array}{rrrr|l}
1 & 0 & -3 & 5 & \underline{1} \\
  & 1 & 1 & -2 & \\
\hline
1 & 1 & -2 & 3 &
\end{array}
\qquad
\begin{array}{rrrr|l}
1 & 0 & -3 & 5 & \underline{2} \\
  & 2 & 4 & 2 & \\
\hline
1 & 2 & 1 & 7 & \text{(nonnegative)}
\end{array}
$$

2 is an upper bound.

$$
\begin{array}{rrrr|l}
1 & 0 & -3 & 5 & \underline{-1} \\
  & -1 & 1 & 2 & \\
\hline
1 & -1 & -2 & 3 &
\end{array}
\quad
\begin{array}{rrrr|l}
1 & 0 & -3 & 5 & \underline{-2} \\
  & -2 & 4 & -2 & \\
\hline
1 & -2 & 1 & 3 &
\end{array}
\quad
\begin{array}{rrrr|l}
1 & 0 & -3 & 5 & \underline{-3} \\
  & -3 & 9 & -18 & \\
\hline
1 & -3 & 6 & -13 & \text{(alternates)}
\end{array}
$$

Thus, -3 is a lower bound.

4.206 $R(x) = x^3 - 2x^2 + 3$

▮

$$
\begin{array}{rrrr|l}
1 & -2 & 0 & 3 & \underline{1} \\
  & 1 & -1 & -1 & \\
\hline
1 & -1 & -1 & 2 &
\end{array}
\qquad
\begin{array}{rrrr|l}
1 & -2 & 0 & 3 & \underline{2} \\
  & 2 & 0 & 0 & \\
\hline
1 & 0 & 0 & 3 & \text{(nonnegative)}
\end{array}
$$

2 is an upper bound.

$$
\begin{array}{rrrr|l}
1 & -2 & 0 & 3 & \underline{-1} \\
  & -1 & 3 & -3 & \\
\hline
1 & -3 & 3 & 0 &
\end{array}
\qquad
\begin{array}{rrrr|l}
1 & -2 & 0 & 3 & \underline{-2} \\
  & -2 & 8 & -16 & \\
\hline
1 & -4 & 8 & -12 & \text{(alternates)}
\end{array}
$$

-2 is a lower bound.

4.207 $M(x) = x^4 - x^2 + 3x + 2$

$$
\begin{array}{rrrrr|r}
1 & 0 & -1 & 3 & 2 & \underline{1} \\
 & 1 & 1 & 0 & 3 & \\
\hline
1 & 1 & 0 & 3 & 8 & \text{(nonnegative)}
\end{array}
$$

1 is an upper bound.

$$
\begin{array}{rrrrr|r}
1 & 0 & -1 & 3 & 2 & \underline{-1} \\
 & 1 & 0 & -3 & & \\
\hline
1 & -1 & 0 & 3 & -1 &
\end{array}
\qquad
\begin{array}{rrrrr|r}
1 & 0 & -1 & 3 & 2 & \underline{-2} \\
 & -2 & 4 & -6 & 6 & \\
\hline
1 & -2 & 3 & -3 & 4 & \text{(alternates)}
\end{array}
$$

-2 is a lower bound.

4.208 $P(x) = x^3 + 7x^2 - x + 2$

$$
\begin{array}{rrrr|r}
1 & 7 & -1 & 2 & \underline{1} \\
 & 1 & 8 & 7 & \\
\hline
1 & 8 & 7 & 9 & \text{(nonnegative)}
\end{array}
$$

1 is an upper bound.

$$
\begin{array}{rrrr|r}
1 & 7 & -1 & 2 & \underline{-1} \\
 & -1 & -6 & 7 & \\
\hline
1 & 6 & -7 & 9 &
\end{array}
\quad
\begin{array}{rrrr|r}
1 & 7 & -1 & 2 & \underline{-2} \\
 & -2 & -10 & 22 & \\
\hline
1 & 5 & 11 & 24 &
\end{array}
\quad
\begin{array}{rrrr|r}
1 & 7 & -1 & 2 & \underline{-8} \\
 & -8 & 8 & -56 & \\
\hline
1 & -1 & 7 & -54 &
\end{array}
$$

-8 is a lower bound. (Do you see why we jumped from dividing by -2 to dividing by -8?)

For Probs. 4.209 to 4.211, show that there is at least 1 real zero between the given values of a and b.

4.209 $P(x) = x^2 - 3x - 2$; $a = 3$, $b = 4$

▮ $P(3) = 3^2 - 3(3) - 2 = -2$, and $P(4) = 4^2 - 3(4) - 2 = 2$; since $P(3)$ and $P(4)$ are opposite in sign, there must be at least 1 real zero between them.

4.210 $P(x) = x^3 - 3x + 5$; $a = -3$, $b = -2$

▮ $P(-3) = -13$, and $P(-2) = 3$. Since $P(-3)$ and $P(-2)$ are opposite in sign, there must be a real root between them.

4.211 $Q(x) = x^3 - 3x^2 - 3x + 9$; $a = 1$, $b = 2$

▮ $Q(a) = Q(1) = 4$, and $Q(b) = Q(2) = -1$. 4 and -1 are opposite in sign, so there is a real root between them.

For Probs. 4.212 to 4.215, write an equation whose roots are the negatives of those of $f(x) = 0$.

4.212 $x^3 - 8x^2 + x - 1 = 0$

▮ We change the signs of alternating terms, beginning with the second: $x^3 8x^2 + x + 1 = 0$.

4.213 $x^4 + 3x^2 + 2x + 1 = 0$

▮ $x^4 + 0x^3 + 3x^2 + 2x + 1 = 0$ is changed to $x^4 - 0x^3 + 3x^2 - 2x + 1 = 0$, or $x^4 + 3x^2 - 2x + 1 = 0$. Be careful with equations that are not in standard form.

4.214 $2x^4 - 5x^2 + 8x - 3 = 0$

▮ This is $2x^4 + 0x^3 - 5x^2 + 8x - 3 = 0$; change it to $2x^4 - 0x^3 - 5x^2 - 8x - 3 = 0$, or $2x^4 - 5x^2 - 8x - 3 = 0$.

4.215 $x^5 + x + 2 = 0$

▮ This is $x^5 + 0x^4 + 0x^3 + 0x^2 + x + 2 = 0$; we change it to $x^5 - 0x^4 + 0x^3 - 0x^2 + x - 2 = 0$, or $x^5 + x - 2 = 0$.

For Probs. 4.216 to 4.218, let $P(x) = x^3 - x^2 - 6x + 6$.

4.216 Discuss the possible number of real zeros using Descartes' rule of signs.

$\blacksquare$ $P(x) = x^3 - x^2 - 6x + 6$ which has two variations. $P(-x) = -x^3 - x^2 + 6x + 6$ which has one variation. Thus, there are 2 or 0 positive zeros and 1 negative zero.

4.217 Find the smallest and largest upper and lower bounds of the zeros of $P(x)$.

$\blacksquare$

$$\begin{array}{rrrr|r} 1 & -1 & -6 & 6 & \underline{1} \\ & 1 & 0 & -6 & \\ \hline 1 & 0 & -6 & 0 & \end{array} \qquad \begin{array}{rrrr|r} 1 & -1 & -6 & 6 & \underline{3} \\ & 3 & 6 & 0 & \\ \hline 1 & 2 & 0 & 6 & \text{(nonnegative)} \end{array}$$

Similarly,

$$\begin{array}{rrrr|r} 1 & -1 & -6 & 6 & \underline{-3} \\ & -3 & 12 & -18 & \\ \hline 1 & -4 & 6 & -12 & \text{(first alternating root)} \end{array}$$

Thus, 3 is an upper bound and -3 is a lower bound.

4.218 Discuss the location of real zeros between the upper and lower bounds found in Prob. 4.217.

$\blacksquare$ If we look at the quotient rows for all synthetic divisions between 1 and -3, we find:

$$\begin{array}{rrrr} 1 & -1 & -6 & 6 \end{array}$$

$$\begin{array}{rrrr|r} 1 & 0 & -6 & 0 & 1 \\ 1 & 1 & -4 & -2 & 2 \\ 1 & 2 & 0 & 6 & 3 \\ 1 & -1 & -6 & 6 & 0 \\ 1 & -2 & -4 & 10 & -1 \\ 1 & -3 & 0 & 6 & -2 \\ 1 & -4 & 6 & -12 & -3 \end{array}$$

1 is a zero. Also, $P(2)$ and $P(3)$ have opposite signs, as do $P(-2)$ and $P(-3)$. There must be roots between 2 and 3, and between -2 and -3.

For Probs. 4.219 to 4.221, let $P(x) = x^4 + 4x^3 - 2x^2 - 12x - 3$.

4.219 Discuss the possible number of real zeros.

$\blacksquare$ $P(x) = x^4 + 4x^3 - 2x^2 - 12x - 3$ which has one variation, and $P(-x) = x^4 - 4x^3 - 2x^2 + 12x - 3$ which has three variations. Thus, there are 3 or 1 negative zeros.

4.220 Find upper and lower bounds for the zeros of $P(x)$.

$\blacksquare$ Using synthetic division, we find:

$$\begin{array}{rrrrr|r} 1 & 4 & -2 & -12 & -3 & \underline{2} \\ & 2 & 12 & 20 & 16 & \\ \hline 1 & 6 & 10 & 8 & 13 & \end{array} \qquad \begin{array}{rrrrr|r} 1 & 4 & -2 & -12 & -5 & \underline{-5} \\ & -5 & 5 & -15 & 135 & \\ \hline 1 & -1 & 3 & -27 & 132 & \end{array}$$

Thus, 2 is an upper bound and -5 is a lower bound.

4.221 Discuss the location of real zeros in the interval found in Prob. 4.220.

$\blacksquare$ Using a synthetic division table, we find:

$$\begin{array}{rrrrr} 1 & 4 & -2 & -12 & -3 \end{array}$$

$$\begin{array}{rrrrr|l} 1 & 5 & 3 & -9 & -12 & 1\; \} \\ 1 & 6 & 10 & 8 & 13 & 2\; \} \;-\text{ to } + \\ 1 & 4 & -2 & -12 & -3 & 0\; \} \\ 1 & 3 & -5 & -7 & 4 & -1\; \} \;-\text{ to } + \\ 1 & 2 & -6 & 0 & -3 & -2\; \} \;+\text{ to } - \\ 1 & 1 & -5 & 3 & -12 & -3\; \} \\ 1 & 0 & -2 & -4 & 13 & -4\; \} \;-\text{ to } + \\ 1 & -1 & 3 & -27 & 132 & -5 \end{array}$$

There are roots between 1 and 2, -1 and 0, -2 and -1, and -4 and -3.

For Probs. 4.222 and 4.223, find all roots for the given polynomial equation.

4.222 $2x^3 - 5x^2 + 1 = 0$

▮ ± 1 and $\pm \frac{1}{2}$ are the possible zeros for $P(x) = 2x^3 - 5x^2 + 1$. $P(x) = 2x^3 - 5x^2 + 1$ which has 2 or 0 positive zeros, and $P(-x) = -2x^3 - 5x^2 + 1$ which has 1 negative zero.

$$
\begin{array}{rrrr|l}
2 & -5 & 0 & 1 & \underline{1} \\
 & 2 & -3 & -3 & \\
\hline
2 & -3 & -3 & -2 &
\end{array}
\qquad
\begin{array}{rrrr|l}
2 & -5 & 0 & 1 & \underline{-1} \\
 & -2 & 7 & -7 & \\
\hline
2 & -7 & 7 & -6 &
\end{array}
\qquad
\begin{array}{rrrr|l}
2 & -5 & 0 & 1 & \underline{\frac{1}{2}} \\
 & 1 & -2 & 0 & \\
\hline
2 & -4 & -2 & 0 &
\end{array}
$$

Thus, $\frac{1}{2}$ is a zero and $P(x) = 2x^3 - 5x^2 + 1 = (x - \frac{1}{2})(2x^2 - 4x - 2)$. Using the quadratic formula, we find $x = 1 \pm \sqrt{2}$. The roots are $\frac{1}{2}$, $1 \pm \sqrt{2}$.

4.223 $x^4 + 4x^3 - x^2 - 20x - 20 = 0$

▮ Possible rational roots are 11, ± 2, ± 4, ± 5, ± 10, ± 20. By Descartes' rule, we find 1 positive zero and 0, 2, or 4 negative zeros. Using a synthetic division table, we find:

$$
\begin{array}{rrrrr}
1 & 4 & -1 & -20 & -20 \\
\end{array}
$$

$$
\begin{array}{rrrrr|rl}
1 & 5 & 4 & -16 & -36 & 1 & \\
1 & 6 & 11 & 2 & -16 & 2 & \\
1 & 8 & 31 & 104 & 396 & 4 & \text{(nonnegative)} \\
1 & 3 & -4 & -16 & -4 & -1 & \\
1 & 2 & -5 & -10 & 0 & -2 &
\end{array}
$$

Thus, -2 is a zero, and 5, 10, and 20 are eliminated since 4 is an upper bound. Consider $x^3 + 2x^2 - 5x - 10$. We find -2 is a zero. Then, $x^3 + 2x^2 - 5x - 10 = (x + 2)(x^2 - 5)$. Then from $x^2 - 5 = 0$, we have $x = \pm\sqrt{5}$. The roots of $P(x)$ are -2, $\sqrt{5}$, $-\sqrt{5}$.

For Probs. 4.224 and 4.225, show that the given number is not rational.

4.224 $\sqrt{6}$

▮ $\sqrt{6}$ is the solution for $x^2 - 6 = 0$. The only possible rational roots for $x^2 - 6 = 0$ are ± 1, ± 3, ± 6. None of these satisfies $x^2 - 6 = 0$. Thus, there are no rational roots, and 6 is irrational.

4.225 $\sqrt[3]{5}$

▮ $\sqrt[3]{5}$ is the solution for $x^3 - 5 = 0$. Then, ± 1 and ± 5 are the only possible rational solutions of $x^3 - 5 = 0$, but, none of them solves $x^3 - 5 = 0$, so $\sqrt[3]{5}$ is irrational.

For Probs. 4.226 and 4.227, find the irrational zero to one decimal place in the given interval.

4.226 $P(x) = x^3 - 5x^2 + 3$; [4, 5]

▮ Find $P(4)$ and $P(5)$: $P(4) = -13$, $P(5) = 3$. The zero lies closer to 5 than to 4, since $|3 - 0| < |-13 - 0|$. Using a calculator, find $P(4.8)$, $P(4.9)$ until a sign change. $P(4.9) = 0.599$, $P(4.8) = -1.608$. The zero lies closer to 4.9 (see above). Repeat this process using $P(4.89)$, $P(4.88)$, etc. We find $P(4.88) = 0.1423$ and $P(4.87) = -0.0832$. There is a sign change. Thus, the zero occurs in the interval $(4.87, 4.88)$ or at ~ 4.9 (to one decimal place).

4.227 $P(x) = x^3 + x - 1$; [0, 1]

▮ $P(0) = -1$, $P(1) = 1$. Since $|1 - 0| = |0 - (-17)|$, the root is near the middle of the interval [0, 1]. $P(0.5) = -0.375$; $P(0.6) = -0.184$; $P(0.7) = 0.043$ (sign change). Also, the root will be nearer to 0.7. $P(0.69) = 0.0185$; $P(0.68) = -0.0006$. The zero is in the interval $(0.68, 0.69)$ or at ~ 0.7.

4.6 PARTIAL FRACTIONS

For Probs. 4.228 to 4.254, express the given fraction as a sum of partial fractions. The method is detailed in Prob. 4.228.

4.228 $\dfrac{5x + 1}{(x - 1)(x + 2)}$

▮ $\dfrac{5x+1}{(x-1)(x+2)} = \dfrac{A}{x-1} + \dfrac{B}{x+2}$. (Note: We use A and B only, i.e., constants only, as numerators since the denominators are linear.) Multiply the equation on both sides by $(x-1)(x+2)$: $5x+1 = A(x+2) + B(x-1)$. Choose 'convenient' values for x. Let $x = -2$; then $5(-2) + 1 = 0 + B(-2-1)$, or $-9 = -3B$. Let $x = 1$; then $5(1) + 1 = A(1+2) + 0$, or $6 = 3A$. Solve the equations obtained: $B = 3$ and $A = 2$. Thus, $\dfrac{5x+1}{(x-1)(x+2)} = \dfrac{2}{x-1} + \dfrac{3}{x+2}$.

4.229 $\dfrac{6x+7}{(2x+1)(3x-1)}$

▮ $\dfrac{6x+7}{(2x+1)(3x-1)} = \dfrac{A}{2x+1} + \dfrac{B}{3x-1}$. Multiply the equation on both sides by $(2x+1)(3x-1)$: $6x + 7 = A(3x-1) + B(2x+1)$. Let $x = \frac{1}{3}$; then $6(\frac{1}{3}) + 7 = A(0) + B[2(\frac{1}{3}) + 1]$, or $9 = \frac{5}{3}B$. Let $x = -\frac{1}{2}$; then $6(-\frac{1}{2}) + 7 = A[3(-\frac{1}{2}) - 1] + B(0)$, or $4 = -\frac{5}{2}A$. Solving the equations obtained: $B = \frac{27}{5}$ and $A = -\frac{8}{5}$. Thus, $\dfrac{6x+7}{(2x+1)(3x-1)} = \dfrac{-\frac{8}{5}}{2x+1} + \dfrac{\frac{27}{5}}{3x-1}$.

4.230 $\dfrac{x-1}{x^2}$

▮ $\dfrac{x-1}{x^2} = \dfrac{A}{x} + \dfrac{B}{x^2}$. (Note: Again, we use constants A and B since a linear term, x, is raised to a power. Notice that the denominators are x and x^2. We use denominators $x, x^2, \dots, x^n$ when the fraction being decomposed has denominator x^n.) Proceed as in the above examples. Multiply by x^2: $x - 1 = Ax + b$. Let $x = 1$; then $0 = A + B$, or $A = -B$. Let $x = 0$; then $-1 = 0 + B$, or $-1 = B$. So $B = -1$ and $A = 1$. Thus, $\dfrac{x-1}{x^2} = \dfrac{1}{x} + \dfrac{-1}{x^2}$.

4.231 $\dfrac{2x^2 - 5x + 7}{(x-1)(x^2+1)}$

▮ $\dfrac{2x^2 - 5x + 7}{(x-1)(x^2+1)} = \dfrac{A}{x-1} + \dfrac{Bx+C}{x^2+1}$. (Notice that we use $Bx + C$ since $x^2 + 1$ is a quadratic.) Multiply the equation on both sides by $(x-1)(x^2+1)$: $2x^2 - 5x + 7 = A(x^2+1) + (Bx+C)(x-1)$. Let $x = 0$; then $7 = A(1) + C(-1)$. Let $x = 1$; then $4 = A(2) + (B+C)(0)$, or $4 = 2A$. Let $x = -1$; then $14 = A(2) + (-B+C)(-2)$. From $4 = 2A$, we have $A = 2$. From $7 = A - C = 2 - C$, we have $C = 2 - 7 = -5$. Substituting these values into $14 = 2A + 2B - 2C$, we have $14 = 4 + 2B + 10$. Therefore, $B = 0$. Thus, $\dfrac{2x^2 - 5x + 7}{(x-1)(x^2+1)} = \dfrac{2}{x-1} + \dfrac{-5}{x^2+1}$.

4.232 $\dfrac{x^3 + 2x^2 - 9x - 3}{(x-2)^2(x^2+1)}$

▮ $\dfrac{x^2 + 2x^2 - 9x - 3}{(x-2)^2(x^2+1)} = \dfrac{A}{x-2} = \dfrac{B}{(x-2)^2} + \dfrac{Cx+D}{x^2+1}$. Then $x^3 + 2x^2 - 9x - 3 = A(x-2)(x^2+1) + B(x^2+1) + (Cx+D)(x-2)^2$. Let $x = 0$: $-3 = -2A + B + 4D$. Let $x = 1$: $-9 = -2A + 2B + (C+D)$. Let $x = 2$: $-5 = 0 + 5B + 0$ or $B = -1$. Let $x = -1$: $7 = -6A + 2B + (D - C)(9)$. Then solving the obtained equations simultaneously, $A = 3$, $B = -1$, $C = -2$, $D = 1$, and, $\dfrac{x^3 + 2x^2 - 9x - 3}{(x-2)^2(x^2+1)} = \dfrac{3}{x-2} + \dfrac{-1}{(x-2)^2} + \dfrac{-2x+1}{x^2+1}$.

4.233 $\dfrac{7x+10}{x^2 + 3x + 2}$

▮ $\dfrac{7x+10}{x^2+3x+2} = \dfrac{7x+10}{(x+2)(x+1)} = \dfrac{A}{x+2} = \dfrac{B}{x+1}$. Then, $7x + 10 = A(x+1) + B(x+2)$. Let $x = -1$: $3 = 0 + B$. Let $x = -2$: $-4 = -A + 0$. Thus, $\dfrac{7x+10}{(x+2)(x+1)} = \dfrac{4}{x+2} + \dfrac{3}{x+1}$.

4.234 $\dfrac{x-19}{x^2 + 4x - 5}$

▮ Since $x^2 + 4x - 5 = (x-1)(x+5)$, we have $\dfrac{x-19}{(x-1)(x+5)} = \dfrac{A}{x-1} + \dfrac{B}{x+5}$. Then $x - 19 = A(x+5) + B(x-1)$. Let $x = -5$: $-24 = 0 + B(-6)$, or $B = 4$. Let $x = 1$: $-18 = 6A + 0$, or $A = -3$. Thus, $\dfrac{x-19}{(x-1)(x+5)} = \dfrac{-3}{x-1} + \dfrac{4}{x+5} = \dfrac{4}{x+5} - \dfrac{3}{x-1}$.

4.235 $\dfrac{x+5}{12x^2+x-1}$

▮ Since $12x^2+x-1=(4x-1)(3x+1)$, we have $\dfrac{x+5}{(4x-1)(3x+1)}=\dfrac{A}{4x-1}+\dfrac{B}{3x+1}$. Then $x+5=A(3x+1)+B(4x-1)$. Let $x=-\frac13$: $4(\frac23)=0+b(-\frac43-1)$. Let $x=\frac14$: $5(\frac14)=A(\frac34+1)+0$. Then solving the equations obtained, $A=3$, and $B=-2$. Thus, $\dfrac{x+5}{12x^2+x-1}=\dfrac{3}{4x-1}-\dfrac{2}{3x+1}$.

4.236 $\dfrac{5x-1}{6x^2-5x+1}$

▮ Since $6x^2-5x+1=(2x-1)(3x-1)$, we have $\dfrac{5x-1}{(2x-1)(3x-1)}=\dfrac{A}{2x-1}+\dfrac{B}{3x-1}$. Then $5x-1=A(3x-1)+B(2x-1)$. Let $x=\frac13$: $\frac53-1=0+B(\frac23-1)$. Let $x=\frac12$: $\frac52-1=A(\frac32-1)+0$. Then, solving the equations obtained, $A=3$, $B=-2$. Thus, $\dfrac{5x-1}{(2x-1)(3x-1)}=\dfrac{3}{2x-1}-\dfrac{2}{3x-1}$.

4.237 $\dfrac{-3x+7}{x^2-2x+1}$

▮ Since $x^2-2x+1=(x-1)^2$, we have $\dfrac{-3x+7}{(x-1)^2}=\dfrac{A}{x-1}+\dfrac{B}{(x-1)^2}$. Then $-3x+7=A(x-1)+B$. Let $x=1$: $-3(1)+7=0+B$. Let $x=0$: $0+7=-A+B$. Then, solving the equations obtained, $A=-3$ and $B=4$. Thus, $\dfrac{-3x+7}{(x-1)^2}=\dfrac{-3}{x-1}+\dfrac{4}{(x-1)^2}$.

4.238 $\dfrac{4x+23}{(x-3)(x+2)(x+4)}$

▮ $\dfrac{4x+23}{(x-3)(x+2)(x+4)}=\dfrac{A}{x-3}+\dfrac{B}{x+2}+\dfrac{C}{x+4}$. Then $4x+23=A(x+2)(x+4)+B(x-3)(x+4)+C(x-3)(x+2)$. Let $x=-2$: $-8+23=0+B(-1)(2)+0$. Let $x=-4$: $-16+23=0+0+C(-7)(-2)$. Let $x=3$: $11+23=A(5)(7)+0+0$. Then $A=1$, $B=-\frac32$, $C=\frac12$, and the solution is $\dfrac{1}{x-3}-\dfrac{\frac32}{x+2}+\dfrac{\frac12}{x+4}$.

4.239 $\dfrac{17x-45}{x^3-2x^2-15x}$

▮ Since $x^3-2x^2-15x=x(x^2-2x-15)=x(x-5)(x+3)$, $\dfrac{17x-45}{x^3-2x^2-15x}+\dfrac{A}{x}+\dfrac{B}{x-5}+\dfrac{C}{x+3}$. Then $17x-45=A(x-5)(x+3)+B(x)(x+3)+C(x)(x-5)$. Let $x=0$: $-45=A(-5)(3)+0+0$. Let $x=5$: $85-45=0+B(5)(8)+0$. Let $x=-3$: $-51-45=0+0+C(-3)(-8)$. Then, $A=3$, $B=1$, $C=-4$, and the solution is $\dfrac{3}{x}+\dfrac{1}{x-5}-\dfrac{4}{x+3}$.

4.240 $\dfrac{2x^2+x+3}{x^2-9}$

▮ $\dfrac{2x^2+x+3}{x^2-9}$ is not a proper fraction, since the degree of the numerator equals the degree of the denominator. Dividing $2x^2+x+3$ by x^2-9, we get $2+\dfrac{x+21}{x^2-9}$. Thus, $\dfrac{2x^2+8x+3}{x^2-9}=2+\dfrac{x+21}{(x+3)(x-3)}$. Solving $\dfrac{x+21}{(x+3)(x-3)}=\dfrac{A}{x+3}+\dfrac{B}{x-3}$, we have $A=-3$ and $B=4$, and the solution is $2-\dfrac{3}{x+3}+\dfrac{4}{x-3}$. *Note:* For Probs. 4.241 through the end of this section, we will condense the algebra. See Probs. 4.228 to 4.237 for detailed algebra.

4.241 $\dfrac{x^2+8}{x^2(x+2)}$

▮ $\dfrac{x^2+8}{x^2(x+2)}=\dfrac{A}{x}+\dfrac{B}{x^2}+\dfrac{C}{x+2}$. Then $x^2+8=A(x)(x+2)+B(x+2)+Cx^2$. Let $x=0$, $x=-2$, and $x=1$. Then $A=-2$, $B=4$, $C=3$, and the solution is $\dfrac{4}{x^2}-\dfrac{2}{x}+\dfrac{3}{x+2}$.

4.242 $\dfrac{x^2 - 1}{(x-2)^3}$

▮ $\dfrac{x^2 - 1}{(x-2)^3} = \dfrac{A}{x-2} + \dfrac{B}{(x-2)^2} + \dfrac{C}{(x-2)^3}$. Then $x^2 - 1 = A(x-2)^2 + B(x-2) + C$. Let $x = 2, 1, 0$.
Then $A = 1$, $B = 4$, $C = 3$, and the solution is $\dfrac{3}{(x-2)^3} + \dfrac{4}{(x-2)^2} + \dfrac{1}{x-2}$.

4.243 $\dfrac{3x^3 - 2x^2 - 10x + 6}{(x+2)^3(2x+1)}$

▮ $\dfrac{3x^3 - 2x^2 - 10x + 6}{(x+2)^3(2x+1)} = \dfrac{A}{x+2} + \dfrac{B}{(x+2)^2} + \dfrac{C}{(x+2)^3} + \dfrac{D}{2x+1}$. Then $3x^3 - 2x^2 - 10x + 6 = A(x+2)^2$
$(2x+1) + B(x+2)(2x+1) + C(2x+1) + D(x+2)^3$. Let $x = -\frac{1}{2}, -2, 0, 1$. Then $A = 0$, $B = -10$,
$C = 2$, $D = 3$, and the solution is $\dfrac{2}{(x+2)^3} - \dfrac{10}{(x+2)^2} + \dfrac{3}{2x+1}$.

4.244 $\dfrac{x^2 + 5x + 3}{x(x+1)^2}$

▮ $\dfrac{x^2 + 5x + 5}{x(x+1)^2} = \dfrac{A}{x} + \dfrac{B}{x+1} \dfrac{C}{(x+1)^2}$. Then $x^2 + 5x + 3 = A(x+1)^2 + Bx(x+1) + Cx$. Let $x = -1, 0, 1$.
Then $A = 3$, $B = -2$, $C = 1$, and the solution is $\dfrac{3}{x} - \dfrac{2}{x+1} + \dfrac{1}{(x+1)^2}$.

4.245 $\dfrac{x^2 - 3x}{(x-4)(x-2)^2}$

▮ $\dfrac{x^2 - 3x}{(x-4)(x-2)^2} = \dfrac{A}{x-4} + \dfrac{B}{(x-2)} + \dfrac{C}{(x-2)^2}$. Then $x^2 - 3x = A(x-2)^2 + B(x-4)(x-2) + C(x-4)$.
Let $x = 2$, $x = 4$, $x = 0$. Then $A = 1$, $B = 0$, $C = 1$, and the solution is $\dfrac{1}{x-4} + \dfrac{1}{(x-2)^2}$.

4.246 $\dfrac{5x^2 - 8}{x(x^2 + 2x - 4)}$

▮ Since $(x^2 + 2x - 4)$ is nonfactorable, $\dfrac{5x^2 - 8}{x(x^2 + 2x - 4)} = \dfrac{A}{x} + \dfrac{Cx + D}{x^2 + 2x - 4}$. Then $5x^2 - 8 =$
$A(x^2 + 2x - 4) + (Cx + D)(x)$. Let $x = 0, 1, -1$. Then $A = 2$, $C = 3$, $D = -4$, and the solution is
$\dfrac{2}{x} + \dfrac{3x - 4}{x^2 + 2x - 4}$.

4.247 $\dfrac{x^2 - 6x - 3}{x^4 + 3x^2 - 18}$

▮ Since $x^4 + 3x^2 - 18 = (x^2 + 6)(x^2 - 3)$, we have $\dfrac{x^2 - 6x - 3}{(x^2 + 6)(x^2 - 3)} = \dfrac{Ax + B}{x^2 + 6} + \dfrac{Cx + D}{x^2 - 3}$. Then $x^2 -$
$6x - 3 = (Ax + B)(x^2 - 3) + (Cx + D)(x^2 + 6)$. Let $x = 0, 1, -1, 2$. Then $A = 2$, $B = 3$, $C = 2$, $D = 0$,
and the solution is $\dfrac{2x + 3}{3(x^2 + 6)} - \dfrac{2x}{3(x^2 - 3)}$.

4.248 $\dfrac{7x - 14}{(x-4)(x+3)}$

▮ $\dfrac{7x - 14}{(x-4)(x+3)} = \dfrac{A}{x-4} + \dfrac{B}{x+3}$. Then $7x - 14 = A(x+3) + B(x-4)$. Let $x = -3, 4$. Then $A = 2$,
$B = 5$, and the solution is $\dfrac{2}{x-4} + \dfrac{5}{x+3}$.

4.249 $\dfrac{17x - 1}{(2x-3)(3x-1)}$

▮ $\dfrac{17x - 1}{(2x-3)(3x-1)} = \dfrac{A}{2x-3} + \dfrac{B}{3x-1}$. Then $17x - 1 = A(3x-1) + B(2x-3)$. Let $x = \frac{1}{3}, \frac{3}{2}$. Then
$A = 7$, $B = -2$, and the solution is $\dfrac{7}{2x-3} - \dfrac{2}{3x-1}$.

4.250 $\dfrac{3x^2+7x+1}{x(x+1)^2}$

▮ $\dfrac{3x^2+7x+1}{x(x+1)^2}=\dfrac{A}{x}+\dfrac{B}{x+1}+\dfrac{C}{(x+1)^2}$. Then $3x^2+7x+1=A(x+1)^2+Bx(x+1)+Cx$. Let $x=-1,1,0$. Then, $A=1$, $B=2$, $C=3$, and the solution is $\dfrac{1}{x}+\dfrac{2}{x+1}+\dfrac{3}{(x+1)^2}$.

4.251 $\dfrac{3x^2+x}{(x-2)(x^2+3)}$

▮ $\dfrac{3x^2+x}{(x-2)(x^2+3)}=\dfrac{A}{x-2}+\dfrac{Bx+C}{x^2+3}$. Then $3x^2+x=A(x^2+3)+(Bx+C)(x-2)$. Let $x=0,2,1$. Then $A=2$, $B=1$, $C=3$, and the solution is $\dfrac{2}{x-2}+\dfrac{x+3}{x^2+3}$.

4.252 $\dfrac{2x^2+4x-1}{(x^2+x+1)^2}$

▮ Be careful! Here we have a quadratic to a power. $\dfrac{2x^2+4x-1}{(x^2+x+1)^2}=\dfrac{Ax+B}{x^2+x+1}+\dfrac{Cx+D}{(x^2+x+1)^2}$. Then $2x^2+4x-1=(Ax+B)(x^2+x+1)+Cx+D$. Let $x=0,-1,1,2$. Then $A=0$, $B=2$, $C=2$, $D=-3$, and the solution is $\dfrac{2}{x^2+x+1}+\dfrac{2x-3}{(x^2+x+1)^2}$.

4.253 $\dfrac{-x+22}{x^2-2x-8}$

▮ Since $x^2-2x-8=(x-4)(x+2)$, $\dfrac{-x+22}{x^2-2x-8}=\dfrac{A}{x-4}+\dfrac{B}{x+2}$. Then $-x+22=A(x+2)+B(x-4)$. Let $x=-2,4$. Then $A=3$, $B=-4$, and the solution is $\dfrac{3}{x-4}-\dfrac{4}{x+2}$.

4.254 $\dfrac{3x-13}{6x^2-x-12}$

▮ Since $6x^2-x-12-(3x+4)(2x-3)$ we have $\dfrac{3x-13}{(3x+4)(2x-3)}=\dfrac{A}{3x+4}+\dfrac{B}{2x-3}$. Then $3x-13=A(2x-3)+B(3x+4)$. Let $x=3/2,-\frac43$. Then $A=3$, $B=-1$, and the solution is $\dfrac{3}{3x+4}-\dfrac{1}{2x-3}$.

For Probs. 4.255 to 4.264, put the given improper fraction in proper form (i.e., write it as the sum of a proper fraction and a polynomial).

4.255 $\dfrac{x^2}{x^2+1}$

$$x^2+1\overline{\smash{)}x^2}$$ with quotient 1, x^2+1, remainder -1

Thus, $\dfrac{x^2}{x^2+1}=1+\dfrac{-1}{x^2+1}$.

4.256 $\dfrac{x^2+1}{x^2+2}$

▮ $x^2+2\overline{\smash{)}x^2+1}$ with quotient 1, x^2+2, remainder -1

Thus, $\dfrac{x^2+1}{x^2+2}=1+\dfrac{-1}{x^2+2}$.

4.257 $\dfrac{x^2 + 3x}{x^3}$

▮

$$x^3 \overline{\smash{\big)}\, x^3 + 3x} \quad {}^{1}$$
$$\underline{x^3}$$
$$3x$$

Thus, $\dfrac{x^2 + 3x}{x^3} = 1 + \dfrac{3x}{x^3} = 1 + \dfrac{3}{x^2}$.

4.258 $\dfrac{x^3}{x^3 + 5}$

▮

$$x^3 + 5 \overline{\smash{\big)}\, x^3} \quad {}^{1}$$
$$\underline{x^3 + 5}$$
$$-5$$

Thus, $\dfrac{x^3}{x^3 + 5} = 1 + \dfrac{-5}{x^3 + 5} = 1 - \dfrac{5}{x^3 + 5}$.

4.259 $\dfrac{2x^2 + 3}{5x^2 + 1}$

▮

$$5x^2 + 1 \overline{\smash{\big)}\, 2x^2 + 3} \quad {}^{\frac{2}{5}}$$
$$\underline{2x^2 + \tfrac{2}{5}}$$
$$\tfrac{13}{5}$$

Thus, $\dfrac{2x^2 + 3}{5x^2 + 1} = \dfrac{2}{5} + \dfrac{\frac{13}{5}}{5x^2 + 1}$.

4.260 $\dfrac{x + 1}{2x - 63}$

▮

$$2x - 63 \overline{\smash{\big)}\, x + 1} \quad {}^{\frac{1}{2}}$$
$$\underline{x - \tfrac{63}{2}}$$
$$-\tfrac{61}{2}$$

Thus, $\dfrac{x + 1}{2x - 63} = \tfrac{1}{2} - \dfrac{\frac{61}{2}}{2x - 63}$.

4.261 $\dfrac{x^3 + x + 1}{x^2 + 3}$

▮

$$x^2 - 3 \overline{\smash{\big)}\, x^3 + x + 1} \quad {}^{x}$$
$$\underline{x^3 - 3x}$$
$$4x + 1$$

Thus, $\dfrac{x^3 + x + 1}{x^2 - 3} = x + \dfrac{4x + 1}{x^2 - 3}$.

4.262 $\dfrac{x^2 + 2x + 5}{x}$

▮

$$x \overline{\smash{\big)}\, x^2 + 2x + 5} \quad {}^{x}$$
$$\underline{x^2}$$
$$2x + 5$$

Thus, $\dfrac{x^2 + 2x + 5}{x} = x + \dfrac{2x + 5}{x}$, but $\dfrac{2x + 5}{x}$ is still improper.

$$x\overline{\smash{\big)}\begin{matrix} 2 \\ 2x + 5 \\ \underline{2x} \\ 5 \end{matrix}}$$

Thus, $\dfrac{x^2 + 2x + 5}{x} = x + 2 + \dfrac{5}{x}$.

4.263 $\dfrac{x^4 + 1}{x^2}$

▮ $\dfrac{x^4 + 1}{x^2} = \dfrac{x^4}{x^2} + \dfrac{1}{x^2} = x^2 + \dfrac{1}{x^2}$.

4.264 $\dfrac{ax + b}{cx + d}$ $(cx + d \neq 0)$

▮ $$cs + d\overline{\smash{\big)}\begin{matrix} a/c \\ ax + b \\ \underline{ax + da/c} \\ b - da/c \end{matrix}}$$

Thus, $\dfrac{ax + b}{cx + d} = \dfrac{a}{c} + \dfrac{b - da/c}{cx + d}$ (when $cx + d \neq 0$).

For Probs. 4.265 to 4.270, the denominator of a proper fraction is given. Write the partial fraction decomposition.

4.265 $(4x^2 + 3x + 5)^2$

▮ This is a quadratic term squared: $\dfrac{Ax + B}{4x^2 + 3x + 5} + \dfrac{Cx + D}{(4x^2 + 3x + 5)^2}$.

4.266 $(3x - 10)^3$

▮ Since $3x - 10$ is linear, we will have only constants in the numerator: $\dfrac{A}{3x - 10} + \dfrac{B}{(3x - 10)^2} + \dfrac{C}{(3x - 10)^3}$.

4.267 $(x^2 + 1)^2(x^3 - x)$

▮ $x^3 - x = x(x^2 - 1) = x(x - 1)(x + 1)$: $\dfrac{Ax + B}{x^2 + 1} + \dfrac{Cx + D}{(x^2 + 1)^2} + \dfrac{E}{x} + \dfrac{F}{x - 1} + \dfrac{G}{x + 1}$.

4.268 $x^3(x^3 + x + 1)$

▮ $\dfrac{A}{x} + \dfrac{B}{x^2} + \dfrac{C}{x^3} + \dfrac{Dx^2 + Ex + G}{x^3 + x + 1}$.

4.269 $x^3 - 1$

▮ $x^3 - 1 = (x - 1)(x^2 + x + 1)$. We can see this from synthetic division.

$$\begin{array}{rrrr|l} 1 & 0 & 0 & -1 & \underline{1} \\ & 1 & 1 & 1 & \\ \hline 1 & 1 & 1 & 0 & \end{array}$$

The fractions will be of the form: $\dfrac{A}{x - 1} + \dfrac{Bx + C}{x^2 + x + 1}$.

4.270 $x^4 + x^3 + 2x^2$

▮ $x^4 + x^3 + 2x^2 = x^2(x^2 + x + 2)$: $\dfrac{A}{x} + \dfrac{B}{x^2} + \dfrac{Cx + D}{x^2 + x + 2}$.

4.7. RATIONAL AND ALGEBRAIC FUNCTIONS

For Probs. 4.271 to 4.281, tell whether the given function is rational and why.

4.271 $f(x) = cx + d$

▮ This is rational, since $f(x) = cx + d = \dfrac{cx + d}{1}$ where $cx + d$ and 1 are polynomials.

4.272 $f(x) = \dfrac{(x-2)^{3/2}}{x-2}$

▮ This is nonrational. The numerator cannot be written as a polynomial. If we divide, we find $f(x) = \sqrt{x-2}$.

4.273 $g(x) = |x|$

▮ This is nonrational. $|x|$ is not a polynomial. Look at Fig. 4.35 which is the graph of $g(x)$. Do you notice the sharp peak? You will learn in calculus that this assures you that the function is nonrational.

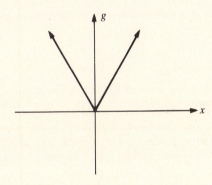

Fig. 4.35

4.274 $f(x) = \dfrac{1}{\pi x}$

▮ 1 and πx are both polynomials. This is a rational function. Do not be mislead by the irrational coefficient of x.

4.275 $h(x) = 3^{x^2} + 2$

▮ This is nonrational. The power to which 3 is raised is variable. Thus, this "numerator" is nonpolynomial.

4.276 $y = \dfrac{x^2 + 3x - 7}{x^2 + 5x - 6}$

▮ The numerator and denominator are each a polynomial of degree 2. This is a rational function.

4.277 $f(x) = (x + 2)^x$

▮ This is nonrational. The exponent x is variable.

4.278 $y = \dfrac{x - 1}{x}$

▮ This function is rational. The numerator and denominator are each a polynomial of degree 1.

4.279 $f(x) = \sqrt{x - 2}$

▮ $\sqrt{x-2} = (x-2)^{1/2}$ which is not a polynomial (the exponent is nonintegral).

4.280 $h(x) = \dfrac{(x+1)^2}{2x + i}$

▮ This is rational. The numerator and denominator are each polynomials.

4.281 $f(x) = \dfrac{x^n}{3^x}$

▮ This is nonrational. The denominator is not a polynomial.

For Probs. 4.282 to 4.290, tell whether the given function is an explicit algebraic function and why.

4.282 $y = x^{2/3} + x^{3/2}$

▮ This is an explicit algebraic function since it is formed by a finite number of algebraic operations. Note that it is not rational.

4.283 $y = |x|$

▮ This is nonalgebraic. $|x|$ is not an algebraic operation.

4.284 $y + \dfrac{x^5}{5} - \dfrac{2ix^3}{3}$

▮ This is algebraic since it is a polynomial and all polynomial and rational functions are algebraic.

4.285 $y = \begin{cases} x, & x \text{ rational} \\ -x, & x \text{ irrational} \end{cases}$

▮ This is nonalgebraic. For any interval (a, b) this y will jump infinitely often between x and $-x$. This is not a finite series of algebraic operations.

4.286 $f(x) = 4^x$

▮ This is nonalgebraic since we are raising 4 to a variable power. That is not an algebraic operation.

4.287 $g(x) = |x^2|$

▮ We need to be very careful here. $|x^2| = x^2$, and x^2 is algebraic. Thus $g(x)$ is an algebraic function.

4.288 $h(x) = \sqrt{\sqrt{x} + 2}$

▮ h is algebraic. This function is evaluated at each x via a finite number of algebraic steps.

4.289 $f(x) = \dfrac{x^2}{12!} + \dfrac{x^6}{16!} + \sqrt{x}$

▮ f is algebraic. $f(x)$ is found by evaluating x^2/k, x^6/l, and $\sqrt{x}$, where k and l are constants.

4.290 $f(x) = |x^3|$

▮ f is nonalgebraic. $|x^3| \neq x^3$ for all $x \in \mathcal{R}$. Compare this with Prob. 4.287.

For Probs. 4.291 to 4.295, tell what kind of function $f(x)$ is, and give its domain.

4.291 $f(x) + x^2 + \sqrt{3}x$

▮ $f(x)$ is a polynomial. Domain = all reals.

4.292 $f(x) = x^3 + \sqrt{3x}$

▮ $f(x)$ is algebraic. Domain of $f = \{x \in \mathcal{R} \,|\, x \geq 0\}$. Notice $\sqrt{3x}$ and compare to $\sqrt{3}x$ in Prob. 4.291.

4.293 $f(x) = \dfrac{\sqrt{x^3 + 3x}}{\sqrt{x + 1}}$

▮ $f(x)$ is algebraic. We need $x^3 + 3x \geq 0$ and $x + 1 \geq 0$. Since $x + 1 \geq 0$, $x \geq -1$; since $x(x^2 + 3) = 0$, $x \geq 0$. Domain = $\{x \in \mathcal{R} \,|\, x \geq 0\}$.

4.294 $f(x) = \dfrac{x^2}{(x+1)(x-6)}$

▮ f here is rational. The domain will be all reals except those making the denominator zero. Domain = $\{x \in \mathcal{R} \mid x \neq 6, \; x \neq -1\}$.

4.295 $f(x) = \sqrt{x-2} - \sqrt{x-4}$

▮ This f is algebraic. We need $x - 2 \geq 0$ and $x - 4 \geq 0$. Since $x - 2 \geq 0$, $x \geq 2$; since $x - 4 \geq 0$, $x \geq 4$. Thus, the domain is $\{x \in \mathcal{R} \mid x \geq 4\}$.

For Probs. 4.296 to 4.304, find all vertical and horizontal asymptotes of the graph of the given function. Do *not* graph the function.

4.296 $f(x) = \dfrac{x}{x+1}$

▮ *Vertical*: Since $f(x) = \dfrac{x}{x+1}$, as $x + 1 \to 0$, f grows without bound. Thus, $x = -1$ is a vertical asymptote. *Horizontal*: As x grows without bound, $f(x) = \dfrac{x}{x+1}$ gets nearer and nearer to 1. $y = 1$ is a horizontal asymptote.

4.297 $f(x) = \dfrac{x+1}{x}$

▮ As $x \to 0$, $\dfrac{x+1}{x}$ grows without bound. $x = 0$ is a vertical asymptote. As $x \to \infty$, $\dfrac{x+1}{x}$ grows closer to 1. $y = 1$ is a horizontal asymptote.

4.298 $f(x) = \dfrac{x^2}{x^2 - 4}$

▮ As $x^2 - 4 \to 0$, $\dfrac{x^2}{x^2-4} \to \infty$, which implies that $x^2 - 4 = 0$ will give us vertical asymptotes. $x = 2$, $x = -2$ are vertical asymptotes. As $x \to \infty$, $\dfrac{x^2}{x^2-4} \to 1$. $y = 1$ is a horizontal asymptote.

4.299 $f(x) = \dfrac{2x^2}{3x^2 + 4}$

▮ Since $3x^2 + 4$ does not approach 0 for any choices of x, there is no vertical asymptote. As $x \to \infty$, $f(x) \to 1$. $y = 1$ is a horizontal asymptote.

4.300 $f(x) = \dfrac{x}{(x-1)(x+2)}$

▮ As $x - 1 \to 0$ or $x + 2 \to 0$, $f(x) \to \infty$. Thus, $x = 1$ and $x = -2$ are vertical asymptotes. As $x \to \infty$, $\dfrac{x}{(x-1)(x+2)} \to 0$. $y = 0$ is a horizontal asymptote.

4.301 $f(x) = \dfrac{x^2 + 2}{x^4 + x}$

▮ $x^4 + x = x(x^3 + 1)$, and $x(x^3+1) \to 0$ if $x \to 0$ or $x^3 + 1 \to 0$. Thus, $x = 0$ and $x = -1$ are vertical asymptotes. As $x \to \infty$, $f(x) \to 0$. $y = 0$ is a horizontal asymptote.

4.302 $f(x) = \dfrac{x+3}{(x-2)(x+1)(x+2)}$

▮ If $x \to 2$, $x \to -1$ or $x \to -2$, the denominator $\to 0$. Thus, $x = 2$, $x = -1$, $x = -2$ are vertical asymptotes. As $x \to \infty$, $f(x) \to 0$, and $y = 0$ is a horizontal asymptote.

4.303 $f(x) = \dfrac{1}{(1-x)^3}$

▮ As $x \to 1$, $(1-x)^3 \to 0$ and $f \to \infty$. Thus, $x = 1$ is a vertical asymptote. As $x \to \infty$, $f(x) \to 0$, and $y = 0$ is a horizontal asymptote.

4.304 $f(x) = 1 + 3/4x$

▮ As $x \to 0$, $3/4x \to \infty$. $x = 0$ is a vertical asymptote. As $x \to \infty$, $1 + 3/4x \to 1$. $y = 1$ is a horizontal asymptote.

For Probs. 4.305 to 4.307, find all zeros, isolated points, and asymptotes.

4.305 $f(x) = x^3 + x$

▮ f is a polynomial, so there are no asymptotes or isolated points. $x = 0$ is the only zero.

4.306 $f(x) = \sqrt{\dfrac{x^3}{2-4}}$

▮ This is an explicit algebraic function. If $x = 0$, $\dfrac{x^3}{2-x} = 0$, and 0 is the only such zero ($x^3 = 0$ if and only if $x = 0$). When $x = 2$, $2 - x = 0$, and thus $x = 2$ is a vertical asymptote. There are no isolated points, and no horizontal asymptotes. As $x \to \infty$, so does $f(x)$.

4.307 $f(x) = \dfrac{2x}{\sqrt{x-1}}$

▮ f here is algebraic. If $2x = 0$, then $x = 0$ and $y = 0$, and 0 is the only zero. As $x \to 1$, $\sqrt{x-1} \to 0$, and $x = 1$ is a vertical asymptote. There are no horizontal asymptotes. Note that $(0, 0)$ is an isolated point. $(0, 0)$ is on the graph, but, except for that point, $x \geq 1$ or f is undefined.

For Probs. 4.308 to 4.313, sketch the graph of the given rational function.

4.308 $f(x) = 1/x$

▮ If $x = 0$, then we have a vertical asymptote. As $x \to \infty$, $1/x \to 0$, so $y = 0$ is a vertical asymptote. There are no x or y intercepts. $y = 1/x$ shows origin symmetry. We plot a few points: $(1, 1)$, $(-1, -1)$, $(2, \frac{1}{2})$. See Fig. 4.36.

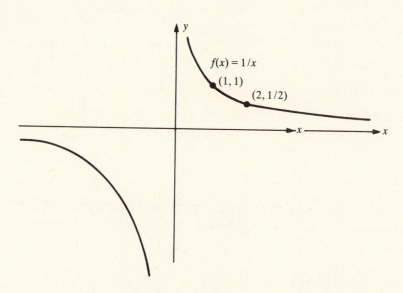

$f(x) = 1/x$
$(1, 1)$
$(2, 1/2)$

Fig. 4.36

4.309 $f(x) = -2/x$

▮ This is similar to Prob. 4.308, but we multiply each ordinate by -2. See Fig. 4.37.

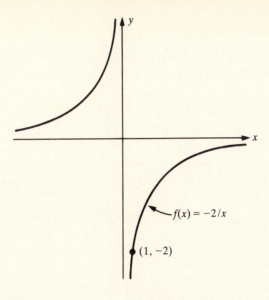

Fig. 4.37

4.310 $f(x) = 2 + 1/x$

$\blacksquare$ If $x \to 0$, $1/x$ grows without bound. $x = 0$ is a vertical asymptote. As $x \to \infty$, $f \to 2$, so $y = 2$ is a horizontal asymptote. There is no symmetry. We plot the points $(1, 3)$ and $(\frac{1}{2}, 4)$. See Fig. 4.38.

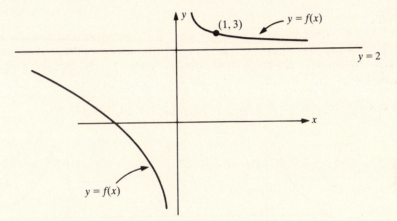

Fig. 4.38

4.311 $f(x) = \dfrac{1}{2(x-1)}$

$\blacksquare$ We find vertical asymptote $x = 1$, horizontal asymptote $y = 0$. There is no symmetry. We plot the points $(2, \frac{1}{2})$, $(-2, -\frac{1}{6})$, $(3, \frac{1}{4})$. See Fig. 4.39.

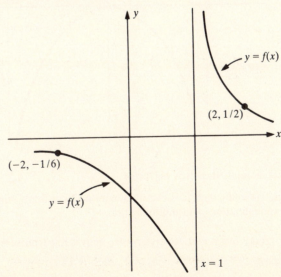

Fig. 4.39

4.312 $f(x) = \dfrac{x}{x+1}$

▮ As $x \to -1$, $x + 1 \to 0$. $x = -1$ is a vertical asymptote. As $x \to \infty$, $f \to 1$, and $y = 1$ is a horizontal asymptote. We plot the points $(1, \frac{1}{2})$, $(-2, 2)$, $(2, \frac{2}{3})$. See Fig. 4.40.

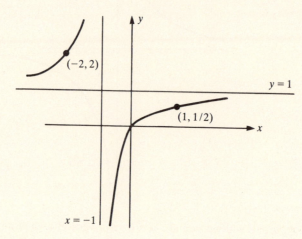

Fig. 4.40

4.313 $f(x) = \dfrac{x^2}{x^2 - 4}$

▮ If $x^2 - 4 = 0$, $x = \pm 2$. As $x \to 2$ or $x \to -2$, f grows without bound; $x = \pm 2$ are vertical asymptotes. As $x \to \infty$, $f \to 1$, so $y = 1$ is a horizontal asymptote. Replacing x by $-x$ leaves equation line unchanged: the graph exhibits y-axis symmetry. We plot the points $(0,0)$, $(1, -\frac{1}{3})$, $(3, \frac{9}{5})$. See Fig. 4.41.

For Probs. 4.314 to 4.321, tell whether the given statement is true or false and why.

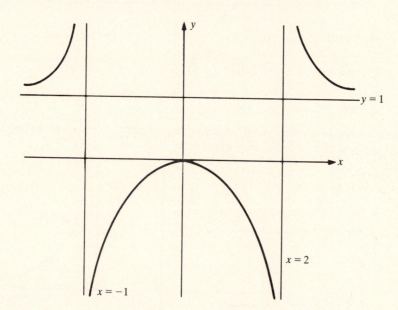

Fig. 4.41

4.314   If $y = f(x)$ is a polynomial, then it is algebraic.

▮ True; if it is a polynomial, then it is clearly generated by a finite number of algebraic operations: addition, integral exponentiation, and multiplication.

4.315   If the graph of $y = f(x)$ has a vertical asymptote, then it must possess a horizontal asymptote.

▮ False; see Fig. 4.42.

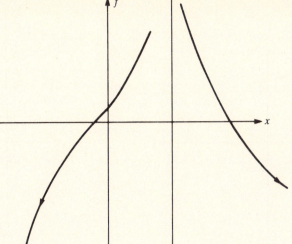

Fig. 4.42

4.316 If the graph of $y = f(x)$ has a vertical asymptote, then $f(x)$ is not a polynomial.

▮ True. If it has a vertical asymptote, then as $x \to a$, f grows without bound. Then f is not a polynomial.

4.317 If the graph of $y = f(x)$ has a horizontal asymptote, then $f(x)$ is not a polynomial.

▮ True. If when $x \to \infty$, $f \to b$, and b is finite, then f is nonpolynomial.

4.318 Only algebraic functions possess asymptotes.

▮ False. Consider $f(x) = \text{Tan } x$. Recall: As $x \to \pi/2$, $f \to \infty$.

4.319 If $y = f(x)$ is rational and the denominator is a cubic polynomial, then the graph of f has three vertical asymptotes.

▮ False; not necessarily. For example, $f(x) = \dfrac{3 + x}{(x - 1)^3}$. The only vertical asymptote is $x = 1$.

4.320 If $f(x)$ and $g(x)$ are algebraic, then so is $f(x) + g(x)$.

▮ True; to find $(f + g)(x)$, we would simply add f to g, and that is simply one more algebraic step. If f and g are algebraic, we would still have a finite number of algebraic manipulations.

4.321 If $y = f(x)$ is a rational function, and has no vertical asymptotes, then $y = kf(x)$ is rational with no vertical asymptotes, where $k \in \mathbb{R}$.

▮ True; clearly $kf(x)$ is rational, since $f(x) = g(x)/h(x)$ where g and h are polynomials. Also, if g is a polynomial, so is kg. If f has no vertical asymptotes, $h(x)$ has no zeros. But $h(x)$ is still the denominator in kg/h.

For Probs. 4.322 to 4.324, refer to Figs. 4.43 to 4.46.

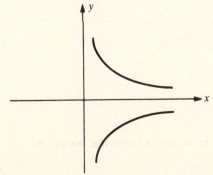

Fig. 4.43

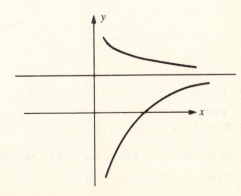

Fig. 4.44

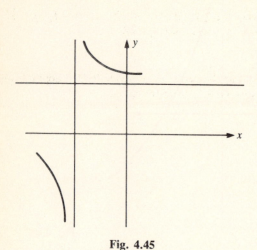

Fig. 4.45

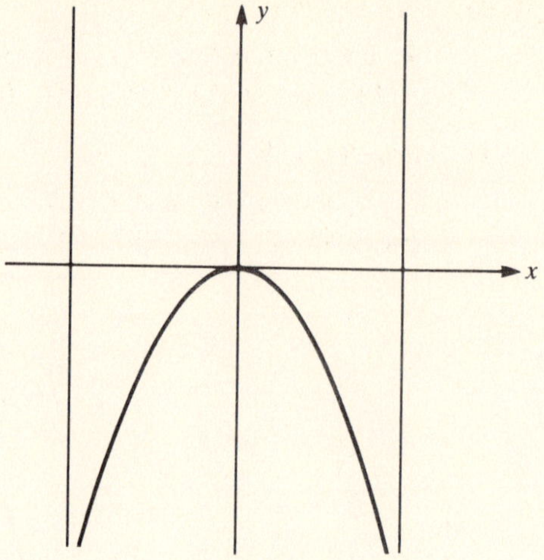

Fig. 4.46

4.322 For which of these relations is it the case that as $x \to \infty$, $f(x) \to 0$?

▮ In Fig. 4.43 note that as we move to the right along the x axis, y gets closer to $y = 0$, which is a horizontal asymptote.

4.323 For which of these relations might the denominator be $(x - a)(x + b)$?

▮ If the denominator is $(x - a)(x + b)$, then as $x \to a$, $f \to \infty$, and as $x \to -b$, $f \to \infty$. The only graph in which this is possible is the one shown in Fig. 4.46.

4.324 For which of these relations do we find x-axis symmetry? y-axis symmetry? Which graphs represent y as a function of x?

▮ We find x symmetry in the graph of Fig. 4.43; y symmetry in the graph of Fig. 4.46. In Figs. 4.44 and 4.45, y is a function of x. That is not the case in the other figures.

For Probs. 4.325 to 4.329, a graph of a function is given. Give an example of a function that might have the given graph.

4.325 See Fig. 4.47.

▮ $y = f(x)$ has asymptotes $y = 0$ and $x = 0$. $f(x) = 1/x$ is one such function. There are others, such as $f(x) = 2/x$, $f(x) = 5/x$, etc.

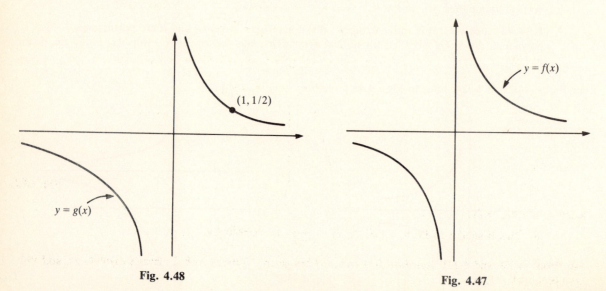

Fig. 4.48

Fig. 4.47

4.326 See Fig. 4.48.

$\blacksquare$ See Prob. 4.325. This is similar except here we know $(1, \frac{1}{2})$ is on the graph. If $x = 1$, $y = \frac{1}{2}$. Thus, $f(x) = 1/2x$ is one possibility, since $x = 0$ and $y = 0$ are again asymptotes.

4.327 See Fig. 4.49.

$\blacksquare$ Again, $x = 0$ and $y = 0$ are asymptotes. This time, however, as x gets close to 0 approaching from the right, $f \to -\infty$. $f(x) = -1/x$ is one possibility since $(1, -1)$ is on the graph.

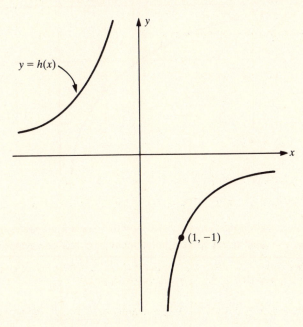

Fig. 4.49

4.328 See Fig. 4.50.

$\blacksquare$ $y = 3$ and $x = -3$ are asymptotes, and $(0, 0)$ is on the curve. Then, $f(x)$ must have $x + 3$ in its denominator and x in its numerator. $f(x) = \dfrac{3x}{x + 3}$ is one possibility.

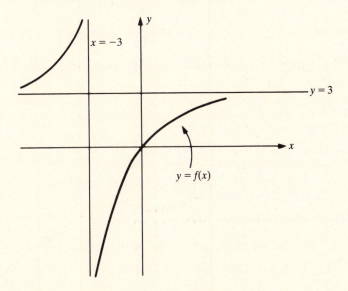

Fig. 4.50

4.329 See Fig. 4.51.

$\blacksquare$ This is similar to Prob. 4.238. $g(x) = \dfrac{x}{x - 2}$ is a possibility.

For Probs. 4.330 and 4.331, functions $f(x)$ and $g(x)$ are given. Discuss each function's asymptotes, and the asymptotes of $f \circ g(x)$.

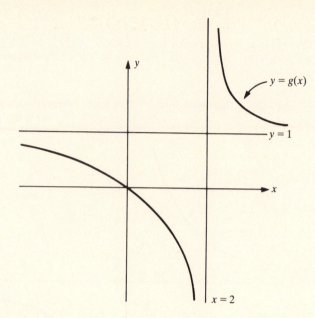

Fig. 4.51

4.330 $f(x) = 1/x$, $g(x) = 2/x$

▐ f and g both have asymptotes $x = 0$ and $y = 0$. See earlier problems for verification. $f \circ g(x) = f(2/x) = 1/(2/x) = \dfrac{x}{2}$. $f \circ g(x) = x/2$ is a polynomial, and has no asymptotes.

4.331 $f(x) = \dfrac{x}{x+2}$, $g(x) = \dfrac{x-2}{x+1}$

▐ f: As $x \to -2$, $x + 2 \to 0$. As $x \to \infty$, $f \to 1$. Thus, f has asymptotes $x = -2$, $y = 1$. g: As $x \to -1$, $x + 1 \to 0$. As $x \to \infty$, $g \to 1$. Thus, g has asymptotes $x = -1$, $y = 1$. Then,

$$f \circ g(x) = f\left(\frac{x-2}{x+1}\right) = \frac{(x-2)/(x+1)}{(x-2)/(x+1)+2} = \frac{(x-2)/(x+1)}{[x-2+2(x+1)]/(x+1)} = \frac{(x-2)/(x+1)}{3x/(x+1)} = \frac{(x-2)(x+1)}{3x(x+1)}$$

$$= \frac{x-2}{3x}$$

$\dfrac{x-2}{3x}$ has asymptotes $x = 0$, and $y = \frac{1}{3}$. Compare this result to Prob. 4.330.

4.332 True or false: If f and g have a horizontal asymptote, then so does $f \circ g$.

▐ False! See Prob. 4.330.

4.333 True or false: If f, g and $f \circ g$ all have horizontal asymptotes, then $f \circ g$ must have the same horizontal asymptote as f or g.

▐ False. See Prob. 4.331.

4.334 True or false: If $f \circ g$ and $g \circ f$ both have horizontal asymptotes, then they have the same horizontal asymptotes.

▐ False. See the functions in Prob. 4.331.

4.335 True or false: If f has no asymptotes, then $f \circ g$ must not have any.

▐ False. Let $f(x) = x^2 + 1$ and $g(x) = 1/x$. Then $f \circ g(x) = f(1/x) = 1/x^2 + 1 = (1 + x^2)/x^2$ which is asymptotic.

CHAPTER 5
Systems of Equations and Inequalities

5.1 SYSTEMS OF LINEAR EQUATIONS

For Probs. 5.1 to 5.10, solve the system of equations by graphing.

5.1 $2x + 3y = 12$, $x - 3y = -3$

▮ See Fig. 5.1. Graph the two lines l_1: $2x + 3y = 12$ and l_2: $x - 3y = -3$. For l_1, if $x = 0$, $y = 4$; if $y = 0$, $x = 6$. For l_2, if $x = 0$, $y = 1$; if $y = 0$, $x = -3$. Next find where they intersect. From our graph, the intersection point is $(3, 2)$. Thus, $x = 3$, $y = 2$ is the solution. *Check*: $2(3) + 3(2) = 12$ and $3 - 3(2) = -3$.

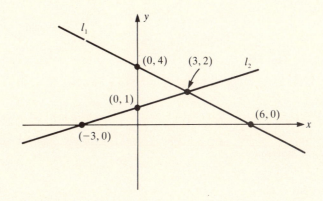

Fig. 5.1

5.2 $x + 2y = 6$, $3x - 2y = 6$

▮ See Fig. 5.2. We graph the two lines l_1: $x + 2y = 6$ and l_2: $3x - 2y = 6$. For l_1; if $x = 0$, $y = 3$; if $y = 0$, $x = 6$. For l_2; if $x = 0$, $y = -3$; if $y = 0$, $x = 2$. The lines intersect at $(3, \frac{3}{2})$: $x = 3$, $y = \frac{3}{2}$. *Check*: $3 + 2(\frac{3}{2}) = 6$, and $3(3) - 2(\frac{3}{2}) = 6$.

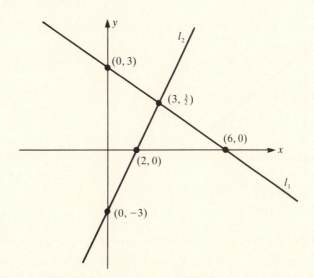

Fig. 5.2

5.3 $2x + 4y = -7$, $x - 3y = 9$

▮ See Fig. 5.3. Let $2x + 4y = -7$ be l_1 and $x - 3y = 9$ be l_2. For l_1, if $x = 0$, $y = -\frac{7}{4}$; if $y = 0$, $x = -\frac{7}{2}$. For l_2, if $x = 0$, $y = -3$; if $y = 0$, $x = 9$. The lines intersect at $(1.5, -1.5)$: $x = 1.5$, $y = -1.5$.

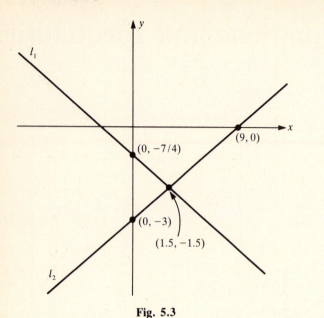

Fig. 5.3

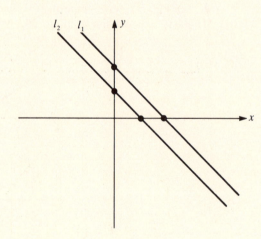

Fig. 5.4

5.4 $x - 2y = 3$, $3x - 6y = 9$

▮ See Fig. 5.4. For l_1 $(x - 2y = 3)$, if $x = 0$, $y = -\frac{3}{2}$; if $y = 0$, $x = 3$. For l_2 $(3x - 6y = 9)$, if $x = 0$, $y = -\frac{9}{6} = -\frac{3}{2}$; if $y = 0$, $x = 3$. Then l_1 and l_2 coincide, and the system is dependent: Every (x, y) will satisfy the system.

5.5 $4x + 5y = 6$, $8x + 10y = 12$

▮ See Fig. 5.5. The system is dependent. See Prob. 5.4.

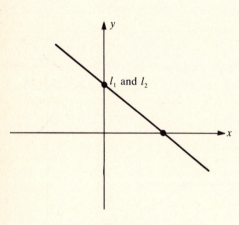

Fig. 5.5

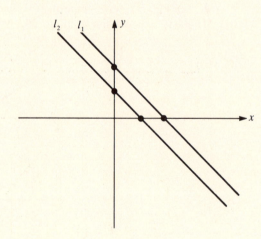

Fig. 5.6

5.6 $2x + y = 2$, $6x + 3y = 5$

▮ See Fig. 5.6. For l_1 $(2x + y = 2)$, if $x = 0$, $y = 2$; if $y = 0$, $x = 1$. For l_2 $(6x + 3y = 5)$, if $x = 0$, $y = \frac{5}{3}$; if $y = 0$, $x = \frac{5}{6}$. Then $l_1 \| l_2$; there is no solution (the system is inconsistent).

5.7 $5x - 3y = 1$, $10x - 6y = -2$

▮ See Fig. 5.7. For l_1 $(5x - 3y = 1)$, if $x = 0$, $y = -\frac{1}{3}$; $y = 0$, $x = \frac{1}{5}$. For l_2 $(10x - 6y = -2)$, if $x = 0$, $y = \frac{1}{3}$; if $y = 0$, $x = -\frac{1}{5}$. Then $l_1 \| l_2$; no solution.

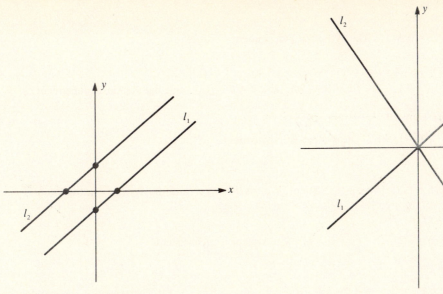

Fig. 5.7

Fig. 5.8

5.8 $x = y$, $x = -y$

▮ See Fig. 5.8. For l_1 $(x = y)$, we plot the points $(1, 1)$ and $(2, 2)$. For l_2 $(x = -y)$ we plot the points $(1, -1)$ and $(2, -2)$. Then $l_1 \cap l_2 = \{(0, 0)\}$, and $x = 0$, $y = 0$.

5.9 $x + y = 1$, $y = 1$

▮ See Fig. 5.9. Let l_1 be $x + y = 1$ and l_2 be $y = 1$. Then l_1 contains the points $(0, 1)$ and $(1, 0)$. Then $l_1 \cap l_2 = \{(0, 1)\}$, and $x = 0$, $y = 1$.

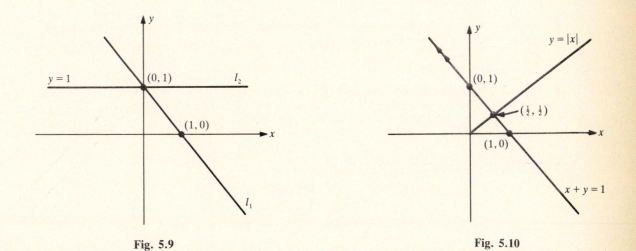

Fig. 5.9

Fig. 5.10

5.10 $x + y = 1$, $y = |x|$

▮ See Fig. 5.10. Here $x = \frac{1}{2}$ and $y = \frac{1}{2}$. Note that, for $x < 0$, $(x + y = 1) \| (y = |x|)$.

For Probs. 5.11 to 5.18, without solving, tell whether the given system is independent, dependent, or inconsistent.

5.11 $2x + y = 6$, $x + y = 10$

▮ Recall that if $ax + by = c$ and $dx + ey = f$, then the system is independent if $d/a \neq e/b$, dependent if $d/a = e/b = f/c$, and inconsistent if $d/a = e/b \neq f/c$. Here, $d/a = 2/1$ and $e/b = 1/1$. The system is independent.

5.12 $2x - y = 6$, $x + y = 6$

▮ $d/a = 2/1$; $e/b = -1/1$.   The system is independent.

5.13 $2x + y = 10$, $-4x - 2y = -20$

▮ $d/a = 2/-4$; $e/b = 1/-2$; $f/c = 10/-20$; $d/a = e/b = f/c$,   and the system is dependent.

5.14 $2x + y = 10$, $-4x + y = 10$

▮ $d/a = 2/-4$, $e/b = 1/-1$.   The system is independent.

5.15 $y + x = 5$, $y + x = 8$

▮ $d/a = 1/1$; $e/b = 1/1$; $f/c = 5/8$.   Then $d/a = e/b \neq f/c$,   and the system is inconsistent.

5.16 $x - 2y - 6 = 0$, $x - 2y = -6$

▮ $d/a = 1/1$; $e/b = -2/-2$; $f/c = 6/-6$.   The system is inconsistent.

5.17 $x = 3y + 8$, $2y = x + 4$

▮   Then $x - 3y = 8$ and $-x + 2y = 4$. $d/a = 1/-1$; $e/b = -3/2$.   The system is independent.

5.18 $y = x$, $2y = x + 3$

▮   Then $-x + y = 0$ and $-x + 2y = 3$. $d/a = -1/-1$; $e/b = 1/2$.   The system is independent.

For Probs. 5.19 to 5.30, solve using the addition/subtraction method.

5.19 $x - y = 10$, $x + y = 14$

▮   Adding, $(x + x) + [y + (-y)] = 24$.   Then $2x = 24$,   or $x = 12$.   Replace x by 12 in either equation. Then $12 - y = 10$,   or $y = 2$.   *Check*: $12 - 2 = 10$,   and $12 + 2 = 14$.

5.20 $2x + y = 20$, $-x - y = 10$

▮   Adding, $(2x - x) + 0 = 30$,   or $x = 30$.   Substituting, $2(30) + y = 20$,   or $y = -40$.   *Check*: $2(30) - 40 = 20$,   and $-30 - (-40) = 10$.

5.21 $x - \frac{1}{2}y = 6$, $2x - \frac{1}{2}y = 10$

▮   Subtract the second equation from the first: $(x - 2x) - \frac{1}{2}y + \frac{1}{2}y = 6 - 10$.   (Note the $+\frac{1}{2}y$ term.   It is positive because we are subtracting a $-\frac{1}{2}y$.)   Then $-x = -4$,   or $x = 4$.   Substituting, $4 - \frac{1}{2}y = 6$, $\frac{1}{2}y = -2$,   or $y = -4$.   *Check*: $4 - \frac{1}{2}(-4) = 6$,   and $8 - \frac{1}{2}(-4) = 10$.

5.22 $3x + y = 7$, $x + 2y = 5$

▮   Multiply the first equation by 2 and subtract:

$$\begin{array}{r} 6x + 2y = 14 \\ x + 2y = 5 \\ \hline 5x = 9 \end{array}$$

Then $x = \frac{9}{5}$.   Substituting $\frac{9}{5}$ for x in $x + 2y = 5$,   we get $\frac{9}{5} + 2y = 5$, $2y = \frac{25}{5} - \frac{9}{15} = \frac{16}{15}$,   or $y = \frac{16}{30}$. Check this.

5.23 $4x - y = 10$, $x + 3y = 8$

▮   Multiply the first equation by 3 and add:

$$\begin{array}{r} 12x - 3y = 30 \\ x + 3y = 8 \\ \hline 13x = 38 \end{array}$$

Then $x = \frac{38}{13}$.   Substituting, $x + 3y = 8$,   so $\frac{38}{13} + 3y = 8$, $3y = 8 - \frac{38}{13} = \frac{104}{13} - \frac{38}{13} = \frac{66}{13}$,   or $y = \frac{66}{39} = \frac{22}{13}$. See Prob. 5.24.

5.24 $4x - y = 10$, $x + 3y = 8$

▮ This time, we multiply the second equation by 4 and subtract:

$$
\begin{array}{r}
4x - y = 10 \\
4x + 12y = 32 \\
\hline
-13y = -22
\end{array}
$$

Then $y = \frac{22}{13}$. Substituting, $x + 3(\frac{22}{13}) = 8$, or $x = 8 - \frac{66}{13} = \frac{38}{13}$. Note that we get the same result as in Prob. 5.23. We should, since these are equivalent techniques.

5.25 $2x - 3y = 10$, $3x - 2y = 5$

▮ Multiply the first equation by 3, the second equation by 2, and subtract:

$$
\begin{array}{r}
6x - 9y = 30 \\
6x - 4y = 10 \\
\hline
-5y = 20
\end{array}
$$

Then $y = -4$. Substituting, $2x + 12 = 10$, $2x = -2$, or $x = -1$.

5.26 $2x + y = 5$, $4x + 2y = 10$

▮ Multiply the first equation by 2 and subtract:

$$
\begin{array}{r}
4x + 2y = 10 \\
4x + 2y = 10 \\
\hline
0 = 0
\end{array}
$$

The system is dependent.

5.27 $3x + y = 20$, $2x + y = 20$

▮ Subtracting, we get $(3x - 2x) + 0 = 0$, or $x = 0$. Then, substituting, $0 + y = 20$, or $y = 20$.

5.28 $3x + y = 20$, $3x + y = 30$

▮ Subtracting, we get $(3x - 3x) + (y - y) = 20 - 30$, or $0 = -10$. The system is inconsistent.

5.29 $0.5x + 0.7y = 10$, $0.1x + y = 15$

▮ Multiply the second equation by 5 and subtract:

$$
\begin{array}{r}
0.5x + 0.7y = 10 \\
0.5x + 5y = 75 \\
\hline
-4.3y = -65
\end{array}
$$

Then $y = 65/4.3$. Substituting into $0.1x + y = 15$, we get $0.1x + 65/4.3 = 15$, $0.1x = 15 - 65/4.3 \approx 15 - 15.12$ (rounding) $= -0.12$. Thus, $x \approx 1.2$.

5.30 $\dfrac{x}{2} + \dfrac{y}{3} = 6$, $\dfrac{-x}{3} + \dfrac{y}{2} = 8$

▮ Multiply both equations by 6: $3x + 2y = 36$ and $-2x + 3y = 48$. Multiply the first equation by 2, the second equation by 3, and add:

$$
\begin{array}{r}
6x + 4y = 72 \\
-6x + 9y = 144 \\
\hline
13y = 216
\end{array}
$$

Then $y = 16.62$ (rounded). Substituting, $3x + 33.24 = 36$, $3x = 2.76$, or $x = 0.92$.

For Probs. 5.31 to 5.41, solve using substitution.

5.31 $x + y = 15$, $x - y = 10$

▮ If $x + y = 15$, then $x = 15 - y$. Substitute $15 - y$ for x in the second equation. Then $(15 - y) - y = 10$, $15 - 2y = 10$, $2y = 5$, or $y = \frac{5}{2}$. Substituting for y, $x + \frac{5}{2} = 15$, or $x = 15 - \frac{5}{2} = 12\frac{1}{2}$. Check this.

5.32 $x + 2y = 10$, $2x + 3y = 8$

▮ $x = 10 - 2y$. Substitute $10 - 2y$ for x in the second equation. Then $2(10 - 2y) + 3y = 8$, $20 - 4y + 3y = 8$, $-y = -12$, or $y = 12$. Substituting for y; $x = 10 - 2y = 10 - 24 = -14$.

5.33 $-x - y = 15$, $-2x - 4y = 4$

▮ $-x = 15 + y$, or $x = -y - 15$. Substituting $-y - 15$ for x in the second equation, $-2(-y - 15) - 4y = 4$, $2y + 30 - 4y = 4$, $-2y = -26$, or $y = 13$. Substituting for y, $x = -13 - 15 = -28$. See Prob. 5.34.

5.34 $-x - y = 15$, $-2x - 4y = 4$

▮ Solve the second equation for x: $-2x = 4 + 4y$, or $x = -2 - 2y$. Substituting $4 + 4y$ for x in the first equation, $-(-2 - 2y) - y = 15$, $2 + 2y - y = 15$, or $y = 13$. Then $x = -29$. See Prob. 5.33.

5.35 $x = 5 - 2y$, $y = 10x - 6$

▮ $y = 10(5 - 2y) - 6$, or $y = 50 - 20y - 6$, $21y = 44$, or $y = \frac{44}{21}$. Substituting $\frac{44}{21}$ for y in the first equation, $x = 5 - 2(\frac{44}{21}) = 5 - \frac{88}{21} = \frac{105}{21} - \frac{88}{21} = \frac{17}{21}$.

5.36 $\frac{x}{2} + y = 6$, $\frac{y}{2} + x = 10$

▮ $y = 6 - x/2$. Substituting $6 - x/2$ for y in the second equation, $\frac{6 - x/2}{3} + x = 10$, $6 - x/2 + 3x = 30$, $5x/2 = 24$, $5x = 48$, or $x = \frac{48}{5}$. Substituting for x, $y = 6 - \frac{48}{5}/2 = 6 - \frac{48}{10} = \frac{60}{10} - \frac{48}{10} = \frac{12}{10} = \frac{6}{5}$. See Prob. 5.37.

5.37 $\frac{x}{2} + y = 6$, $\frac{y}{3} + x = 10$

▮ $x/2 = 6 - y$, or $x = 12 - 2y$. Substituting $12 - 2y$ for x in the second equation, $y/3 + (12 - 2y) = 10$, $y + 36 - 6y = 30$, $-5y = -6$, or $y = \frac{6}{5}$. Substituting for y, $x = 12 - 2(\frac{6}{5}) = 12 - \frac{12}{5} = \frac{48}{5}$. See Prob. 5.36.

5.38 $\frac{x}{2} + \frac{y}{3} = 8$, $\frac{x}{3} + \frac{y}{2} = 10$

▮ $x/2 = 8 - y/3$ or $x = 16 - 2y/3$. Substituting $16 - 2y/3$ for x in the second equation, $x/3 = \frac{16}{3} - 2y/9$, $x/3 + y/2 = 10$, $\frac{16}{3} - 2y/9 + y/2 = 10$, $\frac{16}{3} + 5y/18 = 10$, $5y/18 = \frac{30}{3} - \frac{16}{3} = \frac{14}{3}$, $15y = 252$, or $y = 16.8$. Substituting for y, $x = 16 - \frac{2}{3}(16.8) = 4.8$.

5.39 $0.2x + 0.5y = 10$, $x - 0.3y = 15$

▮ $x = 15 + 0.3y$. Substituting for x, $0.2(15 + 0.3y) + 0.5y = 10$, $3.0 + 0.06y + 0.5y = 10$, $0.56y = 7$, or $y = 12.5$. Substituting for y, $x = 15 + 0.3(12.5) = 18.75$. See Prob. 5.40.

5.40 $0.2x + 0.5y = 10$, $x - 0.3y = 15$

▮ $0.2x = 10 - 0.5y$, $2x = 100 - 5y$, or $x = 50 - 2.5y$. Substituting for x, $50 - 2.5y - 0.3y = 15$, $50 - 2.8y = 15$, $2.8y = 35$, or $y = 12.5$. Substituting for y, $x = 50 - 2.5(12.5) = 18.75$. See Prob. 5.39.

5.41 $0.2x - \frac{1}{2}y = 10$, $0.6y - x/3 = 12$

▮ $0.2x = 10 + \frac{1}{2}y$, $2x = 100 + 5y$, or $x = 50 + \frac{5}{2}y$. Substituting for x, $0.6y - [50 + \frac{5}{2}y]/3 = 12$, $1.8y - 50 - \frac{5}{2}y = 36$, $-0.7y = 86$, or $y = -86/0.7 = -860/7$. Substituting for y, $x = 50 + \frac{5}{2}(-\frac{860}{7}) = 50 + (-\frac{4300}{14}) = \frac{700}{14} - \frac{4300}{14} = \frac{3600}{14} = \frac{1800}{7}$.

For Probs. 5.42 to 5.44, solve the given problem using a system of linear equations.

5.42 The sum of two numbers is 20, and their difference is 10.   Find the two numbers.

▮ Let $x =$ one number, and $y =$ the other.   Then $x + y = 20$ and $x - y = 10$.   Thus, $2x = 30$, or $x = 15$, and $y = x - 10 = 5$.   *Check*: Sum $= 20$; difference $= 10$.

5.43 The length of a rectangle is 10 in more than the width.   The perimeter of the rectangle is 60 in.   Find the dimensions of the rectangle.

▮ Let $l =$ length and $w =$ width.   Then $l = 10 + w$ and $2l + 2w = 60$.   Substituting for l, $2(10 + w) + 2w = 60$, $20 + 2w + 2w = 60$, $4w = 40$,   or $w = 10$ in.   Then $l = 10$ in $+ 10$ in $= 20$ in. *Check*: $20 = 10 + 10$ (length is 10 more),   and $P = 2(20) + 2(10) = 60$.

5.44 The sum of two numbers is 15.   The larger minus thrice the smaller is the negative of the larger.   Find the numbers.

▮ Let $x =$ smaller and $y =$ larger.   Then $x + y = 15$ and $y - 3x = -y$ or $2y - 3x = 0$.   Multiply the first equation by 2 and subtract:

$$\begin{array}{r} 2y + 2x = 30 \\ 2y - 3x = 0 \\ \hline 5x = 30 \end{array}$$

Then $x = 6$ and $y = 9$.

For Probs. 5.45 to 5.47, solve the given system.

5.45 $x + y + z = 5$, $x - y + z = 10$, $x + y = 3$

▮ Add the first two equations: $2x + 2z = 15$.   Add the third and the second: $2x + z = 13$.   Then subtract:

$$\begin{array}{r} 2x + 2z = 15 \\ 2x + z = 13 \\ \hline z = 2 \end{array}$$

Substituting, $2x + 4 = 15$,   or $x = \frac{11}{2}$.   Then $y = 3 - x = 3 - \frac{11}{2} = -\frac{5}{2}$.

5.46 $x + y + 2z = 10$, $x - y - 2z = 5$, $x + y + z = 20$

▮ Add the first two equations: $2x = 15$,   or $x = \frac{7}{2}$.   Subtract the third from the first: $2z - z = -10$,   or $z = -10$.   Then from $x + y + z = 20$, $y = 20 - x - z = 20 - \frac{7}{2} - (-10) = 30 - \frac{7}{2} = \frac{53}{2}$.

5.47 $x + 2y + z = -5$, $2x - 2y + 2z = 10$, $-x - y - z = 5$

▮ Add the first two equations: $3x + 3z = 5$.   Multiply the third by -2, and add to the second:

$$\begin{array}{r} 2x - 2y + 2z = 10 \\ 2x + 2y + 2z = -10 \\ \hline 4x + 4z = 0 \end{array}$$

Then $4x + 4z = 0$ and $3x + 3z = 5$,   so $12x + 12z = 0$ and $12x + 12z = 20$.   No solution; inconsistent.

5.2 MATRICES AND DETERMINANTS

For Probs. 5.48 to 5.51, find a, b, c, d.

5.48 $[a \quad b \quad c \quad d] = [2 \quad 3 \quad -1 \quad 0]$

▮ Two matrices are equal if and only if their corresponding elements are equal.   Thus, $a = 2$, $b = 3$, $c = -1$, $d = 0$.

5.49 $\begin{bmatrix} a & b \\ c & d \end{bmatrix} = \begin{bmatrix} 1 & 5 \\ -2 & 3 \end{bmatrix}$

▮ Corresponding elements must be equal.   Thus, $a = 1$, $b = 5$, $c = -2$, $d = 3$.

5.50 $\begin{bmatrix} a+2 & 3c-1 \\ 2b+3 & d-2 \end{bmatrix} = \begin{bmatrix} 5 & 8 \\ 7 & 0 \end{bmatrix}$

▎ Then $a+2=5$, $a=3$; $3c-1=8$, $c=3$; $2b+3=7$, $b=2$; $d-2=0$, $d=2$.

5.51 $\begin{bmatrix} a+2b & 2a-b & c+2 \\ a & 3d-4 & b+1 \end{bmatrix} = \begin{bmatrix} 5 & 0 & 6 \\ 1 & 2 & 3 \end{bmatrix}$

▎ $a+2b=5$, so $2b=4$, $b=2$, since $a=1$. Since $c+2=6$, $c=4$; since $3d-4=2$, $d=2$.

For Probs. 5.52 to 5.66, refer to the following matrices.

$$A = \begin{bmatrix} 2 & -1 \\ 3 & 0 \end{bmatrix} \quad B = \begin{bmatrix} -3 & 1 \\ 2 & -3 \end{bmatrix} \quad C = \begin{bmatrix} 2 \\ -3 \\ 0 \end{bmatrix}$$

$$D = \begin{bmatrix} 1 \\ 3 \\ 5 \end{bmatrix} \quad E = [-4 \quad 1 \quad 0 \quad -2] \quad F = \begin{bmatrix} 2 & -3 \\ -2 & 0 \\ 1 & 2 \\ 3 & 5 \end{bmatrix}$$

5.52 What are the dimensions of **B**? of **E**?

▎ A matrix is $m \times n$ if it has m rows (horizontal) and n columns (vertical). In this case, **B** has two rows and two columns, so it is 2×2; **E** has one row and four columns, so it is 1×4.

5.53 What are the dimensions of **F**? **D**?

▎ See Prob. 5.52. **F** is 4×2; **D** is 3×1.

5.54 What element is in the second row, first column of **F**?

▎ $[-2 \quad 0]$ is the second row. Then -2 is in the second row, first column position.

5.55 Write a zero matrix of the same dimension as **B**.

▎ A zero matrix has all entries zero. Thus $\begin{bmatrix} 0 & 0 \\ 0 & 0 \end{bmatrix}$ is the matrix we are looking for.

5.56 How many additional columns would **F** need so it would be square?

▎ **F** is 4×2. If it were 4×4, it would be square. It needs two additional columns.

5.57 Find **A** + **B**.

▎ $A + B = \begin{bmatrix} 2 & -1 \\ 3 & 0 \end{bmatrix} + \begin{bmatrix} -3 & 1 \\ 2 & -3 \end{bmatrix} = \begin{bmatrix} -1 & 0 \\ 5 & -3 \end{bmatrix}$.

5.58. Find **C** + **D**.

▎ $C + D = \begin{bmatrix} 2 \\ -3 \\ 0 \end{bmatrix} + \begin{bmatrix} 1 \\ 3 \\ 5 \end{bmatrix} = \begin{bmatrix} 2+1 \\ -3+3 \\ 0+5 \end{bmatrix} = \begin{bmatrix} 3 \\ 0 \\ 5 \end{bmatrix}$.

5.59 Find the negative of matrix **B**.

▎ For any matrix **B**, $-B$ is the matrix such that $B + (-B) =$ the zero matrix. All of the entries of $-B$ must be the negative of the corresponding entries in **B**.

$$-B = \begin{bmatrix} -(-3) & -1 \\ -2 & -(-3) \end{bmatrix} = \begin{bmatrix} 3 & -1 \\ -2 & 3 \end{bmatrix}.$$

Note that $B + (-B) = 0$ ($0 =$ the zero matrix).

5.60   Find $D - C$.

$$D - C = D + (-C) = \begin{bmatrix} 1 \\ 3 \\ 5 \end{bmatrix} + \begin{bmatrix} -2 \\ 3 \\ 0 \end{bmatrix} = \begin{bmatrix} -1 \\ 6 \\ 5 \end{bmatrix}.$$

5.61   Find $A - A$.

$$A - A = A + (-A) = \begin{bmatrix} 2 & -1 \\ 3 & 0 \end{bmatrix} + \begin{bmatrix} -2 & 1 \\ -3 & 0 \end{bmatrix} = \begin{bmatrix} 0 & 0 \\ 0 & 0 \end{bmatrix} = 0.$$

5.62   Find $5B$.

$$5B = 5 \begin{bmatrix} -3 & 1 \\ 2 & -3 \end{bmatrix} = \begin{bmatrix} 5(-3) & 5(1) \\ 5(2) & 5(-3) \end{bmatrix} = \begin{bmatrix} -15 & 5 \\ 10 & -15 \end{bmatrix}.$$

5.63   Find $-2E$.

$$-2E = -2[-4 \quad 1 \quad 0 \quad -2] = [8 \quad -2 \quad 0 \quad 4].$$

5.64   Find $2A + B$

$$2A + B = \begin{bmatrix} 4 & -2 \\ 6 & 0 \end{bmatrix} + \begin{bmatrix} -3 & 1 \\ 2 & -3 \end{bmatrix} = \begin{bmatrix} 1 & -1 \\ 8 & -3 \end{bmatrix}.$$

5.65   Find $B + 2A$.

$$B + 2A = \begin{bmatrix} -3 & 1 \\ 2 & -3 \end{bmatrix} + \begin{bmatrix} 4 & -2 \\ 6 & 0 \end{bmatrix} = \begin{bmatrix} 1 & -1 \\ 8 & -3 \end{bmatrix}. \quad \text{See Prob. 5.64.}$$

5.66   Find $3D - 4C$

$$3D - 4C = 3D + (-4C) = \begin{bmatrix} 3 \\ 9 \\ 15 \end{bmatrix} + \begin{bmatrix} -8 \\ 12 \\ 0 \end{bmatrix} = \begin{bmatrix} -5 \\ 21 \\ 15 \end{bmatrix}.$$

For Probs. 5.67 to 5.74, let $A = \begin{bmatrix} 2 & 3 & 1 \\ 0 & -4 & 5 \end{bmatrix}$ and $B = \begin{bmatrix} -5 & 2 & 4 \\ 3 & 0 & -1 \end{bmatrix}$.

5.67   Find $3B$

$$3B = 3 \begin{bmatrix} -5 & 2 & 4 \\ 3 & 0 & -1 \end{bmatrix} = \begin{bmatrix} -15 & 6 & 12 \\ 9 & 0 & -3 \end{bmatrix}.$$

5.68   Find $0B$.

$$0B = 0 \begin{bmatrix} -5 & 2 & 4 \\ 3 & 0 & -1 \end{bmatrix} = \begin{bmatrix} 0 & 0 & 0 \\ 0 & 0 & 0 \end{bmatrix}.$$

5.69   Find B^T.

B^T = the transpose of B = that matrix which has rows and columns of B interchanged.

$B^T = \begin{bmatrix} -5 & 3 \\ 2 & 0 \\ 4 & -1 \end{bmatrix}.$ Row 1 of B is column 1 of B^T.

5.70   Find $2B^T$.

$$2B^T = 2 \begin{bmatrix} -5 & 3 \\ 2 & 0 \\ 4 & -1 \end{bmatrix} \quad \text{(see Prob. 5.69)} = \begin{bmatrix} -10 & 6 \\ 4 & 0 \\ 8 & -2 \end{bmatrix}.$$

5.71   Find A^T.

See Prob. 5.69. $A^T = \begin{bmatrix} 2 & 0 \\ 3 & -4 \\ 1 & 5 \end{bmatrix}.$

5.72 Find $\mathbf{A}^T - 2\mathbf{B}^T$.

$$\mathbf{A}^T - 2\mathbf{B}^T = \mathbf{A}^T + (-2)\mathbf{B}^T = \begin{bmatrix} 2 & 0 \\ 3 & -4 \\ 1 & 5 \end{bmatrix} + \begin{bmatrix} 10 & -6 \\ -4 & 0 \\ -8 & 2 \end{bmatrix} = \begin{bmatrix} 12 & -6 \\ -1 & -4 \\ -7 & 7 \end{bmatrix}.$$

5.73 Find $\mathbf{A} + \mathbf{B}^T$.

$\mathbf{A}$ is 2×1, and $\mathbf{B}^T$ is 1×2. Thus, $\mathbf{A} + \mathbf{B}^T$ is not defined.

5.74 Find $\mathbf{A}^T + \mathbf{B}^T$.

$$\mathbf{A}^T + \mathbf{B}^T = \begin{bmatrix} 2 & 0 \\ 3 & -4 \\ 1 & 5 \end{bmatrix} + \begin{bmatrix} -5 & 3 \\ 2 & 0 \\ 4 & -1 \end{bmatrix} = \begin{bmatrix} -3 & 3 \\ 5 & -4 \\ 5 & 4 \end{bmatrix}.$$

For Probs. 5.75 to 5.78, perform the operations indicated.

5.75 $[1 \quad 0 \quad 5] + 2[0 \quad 1 \quad 5]$

$[1 \quad 0 \quad 5] + 2[0 \quad 1 \quad 5] = [1 \quad 0 \quad 5] + [0 \quad 2 \quad 10] = [1 \quad 2 \quad 15].$

5.76 $2[1 \quad -1] - 5[0 \quad -1]$

$2[1 \quad -1] - 5[0 \quad -1] = [2 \quad -2] + (-5)[0 \quad -1] = [2 \quad -2] + [0 \quad 5] = [2 \quad 3].$

5.77 $-\begin{bmatrix} 1 \\ 4 \end{bmatrix} + 2[0 \quad 1]^T$

$[0 \quad 1]^T = [i]$. Thus, $-\begin{bmatrix} 1 \\ 4 \end{bmatrix} + 2[0 \quad 1]^T = \begin{bmatrix} -1 \\ -4 \end{bmatrix} + \begin{bmatrix} 0 \\ 2 \end{bmatrix} = \begin{bmatrix} -1 \\ -2 \end{bmatrix}.$

5.78 $-1\begin{bmatrix} 4 & -1 \\ -1 & -2 \end{bmatrix} - (-3)\begin{bmatrix} 0 & 1 \\ 1 & 1 \end{bmatrix}$

$\begin{bmatrix} -4 & 1 \\ 1 & 2 \end{bmatrix} + 3\begin{bmatrix} 0 & 1 \\ 1 & 1 \end{bmatrix} = \begin{bmatrix} -4 & 1 \\ 1 & 2 \end{bmatrix} + \begin{bmatrix} 0 & 3 \\ 3 & 3 \end{bmatrix} = \begin{bmatrix} -4 & 4 \\ 4 & 5 \end{bmatrix}.$

For Probs. 5.79 to 5.81, solve for x and y.

5.79 $[x + 3 \quad 5 + y] = 2\begin{bmatrix} 1 \\ 0 \end{bmatrix}^T$

$2\begin{bmatrix} 1 \\ 0 \end{bmatrix}^T = 2[1 \quad 0] = [2 \quad 0]$. Thus, $x + 3 = 2$, $x = -1$; and $5 + y = 0$, $y = -5$.

5.80 $\begin{bmatrix} 1 & 0 \\ 0 & 1 \end{bmatrix} = -2\begin{bmatrix} x - 3 & 0 \\ 0 & y + 2 \end{bmatrix}$

Then $1 = -2x + 5$, $2x = 5$, or $x = \frac{5}{2}$; and $1 = -2y - 4$, $2y = -5$, or $y = -\frac{5}{2}$.

5.81 $\begin{bmatrix} a & b \\ x & y \end{bmatrix} = -2\begin{bmatrix} e & f \\ 11 & 12 \end{bmatrix}$

Then $-x = -2(11)$, or $x = 22$; and $-y = -2(12)$, or $y = 24$.

For Probs. 5.82 to 5.86, let $\mathbf{A}i = \begin{bmatrix} a_i & b_i & c_i \\ d_i & e_i & f_i \end{bmatrix}$ where $i = 1, 2, 3$ be any three matrices, and $h, k \in \mathcal{R}$.

5.82 Prove $\mathbf{A}_1 + \mathbf{A}_2 = \mathbf{A}_2 + \mathbf{A}_1$

$$\mathbf{A}_1 + \mathbf{A}_2 = \begin{bmatrix} a_1 & b_1 & c_1 \\ d_1 & e_1 & f_1 \end{bmatrix} + \begin{bmatrix} a_2 & b_2 & c_2 \\ d_2 & e_2 & f_2 \end{bmatrix} = \begin{bmatrix} a_1 + a_2 & b_1 + b_2 & c_1 + c_2 \\ d_1 + d_2 & e_1 + e_2 & f_1 + f_2 \end{bmatrix}$$

$$= \begin{bmatrix} a_2 + a_1 & b_2 + b_1 & c_2 + c_1 \\ d_2 + d_1 & e_2 + e_1 & f_2 + f_1 \end{bmatrix} = \mathbf{A}_2 + \mathbf{A}_1$$

(Note that $a_1 + a_2 = a_2 + a_1$ since $a_i \in \mathbb{R}$.)

5.83 $h(\mathbf{A}_1 + \mathbf{A}_3) = h\mathbf{A}_1 + h\mathbf{A}_3$

▮ $h(\mathbf{A}_1 + \mathbf{A}_2) = h\begin{bmatrix} a_1+a_3 & b_1+b_3 & c_1+c_3 \\ d_1+d_3 & e_1+e_3 & f_1+f_3 \end{bmatrix}$

$= \begin{bmatrix} h(a_1+a_3) & h(b_1+b_3) & h(c_1+c_3) \\ h(d_1+d_3) & h(e_1+e_3) & h(f_1+f_3) \end{bmatrix} = \begin{bmatrix} ha_1+ha_3 & hb_1+hb_3 & hc_1+hc_3 \\ hd_1+hd_3 & he_1+he_3 & hf_1+hf_3 \end{bmatrix}$

$= \begin{bmatrix} ha_1 & hb_1 & hc_1 \\ hd_1 & he_1 & hf_1 \end{bmatrix} + \begin{bmatrix} ha_3 & hb_3 & hc_3 \\ hd_3 & he_3 & hf_3 \end{bmatrix} = h\mathbf{A}_1 + h\mathbf{A}_3$

5.84 $(h+k)\mathbf{A}_1 = h\mathbf{A}_1 + k\mathbf{A}_1$

▮ $(h+k)\mathbf{A}_1 = (h+k)\begin{bmatrix} a_1 & b_1 & c_1 \\ d_1 & e_1 & f_1 \end{bmatrix} = \begin{bmatrix} (h+k)a_1 & (h+k)b_1 & (h+k)c_1 \\ (h+k)d_1 & (h+k)e_1 & (h+k)f_1 \end{bmatrix}$

$= \begin{bmatrix} ha_1+ka_1 & hb_1+kb_1 & hc_1+kc_1 \\ hd_1+kd_1 & he_1+he_1 & hf_1+hf_1 \end{bmatrix} = h\mathbf{A}_1 + k\mathbf{A}_1$

5.85 $\mathbf{A}_2 + (-\mathbf{A}_2) = 0$

▮ $\mathbf{A}_2 + (-\mathbf{A}_2) = \begin{bmatrix} a_2 & b_2 & c_2 \\ d_2 & e_2 & f_2 \end{bmatrix} + \begin{bmatrix} -a_2 & -b_2 & -c_2 \\ -d_2 & -e_2 & -f_2 \end{bmatrix} = \begin{bmatrix} 0 & 0 & 0 \\ 0 & 0 & 0 \end{bmatrix} = 0.$

5.86 $\mathbf{A}_1 + 0 = \mathbf{A}_1$

▮ $\mathbf{A}_1 + 0 = \begin{bmatrix} a_1 & b_1 & c_1 \\ d_1 & e_1 & f_1 \end{bmatrix} + \begin{bmatrix} 0 & 0 & 0 \\ 0 & 0 & 0 \end{bmatrix} = \begin{bmatrix} a_1+0 & b_1+0 & c_1+0 \\ d_1+0 & e_1+0 & f_1+0 \end{bmatrix} = \begin{bmatrix} a_1 & b_1 & c_1 \\ d_1 & e_1 & f_1 \end{bmatrix} = \mathbf{A}_1.$

For Probs. 5.87 to 5.89, find the indicated matrix, where $\mathbf{A} = \begin{bmatrix} 1 & 3 \\ 2 & 4 \end{bmatrix}$ and $\mathbf{B} = \begin{bmatrix} 1 & 0 \\ 0 & 1 \end{bmatrix}$.

5.87 Find $\mathbf{B}^T$.

▮ $\mathbf{B} = \begin{bmatrix} 1 & 0 \\ 0 & 1 \end{bmatrix}$ and $\mathbf{B}^T = \begin{bmatrix} 1 & 0 \\ 0 & 1 \end{bmatrix} = \mathbf{B}$. The first row of $\mathbf{B}$ is the first column of $\mathbf{B}^T$.

5.88 Find $(\mathbf{A}^T)^T$.

▮ $\mathbf{A}^T = \begin{bmatrix} 1 & 2 \\ 3 & 4 \end{bmatrix}$ and $(\mathbf{A}^T)^T = \begin{bmatrix} 1 & 3 \\ 2 & 4 \end{bmatrix} = \mathbf{A}$. The first row of $\mathbf{A}^T$ is the first column of $(\mathbf{A}^T)^T$.

5.89 Find $[(\mathbf{A}^T)^T]^T$.

▮ $(\mathbf{A}^T)^T = \begin{bmatrix} 1 & 3 \\ 2 & 4 \end{bmatrix}$ (see Prob. 5.88). Then $[(\mathbf{A}^T)^T]^T = \begin{bmatrix} 1 & 2 \\ 3 & 4 \end{bmatrix} = \mathbf{A}^T$. The first row of $(\mathbf{A}^T)^T$ is the first column of $[(\mathbf{A}^T)^T]^T$.

For Probs. 5.90 to 5.92, find the matrix $\mathbf{X}$.

5.90 $\mathbf{X} - \begin{bmatrix} 2 & 3 \\ -1 & 4 \end{bmatrix} = \begin{bmatrix} 5 & 8 \\ 2 & 1 \end{bmatrix}$

▮ Let $\mathbf{X} = \begin{bmatrix} a & b \\ c & d \end{bmatrix}$. Then $a-2=5$, $a=7$; $b-3=8$, $b=11$; $c+1=2$, $c=1$; $d-4=1$, $d=5$. Thus, $\mathbf{X} = \begin{bmatrix} 7 & 11 \\ 1 & 5 \end{bmatrix}$.

5.91 $\mathbf{X} + \begin{bmatrix} 3 & 2 \\ 0 & 1 \end{bmatrix} = 2\begin{bmatrix} -1 & 3 \\ 2 & 4 \end{bmatrix}$

▮ $\begin{bmatrix} a & b \\ c & d \end{bmatrix} + \begin{bmatrix} 3 & 2 \\ 0 & 1 \end{bmatrix} = \begin{bmatrix} -2 & 6 \\ 4 & 8 \end{bmatrix}$. $a+3=-2$, $a=-5$; $b+2=6$, $b=4$; $c+0=4$, $c=4$; $d+1=8$, $d=7$. Thus, $\mathbf{X} = \begin{bmatrix} -5 & 4 \\ 4 & 7 \end{bmatrix}$.

5.92 $2x + \begin{bmatrix} 2 & 4 \\ 0 & 6 \end{bmatrix} = 3\begin{bmatrix} -2 & 0 \\ 2 & 4 \end{bmatrix}$

▮ $\begin{bmatrix} 2a & 2b \\ 2c & 2d \end{bmatrix} + \begin{bmatrix} 2 & 4 \\ 0 & 6 \end{bmatrix} = \begin{bmatrix} -6 & 0 \\ 6 & 12 \end{bmatrix}$. $2a + 2 = -6$, $2a = -8$, $a = -4$; $2b + 4 = 0$, $2b = -4$, $b = -2$; $2c = 6$, $c = 3$; $2d + 6 = 12$, $2d = 6$, $d = 3$. Thus, $\mathbf{X} = \begin{bmatrix} -4 & -2 \\ 3 & 3 \end{bmatrix}$.

For Probs. 5.93 to 5.102, find the product indicated.

5.93 $\begin{bmatrix} 1 & 3 \\ 2 & 1 \end{bmatrix}\begin{bmatrix} 1 & 2 \\ -1 & 2 \end{bmatrix}$

▮ This product is defined since the number of columns of $\begin{bmatrix} 1 & 3 \\ 2 & 1 \end{bmatrix}$ equals the number of rows of $\begin{bmatrix} 1 & 2 \\ -1 & 2 \end{bmatrix}$. We then find the product as follows:

$$\begin{bmatrix} 1 & 3 \\ 2 & 1 \end{bmatrix}\begin{bmatrix} 1 & 2 \\ -1 & 2 \end{bmatrix} = \begin{bmatrix} 1\cdot1 + 3(-1) & 1\cdot2 + 3\cdot2 \\ 2\cdot1 + 1(-1) & 2\cdot2 + 2\cdot2 \end{bmatrix}$$

In general, $(\mathbf{AB})_{m\times n}$ of the matrices $\mathbf{A}_{m\times s}$ and $\mathbf{B}_{s\times n}$ is the matrix whose ijth element is $a_{i1}b_{1j} + a_{i2}b_{2j} + \cdots + a_{is}b_{sj}$, where i ranges from 1 to m and j ranges from 1 to n. In this case, then, the product is $\begin{bmatrix} 1-3 & 2+6 \\ 2-1 & 4+4 \end{bmatrix} = \begin{bmatrix} -2 & 8 \\ 1 & 8 \end{bmatrix}$.

5.94 $\begin{bmatrix} 2 & 3 \\ 1 & 4 \end{bmatrix}\begin{bmatrix} 0 & 1 \\ 6 & 2 \end{bmatrix}$

▮ The product is $\begin{bmatrix} 2\cdot0 + 3\cdot6 & 2\cdot1 + 3\cdot2 \\ 1\cdot0 + 4\cdot6 & 1\cdot1 + 4\cdot2 \end{bmatrix} = \begin{bmatrix} 18 & 8 \\ 24 & 9 \end{bmatrix}$.

5.95 $\begin{bmatrix} 0 & 1 \\ 6 & 2 \end{bmatrix}\begin{bmatrix} 2 & 3 \\ 1 & 4 \end{bmatrix}$

▮ $\begin{bmatrix} 0\cdot2 + 1\cdot1 & 0\cdot3 + 1\cdot4 \\ 6\cdot2 + 2\cdot1 & 6\cdot3 + 2\cdot4 \end{bmatrix} = \begin{bmatrix} 1 & 4 \\ 14 & 24 \end{bmatrix}$. Look at the result in Prob. 5.94! $\mathbf{AB} \neq \mathbf{BA}$.

5.96 $\begin{bmatrix} 1 \\ 1 \\ 0 \end{bmatrix}[2 \quad 3 \quad 1]$

▮ Let $\mathbf{A} = \begin{bmatrix} 1 \\ 1 \\ 0 \end{bmatrix}$ and $\mathbf{B} = [2 \quad 3 \quad 1]$. Is $\mathbf{AB}$ defined? Yes; the number of columns of $\mathbf{A}$ = the number of rows of $\mathbf{B} = 1$. Then $\begin{bmatrix} 1 \\ 1 \\ 0 \end{bmatrix}[2 \quad 3 \quad 1] = \begin{bmatrix} 1\cdot2 & 1\cdot3 & 1\cdot1 \\ 1\cdot2 & 1\cdot3 & 1\cdot1 \\ 0\cdot2 & 0\cdot3 & 0\cdot1 \end{bmatrix} = \begin{bmatrix} 2 & 3 & 1 \\ 2 & 3 & 1 \\ 0 & 0 & 0 \end{bmatrix}$.

5.97 $[1 \quad 5]\begin{bmatrix} 5 & 6 \\ 0 & 1 \end{bmatrix}$

▮ $[1 \quad 5]$ has two columns; $\begin{bmatrix} 5 & 6 \\ 0 & 1 \end{bmatrix}$ has two rows. The product is $[1\cdot5 + 5\cdot0 \quad 1\cdot6 + 5\cdot1] = [5 \quad 11]$. Be very careful. $1\cdot6 + 5\cdot1$ is the first row, second column position. Since $[1 \quad 5]$ has only one row, the product has only one row.

5.98 $[2 \quad 1 \quad 3]\begin{bmatrix} 3 & 1 & 0 \\ 0 & 0 & 0 \\ 1 & 0 & 1 \end{bmatrix}$

▮ The product is $[2\cdot3 + 1\cdot0 + 3\cdot1 \quad 2\cdot1 + 1\cdot0 + 3\cdot0 \quad 2\cdot0 + 1\cdot0 + 3\cdot1] = [9 \quad 2 \quad 3]$.

5.99 $\begin{bmatrix} 2 \\ 1 \\ 3 \end{bmatrix}\begin{bmatrix} 1 & 0 & 0 \\ 0 & 1 & 0 \\ 0 & 0 & 1 \end{bmatrix}$

▮ The product is not defined. Do you see why? The number of columns of $\begin{bmatrix} 2 \\ 1 \\ 3 \end{bmatrix} \neq$ the number of rows of $\begin{bmatrix} 1 & 0 & 0 \\ 0 & 1 & 0 \\ 0 & 0 & 1 \end{bmatrix}$.

5.100 $\begin{bmatrix} 1 & 2 & 3 \\ 1 & 0 & 1 \end{bmatrix}\begin{bmatrix} 3 & 1 \\ 4 & 2 \\ 6 & 0 \end{bmatrix}$

▮ The product is $\begin{bmatrix} 1\cdot3+2\cdot4+3\cdot6 & 1\cdot1+2\cdot2+3\cdot0 \\ 1\cdot3+0\cdot4+1\cdot6 & 1\cdot1+0\cdot2+1\cdot0 \end{bmatrix}=\begin{bmatrix} 29 & 5 \\ 9 & 1 \end{bmatrix}$.

5.101 $\begin{bmatrix} -6 & 4 \\ 193 & 8 \end{bmatrix}\begin{bmatrix} 1 & 0 \\ 0 & 1 \end{bmatrix}$

▮ The product is $\begin{bmatrix} -6\cdot1+4\cdot0 & -6\cdot0+4\cdot1 \\ 193\cdot1+8\cdot0 & 193\cdot0+8\cdot1 \end{bmatrix}=\begin{bmatrix} -6 & 4 \\ 193 & 8 \end{bmatrix}$. Notice that $\begin{bmatrix} -6 & 4 \\ 193 & 8 \end{bmatrix}$ was unchanged by multiplication by $\begin{bmatrix} 1 & 0 \\ 0 & 1 \end{bmatrix}$. See Prob. 5.102.

5.102 $\begin{bmatrix} 5 & -2 \\ 1 & 6 \end{bmatrix}\begin{bmatrix} 1 & 0 \\ 0 & 1 \end{bmatrix}$

▮ The product is $\begin{bmatrix} 5\cdot1+(-2)\cdot0 & 5\cdot0+(-2)\cdot1 \\ 1\cdot1+6\cdot0 & 1\cdot0+6\cdot1 \end{bmatrix}=\begin{bmatrix} 5 & -2 \\ 1 & 6 \end{bmatrix}$. Do you see that $\begin{bmatrix} 1 & 0 \\ 0 & 1 \end{bmatrix}$ is a good candidate for multiplicative identity?

For Probs. 5.103 to 5.106, find the dot product of **A** and **B**.

5.103 $[1 \quad 0]\cdot\begin{bmatrix} 0 \\ 5 \end{bmatrix}$

▮ Here, $\mathbf{A}\cdot\mathbf{B}=1\cdot0+0.5=0$.

5.104 $[1 \quad -3]\cdot\begin{bmatrix} 1 \\ -6 \end{bmatrix}$

▮ $\mathbf{A}\cdot\mathbf{B}=1\cdot1+(-3)(-6)=1+18=19$.

5.105 $[1 \quad 6]\cdot\begin{bmatrix} 0 \\ 1 \end{bmatrix}$

▮ $\mathbf{A}\cdot\mathbf{B}=1\cdot0+6\cdot1=6$.

5.106 $[1 \quad 2 \quad 3 \ldots n]\cdot\begin{bmatrix} 1 \\ 2 \\ . \\ . \\ . \\ n \end{bmatrix}$

▮ $\mathbf{A}\cdot\mathbf{B}=1\cdot1+2\cdot2+\cdots+n^2=n^2+(n-1)^2+\cdots+2^2+1^2$.

For Probs. 5.107 to 5.110, verify the given statements, where

$$\mathbf{A}=\begin{bmatrix} 1 & 2 \\ 0 & 1 \end{bmatrix} \qquad \mathbf{B}=\begin{bmatrix} 1 & 1 \\ 2 & 3 \end{bmatrix} \qquad \mathbf{C}=\begin{bmatrix} -3 & 1 \\ -1 & 2 \end{bmatrix}$$

5.107 $\mathbf{AB}\neq\mathbf{BA}$.

▮ $\mathbf{AB}=\begin{bmatrix} 1\cdot1+2\cdot2 & 1\cdot1+2\cdot3 \\ 0\cdot1+1\cdot2 & 0\cdot2+1\cdot3 \end{bmatrix}=\begin{bmatrix} 5 & 7 \\ 2 & 3 \end{bmatrix}$

$\mathbf{BA}=\begin{bmatrix} 1\cdot1+1\cdot0 & 1\cdot2+1\cdot1 \\ 2\cdot1+3\cdot0 & 2\cdot2+3\cdot1 \end{bmatrix}\neq\begin{bmatrix} 5 & 7 \\ 2 & 3 \end{bmatrix}$

$\mathbf{AB}\neq\mathbf{BA}$.

5.108 $(AB)C = A(BC)$.

▮ $AB = \begin{bmatrix} 5 & 7 \\ 2 & 3 \end{bmatrix}$ (see Prob. 5.107)

$(AB)C = \begin{bmatrix} 5 & 7 \\ 2 & 3 \end{bmatrix}\begin{bmatrix} -3 & 1 \\ -1 & 2 \end{bmatrix} = \begin{bmatrix} 5(-3) + 7(-1) & 5\cdot 1 + 7\cdot 2 \\ 2(-3) + 3(-1) & 2\cdot 1 + 3\cdot 2 \end{bmatrix} = \begin{bmatrix} -22 & 19 \\ -9 & 8 \end{bmatrix}$

$BC = \begin{bmatrix} 1 & 1 \\ 2 & 3 \end{bmatrix}\begin{bmatrix} -3 & 1 \\ -1 & 2 \end{bmatrix} = \begin{bmatrix} -4 & 3 \\ -9 & 8 \end{bmatrix}$

$A(BC) = \begin{bmatrix} 1 & 2 \\ 0 & 1 \end{bmatrix}\begin{bmatrix} -4 & 3 \\ -9 & 8 \end{bmatrix} = \begin{bmatrix} -22 & 19 \\ -9 & 8 \end{bmatrix} = (AB)C$

5.109 $A(B + C) = AB + AC$.

▮ $A(B + C) = \begin{bmatrix} 1 & 2 \\ 0 & 1 \end{bmatrix}\begin{bmatrix} -2 & 2 \\ 1 & 5 \end{bmatrix} = \begin{bmatrix} 0 & 12 \\ 1 & 5 \end{bmatrix}$. $AB = \begin{bmatrix} 5 & 7 \\ 2 & 3 \end{bmatrix}$. $AC = \begin{bmatrix} -5 & 5 \\ -1 & 2 \end{bmatrix}$. $AB + AC = \begin{bmatrix} 0 & 12 \\ 1 & 5 \end{bmatrix} = A(B + C)$.

5.110 $(B + C)A = BA + CA$

▮ $B + C = \begin{bmatrix} -2 & 2 \\ 1 & 5 \end{bmatrix}$. $(B + C)A = \begin{bmatrix} -2 & 2 \\ 1 & 5 \end{bmatrix}\begin{bmatrix} 1 & 2 \\ 0 & 1 \end{bmatrix} = \begin{bmatrix} -2 & -2 \\ 1 & 7 \end{bmatrix}$. $BA = \begin{bmatrix} 1 & 3 \\ 2 & 7 \end{bmatrix}$

(see Prob. 5.107). $CA = \begin{bmatrix} -3 & 1 \\ -1 & 2 \end{bmatrix}\begin{bmatrix} 1 & 2 \\ 0 & 1 \end{bmatrix} = \begin{bmatrix} -3 & -5 \\ -1 & 0 \end{bmatrix}$. $BA + CA = \begin{bmatrix} -2 & -2 \\ 1 & 7 \end{bmatrix} = (B + C)A$.

For Probs. 5.111 to 5.114, evaluate the given determinants.

5.111 $\begin{vmatrix} 1 & 0 \\ 5 & 6 \end{vmatrix}$

▮ Remember that $\begin{vmatrix} a_{11} & a_{12} \\ a_{21} & a_{22} \end{vmatrix} = a_{11}a_{22} - a_{21}a_{12}$. In this case $\begin{vmatrix} 1 & 0 \\ 5 & 6 \end{vmatrix} = 1\cdot 6 - 0\cdot 5 = 6 - 0 = 6$.

5.112 $\begin{vmatrix} 1 & 4 \\ -3 & 8 \end{vmatrix}$

▮ $\begin{vmatrix} 1 & 4 \\ -3 & 8 \end{vmatrix} = 1\cdot 8 - (4)(-3) = 8 + 12 = 20$.

5.113 $\begin{vmatrix} 1 & \frac{1}{2} \\ \frac{1}{8} & \frac{1}{4} \end{vmatrix}$

▮ $\begin{vmatrix} 1 & \frac{1}{2} \\ \frac{1}{8} & \frac{1}{4} \end{vmatrix} = 1\cdot \frac{1}{4} - \frac{1}{2}\cdot \frac{1}{8} = \frac{1}{4} - \frac{1}{16} = \frac{3}{16}$.

5.114 $\begin{vmatrix} 1 & 4 \\ 0.1 & 0.7 \end{vmatrix}$

▮ $\begin{vmatrix} 1 & 4 \\ 0.1 & 0.7 \end{vmatrix} = 1(0.7) - (4)(0.1) = 0.7 - 0.4 = 0.3$.

For Probs. 5.115 to 5.119, evaluate the given determinant using cofactors.

5.115 $\begin{vmatrix} 3 & -2 & -8 \\ -2 & 0 & -3 \\ 1 & 0 & -4 \end{vmatrix}$

▮ Use column 2 since it has two zero entries.

$\begin{vmatrix} 3 & -2 & -8 \\ -2 & 0 & -3 \\ 1 & 0 & -4 \end{vmatrix} = (-2)(-1)^{1+2}\begin{vmatrix} -2 & -3 \\ 1 & -4 \end{vmatrix} + 0 + 0 = 2\begin{vmatrix} -2 & -3 \\ 1 & -4 \end{vmatrix} = 2(11) = 22$.

5.116 $\begin{vmatrix} 1 & 4 & 1 \\ 1 & 1 & -2 \\ 2 & 1 & -1 \end{vmatrix}$

▮ Using row 1, we get $1(-1)^{1+1}\begin{vmatrix} 1 & -2 \\ 1 & -1 \end{vmatrix} + 4(-1)^{1+2}\begin{vmatrix} 1 & -2 \\ 2 & -1 \end{vmatrix} + 1(-1)^{1+3}\begin{vmatrix} 1 & 1 \\ 2 & 1 \end{vmatrix} = \begin{vmatrix} 1 & -2 \\ 1 & -1 \end{vmatrix} - 4\begin{vmatrix} 1 & -2 \\ 2 & -1 \end{vmatrix} + \begin{vmatrix} 1 & 1 \\ 2 & 1 \end{vmatrix} = -12.$

5.117 $\begin{vmatrix} 1 & 0 & 0 \\ -2 & 4 & 3 \\ 5 & -2 & 1 \end{vmatrix}$

▮ Use row 1 (it has two zeros). $\begin{vmatrix} 1 & 0 & 0 \\ -2 & 4 & 3 \\ 5 & -2 & 1 \end{vmatrix} = 1\begin{vmatrix} 4 & 3 \\ -2 & 1 \end{vmatrix} + 0 + 0 = 4 - (-6) = 10.$

5.118 $\begin{vmatrix} 0 & 1 & 5 \\ 3 & -7 & 6 \\ 0 & -2 & 3 \end{vmatrix}$

▮ Use column 1. Then $\begin{vmatrix} 0 & 1 & 5 \\ 3 & -7 & 6 \\ 0 & -2 & 3 \end{vmatrix} = 0 + 3(-1)^{2+1}\begin{vmatrix} 1 & 5 \\ -2 & 3 \end{vmatrix} + 0 = -21.$

5.119 $\begin{vmatrix} 2 & 6 & 1 & 7 \\ 0 & 3 & 0 & 0 \\ 3 & 4 & 2 & 5 \\ 0 & 9 & 0 & 2 \end{vmatrix}$

▮ Use row 2; then the determinant $= 3(-1)^{2+2}\begin{vmatrix} 2 & 1 & 7 \\ 3 & 2 & 5 \\ 0 & 0 & 2 \end{vmatrix}$.
(The others are all zero.) Now use row 3 (which has two zeros). $\begin{vmatrix} 2 & 1 & 7 \\ 3 & 2 & 5 \\ 0 & 0 & 2 \end{vmatrix} = 2(-1)^{3+3}\begin{vmatrix} 2 & 1 \\ 3 & 2 \end{vmatrix} = 2 \cdot 1 \cdot 1 = 2$; $3(-1)^{2+2} \cdot 2 = 6.$

For Probs. 5.120 to 5.123, show that each statement is true. All letters represent real numbers.

5.120 $\begin{vmatrix} a & b \\ ka & kb \end{vmatrix} = 0$

▮ $\begin{vmatrix} a & b \\ ka & kb \end{vmatrix} = a(kb) - b(ka) = abk - abk = 0.$

5.121 $\begin{vmatrix} a & b \\ c & d \end{vmatrix} = -\begin{vmatrix} b & a \\ d & c \end{vmatrix}$

▮ $\begin{vmatrix} a & b \\ c & d \end{vmatrix} = ad - bc - \begin{vmatrix} b & a \\ d & c \end{vmatrix} = -(bc - ad) = bc + ad = ad - bc = \begin{vmatrix} a & b \\ c & d \end{vmatrix}.$

5.122 $\begin{vmatrix} a & b \\ c & d \end{vmatrix} = \begin{vmatrix} a & c \\ b & d \end{vmatrix}$

▮ $\begin{vmatrix} a & b \\ c & d \end{vmatrix} = ad - bc = ad - cb = \begin{vmatrix} a & c \\ b & d \end{vmatrix}.$

5.123 $\begin{vmatrix} ka & kb \\ c & d \end{vmatrix} = k\begin{vmatrix} a & b \\ c & d \end{vmatrix}$

▮ $\begin{vmatrix} ka & kb \\ c & d \end{vmatrix} = (ka)d - (kb)c = kad - kbc.$ Then $k\begin{vmatrix} a & b \\ c & d \end{vmatrix} = k(ad - bc) = kad - kbc = \begin{vmatrix} ka & kb \\ c & d \end{vmatrix}.$

For Probs. 5.124 to 5.130, state the theorem that can be used to justify the given statement.

5.124 $\begin{vmatrix} 16 & 8 \\ 0 & -1 \end{vmatrix} = 8\begin{vmatrix} 2 & 1 \\ 0 & -1 \end{vmatrix}$

▮ If every element in a column or row of a determinant is multiplied by a constant (here, row 1 is multiplied by 8), the new determinant is the constant times the original.

5.125 $\begin{vmatrix} 1 & 0 & 6 \\ 2 & 0 & 5 \\ 3 & 0 & 8 \end{vmatrix} = 0$

▮ In a given determinant, if any column or row is all zeros, the determinant $= 0$. Here, column 2 is all zeros.

5.126 $\begin{vmatrix} -1 & 4 & 6 & 3 & 8 \\ 0 & 0 & 0 & 0 & 0 \\ 4 & 1 & 6 & 18 & 9 \\ 0 & 1 & 4 & 0 & 1 \\ 1 & 0 & 1 & 1 & 0 \end{vmatrix} = 0$

▮ See Prob. 5.125. All entries here in row 2 are zero. Notice how much work is saved using this theorem.

5.127 $\begin{vmatrix} 4 & 3 \\ 1 & 2 \end{vmatrix} = \begin{vmatrix} 4-4 & 3-8 \\ 1 & 2 \end{vmatrix}$

▮ If a constant multiple of a row or column is added to another row or column, the determinant's value is unchanged. Here, we have multiplied row 2 by -4 and added it to row 1.

5.128 $\begin{vmatrix} 3 & 2 \\ 5 & 1 \end{vmatrix} = \begin{vmatrix} 3+4 & 2 \\ 5+2 & 1 \end{vmatrix}$

▮ See Prob. 5.127. We multiply column 2 by 2 and add it to column 1. The determinant's value does not change.

5.129 $\begin{vmatrix} 1 & 4 & 6 & 1 \\ 2 & 3 & 8 & 2 \\ 1 & 4 & 6 & 1 \\ 2 & 1 & 5 & 2 \end{vmatrix} = 0$

▮ If two rows or columns in a given determinant are identical, then the determinant's value $= 0$. Here, row 1 = row 3.

5.130 $\begin{vmatrix} 2 & 1 & 3 \\ 4 & 6 & 9 \\ 2 & 8 & 1 \end{vmatrix} = \begin{vmatrix} 4 & 6 & 9 \\ 2 & 1 & 3 \\ 2 & 8 & 1 \end{vmatrix}$

▮ If two rows or columns are interchanged in a given determinant, the resulting determinant is the negative of the original. Here, rows 1 and 2 were interchanged.

For Probs. 5.131 and 5.132, prove the given statement.

5.131 $\begin{vmatrix} a & b & c \\ c & a & e \\ a & b & c \end{vmatrix} = 0$

▮ In the given determinant, row 1 = row 3. Thus, the value of the determinant $= 0$.

5.132 $\begin{vmatrix} a & b & c \\ e & f & g \\ q & r & t \end{vmatrix} = -\begin{vmatrix} e & f & g \\ a & b & c \\ q & r & t \end{vmatrix} = \begin{vmatrix} e & g & f \\ a & c & b \\ q & t & r \end{vmatrix}$

▮ Here, A is the first determinant, B is the second, C is the third. $A = -B = C$ since A and B are the same except rows 1 and 2 are interchanged; thus, $A = -B$. B and C have columns 2 and 3 interchanged; thus, $B = -C$ or $-B = C$; $A = -B = C$.

5.133 Show that $(2,5)$ and $(-3,4)$ satisfy the equation $\begin{vmatrix} x & y & 1 \\ 2 & 5 & 1 \\ -3 & 4 & 1 \end{vmatrix} = 0.$

▮ If $x=2$, $y=5$, then we have $\begin{vmatrix} x & y & 1 \\ 2 & 5 & 1 \\ -3 & 4 & 1 \end{vmatrix} = \begin{vmatrix} 2 & 5 & 1 \\ 2 & 5 & 1 \\ -3 & 4 & 1 \end{vmatrix}$ which is 0 since row 1 = row 2.
Similarly, if $x=-3$ and $y=4$, then row 1 = row 3 and the determinant is zero.

5.134 Show that $\begin{vmatrix} x & y & 1 \\ 2 & 3 & 1 \\ -1 & 2 & 1 \end{vmatrix} = 0$ is a line passing through $(2,3)$, $(-1,2)$.

▮ Note that $(2,3)$ and $(-1,2)$ satisfy the equation, since we would have equal rows. Next, expanding, we get $-2\begin{vmatrix} y & 1 \\ 2 & 1 \end{vmatrix} + (-1)\begin{vmatrix} y & 1 \\ 3 & 1 \end{vmatrix} + x\begin{vmatrix} 3 & 1 \\ 2 & 1 \end{vmatrix} = 0.$ Then $-2(y-2)-1(y-3)+x(3-2)=0$, $-2y+4-y+3+x=0$, and $-3y+x=-7$. This is a line.

For Probs. 5.135 to 5.137, solve for x.

5.135 $\begin{vmatrix} 2 & 1 \\ -6 & x \end{vmatrix} = 0$

▮ $2x-(-6)=0$, $2x=-6$, or $x=-3$.

5.136 $\begin{vmatrix} 1 & x \\ 2-x & -3 \end{vmatrix} = 0$

▮ $-3-x(2-x)=0$, $3-2x+x^2=0$, $x^2-2x-3=0$, $(x-3)(x+1)=0$, or $x=3$, $x=-1$.

5.137 $\begin{vmatrix} x & 1 & 3 \\ 1 & x & 2 \\ 1 & 1 & 2 \end{vmatrix} = 0$

▮ Using cofactors, we get $x\begin{vmatrix} x & 2 \\ 1 & 2 \end{vmatrix} - 1\begin{vmatrix} 1 & 3 \\ 1 & 2 \end{vmatrix} + 1\begin{vmatrix} 1 & 3 \\ x & 2 \end{vmatrix} = 0.$ Then $x(2x-2)-(2-3)+(2-3x)=0$, $2x^2-2x+1+2-3x=0$, $2x^2-5x+3=0$, $(2x-3)(x-1)=0$, or $x=\frac{3}{2}$, $x=1$.

For Probs. 5.138 and 5.139, show that the two matrices are each other's inverse.

5.138 $\begin{bmatrix} 3 & -4 \\ -2 & 3 \end{bmatrix} \begin{bmatrix} 3 & 4 \\ 2 & 3 \end{bmatrix}$

▮ The product is $\begin{bmatrix} 3\cdot3+(-4)(2) & 3\cdot4+(-4)(3) \\ (-2)(3)+3\cdot2 & (-2)(4)+3\cdot3 \end{bmatrix} = \begin{bmatrix} 1 & 0 \\ 0 & 1 \end{bmatrix} = I$, the identity matrix. Thus, they are inverses of each other. If $AB=I$, then $A=B^{-1}$.

5.139 $\begin{bmatrix} 1 & -1 & 1 \\ 0 & 2 & -1 \\ 2 & 3 & 0 \end{bmatrix} \begin{bmatrix} 3 & 3 & -1 \\ -2 & -2 & 1 \\ -4 & -5 & 2 \end{bmatrix}$

▮ The product here is
$$\begin{bmatrix} 1\cdot3+(-1)(-2)+(1)(-4) & \cdots \\ 0\cdot3+2(-2)+(-1)(-4) & \cdots \\ \cdots & \cdots \end{bmatrix} = \begin{bmatrix} 1 & 0 & 0 \\ 0 & 1 & 0 \\ 0 & 0 & 1 \end{bmatrix} = I$$
$A=B^{-1}$.

For Probs. 5.140 to 5.143, a matrix M is given. Find M^{-1}.

5.140 $\begin{bmatrix} 1 & 2 \\ 1 & 3 \end{bmatrix}$

▮ We write the augmented matrix $\begin{bmatrix} 1 & 2 & 1 & 0 \\ 1 & 3 & 0 & 1 \end{bmatrix}$ and transform it into a matrix of the form: $\begin{bmatrix} 1 & 0 \\ 0 & 1 \end{bmatrix} S$. When we do this, $\mathbf{S} = \mathbf{A}^{-1}$ where $\mathbf{A} = \begin{bmatrix} 1 & 2 \\ 1 & 3 \end{bmatrix}$. We convert one matrix into the other using elementary row operations: (1) interchange two rows, (2) multiply a row by $k \neq 0$, (3) add $k\mathbf{R}_i$ to $\mathbf{R}_j$, where $\mathbf{R}$ is a row in the matrix. In this case we have the following: Add $-\mathbf{R}_1$ to $\mathbf{R}_2$ and replace $\mathbf{R}_2$ by $\mathbf{R}_2 - \mathbf{R}_1$. Then $\begin{bmatrix} 1 & 2 & 1 & 0 \\ 1 & 3 & 0 & 1 \end{bmatrix} \sim \begin{bmatrix} 1 & 2 & 1 & 0 \\ 0 & 1 & -1 & 1 \end{bmatrix}$. Replace $\mathbf{R}_1$ by $\mathbf{R}_1 + (-2)\mathbf{R}_2$. Then $\begin{bmatrix} 1 & 2 & 1 & 0 \\ 0 & 1 & -1 & 1 \end{bmatrix} \sim \begin{bmatrix} 1 & 0 & 3 & -2 \\ 0 & 1 & -1 & 1 \end{bmatrix}$. Thus, $\mathbf{M}^{-1} = \begin{bmatrix} 3 & -2 \\ -1 & 1 \end{bmatrix}$. Check it! $\begin{bmatrix} 3 & -2 \\ -1 & 1 \end{bmatrix}\begin{bmatrix} 1 & 2 \\ 1 & 3 \end{bmatrix} = \begin{bmatrix} 1 & 0 \\ 0 & 1 \end{bmatrix}$.

5.141 $\begin{bmatrix} 1 & 3 \\ 2 & 7 \end{bmatrix}$

▮ Replace $\mathbf{R}_2$ by $\mathbf{R}_2 + (-2)\mathbf{R}_1$. Then $\begin{bmatrix} 1 & 3 & 1 & 0 \\ 2 & 7 & 0 & 1 \end{bmatrix} \sim \begin{bmatrix} 1 & 3 & 1 & 0 \\ 0 & 1 & -2 & 1 \end{bmatrix}$. Then if we replace $\mathbf{R}_1$ by $\mathbf{R}_1 + (-3)\mathbf{R}_2$, we get $\begin{bmatrix} 1 & 0 & 7 & -3 \\ 0 & 1 & -2 & 1 \end{bmatrix}$. Thus, $\mathbf{M}^{-1} = \begin{bmatrix} 7 & -3 \\ -2 & 1 \end{bmatrix}$.

5.142 $\begin{bmatrix} 1 & -3 & 0 \\ 0 & 3 & 1 \\ 2 & -1 & 2 \end{bmatrix}$

▮ $\begin{bmatrix} 1 & -3 & 0 & 1 & 0 & 0 \\ 0 & 3 & 1 & 0 & 1 & 0 \\ 2 & -1 & 2 & 0 & 0 & 1 \end{bmatrix} \sim \begin{bmatrix} 1 & -3 & 0 & 1 & 0 & 0 \\ 0 & 3 & 1 & 0 & 1 & 0 \\ 0 & 5 & 2 & -2 & 0 & 1 \end{bmatrix}$ (We replaced $\mathbf{R}_3$ by $\mathbf{R}_3 - 2\mathbf{R}_2$.)

$\sim \begin{bmatrix} 1 & -3 & 0 & 1 & 0 & 0 \\ 0 & 1 & \frac{1}{3} & 0 & \frac{1}{3} & 0 \\ 0 & 5 & 2 & -2 & 0 & 1 \end{bmatrix}$ (We replaced $\mathbf{R}_2$ by $\frac{1}{3}\mathbf{R}_2$.)

$\sim \begin{bmatrix} 1 & 0 & 1 & 1 & 1 & 0 \\ 0 & 1 & \frac{1}{3} & 0 & \frac{1}{3} & 0 \\ 0 & 0 & \frac{1}{3} & -2 & -\frac{5}{3} & 1 \end{bmatrix}$ (We replaced $\mathbf{R}_1$ by $\mathbf{R}_1 + 3\mathbf{R}_2$, and $\mathbf{R}_3$ by $\mathbf{R}_3 - 5\mathbf{R}_2$.)

$\sim \begin{bmatrix} 1 & 0 & 1 & 1 & 1 & 0 \\ 0 & 1 & \frac{1}{3} & 0 & \frac{1}{3} & 0 \\ 0 & 0 & 1 & -6 & -5 & 3 \end{bmatrix}$ (We replaced $\mathbf{R}_3$ by $3\mathbf{R}_3$.)

$\sim \begin{bmatrix} 1 & 0 & 0 & 7 & 6 & -3 \\ 0 & 1 & 0 & 2 & 2 & -1 \\ 0 & 0 & 1 & -6 & -5 & 3 \end{bmatrix}$ (We replaced $\mathbf{R}_1$ by $\mathbf{R}_1 - \mathbf{R}_2$, $\mathbf{R}_2$ by $\mathbf{R}_2 - \frac{1}{3}$.)

Then $\mathbf{M}^{-1} = \begin{bmatrix} 7 & 6 & -3 \\ 2 & 2 & -1 \\ -6 & -5 & 3 \end{bmatrix}$.

5.143 Show that if $\mathbf{M} = \begin{bmatrix} 3 & 9 \\ 2 & 6 \end{bmatrix}$, then $\mathbf{M}^{-1}$ does not exist.

▮ Replace $\mathbf{R}_3$ by $\mathbf{R}_3$ by $\mathbf{R}_3 - \frac{2}{3}\mathbf{R}_2$. Then $\begin{bmatrix} 3 & 9 & 1 & 0 \\ 2 & 6 & 0 & 1 \end{bmatrix} \sim \begin{bmatrix} 3 & 9 & 1 & 0 \\ 0 & 0 & -\frac{2}{3} & 1 \end{bmatrix}$. But then $\mathbf{M}^{-1}$ does not exist since we get a row of all zeros on the left-hand side.

5.3 MATRICES, DETERMINANTS, AND LINEAR EQUATIONS

For Probs. 5.144 to 5.154, use Cramer's rule to solve the given system.

5.144 $x + 2y = 6$, $3x - 5y = 10$

▮ By Cramer's rule,

$$x = \frac{\begin{vmatrix} k_1 & a_{12} \\ k_2 & a_{22} \end{vmatrix}}{D} \qquad y = \frac{\begin{vmatrix} a_{11} & k_1 \\ a_{21} & k_2 \end{vmatrix}}{D}$$

where $D = \begin{vmatrix} a_{11} & a_{12} \\ a_{21} & a_{22} \end{vmatrix}$ $(\neq 0)$. Here $a_{11} = 1$, $a_{12} = 2$, $a_{21} = 3$, $a_{22} = -5$, $k_1 = 6$, and $k_2 = 10$.

$$x = \frac{\begin{vmatrix} 6 & 2 \\ 10 & -5 \end{vmatrix}}{D} \qquad y = \frac{\begin{vmatrix} 1 & 6 \\ 3 & 10 \end{vmatrix}}{D} \qquad D = \begin{vmatrix} 1 & 2 \\ 3 & -5 \end{vmatrix} = -5 - 6 = -11$$

Since $\begin{vmatrix} 6 & 2 \\ 10 & -5 \end{vmatrix} = -30 - 20 = -50$, and $\begin{vmatrix} 1 & 6 \\ 3 & 10 \end{vmatrix} = 10 - 18 = -8$, $x = \frac{-50}{-11} = \frac{50}{11}$, and $y = \frac{-8}{-11} = \frac{8}{11}$.

5.145 $x - y = 8$, $2x + 3y = 10$

▮ $a_{11} = 1$, $a_{12} = -1$, $a_{21} = 2$, $a_{22} = 3$, $k_1 = 8$, and $k_2 = 10$.

$$x = \frac{\begin{vmatrix} 8 & -1 \\ 10 & 3 \end{vmatrix}}{D} \qquad y = \frac{\begin{vmatrix} 1 & 8 \\ 2 & 10 \end{vmatrix}}{D} \qquad D = \begin{vmatrix} 1 & -1 \\ 2 & 3 \end{vmatrix} = 3 - (-2) = 5$$

Then $x = \dfrac{24 - (-10)}{5} = \dfrac{34}{5}$; $y = \dfrac{10 - 16}{5} = \dfrac{-6}{5}$.

5.146 $3x - 5y = 8$, $2x + y = 2$

▮ $x = \dfrac{\begin{vmatrix} 8 & -5 \\ 2 & 1 \end{vmatrix}}{D} \qquad y = \dfrac{\begin{vmatrix} 3 & 8 \\ 2 & 2 \end{vmatrix}}{D} \qquad D = \begin{vmatrix} 3 & -5 \\ 2 & 1 \end{vmatrix} = 3 - (-10) = 13$

Then $x = \dfrac{8 + 10}{13} = \dfrac{18}{13}$; $y = \dfrac{6 - 16}{13} = \dfrac{-10}{13}$.

5.147 $2x - 6y = 1$, $3x + y = 2$

▮ $x = \dfrac{\begin{vmatrix} 1 & -6 \\ 2 & 1 \end{vmatrix}}{D} \qquad y = \dfrac{\begin{vmatrix} 2 & 1 \\ 3 & 2 \end{vmatrix}}{D} \qquad D = \begin{vmatrix} 2 & -6 \\ 3 & 1 \end{vmatrix} = 2 - (-18) = 20$

Then $x = \dfrac{1 - (-12)}{20} = \dfrac{13}{20}$; $y = \dfrac{4 - 3}{20} = \dfrac{1}{20}$.

5.148 $-x - y = 40$, $2x - y = 35$

▮ $x = \dfrac{\begin{vmatrix} 40 & -1 \\ 35 & -1 \end{vmatrix}}{D} \qquad y = \dfrac{\begin{vmatrix} -1 & 40 \\ 2 & 35 \end{vmatrix}}{D} \qquad D = \begin{vmatrix} -1 & -1 \\ 2 & -1 \end{vmatrix} = 1 - (-2) = 3$

Then $x = \dfrac{-40 - (-35)}{3} = \dfrac{-5}{3}$; $y = \dfrac{-35 - 80}{3} = \dfrac{-115}{3}$.

5.149 $3x - y = 10$, $40x - 11y = -100$

▮ $x = \dfrac{\begin{vmatrix} 10 & -1 \\ -100 & -11 \end{vmatrix}}{D} \qquad y = \dfrac{\begin{vmatrix} 3 & 10 \\ 40 & -100 \end{vmatrix}}{D} \qquad D = \begin{vmatrix} 3 & -1 \\ 40 & -11 \end{vmatrix} = -33 - (-40) = 7$

Then $x = \dfrac{10(-11) - 100}{7} = \dfrac{-210}{7} = -30$; $y = \dfrac{-300 - 400}{7} = \dfrac{-700}{7} = -100$.

5.150 $2x + y = 5$, $4x + 2y = 10$

▮ $D = \begin{vmatrix} 2 & 1 \\ 4 & 2 \end{vmatrix} = 4 - 4 = 0$. Thus, Cramer's rule will not work. This means that the system is either dependent or inconsistent. Here, we notice that the problem is dependency.

5.151 $x + y = 0$, $2y + z = -5$, $-x + z = -3$

▮ Given a system of equations

$$a_{11}x + a_{12}y + a_{13}z = k_1$$
$$a_{21}x + a_{22}y + a_{23}z = k_2$$
$$a_{31}x + a_{32}y + a_{33}z = k_3$$

then

$$x = \frac{\begin{vmatrix} k_1 & a_{12} & a_{13} \\ \cdot & \cdot & \cdot \\ \cdot & \cdot & \cdot \\ \cdot & \cdot & \cdot \\ k_3 & a_{32} & a_{33} \end{vmatrix}}{D} \qquad y = \frac{\begin{vmatrix} a_{11} & k_1 & a_{13} \\ \cdot & \cdot & \cdot \\ \cdot & \cdot & \cdot \\ \cdot & \cdot & \cdot \\ a_{31} & k_3 & a_{33} \end{vmatrix}}{D}$$

$$z = \frac{\begin{vmatrix} a_{11} & a_{12} & k_1 \\ \cdot & \cdot & \cdot \\ \cdot & \cdot & \cdot \\ a_{31} & a_{32} & k_3 \end{vmatrix}}{D} \qquad D = \begin{vmatrix} a_{11} & \cdots & a_{13} \\ \cdot & & \cdot \\ \cdot & & \cdot \\ a_{31} & \cdots & a_{33} \end{vmatrix} \quad (\neq 0)$$

Here, $\quad D = \begin{vmatrix} 1 & 1 & 0 \\ 0 & 2 & 1 \\ -1 & 0 & 1 \end{vmatrix} = 1; \quad x = \frac{\begin{vmatrix} 0 & 1 & 0 \\ -5 & 2 & 1 \\ -3 & 0 & 1 \end{vmatrix}}{D} = \frac{2}{1} = 2$

$$y = \frac{\begin{vmatrix} 1 & 0 & 1 \\ 0 & -5 & 1 \\ -1 & -3 & 1 \end{vmatrix}}{D} = \frac{-2}{1} = -2 \qquad z = \frac{\begin{vmatrix} 1 & 1 & 0 \\ 0 & 2 & -5 \\ -1 & 0 & -3 \end{vmatrix}}{D} = \frac{-1}{1} = -1$$

5.152 $x + y = 1$, $2y + z = 0$, $-x + z = 0$

▮ $D = \begin{vmatrix} 1 & 1 & 0 \\ 0 & 2 & 1 \\ -1 & 0 & 1 \end{vmatrix} = 1$. Then $\quad x = \frac{\begin{vmatrix} 1 & 1 & 0 \\ 0 & 2 & 0 \\ 0 & 0 & 1 \end{vmatrix}}{1} = 2; \quad y = \frac{\begin{vmatrix} 1 & 1 & 0 \\ 0 & 0 & 1 \\ -1 & 0 & 1 \end{vmatrix}}{1} = -1; \quad z = \frac{\begin{vmatrix} 1 & 1 & 1 \\ 0 & 2 & 0 \\ -1 & 0 & 0 \end{vmatrix}}{1} = 2$

5.153 $y + z = -4$, $x + 2z = 0$, $x - y = 5$

▮ $D = \begin{vmatrix} 0 & 1 & 1 \\ 1 & 0 & 2 \\ 1 & -1 & 0 \end{vmatrix} = 1$. Then

$$x = \begin{vmatrix} -4 & 1 & 1 \\ 0 & 0 & 2 \\ 5 & -1 & 0 \end{vmatrix} = 2; \quad y = \begin{vmatrix} 0 & -4 & 1 \\ 1 & 0 & 2 \\ 1 & 5 & 0 \end{vmatrix} = -3; \quad z = \begin{vmatrix} 0 & 1 & -4 \\ 1 & 0 & 0 \\ 1 & -1 & 5 \end{vmatrix} = -1$$

5.154 $2y - z = -4$, $x - y - z = 0$, $x - y + 2z = 6$

▮ $D = \begin{vmatrix} 0 & 2 & -1 \\ 1 & -1 & -1 \\ 1 & -1 & 2 \end{vmatrix} = -6$. Then

$$x = \frac{\begin{vmatrix} -4 & 2 & -1 \\ 0 & -1 & -1 \\ 6 & -1 & 2 \end{vmatrix}}{-6} = \frac{-6}{-6} = 1; \quad y = \frac{\begin{vmatrix} 0 & -4 & -1 \\ 1 & 0 & -1 \\ 1 & 6 & 2 \end{vmatrix}}{-6} = -1; \quad z = \frac{\begin{vmatrix} 0 & 2 & -4 \\ 1 & -1 & 0 \\ 1 & -1 & 6 \end{vmatrix}}{-6} = -1$$

For Probs. 5.155 to 5.158, perform the indicated row operations on the matrix $\mathbf{N} = \begin{bmatrix} 1 & -3 & 2 \\ 4 & -6 & -8 \end{bmatrix}$.

5.155 $\mathbf{R}_1 \leftrightarrow \mathbf{R}_2$

$\blacksquare$ The symbol $\leftrightarrow$ means $\mathbf{R}_1$ and $\mathbf{R}_2$ are interchanged. Thus, $\mathbf{N}$ becomes $\begin{bmatrix} 4 & -6 & -8 \\ 1 & -3 & 2 \end{bmatrix}$.

5.156 $-4\mathbf{R}_1 \rightarrow \mathbf{R}_1$

$\blacksquare$ The symbol $\rightarrow$ means "replaces." Thus, $-4\mathbf{R}_1$ replaces $\mathbf{R}_1$: We multiply each entry in $\mathbf{R}_1$ by -4. Thus, $\mathbf{N}$ becomes $\begin{bmatrix} 16 & 12 & -8 \\ 4 & -6 & -8 \end{bmatrix}$.

5.157 $\mathbf{R}_2 + (-4)\mathbf{R}_1 \rightarrow \mathbf{R}_2$

$\blacksquare$ This means $\mathbf{R}_2 + (-4)\mathbf{R}_1$ replaces $\mathbf{R}_2$. $\mathbf{N}$ becomes $\begin{bmatrix} 1 & -3 & 2 \\ 0 & -18 & -16 \end{bmatrix}$ {where $(-4)\mathbf{R}_1$ is $[-4 \quad -12 \quad -8]$}. Then $0 = -4 + 4$; $-18 = -12 - 6$; $-16 = -8 - 8$.

5.158 $\mathbf{R}_2 + (-1)\mathbf{R}_1 \rightarrow \mathbf{R}_2$

$\blacksquare$ Multiply $\mathbf{R}_1$ by -1, and add this result to $\mathbf{R}_2$. This sum replaces $\mathbf{R}_2$. Then $\begin{bmatrix} 1 & -3 & 2 \\ 4 & -6 & -8 \end{bmatrix}$ becomes $\begin{bmatrix} 1 & -3 & 2 \\ 3 & -3 & -10 \end{bmatrix}$ since $-1 + 4 = 3$; $3 + (-6) = -3$; $-2 - 8 = -10$.

For Probs. 5.159 to 5.163, solve the given system using augmented matrix methods.

5.159 $x_1 + x_2 = 5$, $x_1 - x_2 = 1$

$\blacksquare$ Remember that the technique here is to write the augmented matrix and use elementary row operations to write the row-equivalent matrix having the identity on the left-hand side. In this problem $\begin{bmatrix} 1 & 1 & 5 \\ 1 & -1 & 1 \end{bmatrix}$ is the augmented matrix. We want 0 in the a_{21} spot. Perform the operation $\mathbf{R}_2 + (-1)\mathbf{R}_1 \rightarrow \mathbf{R}_2$. Then $a_{21} = 0$ since $1 + (-1)(1) = 0$. We get $\begin{bmatrix} 1 & 1 & 5 \\ 0 & -2 & -4 \end{bmatrix}$. Now we want a 1 in the a_{22} spot. Multiply $\mathbf{R}_2$ by $-\frac{1}{2}$. $(-\frac{1}{2})\mathbf{R}_2 \rightarrow \mathbf{R}_2$, and we get $\begin{bmatrix} 1 & 1 & 5 \\ 0 & 1 & 2 \end{bmatrix}$. Finally, we need 0 in the a_{12} spot. Use $\mathbf{R}_1 + (-1)\mathbf{R}_2 \rightarrow \mathbf{R}_1$ and get $\begin{bmatrix} 1 & 0 & 3 \\ 0 & 1 & 2 \end{bmatrix} \sim \begin{bmatrix} 1 & 1 & 5 \\ 1 & -1 & 1 \end{bmatrix}$ and $x_1 = 3$, $x_2 = 2$.

5.160 $x_1 - 2x_2 = 1$, $2x_1 - x_2 = 5$

$\blacksquare$ $\begin{bmatrix} 1 & -2 & 1 \\ 2 & -1 & 5 \end{bmatrix} \sim \begin{bmatrix} 1 & -2 & 1 \\ 0 & 3 & 3 \end{bmatrix}$ (We performed the operation $\mathbf{R}_2 + (-2)\mathbf{R}_1 \rightarrow \mathbf{R}_2$.)

$\sim \begin{bmatrix} 1 & -2 & 1 \\ 0 & 1 & 1 \end{bmatrix}$ (We performed the operation $\frac{1}{3}\mathbf{R}_2 \rightarrow \mathbf{R}_2$.)

$\sim \begin{bmatrix} 1 & 0 & 3 \\ 0 & 1 & 1 \end{bmatrix}$ (We performed the operation $\mathbf{R}_1 + 2\mathbf{R}_2 \rightarrow \mathbf{R}_1$.)

Thus, $x_1 = 3$, $x_2 = 1$.

5.161 $x_1 - 4x_2 = -2$, $2x_1 + x_2 = -3$

$\blacksquare$ $\begin{bmatrix} 1 & -4 & -2 \\ 2 & 1 & -3 \end{bmatrix} \sim \begin{bmatrix} 1 & -4 & -2 \\ 0 & -7 & -7 \end{bmatrix}$ ($\mathbf{R}_2 + 2\mathbf{R}_1 \rightarrow \mathbf{R}_2$)

$\sim \begin{bmatrix} 1 & -4 & -2 \\ 0 & 1 & 1 \end{bmatrix}$ ($-\frac{1}{7}\mathbf{R}_2 \rightarrow \mathbf{R}_2$)

$\sim \begin{bmatrix} 1 & 0 & 2 \\ 0 & 1 & 1 \end{bmatrix}$ ($\mathbf{R}_1 + 4\mathbf{R}_2 \rightarrow \mathbf{R}_1$)

Thus, $x_1 = 2$, $x_2 = 1$.

5.162 $x_1 + 2x_2 = 4$, $2x_1 + 4x_2 = -8$

$\blacksquare$ $\begin{bmatrix} 1 & 2 & 4 \\ 2 & 4 & -8 \end{bmatrix} \sim \begin{bmatrix} 1 & 2 & 4 \\ 0 & 0 & -16 \end{bmatrix}$ $[\mathbf{R}_2 + (-2)\mathbf{R}_1 \rightarrow \mathbf{R}_2]$

The $[0 \quad 0]$ row signals no solution or dependency. Here, $x_1 + 2x_2 = 4$ and $0x_1 + 0x_2 = -16$ which means no solution.

5.163 $3x_1 - 6x_2 = -9, \quad -2x_1 + 4x_2 = 6$

$\begin{bmatrix} 3 & -6 & | & -9 \\ -2 & 4 & | & 6 \end{bmatrix} \sim \begin{bmatrix} 1 & -2 & | & -3 \\ -2 & 4 & | & 6 \end{bmatrix} \quad (\frac{1}{3}\mathbf{R}_1 \to \mathbf{R}_1)$

$\sim \begin{bmatrix} 1 & -2 & | & -3 \\ 0 & 0 & | & 0 \end{bmatrix} \quad (2\mathbf{R}_1 - \mathbf{R}_2 \to \mathbf{R}_2)$

$0x_1 + 0x_2 = 0$ signals that there are infinitely many solutions. Since $x_1 - 2x_2 = -3$, any pair (x_1, x_2) having the property $x_1 - 2x_2 = -3$ solves the system.

For Probs. 5.164 to 5.168, tell whether the given matrix is reduced or not, and explain why.

5.164 $\begin{bmatrix} 1 & 0 & 2 & | & 3 \\ 0 & 0 & 0 & | & 0 \\ 0 & 1 & -1 & | & 4 \end{bmatrix}$

▮ Not reduced: In a reduced matrix, an all-zero row must not be above a non-all-zero row. Here, it is.

5.165 $\begin{bmatrix} 0 & 1 & 0 & | & 2 \\ 0 & 0 & 3 & | & -1 \\ 0 & 0 & 0 & | & 0 \end{bmatrix}$

▮ Not reduced: In row 2, the left-most nonzero element is 3. In a reduced matrix, it must be 1.

5.166 $\begin{bmatrix} 1 & 1 & | & 1 \\ 0 & 1 & | & 1 \\ 1 & 0 & | & 1 \end{bmatrix}$

▮ Not reduced: The left-most 1 in row 3 is to the left of the left-most 1 in row 2. But row 2 is above row 1.

5.167 $\begin{bmatrix} 1 & 2 & 0 & 3 & | & 2 \\ 0 & 0 & 1 & -1 & | & 0 \end{bmatrix}$

▮ This is reduced. All four conditions are satisfied.

5.168 $\begin{bmatrix} 1 & 0 & | & 5 \\ 0 & 1 & | & 3 \end{bmatrix}$

▮ This is reduced. All four conditions are satisfied.

For Probs. 5.169 to 5.171, write the linear system corresponding to the given reduced matrix, and solve the system.

5.169 $\begin{bmatrix} 1 & 0 & 0 & | & -2 \\ 0 & 1 & 0 & | & 3 \\ 0 & 0 & 1 & | & 0 \end{bmatrix}$

▮ $x_1 = -2, \quad x_2 = -3, \quad x_3 = 0$. Notice that the system is solved trivially. That is the strength of this method.

5.170 $\begin{bmatrix} 1 & 0 & -2 & | & 3 \\ 0 & 1 & 1 & | & -5 \\ 0 & 0 & 0 & | & 0 \end{bmatrix}$

▮ $x_1 - 2x_3 = 3$ and $x_2 + x_3 = -5$. Suppose x_3 has some value s. Then $x_1 = 3 + 2x_3 = 3 + 2s$, and $x_2 = -5 - x_3 = -5 - s$. The solution is: For any value s of x_3, $x_2 = -5 - s$ and $x_1 = 3 + 2s$.

5.171 $\begin{bmatrix} 1 & 0 & | & 0 \\ 0 & 1 & | & 0 \\ 0 & 0 & | & 1 \end{bmatrix}$

▮ $x_1 = 0, \quad x_2 = 0,$ and $0x_1 + 0x_2 = 1$. But $0 \neq 1$. No solution.

For Probs. 5.172 to 5.174, change the given matrix to one in reduced form.

5.172 $\begin{bmatrix} 1 & 2 & | & -1 \\ 0 & 1 & | & 3 \end{bmatrix}$

▮ $\begin{bmatrix} 1 & 2 & | & -1 \\ 0 & 1 & | & 3 \end{bmatrix} \sim \begin{bmatrix} 1 & 0 & | & -7 \\ 0 & 1 & | & 3 \end{bmatrix}$ $[\mathbf{R}_1 + (-2)\mathbf{R}_2 \rightarrow \mathbf{R}_1]$

5.173 $\begin{bmatrix} 1 & 0 & -3 & | & 1 \\ 0 & 1 & 2 & | & 0 \\ 0 & 0 & 3 & | & -6 \end{bmatrix}$

▮ $\begin{bmatrix} 1 & 0 & -3 & | & 1 \\ 0 & 1 & 2 & | & 0 \\ 0 & 0 & 3 & | & -6 \end{bmatrix} \sim \begin{bmatrix} 1 & 0 & -3 & | & 1 \\ 0 & 1 & 2 & | & 0 \\ 0 & 0 & 1 & | & -2 \end{bmatrix}$ $(\frac{1}{3}\mathbf{R}_3 \rightarrow \mathbf{R}_3)$

$\sim \begin{bmatrix} 1 & 0 & 0 & | & -5 \\ 0 & 1 & 0 & | & 4 \\ 0 & 0 & 1 & | & 2 \end{bmatrix}$ $[\mathbf{R}_1 + 3\mathbf{R}_3 \rightarrow \mathbf{R}_1 \quad \text{and} \quad \mathbf{R}_2 + (-2)\mathbf{R}_3 \rightarrow \mathbf{R}_2]$

5.174 $\begin{bmatrix} 1 & 2 & -2 & | & -1 \\ 0 & 3 & -6 & | & 1 \\ 0 & -1 & 2 & | & -\frac{1}{3} \end{bmatrix}$

▮ $\sim \begin{bmatrix} 1 & 2 & -2 & | & -1 \\ 0 & 1 & -2 & | & \frac{1}{3} \\ 0 & -1 & 2 & | & -\frac{1}{3} \end{bmatrix}$ $(\frac{1}{3}\mathbf{R}_2 \rightarrow \mathbf{R}_2)$

$\sim \begin{bmatrix} 1 & 0 & 2 & | & -\frac{5}{3} \\ 0 & 1 & -2 & | & \frac{1}{3} \\ 0 & 0 & 0 & | & 0 \end{bmatrix}$ $[\mathbf{R}_1 + (-2)\mathbf{R}_2 \rightarrow \mathbf{R}_1 \quad \text{and} \quad \mathbf{R}_2 + \mathbf{R}_3 \rightarrow \mathbf{R}_2]$

For Probs. 5.175 to 5.177, solve using Gauss-Jordan elimination.

5.175 $2x_1 + 2x_2 = 2, \quad x_1 + 2x_2 = 3, \quad -3x_2 = -6$

▮ $\begin{bmatrix} 2 & 2 & | & 2 \\ 1 & 2 & | & 3 \\ 0 & -3 & | & -6 \end{bmatrix} \sim \begin{bmatrix} 1 & 2 & | & 3 \\ 2 & 2 & | & 2 \\ 0 & -3 & | & -6 \end{bmatrix}$ $(\mathbf{R}_1 \leftrightarrow \mathbf{R}_2) \sim \begin{bmatrix} 1 & 2 & | & 3 \\ 0 & -2 & | & -4 \\ 0 & -3 & | & -6 \end{bmatrix}$ $[\mathbf{R}_2 + (-2)\mathbf{R}_1 \rightarrow \mathbf{R}_2]$

$\sim \begin{bmatrix} 1 & 2 & | & 3 \\ 0 & 1 & | & 2 \\ 0 & -3 & | & -6 \end{bmatrix}$ $(-\frac{1}{2}\mathbf{R}_2 \rightarrow \mathbf{R}_2)$

$\sim \begin{bmatrix} 1 & 0 & | & -1 \\ 0 & 1 & | & 2 \\ 0 & 0 & | & 0 \end{bmatrix}$ $[\mathbf{R}_1 + (-2)\mathbf{R}_2 \rightarrow \mathbf{R}_1 \quad \text{and} \quad \mathbf{R}_3 + 3\mathbf{R}_2 \rightarrow \mathbf{R}_3]$

Thus, $x_1 = -1, \quad x_2 = 2.$

5.176 $3x_1 - 4x_2 - x_3 = 1, \quad 2x_1 - 3x_2 + x_3 = 1, \quad x_1 - 2x_2 + 3x_3 = 2$

▮ $\begin{bmatrix} 2 & -1 & | & 0 \\ 3 & 2 & | & 7 \\ 1 & -1 & | & -2 \end{bmatrix} \sim \begin{bmatrix} 1 & -1 & | & -2 \\ 3 & 2 & | & 7 \\ 2 & -1 & | & 0 \end{bmatrix} \sim \begin{bmatrix} 1 & -1 & | & -2 \\ 0 & 5 & | & 13 \\ 0 & 1 & | & 4 \end{bmatrix} \sim \begin{bmatrix} 1 & -1 & | & -2 \\ 0 & 1 & | & 4 \\ 0 & 5 & | & 13 \end{bmatrix} \sim \begin{bmatrix} 1 & -1 & | & -2 \\ 0 & 1 & | & 4 \\ 0 & 0 & | & -7 \end{bmatrix}$. But

$0x_1 + 0x_2 = -7$ means $0 = 7$. No solution.

5.177 $2x_1 - x_2 - 3x_3 = 8, \quad x_1 - 2x_2 = 7$

▮ $\begin{bmatrix} 2 & -1 & -3 & | & 8 \\ 1 & -2 & 0 & | & 7 \end{bmatrix} \sim \begin{bmatrix} 1 & -2 & 0 & | & 7 \\ 2 & -1 & -3 & | & 8 \end{bmatrix} \sim \begin{bmatrix} 1 & -2 & 0 & | & 7 \\ 0 & 3 & -3 & | & -6 \end{bmatrix} \sim \begin{bmatrix} 1 & -2 & 0 & | & 7 \\ 0 & 1 & -1 & | & -2 \end{bmatrix}$

$\sim \begin{bmatrix} 1 & 0 & -2 & | & 3 \\ 0 & 1 & -1 & | & -2 \end{bmatrix}$. Then, if x_3 has some value, say s, then $x_1 + 0x_2 - 2x_3 = x_1 - 2s = 3$ or $x_1 = 3 + 2s$, and $0x_1 + x_2 - x_3 = x_2 - s = -2$ or $x_2 = -2 + s$.

5.4 SYSTEMS OF NONLINEAR EQUATIONS

For Probs. 5.178 to 5.203, solve the given system algebraically.

5.178 $3x - y = 8$, $3x^2 - y^2 = 26$

▮ $y = 3x - 8$. Substituting, we get $26 = 3x^2 - (3x - 8)^2 = 3x^2 - (9x^2 - 48x + 64) = -6x^2 + 48x - 64 = 26$. Then $-6x^2 + 48x - 90 = 0$, $x^2 - 8x + 15 = 0$, $(x - 5)(x - 3) = 0$, and $x = 5$, $x = 3$. If $x = 5$, $y = 3(5) - 8 = 7$. If $x = 3$, $y = 3(3) - 8 = 1$. Check to see that these solutions are correct. They are!

5.179 $3x - 2y = 5$, $3x^2 - 2y^2 = 19$

▮ $2y = 3x - 5$, or $y = \frac{3}{2}x - \frac{5}{2}$. Substituting, $19 = 3x^2 - 2(\frac{3}{2}x - \frac{5}{2})^2 = 3x^2 - 2(\frac{9}{4}x^2 - \frac{30}{4}x + \frac{25}{4}) = 3x^2 - \frac{18}{4}x^2 + \frac{60}{4}x - \frac{50}{4} = -\frac{3}{2}x^2 + \frac{60}{4}x - \frac{50}{4} = -\frac{3}{2}x^2 + 15x - \frac{25}{2}$. Multiplying both sides by 2, we get $-3x^2 + 30x - 25 = 38$. Subtracting 38 from both sides, we get $3x^2 - 30x + 63 = 0$, $x^2 - 10x + 21 = 0$, $(x - 7)(x - 3) = 0$, and $x = 7$, $x = 3$. If $x = 7$, $y = \frac{3}{2} \cdot 7 - \frac{5}{2} = 8$. If $x = 3$, $y - \frac{3}{2} \cdot 3 - \frac{5}{2} = 2$.

5.180 ▮ $2x - y = -1$, $16x^2 - 3y^2 = -11$

▮ $y = 2x + 1$. Substituting, $-11 = 16x^2 - 3(2x + 1)^2 = 16x^2 - 3(4x^2 + 4x + 1) = 16x^2 - 12x^2 - 12x - 3$. Adding 11 to both sides, $4x^2 - 12x + 8 = 0$, $x^2 - 3x + 2 = 0$, $(x - 2)(x - 1) = 0$, and $x = 2$, $x = 1$. If $x = 2$, $y = 2(2) + 1 = 5$. If $x = 1$, $y = 2(1) + 1 = 3$.

5.181 $4x - y = 11$, $8x^2 + 5y^2 = 77$

▮ $y = 4x - 11$. Substituting $77 = 8x^2 + 5(4x - 11)^2 = 8x^2 + 5(16x^2 - 88x + 121) = 8x^2 + 80x^2 - 440x + 605$. Subtracting 77 from both sides, $88x^2 - 440x + 528 = 0$, $x^2 - 5x + 6 = 0$, $(x - 6)(x + 1) = 0$, and $x = 6$, $x = -1$. If $x = 6$, $y = 24 - 11 = 13$. If $x = -1$, $y = -4 - 11 = -15$.

5.182 $x^2 + 2y^2 = 9$, $xy = 2$

▮ $x = 2y$. Substituting, $(2/y)^2 + 2y^2 = 9$, $4/y^2 + 2y^2 = 9$, $4 + 2y^4 = 9y^2$, $2y^4 - 9y^2 + 4 = 0$. Let $S = y^2$; then $2S^2 - 9S + 4 = 0$. $S = \dfrac{9 \pm \sqrt{81 - 4(2)(4)}}{4} = \dfrac{9 \pm \sqrt{49}}{4} = \dfrac{9 \pm 7}{4}$. Thus, $S = \dfrac{9 + 7}{4} = 4$ or $S = \dfrac{9 - 7}{4} = \dfrac{2}{4} = \dfrac{1}{2}$. Then $y^2 = S = 4$, and $y = \pm 2$; or $y^2 = S = \frac{1}{2}$, and $y = \pm\sqrt{2}/\sqrt{2}$. If $y = 2$, $x = 1$; if $y = -2$, $x = -1$; if $y = \sqrt{2}/2$, $x = 2\sqrt{2}$; if $y = -\sqrt{2}/2$, $x = -2\sqrt{2}$.

5.183 $2x^2 + 4y^2 = 19$, $3x^2 - 8y^2 = 25$

▮ Multiply the first equation by 2, and add:

$$\begin{array}{r} 4x^2 + 8y^2 = 38 \\ 3x^2 - 8y^2 = 25 \\ \hline 7x^2 \qquad\;\; = 63 \end{array}$$

Then $x^2 = \frac{63}{7} = 9$, $x = \pm 3$. If $x = 3$, $2(9) + 4y^2 = 19$, $4y^2 = 1$, and $y = \pm 1$. If $x = -3$, $2(9) + 4y^2 = 19$, and $y = \pm 1$. Thus, the solutions are $(3, \pm 1)$, $(-3, \pm 1)$.

5.184 $x^2 + 2y^2 = 3$, $3x^2 - y^2 = 2$

▮ Subtracting,

$$\begin{array}{r} 3x^2 + 6y^2 = 9 \\ 3x^2 - \;\; y^2 = 2 \\ \hline 7y^2 = 7 \end{array}$$

Then $y = \pm 1$. If $y = 1$, $x^2 + 2 = 3$, and $x = \pm 1$. If $y = -1$, $x = \pm 1$. Thus, the solutions are $(\pm 1, 1)$, $(\pm 1, -1)$.

5.185 $x^2 + 2y^2 = 3$, $3x^2 - y^2 = 2$

▎ Multiply the second equation by 2, and add:

$$\begin{array}{r} x^2 + 2y^2 = 3 \\ 6x^2 - y^2 = 4 \\ \hline 7x^2 \quad\;\; = 7 \end{array}$$

Then $x^2 = 1$, or $x = \pm 1$. If $x = 1$, $2y^2 = 2$, and $y = \pm 1$. If $x = -1$, $2y^2 = 2$, and $y = \pm 1$. Thus, the solutions are $(1, \pm 1)$, $(-1, \pm 1)$.

5.186 $4x^2 + 3y^2 = 4$, $8x^2 + 5y^2 = 7$

▎ Multiply the first equation by 2, and subtract:

$$\begin{array}{r} 8x^2 + 6y^2 = 8 \\ 8x^2 + 5y^2 = 7 \\ \hline y^2 = 1 \end{array}$$

Then $y = \pm 1$. If $y = 1$, $4x^2 = 1$, and $x = \pm \frac{1}{2}$; If $y = -1$, $x = \pm \frac{1}{2}$. Thus, the solutions are $(\pm \frac{1}{2}, 1)$, $(\pm \frac{1}{2}, -1)$.

5.187 $2x^2 + 3y^2 = 7$, $3x^2 + 2y^2 = 8$

▎ Multiply the first equation by 3, the second equation by 2, and subtract:

$$\begin{array}{r} 6x^2 + 9y^2 = 21 \\ 6x^2 + 4y^2 = 16 \\ \hline 5y^2 = 5 \end{array}$$

Then $y = \pm 1$. If $y = 1$, $6x^2 = 12$, $x^2 = 2$, and $x = \pm \sqrt{2}$. If $y = -1$, $x = \pm \sqrt{2}$. Thus, the solutions are $(\pm \sqrt{2}, 1)$, $(\pm \sqrt{2}, -1)$.

5.188 $\dfrac{1}{x} + \dfrac{2}{y} = 1$, $\dfrac{3}{x} + \dfrac{10}{y} = 4$

▎ If $1/x = 1 - 2/y$, then $3/x = 3 - 6/y$. Thus, $3 - 6/y + 10/y = 4$, $4/y = 1$, or $y = 4$. If $y = 4$, $1/x + \frac{1}{2} = 1$, $1/x = \frac{1}{2}$, or $x = 2$. The solution is $(2, 4)$.

5.189 $x^2 + 4y^2 - 2x = 1$, $3x^2 + 8y^2 - 6x = 2$

$$\begin{array}{r} 3x^2 + 12y^2 - 6x = 3 \\ 3x^2 + 8y^2 - 6x = 2 \\ \hline 4y^2 \qquad\quad = 1 \end{array}$$

Then $y = \pm \frac{1}{2}$. If $y = \frac{1}{2}$, $x^2 - 2x = 1 - 1 = 0$, $x(x - 2) = 0$, and $x = 0, 2$. If $y = -\frac{1}{2}$, $x = 0, 2$. Thus, the solutions are $(0, \frac{1}{2})$, $(2, \frac{1}{2})$, $(0, -\frac{1}{2})$, $(2, -\frac{1}{2})$.

5.190 $9x^2 - 41y^2 + 60y = 100$, $x^2 - 5y^2 + 8y = 12$

▎
$$\begin{array}{r} 9x^2 - 41y^2 + 60y = 100 \\ 9x^2 - 45y^2 + 72y = 108 \\ \hline 4y^2 - 12y = -8 \end{array}$$

Then $y^2 - 3y + 2 = 0$, $y = \dfrac{3 \pm \sqrt{9 - 4(1)(2)}}{2} = \dfrac{3 \pm \sqrt{9 - 8}}{2} = \dfrac{3 \pm 1}{2}$, and $y = 2$, $y = 1$. If $y = 2$, $x^2 - 20 + 16 = 12$, $x^2 = 16$, and $x = \pm 4$. If $y = 1$, $x^2 - 5 + 8 = 12$, $x^2 = 9$, and $x = \pm 3$. Thus, the solutions are $(3, 1)$, $(-3, 1)$, $(\pm 4, 2)$.

5.191 $2x^2 - y^2 = 14$, $x - y = 1$

▎ Substituting for y, $2x^2 - (x - 1)^2 = 14$, $2x^2 - (x^2 - 2x + 1) = 14$, $x^2 + 2x - 15 = 0$, $(x - 3)(x + 5) = 0$, and $x = 3$, $x = -5$. If $x = 3$, $y = 2$. If $x = -5$, $y = -6$.

5.192 $xy + x^2 = 24$, $y - 3x + 4 = 0$

⬛ $y = 3x - 4$. Substituting for y, $x(3x - 4) + x^2 = 24$, $3x^2 - 4x + x^2 = 24$, $4x^2 - 4x - 24 = 0$, $x^2 - x - 6 = 0$, $(x - 3)(x + 2) = 0$,   and $x = 3$, $x = -2$. If $x = 3$, $y = 5$; if $x = -2$, $y = -10$.

5.193 $3xy - 10x = y$, $2 - y + x = 0$

⬛ $y = 2 + x$. Substituting for y, $3x(2 + x) - 10x = 2 + x$, $6x + 3x^2 - 10x = 2 + x$, $3x^2 - 5x - 2 = 0$, $(3x + 1)(x - 2) = 0$,   and $x = -\frac{1}{3}$, $x = 2$. If $x = -\frac{1}{3}$, $y = \frac{5}{3}$; if $x = 2$, $y = 4$.

5.194 $xy = -3$, $xy = -6$

⬛   If $xy = -3$,   then, since $xy = -6$,   we get $-3 = -6$. No solution.   (Graphically, what does this say about these two hyperbolas?)

5.195 $4x + 5y = 6$, $xy = -2$

⬛ $y = -2/x$. Substituting for y, $4x + 5(-2/x) = 6$, $4x - 10/x = 6$, $4x^2 - 10 = 6x$, $2x^2 - 3x - 5 = 0$, $(2x - 5)(x + 1)$,   and $x = \frac{5}{2}$, $x = -1$. If $x = \frac{5}{2}$, $y = -\frac{4}{5}$; if $x = -1$, $y = 2$.

5.196 $\dfrac{9}{x^2} + \dfrac{16}{y^2} = 5$, $\dfrac{18}{x^2} + \dfrac{12}{y^2} = -1$

⬛ $9/x^2 = 5 - 16/y^2$. Thus, $18/x^2 = 10 - 32/y^2$. But $18/x^2 = -1 + 12/y^2$. Then, $-1 + 12/y^2 = 10 - 32/y^2$, $44/y^2 = 11$, $11y^2 = 44$,   and $y = \pm 2$. If $y = 2$, $9/x^2 + \frac{16}{4} = 5$, $9/x^2 = 1$, $x^2 = 9$,   and $x = \pm 3$. If $y = -2$, $x = \pm 3$. Thus, the solutions are $(3, 2)$, $(-3, 2)$, $(3, -2)$, $(-3, -2)$.

5.197 $x^2 - xy = 12$, $xy - y^2 = 3$

⬛ $x(x - y) = 12$ and $y(x - y) = 3$. Then $x/y = 4$ $(x \neq y)$ and $(4y)^2 - (4y)(y) = 12$. So $16y^2 - 4y^2 = 12$, $12y^2 = 12$,   and $y = \pm 1$. If $y = 1$, $x = 4$; if $y = -1$, $x = -4$. Since we divided by $x - y$, check both solutions.

5.198 $x^3 - y^3 = 9$, $x - y = 3$

⬛ $x = y + 3$. Substituting for x, $(y + 3)^3 - y^3 = 9$, $y^3 + 9y^2 + 27y + 27 - y^3 = 9$, $9y^2 + 27y + 18 = 0$, $y^2 + 3y + 2 = 0$, $(y + 2)(y + 1) = 0$,   and $y = -1$, $y = -2$. If $y = -1$, $x = 2$; if $y = -2$, $x = 1$.

5.199 $9x^2 + y^2 = 90$, $x^2 + 9y^2 = 90$

⬛

$$\begin{array}{r} 9x^2 + \ \ y^2 = \ \ \ 90 \\ 9x^2 + 81y^2 = \ \ 810 \\ \hline -80y^2 = -720 \end{array}$$

Then $y^2 = 9$,   and $y = \pm 3$. If $y = 3$, $x = \pm 3$. If $y = -3$, $x = \pm 3$.

5.200 $\dfrac{2}{x^2} - \dfrac{3}{y^2} = 5$, $\dfrac{1}{x^2} + \dfrac{2}{y^2} = 6$

⬛ $1/x^2 = 6 - 2/y^2$. Thus, $2/x^2 = 12 - 4/y^2$, $12 - 4/y^2 = 5 + 3/y^2$, $12y^2 - 4 = 5y^2 + 3$, $7y^2 = 7$,   and $y = \pm 1$. If $y = 1$, $x = \pm \frac{1}{2}$; if $y = -1$, $x = \pm \frac{1}{2}$.

5.201 $y^2 = 4x - 8$, $y^2 = -6x + 32$

⬛   Then $-6x + 32 = 4x - 8$, $10x = 40$,   and $x = 4$. Then $y^2 = 16 - 8 = 8$,   and $y = \pm 2\sqrt{2}$. Thus, the solutions are $(4, \pm 2\sqrt{2})$.

5.202 $x^2 - y^2 = 16$, $y^2 = 2x - 1$

⬛   From the first equation, $y^2 = x^2 - 16$. Then $y^2 = x^2 - 16 = 2x - 1$, $x^2 - 2x - 15 = 0$, $x = \dfrac{2 \pm \sqrt{4 + 4(1)(15)}}{2} = \dfrac{2 \pm 8}{2}$,   and $x = 5$ or $x = -3$. [We also could have factored: $(x - 5)(x + 3) = 0$.] If $x = 5$, $y^2 = 10 - 1 = 9$,   and $y = \pm 3$. If $x = -3$, $y^2 = -6 - 1 = -7$,   and $y = \pm i\sqrt{7}$. Notice the imaginary solution here.

5.203 $2x^2 + y^2 = 6$, $x^2 + y^2 + 2x = 3$

▮ Subtract the two equations:

$$\begin{array}{r} 2x^2 + y^2 = 6 \\ x^2 + y^2 + 2x = 3 \\ \hline x^2 - 2x = 3 \end{array}$$

Then $x^2 - 2x - 3 = 0$, $(x-3)(x+1) = 0$ and $x = 3$ or $x = -1$. If $x = -1$, $y = \pm\sqrt{3}$. If $x = 3$, $y = \pm\sqrt{6 - 2x^2} = \pm 2i\sqrt{3}$.

For Probs. 5.204 to 5.208, solve the given problem using a system of equations.

5.204 Two numbers differ by 2 and their squares differ by 48. Find the numbers.

▮ Let x and y be the numbers. Then, $x - y = 2$ (differ by 2) and $x^2 - y^2 = 48$ (squares differ by 48). Thus, $x = y + 2$. Substituting for x, $(y+2)^2 - y^2 = 48$, $y^2 + 4y + 4 - y^2 = 48$, $4y = 44$, and $y = 11$. If $y = 11$, $x = 13$.

5.205 The sum of the circumferences of two circles is 88 in and the sum of their areas is $\frac{2200}{7}$ in^2 when $\pi \approx \frac{22}{7}$. Find the radius of each circle.

▮ Let r_1 and r_2 be the radii. Thus, $2(\frac{22}{7})r_1 + 2(\frac{22}{7})r_2 = 88$ and $\frac{22}{7}r_1^2 + \frac{22}{7}r_2^2 = \frac{2200}{7}$. Then $44r_1 + 44r_2 = 616$, $r_1 + r_2 = 14$, $r_1 = 14 - r_2$. Also, $r_1^2 + r_2^2 = 100$. Substituting for r_1 in this equation, $(14 - r_2)^2 + r_2^2 = 100$, $196 + r_2^2 - 28r_2 + r_2^2 = 100$, $2r_2^2 - 28r + 96 = 0$, $r_2^2 - 14r + 48 = 0$, and $r = 6$ in, 8 in.

5.206 A party costing \$30 is planned. It is found that by adding 3 more to the group, the cost per person would be reduced by 50 ¢. For how many people was the party originally planned?

▮ Let $x = $ cost per person and $y = $ number of people. Then $xy = 30$ and $x = \dfrac{30}{y}$. Thus, $x - 0.5 = \dfrac{30}{y+3}$ (since the cost per person was reduced by 50¢ and the number of people was increased by 3). Then $\dfrac{30}{y} - 0.5 = \dfrac{30}{y+3}$, and $y = 12$.

5.207 The square of a certain number exceeds twice the square of another number by 16. Find the numbers if the sum of their squares is 208.

▮ Let x and y be the numbers. Then $x^2 = 2y^2 + 16$ and $x^2 + y^2 = 208$. Thus, $x^2 = 208 - y^2 = 2y^2 + 16$, $2y^2 + 16 = 208 - y^2$, $3y^2 = 192$, $y^2 = 64$, and $y = \pm 8$. If $y = 8$, $x = \pm 12$. If $y = -8$, $x = \pm 12$. Thus the solutions are $x = 12$, $y = 8$; $x = 12$, $y = -8$; $x = -12$, $y = 8$; $x = -12$, $y = -8$.

5.208 The diagonal of a rectangle is 85 ft. If the short side is increased by 11 ft and the long side decreased by 7 ft, the length of the diagonal remains the same. Find the original dimensions.

▮ See Fig. 5.11a. Let $x = $ width and $y = $ length. Then $d = \sqrt{x^2 + y^2}$, and $d' = \sqrt{(x+11)^2 + (y-7)^2}$ (see Fig. 5.11b). If $d = d'$, then $\sqrt{x^2 + y^2} = \sqrt{(x+11)^2 + (y-7)^2}$. Since $\sqrt{x^2 + y^2} = 85$, $\sqrt{(x+11)^2 + (y-7)^2} = 85$. Then we have the equations $(x+11)^2 + (y-7)^2 = 7225$ and $x^2 + y^2 = 7225$. Solving these two equations, we get $x = 40$ ft and $y = 75$ ft.

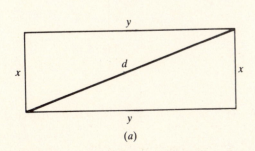

(a)

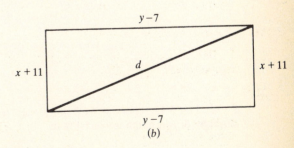

(b)

Fig. 5.11

For Probs. 5.209 to 5.214, solve the given system graphically.

5.209 $y = x^2$, $y = x^3$

▮ See Fig. 5.12. For both equations if $x = 1$, $y = 1$.

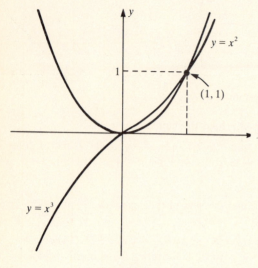

Fig. 5.12

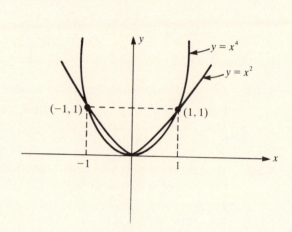

Fig. 5.13

5.210 $y = x^2$, $y = x^4$

▮ See Fig. 5.13, For $y = x^2$, if $x = 1$, $y = 1$. For $y = x^4$, if $x = -1$, $y = 1$.

5.211 $y = x^2$, $y = |x|$

▮ See Fig. 5.14. For $y = x^2$, if $x = 1$, $y = 1$. For $y = |x|$, if $x = -1$, $y = 1$.

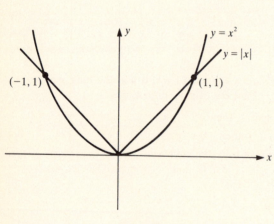

Fig. 5.14

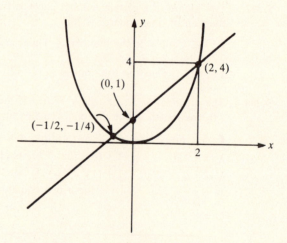

Fig. 5.15

5.212 $y = x^2$, $2y = 3x + 2$

▮ See Fig. 5.15. For $y = x^2$, if $x = 2$, $y = 4$. For $2y = 3x + 2$, if $x = -\frac{1}{2}$, $y = \frac{1}{4}$. Notice that a carefully drawn graph is crucial here.

5.213 $xy = 1$, $x = y$

▮ See Fig. 5.16. For $xy = 1$, if $x = 1$, $y = 1$. For $x = y$, if $x = -1$, $y = -1$.

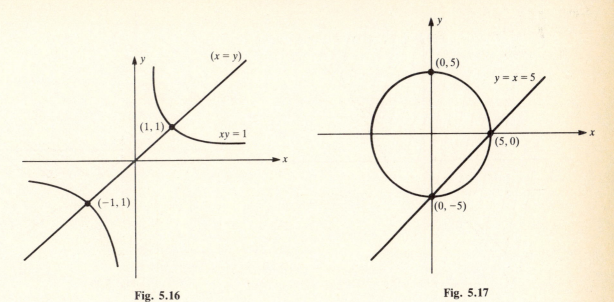

Fig. 5.16   **Fig. 5.17**

5.214 $x^2 + y^2 = 25$, $y = x - 5$

▎ See Fig. 5.17. For $x^2 + y^2 = 25$, if $x = 5$, $y = 0$. For $y = x - 5$, if $x = 0$, $y = -5$.

5.5   SYSTEMS OF INEQUALITIES

For Probs. 5.215 to 5.222, graph the given inequality.

5.215 $2x - 3y < 6$

▎ See Fig. 5.18. We graph the line $2x - 3y = 6$. Then, we find which "side" of the plane satisfies the inequality by testing a point. Dot the line $2x - 3y = 6$ since the inequality is $<$, not $\leq$. Test $(0, 0)$: $2(0) - 3(0) < 6$. Thus, the $(0, 0)$ side of the plane is shaded.

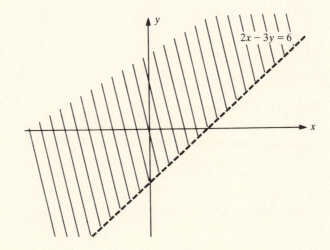

Fig. 5.18

5.216 $x \leq y$

▎ See Fig. 5.19. Test $(1, 0)$: $1 \leq 0$? No! Also, use a solid line since it is $\leq$.

5.217 $x + y \geq 0$

▎ See Fig. 5.20. Test $(1, 0)$: $1 + 0 \geq 0$? Yes!

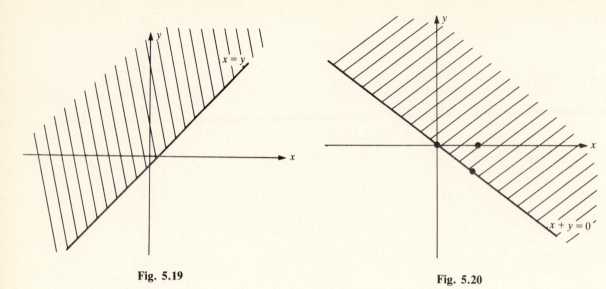

Fig. 5.19

Fig. 5.20

5.218 $4x - y > 8$

▮ See Fig. 5.21. Use a dotted line. Check $(0,0)$: $0 - 0 > 8$? No!

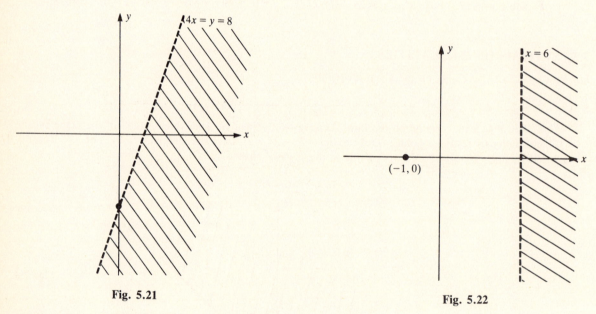

Fig. 5.21

Fig. 5.22

5.219 $x > 6$

▮ See Fig. 5.22. $x > 6$ is the region to the right of $x = 6$. Test $(-1, 0)$.

5.220 $y \le -1$

▮ See Fig. 5.23. Use a solid line here; all points below $y = -1$ satisfy $y \le -1$.

5.221 $-5 < x \le 1$

▮ See Fig. 5.24. Be careful! One line is dotted, and the other is not.

5.222 $1 \le y < 0$

▮ There is no region satisfying this inequality. If $1 \le y$, then $y \ge 1$; thus, y is not less than zero. No solution.

For Probs. 5.223 to 5.237, find the solution set of each system graphically.

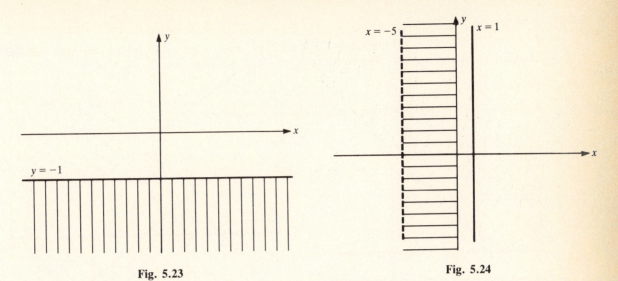

Fig. 5.23

Fig. 5.24

5.223 $-2 \le x < 2$, $-1 < y \le 6$

▮ See Fig. 5.25.   We graph $2 \le x < 2$ and $-1 < y \le 6$ on the same set of axes and find where they intersect.   Notice the cross-hatched region: That is the solution set.

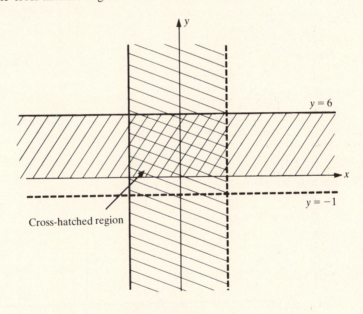

Fig. 5.25

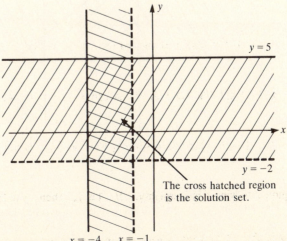

Fig. 5.26

5.224 $-4 \le x < -1$, $-2 < y \le 5$

┃ See Fig. 5.26. We sketch the regions $-4 \le x < -1$ and $-2 < y \le 5$ on the same axes and find the intersection of these regions. The crosshatched region is the solution set.

5.225 $x < 5$, $y > 2$

┃ See Fig. 5.27. The crosshatched quadrant is the solution set.

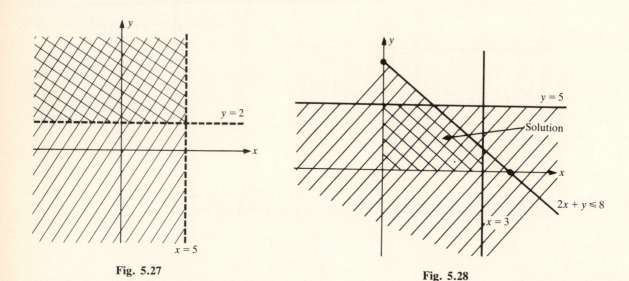

Fig. 5.27   Fig. 5.28

5.226 $2x + y \le 8$, $0 \le x \le 3$, $0 \le y \le 5$

┃ See Fig. 5.28. Use the same technique as in the case of two inequalities. Use all solid lines, and use $(0,0)$ as a test for $2x + y \le 8$. The solution set is the crosshatched area.

5.227 $x + 3y \le 12$, $0 \le x \le 8$, $0 \le y \le 3$

See Fig. 5.29. All lines are solid. Use $(0,0)$ as a test for $x + 3y \le 12$.

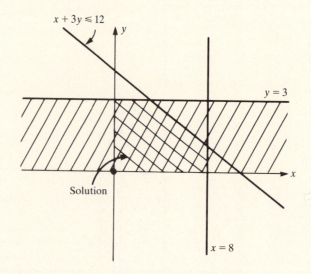

Fig. 5.29

5.228 $2x + y \le 8$, $x + 3y \le 12$, $x \ge 0$, $y \ge 0$

┃ See Fig. 5.30. The solution set is in quadrant I since x, $y \ge 0$. Use $(0,0)$ as a test point for both lines.

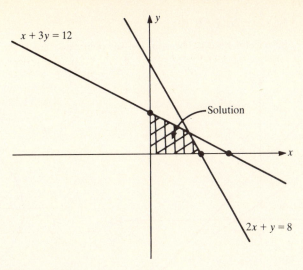

Fig. 5.30

5.229 $x + y \geq 1$, $2x - 3y \leq 6$, $x \geq 0$, $y \geq 0$

▮ See Fig. 5.31. Use $(0,0)$ as a test point for both lines. Use quadrant I only!

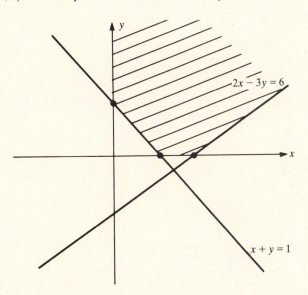

Fig. 5.31

5.230 $x + 2y \leq 10$, $3x + y \leq 15$, $x \geq 0$, $y \geq 0$

▮ See Fig. 5.32. Once again use solid lines and $(0,0)$ as a test point.

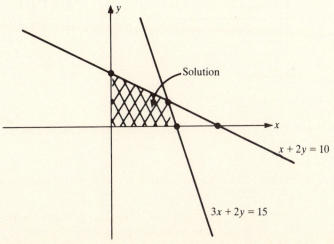

Fig. 5.32

5.231 $3x + 4y \geq 8$, $4x + 3y \geq 24$, $x, y \geq 0$

▮ See Fig. 5.33.   Be careful here!   The solution set is in quadrant I.

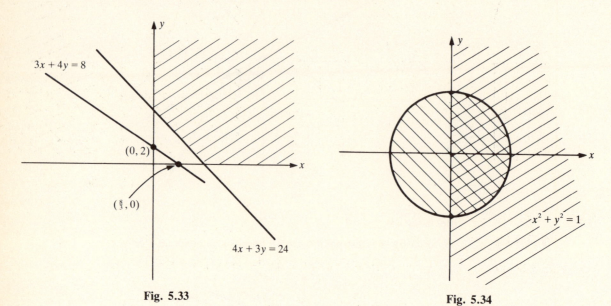

Fig. 5.33   Fig. 5.34

5.232 $x^2 + y^2 \leq 1$, $x \geq 0$

▮ See Fig. 5.34.   Test $(0, 0)$.   The inside of the circle satisfies the inequality.   The crosshatched region is the solution set.

5.233 $y \geq x^2$, $x \geq 0$

▮ See Fig. 5.35.   Use $(0, 1)$ as a test. $1 > 0^2$?   Yes!   The crosshatched region is the solution set.

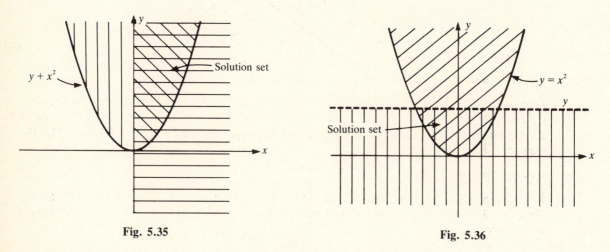

Fig. 5.35   Fig. 5.36

5.234 $y \geq x^2$, $y < 1$

▮ See Fig. 5.36 and Prob. 5.233.   The crosshatched region is the solution set.

5.235 $y > x^2$, $y \leq 1$

▮ See Fig. 5.37 and Prob. 5.234.   The crosshatched region is the solution set.

5.236 $y \leq x^2$, $y \leq 0$

▮ See Fig. 5.38.   The solution set here is the half plane below $y = 0$.

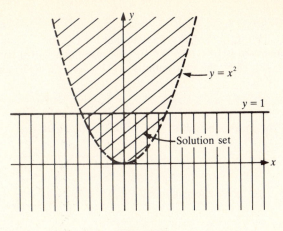

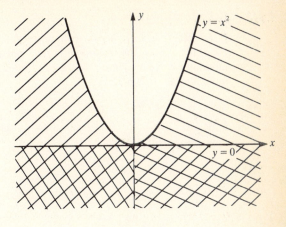

Fig. 5.37   **Fig. 5.38**

5.237 $y \le x^2$, $y > 1$

    ▮   See Fig. 5.39.

For Probs. 5.238 to 5.241, find a parametric representation for the line segment $\overline{P_1 P_2}$.

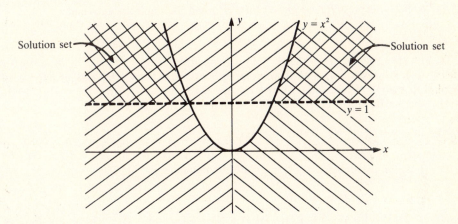

Fig. 5.39

5.238 $P_1(2, 3)$, $P_2(5, 8)$

    ▮   If the line l connects $P_1(x_1, y_1)$ and $P_2(x_2, y_2)$, then $x = x_1 + t(x_2 - x_1)$ and $y = y_1 + t(y_2 - y_1)$ for any $t \in \mathcal{R}$ represents (x, y) on l. In this case, $x = 2 + t(5 - 2)$ and $y = 3 + t(8 - 3)$, or $x = 2 + 3t$ and $y = 3 + 5t$. These last two equations represent (x, y) on line $\overleftrightarrow{P_1 P_2}$. If $0 \le t \le 1$, the segment $\overline{P_1 P_2}$ is represented.

5.239 $P_1(3, 5)$, $P_2(-1, 2)$

    ▮   See Prob. 5.238. $x_1 = 3$, $x_2 = -1$, $y_1 = 5$, $y_2 = 2$. Then $x = 3 + t(-1 - 3)$ and $y = 5 + t(2 - 5)$. Then $x = 3 - 4t$ and $y = 5 - 3t$. These equations represent the line. If $0 \le t \le 1$, they represent the line segment.

5.240 $P_1(-4, 0)$, $P_2(2, -5)$

    ▮   Here, $x_1 = -4$, $y_1 = 0$, $x_2 = 2$, $y_2 = -5$. Then $x = -4 + t[2 - (-4)]$ and $y = 0 + t(-5 - 0)$, or $x = -4 + 6t$ and $y = -5t$. If $0 \le t \le 1$, these equations represent the line segment.

5.241 $P_1(7, 3)$, $P_2(-2, -3)$

    ▮ $x = 7 + t(-2 - 7)$ and $y = 3 + t(-3 - 3)$, or $x = 7 - 9t$ and $y = 3 - 6t$. If $0 \le t \le 1$, these equations represent the line segment.

For Probs. 5.242 to 5.246, find the maximum and minimum values for $f(t)$ in the given range.

5.242 $f(t) = 2t + 5, \quad 0 \le t \le 4$

▮ If $f(t) = at + b, \quad a \ne 0$ (where $a, \ b \in \mathcal{R}$), then if $c \le t \le d$, the extrema for f occur at c and d. Also, if $a > 0$, the maximum is $f(d)$ and the minimum is $f(c)$. If $a < 0$, the maximum is $f(c)$ and the minimum is $f(d)$. Here, $a = 2 > 0$; $c = 0$, $d = 4$. Thus, the maximum is $f(4) = 4(2) + 5 = 13$ and the minimum is $f(0) = 5$.

5.243 $f(t) = 3t - 2, \quad -2 \le t \le 3$

▮ Here, $a = 3 > 0$ (see Prob. 5.242). Thus, the maximum $= f(d) = f(3) = 9 - 2 = 7$, and the minimum $= f(c) = f(-2) = -6 - 2 = -8$.

5.244 $f(t) = -3t + 2, \quad 1 \le t \le 6$

Then $a = -3 < 0$. The maximum $= f(c) = f(1) = -1$, and the minimum $= f(d) = f(6) = -18 + 2 = -16$.

5.245 $f(t) = -t - 4, \quad -5 \le t \le 2$

▮ $a = -1 < 0$. The maximum $= f(c) = f(-5) = 1$, and the minimum $= f(d) = f(2) = -6$.

5.246 $f(t) = 7, \quad 0 \le t \le 5$

▮ The above theorem does not apply since $a = 0$. However, if $f(t) = 7$, then $f = 7$ for all t, and f is constant.

For Probs. 5.247 to 5.249, express $f(x, y)$ as $g(t)$ for $\overline{P_1 P_2}$, and then find the extrema.

5.247 $f(x, y) = 3x + 2y - 5; \quad P_1 = (2, 1), \quad P_2 = (8, 6)$

▮ $f(x, y) = f[x_1 + t(x_2 - x_1), \ y_1 + t(y_2 - y_1)] = ax_1 + by_1 + c + [a(x_2 - x_1) + b(y_2 - y_1)]t = g(t)$, where $P_1 = (x_1, y_1)$, $P_2 = (x_2, y_2)$, and $f(x, y) = ax + by + c$. Then $g(t) = ak_1 + by_1 + c + [a(x_2 - x_1) + b(y_2 - y_1)]t = g(t)$. Thus, $g(t) = 3(2) + 2(1) + (-5) + [3(x - 2) + 2(6 - 1)]t = 3 + [18 + 10]t = 3 + 28t$. Since $28 > 0$, the maximum is $g(1) = 31$ and the minimum is $g(0) = 3$.

5.248 $f(x, y) = 2x - y + 3; \quad P_1 = (-3, 0), \quad P_2 = (2, 3)$

▮ $g(t) = ax_1 + by_1 + c + t[a(x_2 - x_1) + b(y_2 - y_1)] = 2(-3) + (-1)(0) + 3 + t[2(2 + 3) + (-1)(3 - 0)] = -3 + 7t$; $7 > 0$. Thus, the maximum is $g(1) = 4$ and the minimum is $g(10) = -3$.

5.249 $f(x, y) = -x + 4y + 2; \quad P_1 = (4, -3), \quad P_2 = (-1, 4)$

▮ $g(t) = (-1)(4) + 4(-3) + 2 + t[-1(-1 - 4) + 4(4 + 3)] = -14 + 33t$. The maximum $= 19 = g(1)$ and the minimum $= -14 = g(0)$.

For Probs. 5.250 and 5.251, find the extrema for f on $\overline{P_1 P_2}$.

5.250 $f(x, y) = 5x + 2y - 3; \quad P_1(2, 3), \quad P_2(5, -1)$

▮ The extrema must occur at P_1 and P_2, the endpoints of $\overline{P_1 P_2}$. Thus, the maximum is $f(P_1)$ or $f(P_2)$, and the minimum is $f(P_1)$ or $f(P_2)$. $f(P_1) = f(2, 3) = 5(2) + 2(3) - 3 = 13$, and $f(P_2) = f(5, -1) = 5(5) + 2(-1) - 3 = 20$. Since $20 > 13$, the maximum $= 20$ and the minimum $= 13$.

5.251 $f(x, y) = -4x - 2y + 2; \quad P_1(3, 2), \quad P_2(-2, -4)$

▮ $f(P_1) = -4(3) - 2(2) + 2 = -14$, and $f(P_2) = -4(-2) - 2(-4) + 2 = 18$. Since $18 > -14$, $f(P_1) =$ the minimum $= -14$ and $f(P_2) =$ the maximum $= 18$.

For Probs. 5.252 to 5.257, find the maximum and minimum values of f on the set S determined by the given linear inequalities.

5.252 $-2x + 3y - 6 \le 0, \quad y \ge 0, \quad x \le 0; \quad f(x, y) = 2x + y - 1$

▮ In Fig. 5.40, we notice that S is convex. Thus, the maximum and minimum of f on S occur at the vertices. $f(0, 0) = 0 + 0 - 1 = -1$. $f(-3, 0) = -6 + 0 - 1 = -7 =$ minimum. $f(0, 2) = 0 + 2 - 1 = 1 =$ maximum.

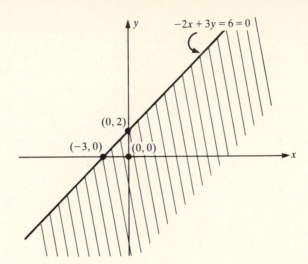

Fig. 5.40

5.253 $2x - y + 2 \geq 0$, $\quad x + y - 2 \leq 0$, $\quad y \geq 0$, $\quad f(x, y) = -x + 3y - 5$

$\quad$ **▐** See Fig. 5.41. S has vertices $(-1, 0)$, $(0, 2)$, $(2, 0)$ and is convex. Then $f(-1, 0) = 1 - 5 = -4$. $f(2, 0) = -2 - 5 = -7 = $ minimum. $f(0, 2) = 6 - 5 = 1 = $ maximum.

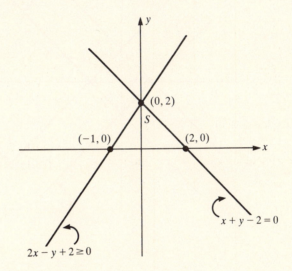

Fig. 5.41

5.254 $2x - y + 2 \geq 0$, $\quad x + y - 2 \leq 0$, $\quad y \geq 0$; $\quad f(x, y) = 2x - 6y + 8$

$\quad$ **▐** See Prob. 5.253. This is the same region. Then $f(-1, 0) = -2 + 8 = 6$. $f(2, 0) = 4 + 8 = 12 = $ maximum. $f(0, 2) = -12 + 8 = -4 = $ minimum.

5.255 $3x - 2y + 6 \geq 0$, $\quad x + y + 2 \geq 0$, $\quad x - y - 3 \leq 0$, $\quad x + y - 3 \leq 0$; $\quad f(x, y) = x + y - 8$

$\quad$ **▐** See Fig. 5.42. Then $f(0, 3) = 0 + 0 - 8 = -8$. $f(-2, 0) = -2 + 0 - 8 = -6$. $f(3, 0) = 3 + 0 - 8 = -5 = $ maximum. $f(\frac{1}{2}, -2\frac{1}{2}) = \frac{1}{2} - 2\frac{1}{2} - 8 = -10 = $ minimum.

5.256 $x - y + 1 \geq 0$; $\quad x + y + 1 \geq 0$, $\quad -x + y + 1 \geq 0$, $\quad -x - y + 1 \geq 0$; $\quad f(x, y) = -4x + 3y + 8$

$\quad$ **▐** See Fig. 5.43. Then $f(0, 1) = 8 + 3 = 11$. $f(1, 0) = 8 - 4 = 4 = $ minimum. $f(-1, 0) = 8 + 4 = 12 = $ maximum. $f(0, -1) = 8 - 3 = 5$.

5.257 $x - y + 1 \geq 0$, $\quad x + y + 1 \geq 0$, $\quad -x + y + 1 \geq 0$, $\quad -x - y + 1 \geq 0$; $\quad f(x, y) = x$

$\quad$ **▐** S here is the same region as in Prob. 5.256 above. Then $f(0, 1) = x = 0$. $f(-1, 0) = x = 1 = $ maximum. $f(-1, 0) = x = -1 = $ minimum. $f(0, -1) = x = 0$.

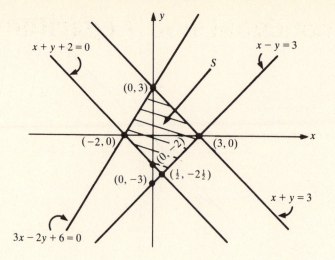

Fig. 5.42

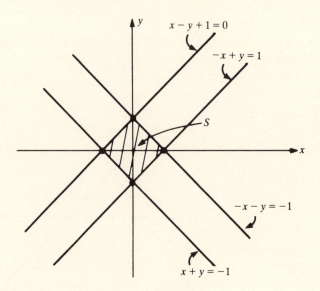

Fig. 5.43

CHAPTER 6
Exponential and Logarithmic Functions

6.1 EXPONENTIAL FUNCTIONS

For Probs. 6.1 to 6.20, sketch the graph of the given equation.

6.1 $y = -2^x$

▮ See Fig. 6.1. If $x = 0$, $y = -2^0 = -1$; $-2^x \neq 0$ implies that y is never zero (the x axis is an asymptote); as x gets large, y decreases.

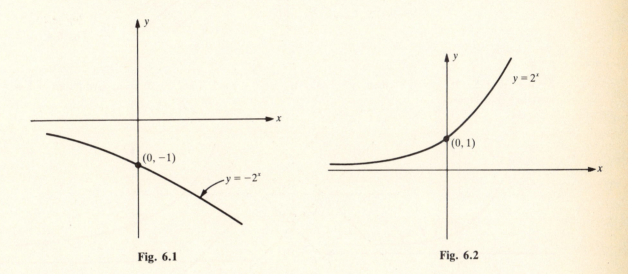

Fig. 6.1   Fig. 6.2

6.2 $y = 2^x$

▮ See Fig. 6.2. If $x = 0$, $y = 2^0 = 1$; x axis is an asymptote; as x increases, y increases.

6.3 $y = 2^x$, $y = 3^x$ on the same axes.

▮ See Fig. 6.3. $y = 3^x$ is steeper; both equations increase as x increases, but 3^x increases more dramatically.

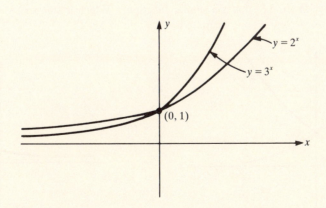

Fig. 6.3

6.4 $y = -2^x$, $y = -3^x$ on the same axes.

▮ See Fig. 6.4. The reasoning is the same as in Prob. 6.4. $y = -3^x$ will decrease more dramatically.

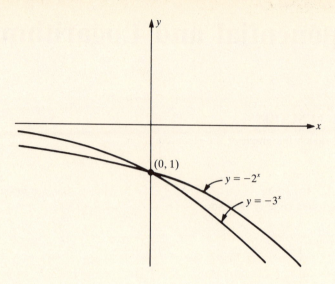

Fig. 6.4

6.5 $y = 3^{-x}$

▮ See Fig. 6.5. If $y = 3^{-x}$, then $y = 1/3^x$. If $x = 0$, $y = 1$. The x axis is an asymptote; and as x increases, y decreases.

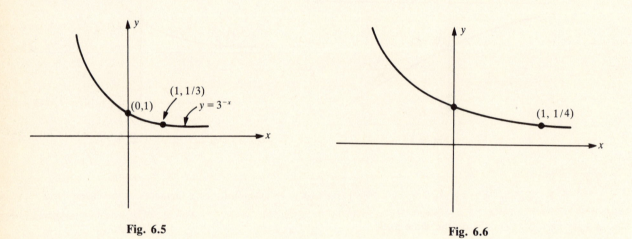

Fig. 6.5

Fig. 6.6

6.6 $y = 4^{-x}$

▮ See Fig. 6.6. See Prob. 6.5; this is a similar function. Which is steeper?

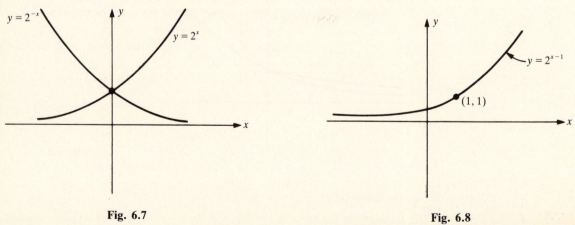

Fig. 6.7

Fig. 6.8

6.7 $y = z^x$ and $y = 2^{-x}$ on the same axes.

▮ See Fig. 6.7. Since $2^x \cdot 2^{-x} = 1$,   we suspect that something interesting will occur. Notice that one curve is the image of the other in the y axis.

6.8 $y = 2^{x-1}$

▮ See Fig. 6.8. Here, if $x = 1$, $y = 2^0 = 1$. Otherwise, this graph is very similar to that of $y = 2^x$. Since y is never 0, and 2^{x-1} gets close to 0 as x gets large through the negatives, the x axis is an asymptote.

6.9 $y = -2^{-x}$

▮ See Fig. 6.9. Look at the graph for Prob. 6.5 (Fig. 6.5). The graph of $y = 2^{-x}$ is similar. The graph of $y = 2^{-x}$ is the ordinate of each point in $y = 2^{-x}$ negated.

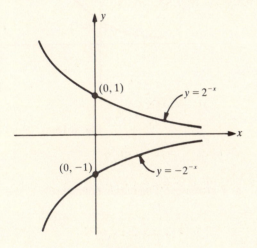

Fig. 6.9

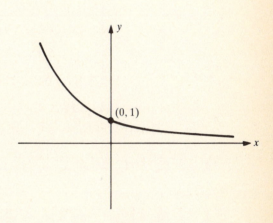

Fig. 6.10

6.10 $y = 3^{-0.5x}$

▮ See Fig. 6.10. $3^{-0.5x} = 1/3^{0.5x}$. If $x = 0$, $y = 1$. The x axis is again an asymptote, since y approaches 0 as x grows large.

6.11 $y = 2 + 2^x$

▮ See Fig. 6.11. We take the graph of 2^x and add 2 to each ordinate.

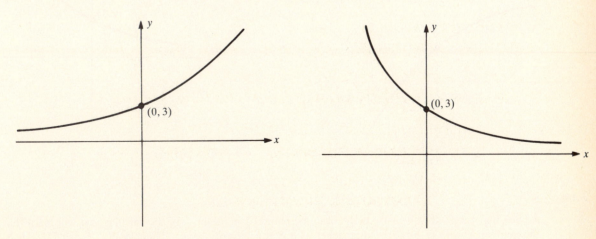

Fig. 6.11   Fig. 6.12

6.12 $y = 3^{1-x}$

❚ See Fig. 6.12. If $x = 3$, $y = 3^0 = 1$. We notice that the x axis is again an asymptote. Also, as x takes on large negative values, y gets large.

6.13 $y = 2^{|x|}$

See Fig. 6.13. If $x = 0$, $y = 1$. For $x > 0$, $2^{|x|} = 2^x$; for $x < 0$, $2^{|x|} = 2^{-x}$. We combine these two graphs.

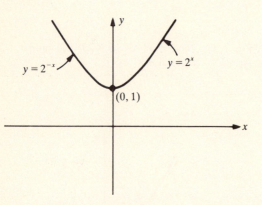

Fig. 6.13

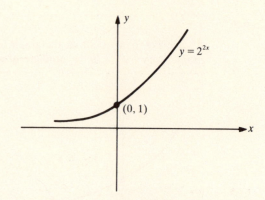

Fig. 6.14

6.14 $y = 2^{2x}$

❚ See Fig. 6.14. When $x = 0$, $y = 1$; besides that, as x increases, y increases dramatically.

6.15 $y = -2^{2x}$

❚ See Fig. 6.15 and Prob. 6.14; we negate each ordinate.

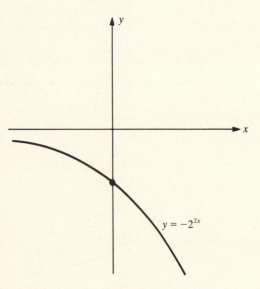

Fig. 6.15

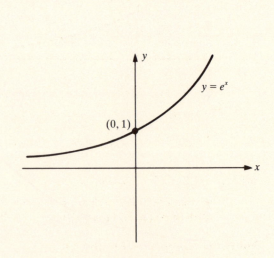

Fig. 6.16

6.16 $y = e^x$

❚ See Fig. 6.16. Recall that e is the base for natural logarithms. In the calculus you will learn that e is the number which $(1 + 1/b)^n$ approaches as n gets arbitrarily large. Numerically, $e \approx 2.718$; it is an irrational number. Thus, when $x = 0$, $e^0 = 1$; as x increases, so does e^x.

6.17 $y = 2e^{-x}$

▍ See Fig. 6.17. $2e^{-x} = 2(1/e^x)$.   If $x = 0$, $1/e^x = 1$;   then $2(1/e^x) = 2(1) = 2$.   As x gets large, $2/e^x$ decreases.

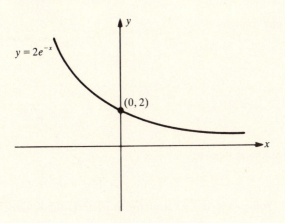

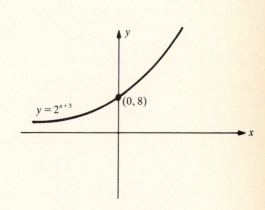

Fig. 6.17   **Fig. 6.18**

6.18 $y = 2^{x+3}$

▍ See Fig. 6.18.   There are two ways (at least!) in which we can obtain the graph.   Notice that $2^{x+3} = 2^x 2^3 = 8 \cdot 2^x$.   We can get the same graph by noting directly that when $x = 0$, $y = 2^3 = 8$.

6.19 $y = 2e^{-x} + 5$

▍ See Fig. 6.19.   We look at Fig. 6.17, and notice that if we add 5 to each ordinate, we get our function.

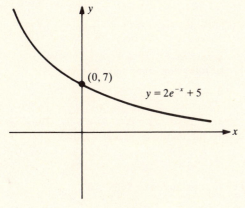

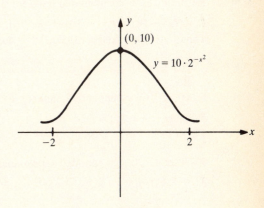

Fig. 6.19   **Fig. 6.20**

6.20 $y = 10 \cdot 2^{-x^2}$, $-2 \le x \le 2$.

▍ See Fig. 6.20.   When $x = 2$, $y = 10 \cdot (1/2^4) = \frac{10}{16}$.   When $x = -2$, $y = \frac{10}{16}$.   When $x = 0$, $y = 10 \cdot 1/2^0 = 10$.   As x goes from -2 to 0, y increases; as x goes from 0 to 2, y decreases.

For Probs. 6.21 to 6.27, let $f(x) = a^x$.

6.21   Prove that $f(x + 2) = f(x) \cdot f(2)$.

▍ If $f(x) = a^x$,   then $f(x + 2) = a^{x+2} = a^x a^2$.   Since $a^x = f(x)$ and $a^2 = f(2)$, $f(x + 2) = a^x a^2 = f(x) \cdot f(2)$.

6.22   Generalize the statement in Prob. 6.21, and prove your generalization.

▍ Prove that $f(x + y) = f(x)f(y)$.   Proof: $f(x + y) = a^{x+y} = a^x a^y = f(x)f(y)$.

6.23 Prove that $f(x-y)=f(x)/f(y)$.

▮ $f(x-y)=a^{x-y}=a^x/a^y=f(x)/f(y)$.

6.24 Prove that f is one-to-one.

▮ Suppose that $m\neq n$. Then $a^m\neq a^n$; f is one-to-one.

6.25 Prove that $f(-x)=1/f(x)$.

▮ $f(x)=a^x$; then $f(-x)=a^{-x}=1/a^x=1/f(x)$.

6.26 Prove that $f(2+x)=a^2f(x)$.

▮ $f(2+x)=a^{2+x}=a^2a^x=a^2a^x$.

6.27 Generalize the statement in Prob. 6.26, and prove your statement.

▮ Prove that $f(b+x)=a^bf(x)$. *Proof*: $f(b+x)=a^{b+x}=a^ba^x=a^bf(x)$.

For Probs. 6.28 to 6.30, compute the compound amount. You should use a calculator to perform the arithmetic.

6.28 \$2000 at 12% compounded semiannually for 3 years.

▮ $A(n)=P(1+i)^n$, where $A(n)=$ compound amount at the end of n interest periods, $i=$ rate per interest period, $P=$ amount invested. Here, $n=6$, $i=0.12/2=0.06$. Then $A(6)=\$2000(1+0.06)^6=\$2000(1.06)^6\approx\$2000(1.42)$ (rounded to two places on the calculator) $=\$2840$.

6.29 \$6000 at 15% compounded quarterly for 4 years, 6 months.

▮ $A(n)=P(1+i)^n$, where $n=18$, $P=\$6000$, $i=0.15/4=0.0375$. Then $A(18)=\$6000(1+0.0375)^{18}=\$6000(1.0375)^{18}\approx\$6000(1.94)$ (rounded to the hundreds place) $=\$11\,640$.

6.30 \$5050 at $11\frac{3}{4}$% compounded daily for 2 years.

▮ $A(n)=P(1+i)^n$, where $n=730$, $P=\$5050$, $i=0.1175/365\approx0.000\,32$ (rounded to five places). Then $A(730)=\$5050(1.000\,32)^{730}=\$6379$.

For Probs. 6.31 to 6.33, compute the principal P invested to yield the following compound amounts A.

6.31 \$5000 of 10% compounded annually for 5 years.

▮ Since $A=P(1+i)^n$, $P=A(1+i)^{-n}$; here, $A=\$5000$, $i=0.1/1=0.1$, $n=5.0$. Then $P=\$5000(1+0.1)^{-5}=\$5000(1.1)^{-5}\approx\$5000(0.62)$ (rounded to the hundreds place) $=\$3100$.

6.32 \$6750 at $12\frac{1}{2}$% compounded semiannually for 3 years.

▮ $P=\$6750(1+0.0625)^{-6}=\$6750(1.0625)^{-6}\approx\$6750(0.69)$ (rounded) $=\$4657.50$.

6.33 \$10 000 at $14\frac{3}{4}$% compounded quarterly for $6\frac{1}{2}$ years.

▮ $P=\$10\,000(1+0.036\,875)^{-26}=\$10\,000(1.036\,875)^{-26}\approx\$10\,000(0.39)$ (rounded) $=\$3900$.

For Probs. 6.34 to 6.36, compute the amount due in each, given that the interest is compounded continuously.

6.34 \$3000 at 10% for 5 years.

▮ $A=Pe^{rt}=\$3000(e^{5(0.1)})=\$3000(e^{0.5})\approx\$4946$ (rounded to the nearest dollar).

6.35 \$4550 at $12\frac{1}{2}$% for 3 years.

▮ $A=\$4550(e^{(0.125)(3)})=\$4550(e^{0.375})\approx\$6620$.

6.36 \$7500 at $16\frac{1}{4}$% for $4\frac{1}{2}$ years.

▮ $A=\$7500(e^{0.731\,25})\approx\$15\,581$. (Does that amount shock you?)

For Probs. 6.37 to 6.40, answer true or false, and explain your answer.

6.37 If $a > b$, and $n \in \mathscr{L}$, then $a^n > b^n$.

$\blacksquare$ False. $4 > 2$, but $4^{-1} = \frac{1}{4}$, $2^{-1} = \frac{1}{2}$, and $\frac{1}{4} < \frac{1}{2}$.

6.38 If $a > b$, and $n \in \mathscr{N}$, then $a^{n^2} > b^{n^2}$.

$\blacksquare$ True. $a^n > b^n \forall n \in \mathscr{L}$, but $n^2 \in \mathscr{N}$. Thus, $a^{n^2} > b^{n^2}$.

6.39 The graphs of $y = a^x$ and $y = b^x$ (where $a \neq b$; $a, b \in \mathscr{N}$) *must* intersect $\forall a, b$ so chosen.

$\blacksquare$ True. If $x = 0$, $a^0 = b^0 = 1$; they intersect at $(0, 1)$.

6.40 The graph of $y = a^x$ and $y = 2a^x$ have no points of intersection.

$\blacksquare$ True. Suppose that $a^x = 2a^x$; $a^x \neq 0$, so we can divide by a^x. Then $1 = 2$.

6.2 LOGARITHMIC FUNCTIONS

For Probs. 6.41 to 6.45, rewrite in an equivalent exponential form.

6.41 $\log_{10} 100 = 2$

$\blacksquare$ Remember that if $\log_a b = x$, then $a^x = b$, and conversely. Thus, if we let $a = 10$, $b = 100$, $x = 2$, then $\log_{10} 100 = 2$ and $10^2 = 100$.

6.42 $\log_{10} 10\,000 = 4$

$\blacksquare$ If $\log_{10} 10\,000 = 4$, then $10^4 = 10\,000$. Remember: A logarithm is an exponent.

6.43 $\log_2 8 = 3$

$\blacksquare$ See Prob. 6.41. Here, $a = 2$, $b = 8$, $x = 2$, and $2^3 = 8$.

6.44 $\log_4 64 = 3$

$\blacksquare$ Using the formula in Prob. 6.41, where $a = 4$, $b = 64$, $x = 3$, we have $4^3 = 64$.

6.45 $\log_{14} 1 = 10$

$\blacksquare$ $a = 14$, $b = 1$, $x = 0$, and $14^0 = 1$.

For Probs. 6.46 to 6.51, rewrite in an equivalent logarithmic form.

6.46 $7^2 = 49$

$\blacksquare$ Again, we use the formula $a^x = b \Leftrightarrow \log_a b = x$. Thus, $7^2 = 49 \Leftrightarrow \log_7 49 = 2$.

6.47 $4^c = 256$

$\blacksquare$ Here, $a = 4$, $b = 256$, $x = c$, and $\log_4 256 = c$.

6.48 $u = v^x$

$\blacksquare$ If $u = v^x$, then $a = v$, $b = u$, $x = x$, and $\log_v u = x$.

6.49 $9 = 27^{2/3}$

$\blacksquare$ Then $\log_{27} 9 = \frac{2}{3}$.

6.50 $625^{0.25} = 5$

$\blacksquare$ Do not be fooled or misled by decimals (or anything else!). The logarithm-exponential conversion still holds; $\log_{625} 5 = 0.25$.

6.51 $729^{1/6} = 3$

$\blacksquare$ $\log_{729} 3 = \frac{1}{6}$.

For Probs. 6.52 to 6.66, evaluate the given expression.

6.52 $\log_2 8$

▮   We are looking for x such that $\log_2 8 = x$.   Then $2^x = 8$,   but $x = 3$,   so $\log_2 8 = 3$.

6.53 $\log_{10} 3$

▮   Remember that $\log a$ means $\log_{10} a$, and $\ln a$ means $\log_e a$.   We let $\log_{10} 3 = x$ which means $10^x = 10^3$.   Then $x = 3$ and $\log_{10} 3 = 3$.   Also, remember that $\log_q r = s$ means "s is the power to which q is raised to yield r."   To what power must 10 be raised to give 10^3?

6.54 $\log_5 125$

▮   If $t = \log_5 125$,   then $5^t = 125$; $t = 3$.

6.55 $\log_a a^z$

▮   To yield a^2, raise a to the second power. $\log_a a^2 = 2$.

6.56 $\log_2 \frac{1}{16}$

▮   If $2^x = \frac{1}{16}$,   then $2^{-x} = 16$,   and $-x = 4$,   or $x = -4$.   See Prob. 6.57.

6.57 $\log_3 \frac{1}{27}$

▮   Compare this to Prob. 6.56.   If $\log_3 \frac{1}{27} = x$,   then $3^x = \frac{1}{27}$,   which means $\left(\frac{1}{3}\right)^{-x} = \left(\frac{1}{3}\right)^3$,   and $-x = 3$,   or $x = -3$.

6.58 $\log_3 1$

▮   If $\log_3 1 = y$,   then $3^y = 1$,   or $y = 0$.

6.59 $\log_{4000} 1$

▮   If $4000^x = 1$, $x = 0$.

6.60 $\ln e^{-2}$

▮ $\ln e^{-2} = \log_e e^{-2}$.   Then observe directly that $\ln e^{-2} = -2$,   or that $\log_e 2^{-2} = x$ means $e^x = e^{-2}$,   or $x = -2$.

6.61 $\log_{32} 2$

▮   If $\log_{32} 2 = p$,   then $32^p = 2$,   or $(2^5)^p = 2$.   Then $5p = 1$,   and $p = \frac{1}{5}$.

6.62 $\log_2 2^{-4}$

▮   Observe directly that $\log_2 2^{-4} = -4$,   or that if $\log_2 2^{-4} = 5$,   then $2^s = 2^{-4}$,   or $s = -4$.

6.63 $\log_b b^u$

▮   If $\log_b b^u = k$,   then $b^k = b^u$,   or $k = u$.   Then $\log_b b^u = u$.   Think about it!   Doesn't this make sense?

6.64 $\log_b b^{uv}$

▮   If $\log_b b^{uv} = k$,   then $b^k = b^{uv}$,   or $k = uv$.   Notice that $k = uv$ also follows directly from the result in Prob. 6.63.

6.65 $\log_2 \sqrt{8}$

▮   If $\log_2 \sqrt{8} = x$,   then $2^x = \sqrt{8} = \sqrt{2^3} = 2^{3/2}$,   and $x = \frac{3}{2}$.

6.66 $\log_5 \sqrt[3]{5}$

▮   If $\log_5 \sqrt[3]{5} = x$,   then $5^x = \sqrt[3]{5} = 5^{1/3}$,   and $x = \frac{1}{3}$.

For Probs. 6.67 to 6.82, solve the given equation.

6.67 $\log 100 = x$

▮ If $\log 100 = x$, then $\log_{10} 100 = x$. Thus, $10^x = 100 = 10^2$ and $x = 2$.

6.68 $\log x = 2$

▮ If $\log x = 2$, then $\log_{10} x = 2$. Thus, $10^2 = x$, and $x = 100$.

6.69 $\log_2 x = 1$

▮ If $\log_2 x = 1$, then $2^1 = x$, and $x = 2$.

6.70 $\log_3 3 = x$

▮ $3^x = 3$ (Ask yourself, "which 3 in the equation is the base?") Then $3^x = 3^1$, and $x = 1$.

6.71 $\log x = 0$

▮ If $\log x = 0$, then $\log_{10} x = 0$, and $10^0 = x$, or $x = 1$.

6.72 $\log_x 81 = 4$

▮ $\log_x 81 = 4$. Then $x^4 = 81$, or $x = 3$ ($3^4 = 81$).

6.73 $\log_x \frac{1}{27} = -3$

▮ $\log_x \frac{1}{27} = -3$. Then $x^{-3} = \frac{1}{27} = 1/3^3 = 3^{-3}$, or $x = 3$.

6.74 $\ln e^x = 5$

▮ If $\ln e^x = 5$, then $\log_e e^x = 5$. Thus, $e^5 = e^x$, or $x = 5$.

6.75 $\ln e^{x+2} = 7$

▮ Then $\log_e e^{x+2} = 7$. Thus, $e^7 = e^{x+2}$, $x + 2 = 7$, or $x = 5$.

6.76 $\log_x 27x = 4$

▮ Then $x^4 = 27x$, $x^4 - 27x = 0$, $x(x^3 - 27) = 0$, and $x = 0$, $x = 3$. But $x = 0$ is extraneous (0 is not a logarithmic base), so $x = 3$.

6.77 $\log_{49} \frac{1}{7} = y$

▮ Then $49^y = \frac{1}{7}$, $(7^2)^y = \frac{1}{7}$, $7^{2y} = \frac{1}{7} = 7^{-1}$, $2y = -1$, or $y = -\frac{1}{2}$.

6.78 $\log_b 1000 = \frac{3}{2}$

▮ Then $b^{3/2} = 1000$, but $1000 = 10^3$. Thus $1000 = 100^{3/2}$, $b^{3/2} = 100^{3/2}$, or $b = 100$.

6.79 $\log_b 4 = \frac{2}{3}$

▮ Then $b^{2/3} = 4$. Raise each side to the power $\frac{3}{2}$. Then $(b^{2/3})^{3/2} = 4^{3/2}$ and $b = 8$. Compare this with Prob. 6.77, which can be done using this technique as well.

6.80 $\log_b b = 1$

▮ If $\log_b b = 1$, then $b^1 = b$; b can be any positive real except 1. Thus, $b \neq 1$, $b > 0$ since b is the logarithm base.

6.81 $\log_b 1 = 0$

▮ Then $b^0 = 1$. This is true for all b. However, b must be positive and not 1 since b is the logarithm base.

6.82 $\log_{e^2} x = 10$

▮ Here, the base is e^2. Then $(e^2)^{10} = x$, or $x = e^{20}$.

For Probs. 6.83 to 6.87, evaluate the given expression where $f(x) = \log x$, $g(x) = 10^x$, $h(x) = \ln x$, $k(x) = e^x$, $l(x) = x^2$.

6.83 $f \circ g(x)$

▍ $f \circ g(x) = f(g(x)) = \log(10^x)$.   But $\log 10^x = x$. $f \circ g(x) = x$.

6.84 $g \circ f(x)$

▍ $g \circ f(x) = 10^{\log x} = x$ $(x > 0)$.

6.85 $h \circ k(x)$

▍ $h \circ k(x) = \ln e^x = \log_e e^x = x$.   Thus, $h \circ k(x) = x$.

6.86 $f \circ l(10)$

▍ $l(x) = x^2$;   thus, $l(10) = 10^2 = 100$.   Then $f \circ l(10) = f(l(10)) = \log 10^2 = 2$.

6.87 $l \circ h(3)$

▍ $h(x) = \ln x$;   thus, $h(3) = \ln 3$.   Then $l \circ h(3) = l(h(3)) = l(\ln 3) = (\ln 3)^2$.

For Prob. 6.88 to 6.97, sketch the given relation.

6.88 $y = \log_2 x$

▍ See Fig. 6.21. When $x = 1$, $y = \log_2 1 = 0$;   when $x = 2$, $y = \log_2 2 = 1$.   Also, as x takes on values between 0 and 1, y decreases: $\log_2 \frac{1}{2} = -1$, $\log_2 \frac{1}{4} = -2$, $\log_2 \frac{1}{8} = -3$,   etc.   The y axis is a vertical asymptote.

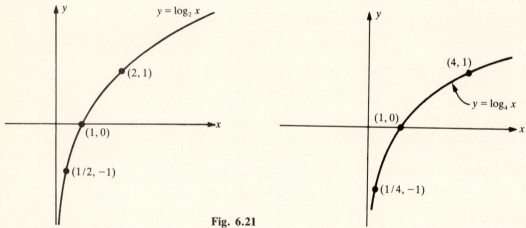

Fig. 6.21   Fig. 6.22

6.89 $y = \log_4 x$

▍ See Fig. 6.22 and Prob. 6.88. $\log_4 1 = 0$, $\log_4 4 = 1$, $\log_4 \frac{1}{4} = -1$.   The y axis is a vertical asymptote.

6.90 $y = \log_2 x$, $y = \log_4 x$ on the same axes.

See Fig. 6.23.

6.91 $y = \log_2 x$, $y = 2^x$ on the same axes.

▍ See Fig. 6.24. Since $y = \log_2 x$ and $y = 2^x$ are inverse functions, their graphs are mirror images about $y = x$.

6.92 $y = \log_3 |x|$

▍ See Fig. 6.25. $y = \log_3 x$ is defined $\forall x \in \mathcal{R}^+$.   However, $|x| \geq 0$ $\forall x \in \mathcal{R}$.   Thus, while the domain of $\log_3 x$ is $\{x | x \in \mathcal{R}^+\}$,   the domain of $\log_3 |x|$ is $\mathcal{R}$.   Since $\log_x |-x| = \log_3 |x| = y$, $f(x) = \log_x |x|$ is symmetric about the y axis.

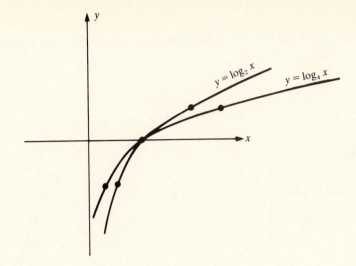

Fig. 6.23

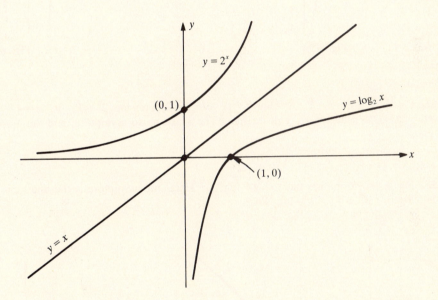

Fig. 6.24

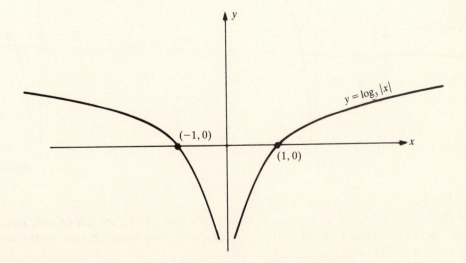

Fig. 6.25

6.93 $y = \log_5 (-x)$

▮ See Fig. 6.26. $y = \log_5 (-x)$ is defined only when $-x > 0$; $-x > 0$ implies that $x < 0$.   The domain of $f(x) = \log_5 (-x)$ is $\mathscr{R}^-$.

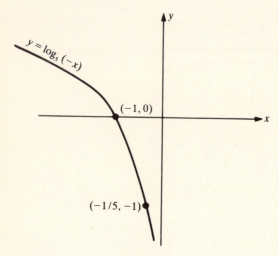

Fig. 6.26

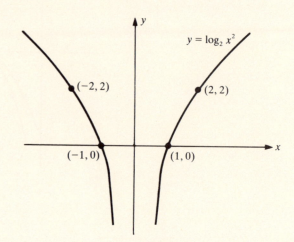

Fig. 6.27

6.94 $y = \log_2 x^2$

▮ See Fig. 6.27.   If $x = 1$, $y = \log_2 1 = 0$;   if $x = 2$, $y = \log_2 2^2 = 2$;   if $x = 4$, $y = \log_2 4^2 = \log_2 2^4 = 4$,   etc.   Also, as x decreases, y decreases.   The y axis as an asymptote, and the graph is symmetric about the y axis.

6.95 $y = \log_2 (x - 2)$

▮ See Fig. 6.28.   Compare this to $y = \log_2 x$.   Here, the asymptote is the line $x = 2$.   Also, if $x = 3$, $y = \log_2 1 = 0$.   If $x = 6$, $y = \log_2 4 = 2$.

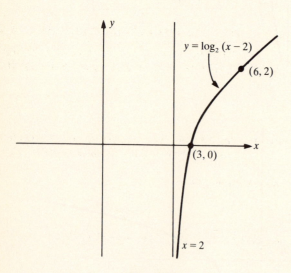

Fig. 6.28

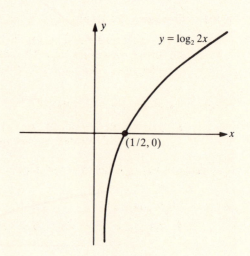

Fig. 6.29

6.96 $y = \log_2 2x$

▮ See Fig. 6.29.   Again, we look to $y = \log_2 x$ for help.   Here, we double each x value.   If $x = \frac{1}{2}$, then $y = \log_2 2x = \log_2 1 = 0$;   if $x = 1$, $y = \log_2 2 = 1$.   The asymptote remains the same.

6.97 $y = \log_{1/2} x$

▮ See Fig. 6.30. Be very careful here. The base $b < 1$. If $x = 1$, $y = \log_{1/2} 1 = 0$; if $x = \frac{1}{2}$, $y = \log_{1/2} \frac{1}{2} = 1$; if $x = 2$, $y = \log_{1/2} 2 = -1$. The y axis is an asymptote since as x decreases, y increases.

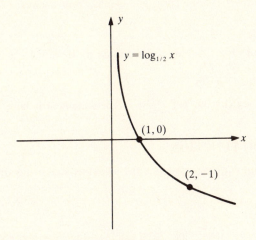

Fig. 6.30

6.98   Rewrite the equation $y = \log |x|$ without using absolute value signs.

▮
$$|x| = \begin{cases} x, x \ge 0 \\ -x, x < 0 \end{cases}$$

But $x = 0$ is not in the domain of this function. Thus, an equivalent definition is

$$y = \begin{cases} \log x, x > 0 \\ \log(-x), x < 0 \end{cases}$$

6.99   Prove that $g(x) = \log_a x$ is one-to-one.

▮ To prove this, we will use the definition of one-to-one. If $\log_a x = \log_a y$ and $\log_a x = p$, then $\log_a y = p$, and $a^p = x$, $a^p = y$. Thus, $x = y$. $g(x)$ is one-to-one.

6.100   Prove that the graphs of $y = \log_a x$ and $y = \log_{1/a} x$ are mirror images of each other about the x axis.

▮ We need to show that, for all x, $\log_a x = -\log_{1/a} x$. Then the y values are the same and the graphs will be mirror images. Let $y = \log_a x$; then, $a^y = x$, $1/a^y = 1/x$, $a^{-y} = 1/x$, $1/a^{-y} = 1/x$, $(1/a)^{-y} = x$, $-y = \log_{1/a} x$, and the proof is complete.

6.101   Prove that $\log 3$ is irrational.

▮ This will be a proof by contradiction. Suppose that $\log 3 = a/b$ with $b, a \in \mathscr{Z}$, $b \ne 0$, $(a, b) = 1$. Then $10^{a/b} = 3$, which implies that $10^a = 3^b$. Then, since $5|10^a, 5|3^b$. But 5 is *not* a factor of 3^b.

For Probs. 6.102 to 6.105, find the domain and range of the given function.

6.102 $y = \log_3(x + 1)$

▮ We know $x + 1 \ne 0$; thus, the domain $= \{x \in \mathscr{R} | x > -1\}$, and the range $= \mathscr{R}$.

6.103 $y = \log_3(2x - 5)$

▮ We must have $2x - 5 > 0$; thus, the domain $= \{x \in \mathscr{R} | x > \frac{5}{2}\}$, and the range $= \mathscr{R}$.

6.104 $y = \log_5(x^2 + 1)$

▮ Since $x^2 + 1 > 0 \forall x \in \mathscr{R}$, the domain $= \mathscr{R}$, and the range $= \mathscr{R}$.

6.105 $y = |\log_6 x|$

    ▮ $x > 0$ for $\log_6 x$ to be defined. But $|\log_6 x| \geq 0 \, \forall x \in \mathcal{R}$; thus, the domain $= \{x \in \mathcal{R} \,|\, x > 0\}$, and the range $= \{y \in \mathcal{R} \,|\, y \geq 0\}$.

For Probs. 6.106 and 6.107, answer true or false, and explain your answer.

6.106   If $x > y$, then $\log_a x > \log_a y$.

    ▮   False; $5 > 4$, but $\log_{1/2} 5 < \log_{1/2} 4$. In fact if $0 < a < 1$, $y = \log_a x$ is a decreasing function.

6.107   If $\log_a x = \log_b x$, then $a = b$.

    ▮ $\log_5 1 = 0$, and $\log_6 1 = 0$; but $5 \neq 6$. The statement is false.

6.3   PROPERTIES OF LOG FUNCTIONS

For Probs. 6.108 to 6.116, write each expression as the algebraic sum of logarithms. The base is any positive real except 1.

6.108 $\log (251)(46)(18)$

    ▮ $\log xy = \log x + \log y$; thus, $\log (251)(46)(18) = \log 251 + \log (46)(18) = \log 251 + \log 46 + \log 18$.

6.109 $\log (34)^2 (2.7)$

    ▮ $\log xy = \log x + \log y$, and $\log z^2 = 2 \log z$; thus, $\log (34)^2 (2.7) = \log (34)^2 + \log (2.7) = 2 \log 34 + \log 2.7$.

6.110 $\log (24)^{1/2} (35)^3$

    ▮   In general, $\log a^x = x \log a$; thus, $\log (24)^{1/2} (35)^3 = \log (24)^{1/2} + \log (35)^3 = \frac{1}{2} \log 24 + 3 \log 35$.

6.111 $\log \dfrac{(83)(41)}{29}$

    ▮ $\log (x/y) = \log x - \log y$; thus, $\log \dfrac{(83)(41)}{29} = \log (83)(41) - \log 29 = (\log 83 + \log 41) - \log 29$.

6.112 $\log \dfrac{(49)(65)}{(71)(86)}$

    ▮ $\log \dfrac{(49)(65)}{(71)(86)} = \log (49)(65) - \log (71)(86) = (\log 49 + \log 65) - (\log 71 + \log 86) = \log 49 + \log 65 -$ $\log 71 - \log 86$.

6.113 $\log \dfrac{(2.7)^2 (58)^{1/3}}{(75)(89)^2}$

    ▮ $\log \dfrac{(2.7)^2 (58)^{1/3}}{(75)(89)^2} = \log (2.7)^2 (58)^{1/3} - \log (75)(89)^2 = (2 \log 2.7 + \frac{1}{3} \log 58) - (\log 75 + 2 \log 89) = 2 \log 2.7$ $+ \frac{1}{3} \log 58 - \log 75 - 2 \log 89$.

6.114 $\log \sqrt{\dfrac{(87)(28)}{15}}$

    ▮ $\log \sqrt{\dfrac{(87)(28)}{15}} = \log \left[\dfrac{(87)(28)}{15} \right]^{1/2} = \frac{1}{2} \log \dfrac{(87)(28)}{15} = \frac{1}{2} (\log 87 + \log 28 - \log 15)$.

6.115 $\log a^n b^m$

    ▮ $\log a^n b^m = \log a^n + \log b^m = n \log a + m \log b$.

6.116 $\log \sqrt[n]{a^{n-1}p}$

$\blacksquare$ $\log \sqrt[n]{a^{n-1}p} = \log (a^{n-1}p)^{1/n} = (1/n)\log a^{n-1}p = (1/n)(\log a^{n-1} + \log p) = (1/n)[(n-1)\log a + \log p]$

$= \dfrac{n-1}{n}\log a + \dfrac{\log p}{n}$.

For Probs. 6.117 to 6.122, express each as a single logarithm.

6.117 $\log a + \log b + \log c$

$\blacksquare$ $\log a + \log b = \log ab$. Thus, $\log a + \log b + \log c = \log ab + \log c = \log (ab)(c) = \log abc$.

6.118 $\log a + \log b + \log c + \log d$

$\blacksquare$ $\log a + \log b = \log ab$; $\log c + \log d = \log cd$. Thus, $\log a + \log b + \log c + \log d = \log ab + \log cd = \log abcd$.

6.119 $\log a - \log b - \log c + \log d$

$\blacksquare$ $\log a - \log b - \log c = \log a/b - \log c = \log \dfrac{a/b}{c} = \log \dfrac{a}{bc}$. Thus, $\log a - \log b - \log c + \log d = \log \left(\dfrac{a}{bc}\cdot d\right) = \log \dfrac{ad}{bc}$.

6.120 $2\log x - 3\log y + \log z$

$\blacksquare$ $2\log x = \log x^2$; $3\log y = \log y^3$. Thus, $2\log x - 3\log y + \log z = \log x^2 - \log y^3 + \log z = \log \dfrac{x^2}{y^3}\cdot z = \log \dfrac{x^2 z}{y^3}$.

6.121 $3\log x - \log (x-2)$

$\blacksquare$ $3\log x = \log x^3$; thus $3\log x - \log (x-2) = \log \dfrac{x^3}{x-2}$.

6.122 $\log 1 + \log 5376$

$\blacksquare$ For any base, $\log 1 = 0$. Thus, $\log 1 + \log 5376 = \log 5376$.

For Probs. 6.123 to 6.133, evaluate the given expression given that $\log_{10} 2 = 0.3010$, $\log_{10} 3 = 0.4771$.

6.123 $\log_{10} 8$

$\blacksquare$ $\log_{10} 8 = \log_{10} 2^3 = 3\log_{10} 2 = 3(0.3010) = 0.9030$.

6.124 $\log_{10} 32$

$\blacksquare$ $32 = 2^5$; thus, $\log_{10} 32 = \log_{10} 2^5 = 5\log_{10} 2 = 5(0.3010) = 1.5050$.

6.125 $\log_{10} 60$

$\blacksquare$ $\log_{10} 60 = \log_{10} (6\cdot 10) = \log_{10} 6 + \log_{10} 10 = (\log_{10} 3 + \log_{10} 2) + \log_{10} 10 = 0.4771 + 0.3010 + 1 = 1.7781$.

6.126 $\log_{10} 600$

$\blacksquare$ $\log_{10} 600 = \log_{10} (100\cdot 6) = \log_{10} 100 + \log_{10} 6 = 2 + \log_{10} 3 + \log_{10} 2 = 2 + 0.4771 + 0.3010 = 2.7781$.

6.127 $\log_{10} \underbrace{60\cdots 000}_{n\text{ zeros}}$

$\blacksquare$ See Probs. 6.125 and 6.126 above. Then for each extra zero, we will add 1 to the result. $\log_{10} 6000 = 1 + \log_{10} 600 = 3.7781$, etc. $\log_{10} \underbrace{60\cdots 000}_{n\text{ zeros}} = n.7781$.

6.128 $\log_{10} 54$

$\blacksquare$ $\log_{10} 54 = \log_{10} (9\cdot 6) = \log_{10} (3^2\cdot 3\cdot 2) = \log_{10} (3^3\cdot 2) = 3\log_{10} 3 + \log_{10} 2 = 3(0.4771) + 0.3010 = 1.7323$.

6.129 $\log_{10} 540$

▮ Since $540 = 54 \cdot 10$, $\log_{10} 540 = 1 + \log_{10} 54 = 1.7323 + 1 = 2.7323$.

6.130 $\log_{10} \sqrt{12}$

▮ $\log_{10} \sqrt{12} = \log_{10} (12)^{1/2} = \frac{1}{2} \log_{10} 12 = \frac{1}{2} \log(2^2 \cdot 3) = \frac{1}{2} \log 2^2 + \frac{1}{2} \log 3 = 2 \cdot \frac{1}{2} \log 2 + \frac{1}{2} \log 3$
$= 0.3010 + 0.4771/2 = 0.539\,55$

6.131 $\log_{10} \sqrt[3]{18}$

▮ $18 = 3^2 \cdot 2$; thus, $\log_{10} (18)^{1/3} = \frac{1}{3} \log_{10} 18 = \frac{1}{3} \log_{10} (3^2 \cdot 2) = \frac{1}{3}(2 \log_{10} 3 + \log_{10} 2) = \frac{2}{3} \log_{10} 3 + \frac{1}{3} \log_{10} 2 = 0.3181 + 0.1003$ (rounding off) $= 0.4184$.

6.132 $\log_{10} \frac{64}{9}$

▮ $\log_{10} \frac{64}{9} = \log_{10} (2^6/3^2) = \log_{10} 2^6 - \log_{10} 3^2 = 6 \log_{10} 2 - 2 \log_{10} 3 = 1.806 - 0.9542 = 0.8518$.

6.133 $\log_{10} \sqrt[4]{72}$

▮ $\log_{10}(72)^{1/4} = \frac{1}{4} \log_{10} 72 = \frac{1}{4} \log_{10} (3^2 \cdot 2^3) = \frac{1}{4} \log_{10} 3^2 + \frac{1}{4} \log_{10} 2^3 = \frac{1}{2} \log_{10} 3 + \frac{3}{4} \log_{10} 2$
$= \frac{1}{2}(0.4771) + \frac{3}{4}(0.3010) = 0.238\,55 + 0.225\,75 = 0.4643$.

For Probs. 6.134 to 6.139, obtain the required logarithm.

6.134 $\log_2 (16)(1024)$

▮ $\log_2 (16)(1024) = \log_2 16 + \log_2 1024 = 4 + 10 = 14$.   (Note: $2^{10} = 1024$.)

6.135 $\log_2 (8)(16\,384)$

▮ $\log_2 (8)(16\,384) = \log_2 8 + \log_2 16\,384$.   But $2^{14} = 16\,384$.   Thus, $\log_2 8 + \log_2 16\,384 = 3 + 14 = 17$.

6.136 $\log_2 (256)(4096)$

▮ $\log_2 (256)(4096) = \log_2 256 + \log_2 4096 = 8 + 12 = 20$.

6.137 $\log_2 (1024)^4$

▮ $\log_2 (1024)^3 = 3 \log_2 1024 = 3 \cdot 10 = 30$.

6.138 $\log_2 (16\,384)^{-2}$

▮ $\log_2 (16\,384)^{-2} = -2 \log 16\,384 = -2 \cdot 14 = -28$.

6.139 $\log_3 \sqrt[4]{65\,536}$

▮ $2^{16} = 65\,536$.   Thus, $\log_2 (65\,536)^{1/4} = \frac{1}{4} \log_2 65\,536 = \frac{1}{4} \cdot 16 = 4$.

For Probs. 6.140 to 6.146, write each expression in terms of a single logarithm with a coefficient of 1.

6.140 $2 \log_b x - \log_b y$

▮ $2 \log_b x = \log_b x^2$;   thus, $2 \log_b x - \log_b y = \log_b x^2 - \log_b y = \log_b (x^2/y)$.

6.141 $\log_b m = \frac{1}{2} \log_b n$

▮ $\log_b m - \frac{1}{2} \log_b n = \log_b m - \log_b n^{1/2} = \log_b (m/n^{1/2})$.

6.142 $3 \log_b x + 2 \log_b y - 4 \log_b z$

▮ $3 \log_b x + 2 \log_b y - 4 \log_b z = \log_b x^3 + \log_b y^2 - \log_b z^4 = \log_b (x^3 y^3/z^4)$.

6.143 $\frac{1}{3} \log_b w - 3 \log_b x - 5 \log_b y$

▮ $\frac{1}{3} \log_b w - 3 \log_b x - 5 \log_b y = \log_b w^{1/3} - \log_b x^3 - \log_b y^5 = \log_b (w^{1/3}/x^{1/3}) - \log_b y^5 = \log_b \dfrac{w^{1/3}/x^{1/3}}{y^5}$
$= \log_b \dfrac{w^{1/3}}{x^{1/3}y^5}$.

6.144 $\frac{1}{5}(2\log_b x + 3\log_b y)$

▮ $\frac{1}{5}(2\log_b x + 3\log_b y) = \frac{2}{5}\log_b x + \frac{3}{5}\log_b y = \log_b x^{2/5} + \log_b y^{3/5} = \log_b x^{2/5}y^{3/5}$. See Prob. 6.145.

6.145 $\frac{1}{7}(3\log_b x + 4\log_b y)$

▮ $3\log_b x + 4\log_b y = \log_b x^3 + \log_b y^4 = \log_b x^3 y^4$; thus, $\frac{1}{7}(3\log_b x + 4\log_b y) = \frac{1}{7}(\log_b x^3 y^4) = \log_b (x^3 y^4)^{1/7}$. Compare this to the method used in Prob. 6.144.

6.146 $\frac{1}{3}(\log_b x - \log_b y)$

▮ $\log_b x - \log_b y = \log_b(x/y)$; thus $\frac{1}{3}[\log_b (x/y)] = \log_b (x/y)^{1/3}$.

For Probs. 6.147 to 6.149, prove the stated logarithm property.

6.147 $\log_b xy = \log_b x + \log_b y$

▮ Assume that $x = b^t$, $y = b^s$; then $\log_b xy = \log_b (b^t b^s = \log_b b^{t+s}$; but $\log_b b^z = z$ for all b, z; thus, $\log_b b^{t+s} = t + s = \log_b x + \log_b y$. The proof is complete.

6.148 $\log_b (x/y) = \log_b x - \log_b y$

▮ Assume that $x = b^y$, $y = b^s$; then $\log_b(x/y) = \log_b (b^t/b^s) = \log_b (b^{t-s}) = t - s = \log_b x - \log_b y$.

6.149 $\log_b x^s = s\log_b x$

▮ Assume that $x = b^t$. Then $\log_b x^s = \log_b (b^t)^s = \log_b b^{ts} = ts = st = s\log_b x$.

For Probs. 6.150 to 6.156, rewrite the given expression in terms of common logarithms.

6.150 $\log_6 7$

▮ Using the property $\log_b x = \dfrac{\log_a x}{\log_a b}$, we want $b = 6$, $x = 7$, $a = 10$ (this is to be in terms of common logarithms). Then $\log_6 7 = \dfrac{\log 7}{\log 6}$ ($\log 7 = \log_{10} 7$).

6.151 $\log_4 15$

▮ $\log_4 15 - \log_b x$ where $b = 4$, $x = 15$. Then $\log_b x = \log_4 15 = \dfrac{\log_a x}{\log_a b} = \dfrac{\log 15}{\log 4}$.

6.152 $\log_{1/2} 10$

▮ $\log_{1/2} 10 = \log_b x$ where $b = \frac{1}{2}$, $x = 10$. Then $\log_b x = \log_{1/2} 10 = \dfrac{\log_a x}{\log_a b} = \dfrac{\log 10}{\log \frac{1}{2}} = \dfrac{1}{\log \frac{1}{2}} = \dfrac{-1}{\log 2}$.

6.153 $\log_{1/3} 30$

▮ $\log_{1/3} 30 = \log_b x$ where $b = \frac{1}{3}$, $x = 30$. Then $\log_b x = \log_{1/3} 30 = \dfrac{\log_a x}{\log_a b} = \dfrac{\log 30}{\log \frac{1}{3}} = \dfrac{\log 10 \cdot 3}{-\log 3} = \dfrac{\log 10 + \log 3}{-\log 3} = \dfrac{1 + \log 3}{-\log 3}$.

6.154 $\log_{100} 10$

▮ $\log_{100} 10 = \dfrac{\log 10}{\log 100} = \dfrac{1}{2}$. Alternately, $\log_{100} 10 = \log_{100} (100)^{1/2} = \frac{1}{2}\log_{100} 100 = \frac{1}{2} \cdot 1 = \frac{1}{2}$.

6.155 $\log_{20} e$

▮ Here, e is *not* the base; 20 is. $\log_{20} e = \dfrac{\log e}{\log 20} = \dfrac{\log e}{\log(10 \cdot 2)} = \dfrac{\log e}{\log 10 + \log 2} = \dfrac{\log e}{1 + \log 2}$.

6.156 $\log_{20} e^3$

▮ See Prob. 6.155. $\log_{20} e^3 = 3\log_{20} e = 3\dfrac{\log e}{1 + \log 2} = \dfrac{3\log e}{1 + \log 2}$.

For Probs. 6.157 to 6.161, rewrite the given expression in terms of logarithms with a base of 13.

6.157 $\log_4 8$

▮ $\log_4 8 = \log_b x$ where $b = 4$, $x = 8$. Here, $a = 13$; then $\log_b x = \log_4 8 = \dfrac{\log_a x}{\log_a b} = \dfrac{\log_{13} 8}{\log_{13} 4}$.

6.158 $\log_{12} 15$

▮ $\log_{12} 15 = \log_b x$ where $b = 12$, $x = 15$. Then $\log_{12} 15 = \dfrac{\log_{13} 15}{\log_{13} 12}$. Which is larger, $\log_{12} 15$ or $\log_{13} 15$?

6.159 $\ln 10$

▮ $\ln 10 = \log_e 10 = \dfrac{\log_{13} 10}{\log_{13} e}$.

6.160 $\ln e^3$

▮ $\ln e^3 = \log e^3 = \dfrac{\log_{13} e^3}{\log_{13} e}$. But $\ln e^3 = \log_e e^3 = 3$. Thus, $\dfrac{\log_{13} e^3}{\log_{13} e} = 3$.

6.161 $\log_{11} 121$

▮ $\log_{11} 121 = \dfrac{\log_{13} 121}{\log_{13} 11}$. But $\log_{11} 121 = \log_{11}(11)^2 = 2$. Thus, $\dfrac{\log_{13} 121}{\log_{13} 11} = 2$. Compare this with Prob. 6.159. Do you notice a pattern?

For Probs. 6.162 to 6.174, tell whether the given statement is true or false, and explain your answer.

6.162 $(\log 15)^0 = 1$

▮ This is true; $(\log 15) \in \mathbb{R}$, and $a^0 = 1 \quad \forall a \in \mathbb{R}$. Do not be tricked by the fact that the real number here is a logarithm. *All* reals can be written in logarithmic form!

6.163 If $f(x) = \log_8 x$, then the range of f is $\mathbb{R}$.

▮ True. Look at their graphs, or recall that the logarithm is the exponent. In $\log_a b = q$, a and b are the numbers which are restricted, not q!

6.164 The domain of $y = \log_x b$ is $\mathbb{R}$.

▮ False. For example, if $x = 1$, then 1 is the logarithmic base, which is impossible.

6.165 $\log_x 100 = -\log_{1/x} \frac{1}{100}$

▮ Let $x = 10$; then $\log_{10} 100 = 2$, but $\log_{1/10} \frac{1}{100} = 2$. The statement is false.

6.166 $\log_x a = \log_{1/x} 1/a$

▮ If $\log_x a = b$, then $x^b = a$, $1/x^b = 1/a$, $(1/x)^b = 1/a$, and $\log_{1/x} 1/a = b$. The statement is true.

6.167 $f(x) = \log_b x$ is an always-decreasing function.

▮ False. $f(x) = \log_b x$ is decreasing for $0 < b < 1$. If $b = 10$, for example, $\log_{10} x_1 > \log_{10} x_2$ when $x_1 > x_2$.

6.168 $\log_k ab = (\log_k a)(\log_k b)$

▮ Clearly, this is a false statement. For all k, a, b, $\log_k ab = \log_k a + \log_k b$. When $k = e$, $a = e$, $b = e$, for example, $\ln(e \cdot e) = \ln e^2 = 2$, but $\ln e \ln e = 1 \cdot 1 = 1$; $2 \neq 1$.

6.169 $g(x) = \ln |x|$ has domain $= \mathbb{R}$.

▮ False; $0 \notin$ domain of g. However, all other reals are in the domain.

6.170 Let $g(x) = \log_a x$, and $h(x) = \log_b x$; then $g \circ h(ab) = \log_a (1 + \log a)$.

▮ $h(ab) = \log_b ab = \log_b a + \log_b b = 1 + \log_b a$. $g(1 + \log_b a) = \log_a (1 + \log_b a)$. The statement is true.

6.171 Using functions g and h in Prob. 6.170, $g \circ h(ab) = \log_a 1 + \log_a(\log_b a)$.

▮ False. For this to be true, we would be saying that "the log of a sum is the sum of the logs." But this is false since, for example, $\log_2 8 = \log_2 2^3 = 8$, but $\log_2(4+4) = 8$, not $2+2$.

6.172 $\log_a 1/x = -\log_a x$

▮ $\log_a 1/x = \log_a(x^{-1}) = -\log_a x$. True.

6.173 $\log_b a = \dfrac{1}{\log_a b}$

▮ $\log_b a = \dfrac{\log_k a}{\log_k b}$ for all log bases k. Let $k = a$; $\log_b a = \dfrac{\log_a a}{\log_a b} = \dfrac{1}{\log_a b}$. True.

6.174 The graph of $h(x) = \ln e^x$ is a straight line.

▮ $h(x) = \ln e^x = x$; thus $h(x)$ is the straight line with slope 1 passing through $(0,0)$. True.

6.175 Find x so that $\frac{3}{2}\log_b 4 - \frac{2}{3}\log_b 8 + 2\log_b 2 = \log_b x$.

▮ Then $\log_b x = \log_b 4^{3/2} - \log_b 8^{2/3} + \log_b 2^2 = \log_b 8 - \log_b 4 + \log_b 4 = \log_b(\frac{8}{4} \cdot 4)$, or $x = 8$.

6.176 Find x so that $3\log_b 2 + \frac{1}{2}\log_b 25 - \log_b 20 = \log_b x$.

▮ $\log_b x = \log_b 8 + \log_b 5 - \log_b 20 = \log_b(8 \cdot 5/20) = \log_b x$, or $x = 2$.

6.177 Write $\log_b y - \log_b c + kt = 0$ in exponential form free of logarithms.

▮ Then $\log_b y - \log_b c = -kt$, and $b^{(\log_b y - \log_b c)} = b^{-kt}$, $\dfrac{b^{\log_b y}}{b^{\log_b c}} = b^{-kt}$, $\dfrac{y}{c} = b^{-kt}$, or $y = cb^{-kt}$.

6.178 Prove that $\log_a xyz = \log_a x + \log_a y + \log_a z$

▮ $\log_a xy + \log_a z = (\log_a x + \log_a y) + \log_a z = \log_a x + \log_a y + \log_a z$, and the proof is complete.

6.179 Let $f(x) = 7^x$. Find the domain and range of $f^{-1}(x)$.

▮ If $f(x) = 7^x$, then $f^{-1}(x) = \log_7 x$. Then the domain of f is $\mathbb{R}^+$, and the range (i.e., the f^{-1} values) is $\mathbb{R}$.

6.180 Solve the equation $A = Pe^{rt}$ for r.

▮ If $A = Pe^{rt}$, then $\ln A = \ln Pe^{rt} = \ln P + \ln e^{rt} = \ln P + rt\ln e = \ln P + rt$ (since $\ln e = 1$). Thus, $rt = \ln A - \ln P$, and $r = \dfrac{\ln A - \ln P}{t}$.

6.181 Solve the equation $\ln(\log x) = 1$ for x.

▮ If $\ln(\log x) = 1$, then $e^{\ln(\log x)} = e^1$. But $e^{\ln P} = P$ for all P. Thus, $e^{\ln(\log x)} = \log x$. Then $\log x = e$, $10^{\log x} = 10^e$, and $x = 10^e$.

6.182 Check the result in Prob. 6.181.

▮ $x = 10^e$ where $\ln(\log x) \overset{?}{=} 1$. $\ln(\log 10^e) = \ln(e\log 10) = \ln(e \cdot 1) = \ln e = 1$. The result checks.

6.4 NATURAL AND COMMON LOGARITHMS

Note: Many problems in this section involve the use of a calculator. If you do not know how to use your calculator for logarithm computations, check the instruction manual.

For Probs. 6.183 to 6.187, express the given number using scientific notation.

6.183 137.65

▮ Scientific notation is of the numerical form $N \times 10^M$, $1 \le N \le 10$, $M \in \mathbb{Z}$. Thus, for 137.65 we move the decimal point over two spaces, so $137.65 = 1.3765 \times 10^2$. Here, $M = 2$, $N = 1.3765$.

6.184   14 632.600 040 00

    ▍ Let $M = 4$, $N = 1.463\,260\,004\,000$.   Then $N \times 10^M = 1.463\,260\,004\,000 \times 10^4 = 14\,632.600\,040\,00$.   (We have moved the decimal point over four spaces.)

6.185   1.6843

    ▍ $1.6843 = 1.6843 \times 10^0$.   Notice that, in essence, the number given was in scientific notation.   The 0 exponent comes about since we moved the decimal over "0" spaces.

6.186   0.006 84

    ▍ $0.006\,84 = 6.84 \times 10^{-3}$.   We can easily see the -3 exponent by observing that we have moved the decimal point three spaces to the right.

6.187 $4.26 \cdot 9^2$

    ▍ $4.26 \times 9^2 = 345.06 = 3.4506 \times 10^2$.

For Probs. 6.188 and 6.189, rewrite the numbers using ordinary notation.

6.188 $4.63 \cdot 10^{-4}$

    ▍ $4.63 \times 10^{-4} = 4.63 \times 0.0001 = 0.000\,463$.   Another, and easier, way to do this is to observe that a 0.0001 multiplier moves the decimal point four spaces to the left ($M < 0$).

6.189 $4.5686 \cdot 10^3$

    ▍ Simply move the decimal three places to the right   ($M > 0$).   Then $4.5686 \times 10^3 = 4568.6$.

For Probs. 6.190 to 6.193, find the characteristic of the logarithm (base 10) of the given number.

6.190   146

    ▍ To determine the characteristic of a number P, put P into scientific notation: $P = N \times 10^M$, $1 \le N < 10$, $M \in \mathbb{Z}$.   Then $M = $ the characteristic.   Thus, $146 = 1.46 \times 10^2$,   and $2 = $ characteristic.

6.191   14 673.1

    ▍ $14\,673.1 = 1.467\,31 \times 10^4$. $4 = $ characteristic.

6.192   0.003 634

    ▍ $0.003\,634 = 3.634 \times 10^{-3}$ and the characteristic $= -3$.

6.193   4.638

    ▍ $4.638 = 4.638 \times 10^0$;   and the characteristic is 0.

For Probs. 6.194 to 6.199, use a logarithm table (base 10) to find the common logarithm of each number.

6.194   log 5.73

    ▍ First notice that the characteristic is 0, since $5.73 = 5.73 \times 10^0$.   Looking up 573 in the table, we find 7582.   Thus, $\log 5.73 = 0.7582$ where 0 is the characteristic and 7582 is the mantissa.

6.195   log 57.3

    ▍ $57.3 = 5.73 \times 10^1$.   We find 7582 in the table for 573.   Thus, $\log 57.3 = 1.7582$.   The 1 in 1.7582 is the characteristic found using scientific notation.

6.196   0.005 73

    ▍ We again use the 7582 found in the table for 573.   Then, $\log 0.005\,73 = \log(5.73 \times 10^{-3}) = \log 5.73 + \log 10^{-3} = 0.7582 + (-3) = 0.7582 - 3$.   You can, if you like, rewrite $0.7582 - 3$ as -2.2418, but $0.7582 - 3$ is a perfectly acceptable form.

6.197   0.0101

▮ $0.0101 = 1.01 \times 10^{-2}$. Look up 101 in your table, and find 0043. Thus, $\log(0.0101) = \log(1.01 \times 10^{-2}) = \log 1.01 + \log 10^{-2} = 0.0043 - 2$ ($= -1.9957$).

6.198 5.13×10^{-4}

▮ $\log(5.13 \times 10^{-4}) = \log(5.13) + \log 10^{-4} = 0.7101 - 4$.

6.199 51.3×10^{-7}

▮ $\log(51.3 \times 10^{-7}) = \log(5.13 \times 10^{-6}) = 0.7101 - 6$.

For Probs. 6.200 to 6.202, rewrite the given log values in standard logarithmic form.

6.200 $5.1938 - 8$

▮ Standard log form is $\log n = $ mantissa + characteristic where the mantissa is nonnegative and less than 1. Thus, $5.1983 - 8$ has a "mantissa" of 5.1983; we subtract 5 from 5.1983, and add 5 to (-8) to obtain: $(5.1983 - 5) + [(-8) + 5] = 0.1983 - 3$. Do you see the power of this method? We now know that, in scientific notation, n is of the form $a.bc \times 10^{-3}$.

6.201 -1.1776

▮ -1.1776 is not in standard log form. $-1.1776 = (-1.1776 + 2)$ (adding 2 is the least you can add to -1.1776 to get it to be nonnegative) $= 2$. Then, $(-1.1776 + 2) - 2 = 0.8224 - 2$.

6.202 -3.2076

▮ $-3.2076 = (-3.2076 + 4) - 4 = 0.7924 - 4$.

For Probs. 6.203 to 6.206, use a common logarithm table to find the antilogarithm of the given number.

6.203   0.7993

▮ Look for 0.7993 in your table. It is going to correspond to 630. Since 0.7993 has a 0 characteristic, antilog $0.7993 = 6.30 \times 10^0 = 6.30$.

6.204   2.5729

▮ We find, in our table, that 0.5729 corresponds to 374. Since the characteristic is 2, antilog $2.5729 = 3.74 \times 10^2 = 374$.

6.205 $0.9460 - 2$

▮ Then antilog $(0.9460 - 2) = 8.83 \times 10^{-2} = 0.0883$.

6.206 -4.0052

▮ $-4.0052 = (-4.0052 + 5) - 5 = 0.9948 - 5$. Since 0.9948 corresponds to 988 in the table, antilog $(-4.0052) = $ antilog $(0.9948 - 5) = 9.88 \times 10^{-5}$.

For Probs. 6.207 to 6.210, use a calculator to find the indicated quantity rounded to six decimal places.

6.207   (a) log 82 734, (b) ln 82 734

▮ (a) If you enter 82 734 and press the log key, you will get log 82 734. What could be easier? *Press*: 8 2 7 3 4 log. *On screen*: 4.917684.
(b) We repeat the same procedure, but use the ln key instead. *Press*: 8 2 7 3 4 ln. *On screen*: 11.323386.

6.208   (a) log 843 250, (b) ln 843 250

▮ (a) *Press*: 8 4 3 2 5 0 log. *On screen*: 5.9259563. Rounding off we get 5.925 956.
(b) *Press*: 8 4 3 2 5 0 ln. *On screen*: 13.645019.

6.209   log 0.0103

▮ *Press*: $\boxed{\cdot}\ \boxed{0}\ \boxed{1}\ \boxed{0}\ \boxed{3}\ \boxed{\log}$. *On screen*: −1.9871628.   Rounding off we get −1.987 163.   Then −1.987 163 = (−1.987 163 + 2) − 2 = 0.012 837 − 2.

6.210   ln 0.081 043

▮ *Press*: $\boxed{\cdot}\ \boxed{0}\ \boxed{8}\ \boxed{1}\ \boxed{0}\ \boxed{4}\ \boxed{3}\ \boxed{\ln}$. *On screen*: −2.5127754.   Rounding off we get −2.512 775.   Then (−2.512 775 + 3) − 3 = 0.487 225 − 3.

For Probs. 6.211 to 6.216, use a calculator to find x to four significant digits.

6.211   log x = 5.3027

▮ Recall that $f(x) = \log x$ and $g(x) = 10^x$ are inverse functions.   Thus, to go from log x to x, we find $10^{\log x}$.   *Press*: 5.3027 $\boxed{10^x}$.   *On screen*: 200770.55.   Then $x = 200\,770.55 \approx 200\,800$ to four significant digits.

6.212   log x = 1.9168

▮ *Press*: 1.9168 $\boxed{10^x}$.   *On screen*: 82.565763.   Then $x = 82.565\,763 \approx 82.57$ to four significant digits.

6.213   log x = −3.1773

▮ *Press*: 3.1773 $\boxed{\pm}$ $\boxed{10^x}$.   *On screen*: 0.0006648.   Then $x = 0.000\,664\,8$ (already is to four significant digits).

6.214   log x = −2.0411

▮ *Press*: 2.0411 $\boxed{\pm}$ $\boxed{10^x}$.   *On screen*: 0.009097.   Then $x = 0.009\,097 \approx 0.0091$ to four significant digits.

6.215   ln x = 3.8655

▮ *Press*: 3.8655 e^x.   *On screen*: 47.72713.   Then $x = 47.727\,13 \approx 47.73$ to four significant digits.

6.216   ln x = −0.3916

▮ *Press*: 0.3916 $\boxed{\pm}$ $\boxed{e^x}$.   *On screen*: 0.6759745.   Then $x = 0.675\,974\,5 \approx 0.6760$ to four significant digits.

For Probs. 6.217 to 6.220, find x to five significant digits using a calculator.   Your calculator may operate slightly differently than exhibited here.

6.217 $x = \log{(5.3147 \times 10^{12})}$

▮ $\log{(5.3147 \times 10^{12})} = \log 5.3147 + \log 10^{12} = \log 5.3147 + 12$.   Find log 5.3147 on the calculator.   *Press*: 5.3147 $\boxed{\log}$.   *On screen*: 12.7254788.   Then $x = 12.725\,478\,8 \approx 12.725$ to five significant digits.   (*Note*: We did not enter 5.3147×10^{12} on the calculator.   Why?)

6.218 $x = \log{(2.0991 \times 10^{17})}$

▮ $\log{(2.0991 \times 10^{17})} = \log 2.0991 + \log 10^{17} = 0.322\,033 + 17$ (0.322 033 is obtained from the calculator) $= 17.322\,031 \approx 17.322$.

6.219 $x = \ln{(6.7917 \times 10^{-12})}$

▮ $x = \ln{(6.7917 \times 10^{-12})} = \ln 6.7917 + \ln 10^{-12} = \ln 6.7917 + \dfrac{\log_{10} 10^{-12}}{\log_{10} e} = 1.915\,701\,3 + \dfrac{-12}{0.434\,297\,4} \approx -25.715.$

6.220   log x = 32.065 23

▮ Enter 32.068 523 into your calculator, and press $\boxed{10^x}$.   Then $x = 1.1709 \times 10^{32}$ is what you will see on the screen (actually, you will see 1.1709 32).   Alternately, if   log x = 32.068 523, $x = 10^{32} \times$ antilog 0.068 523 $= 10^{32} \times 1.170\,908\,6$.   (To find the antilog 0.068 523 enter 0.068523 $\boxed{10^x}$ into your calculator to obtain $\approx 1.1709 \times 10^{32}$.)

For Probs. 6.221 to 6.223, use the change of base formula and a calculator to find the indicated quantity to four significant places.

6.221 $\log_5 372$

▮ Recall that $\log_b M = \dfrac{\log_k M}{\log_k b}$. Then, $\log_5 372 = \dfrac{\ln 372}{\ln 5}$. *Press*: 372 $\boxed{\ln}$ ÷ 5 $\boxed{\ln}$. *On screen*: 3.67765155. Rounding off we get 3.6776.

6.222 $\log_8 0.0352$

▮ $\log_8 0.0352 = \dfrac{\ln 0.0352}{\ln 8}$. *Press*: 0.0352 $\boxed{\ln}$ ÷ 8 $\boxed{\ln}$. *On screen*: −1.6094269. Rounding off we get −1.6094.

6.223 $\log_3 0.1483$

▮ $\log_3 0.1483 = \dfrac{\ln 0.1483}{\ln 3}$. Using the key steps in Probs. 6.221 and 6.222, we get $-1.737\,208 \approx -1.7372$.

For Probs. 6.224 to 6.231, perform the indicated computations. Obtain answers to three significant digits. Use a common logarithm table.

6.224 $(3.86)(59.1)$

▮ $\log 3.86 = 0.5866$ and $\log 59.1 = 1.7716$. Then since $\log (3.86)(59.1) = 2.3582$, $(3.86)(59.1) = 228$. (We find 2.28 in the table, and use the characteristic of 2 to get 2.28. We do not interpolate since we only want the result to three significant digits. 2.28 corresponds to 0.3579 in the table, which is closest to 0.3582.)

6.225 $(807)(0.276)$

▮ Let $P = (807)(0.276)$. Then $\log P = \log 807 + \log 0.276 = 0.9069 + 0.4409 = 1.3478$. Thus, $P = 22.3$ (0.3483 is closest to 0.3478). We use $10^1 \times 2.23$ since the characteristic is 1.

6.226 $(6.23)(4.71)(32.8)$

▮ Let $Q =$ the product. Then $\log Q = \log 6.23 + \log 4.71 + \log 32.8 = 0.7945 + 0.6730 + 1.5159 = 2.9834$. Then since 0.9834 is closest to 0.9836 which corresponds to 9.63, $\log Q = 2.9834$ and $Q = 96.3$.

6.227 $984 \div 237$

▮ Let $S = \frac{948}{237}$. Then $\log S = \log 948 - \log 237 = 2.9768 - 2.3747 = 0.6021$, and $S = 4.00$. (Actually, 0.6021 is in the table!)

6.228 $\dfrac{(24.7)(3.28)}{47.6}$

▮ Let T represent the solution. Then $\log T = \log 24.7 + \log 3.28 - \log 47.6 = 1.3927 + 0.5159 - 1.6776 = 0.231$, which is closest to 0.2304, which corresponds to 1.70. $T = 1.70$.

6.229 $(7.25)^2$

▮ If $P = (7.25)^2$, then $\log P = 2 \log 7.25 = 2 \times 0.8603 = 1.7206$. Then since 0.7206 corresponds to 5.25 in the table (that is the closest), $P = 52.5$.

6.230 $\sqrt[3]{9.84}$

▮ Let $R = \sqrt[3]{9.84}$. Then $R = (9.84)^{1/3}$, and $\log R = \log[(9.84)^{1/3}] = \frac{1}{3} \log 9.84 = \frac{1}{3} \times 0.9930 = 0.331$. Thus, $R = 2.14$ (0.331 is closest to 0.3304 in the table).

6.231 $\sqrt{27.3}\sqrt[3]{0.629}$

▮ Let $T =$ the product. Then $\log T = \frac{1}{2} \log 27.3 + \frac{1}{3} \log 0.629 = \frac{1}{2} \times 1.4362 + \frac{1}{3} \times 0.7987 = 0.7181 + 0.2662 = 0.9843$. Thus, $T = 9.64$ (closest value from table).

For Probs. 6.232 and 6.233, find N to four significant digits if $\log N$ is the number given. Use the common log table.

6.232 $\log N = 0.5801$

▮ We locate two numbers in the table r and s such that $r < 0.5801 < s$; they are 0.5798 and 0.5809

$$0.0011 \begin{bmatrix} 0.5798 \\ 0.5801 \\ 0.5809 \end{bmatrix} 0.0008 \quad \begin{bmatrix} 3.80 \\ x \\ 3.81 \end{bmatrix} 3.81 - x \Bigg] 0.01$$

Then, $\dfrac{0.0011}{0.0008} = \dfrac{0.01}{3.81 - x}$, $(0.0008)(0.01) = [(0.0011)(3.81) - x]$, $8(0.01) = 11(3.81 - x)$, $0.08 = 41.91 - 11x$, $11x = 41.83$, and $x \approx 3.802\,727 \approx 3.803$ (to four significant places).

6.233 $\log N = 0.4063$

▮ From the table, $0.4048 < 0.4063 < 0.4065$.

$$0.0017 \begin{bmatrix} 0.4065 \\ 0.4063 \\ 0.4048 \end{bmatrix} 0.0015 \quad \begin{bmatrix} 2.55 \\ x \\ 2.54 \end{bmatrix} x - 2.54 \Bigg] 0.01$$

Then $\dfrac{0.0017}{0.0015} = \dfrac{0.01}{x - 2.54}$, $\dfrac{17}{15} = \dfrac{0.01}{x - 2.54}$, $17x = (15)(0.01) + (2.54)(17)$, $17x = 43.33$, and $x \approx 2.548\,823 \approx 2.549$.

For Probs. 6.234 and 6.235, find $\log N$ using interpolation.

6.234 $N = 53.62$

▮ $\log 53.6 = 1.7292$ and $\log 53.7 = 1.7300$. Interpolating,

$$0.1 \begin{bmatrix} & 53.7 \\ 0.02 \begin{bmatrix} 53.62 \\ 53.6 \end{bmatrix} \end{bmatrix} \begin{matrix} 1.7300 \\ x \\ 1.7292 \end{matrix} \Bigg] x - 1.7292 \Bigg] 0.0008$$

Then $\dfrac{0.1}{0.02} = \dfrac{0.0008}{x - 1.7292}$, $10(x - 1.7292) = 0.0008(2)$, $10x - 17.292 = 0.0016$, $10x = 0.0016 + 17.292$, and $x = 1.729\,36 \approx 1.7294$. (The accuracy matches the table's accuracy.)

6.235 $N = 141.6$

▮ $\log 141 = 2.1492$ and $\log 142 = 2.1523$. Interpolating,

$$0.6 \begin{bmatrix} 142 \\ 141.6 \\ 141 \end{bmatrix} 1 \quad \begin{bmatrix} 2.1523 \\ x \\ 2.1492 \end{bmatrix} x - 2.1492 \Bigg] 0.0031$$

Then $\dfrac{1}{0.6} = \dfrac{0.0031}{x - 2.1492}$, $x - 2.1492 = 0.00186$, and $x = 2.151\,06 \approx 2.1511$.

6.5 LOGARITHMIC AND EXPONENTIAL EQUATIONS

For Probs. 6.236 to 6.242, solve the given equation for x in terms of y.

6.236 $y = 10^x$

▮ If $y = 10^x$, then $x = \log y$. You can get that answer by remembering that $\log_a x$ and a^x are inverse functions, or by taking the log of both sides of the equation $y = 10^x$. Then $\log y = \log 10^x$ and $\log y = x$.

6.237 $y = 10^{-x}$

▮ If $y = 10^{-x}$, then $\log y = \log 10^{-x}$ and $\log y = -x$. $x = -\log y = \log y^{-1} = \log \frac{1}{4}$.

6.238 $y = 5(10^{-2x})$

▮ $\log y = \log 5(10^{-2x}) = \log 5 + \log 10^{-2x} = \log 5 + (-2x)$. Then $x = \dfrac{\log 5 - \log y}{2} = \frac{1}{2} \log (5/y) = \log \sqrt{5/y}$.

6.239 $y = 3(10^{2x})$

▮ $\log y = \log 3(10^{2x}) = \log 3 + \log 10^{2x} = \log 3 + 2x$. Then $2x = \log y - \log 3 = \log (y/3)$ and $x = \frac{1}{2} \log (y/3) = \log (y/3)^{1/2}$.

6.240 $y = \dfrac{e^x + e^{-x}}{2}$

▮ Let $u = e^x$. (This is a common trick.) Then $e^{-x} = 1/u$, so $2y = u + \dfrac{1}{u} = \dfrac{u^2 + 1}{u}$ and $u^2 - 2uy + 1 = 0$, a quadratic in u. Then $u = \dfrac{2y + \sqrt{4y^2 - 4}}{2} = y + \sqrt{y^2 - 1}$, and $e^x = y \pm \sqrt{y^2 - 1}$. Thus, $x = \ln (y \pm \sqrt{y^2 - 1})$.

6.241 $y = \dfrac{e^x - e^{-x}}{2}$

▮ Using the procedure in Prob. 6.240, let $u = e^x$. Then $1/u = e^{-x}$, and $2y = u - \dfrac{1}{u} = \dfrac{u^2 - 1}{u}$. Then $u^2 - 1 = 2yu$, $u^2 - 2yu - 1 = 0$, $u = y + \sqrt{y^2 + 1}$, and we obtain $x = \ln (y + \sqrt{y^2 + 1})$ by solving the quadratic.

6.242 $y = \ln (\sqrt{x^2 + 1} + x)$

▮ This is the solution from Prob. 6.241; thus $x = \dfrac{e^y - e^{-y}}{2}$.

For Probs. 6.243 to 6.279, solve the given equation; when appropriate, give answers to three decimal places.

6.243 $5^x = 625$

▮ $5^x = 5^3$, or $x = 3$. (If $5^x = 5^3$, then the exponents must be equal; equivalently, if $\log_5 5^x = \log_5 5^3$, $x = 3$.)

6.244 $3^x = 27$

▮ $3^x = 3^3$, $\log_3 3^x = \log_3 3^3$, or $x = 3$.

6.245 $3^{x^2 - 1} = 27$

▮ $3^{x^2 - 1} = 3^3$, $\log_3 3^{x^2 - 1} = \log_3 3^3$, $x^2 - 1 = 3$, $x^2 = 4$, or $x = \pm 2$. (Check both results!)

6.246 $5^{x^2 - 3x} = 625$

▮ $5^{x^2 - 3x} = 5^4$, $x^2 - 3x = 4$, $x^2 - 3x - 4 = 0$, $(x - 4)(x + 1) = 0$, or $x = 4, -1$. (Check these!)

6.247 $5^{x^2 - 4x + 3} = 1$

▮ $5^{x^2 - 4x + 3} = 5^0$, $\log_5 5^{x^2 - 4x + 3} = \log_5 5^0$, $x^2 - 4x + 3 = 0$, $(x - 1)(x - 3) = 0$, or $x = 1, 3$.

6.248 $4^{x^2 + 3x} = 256$

▮ $4^{x^2 + 3x} = 4^4$, $x^2 + 3x = 4$, $x^2 + 3x - 4 = 0$, $(x + 4)(x - 1) = 0$, or $x = -4, 1$.

6.249 $2^x = 41$

▮ $\log 2^x = \log 41$, $x \log 2 = \log 41$, or $x = \dfrac{\log 41}{\log 2} = \dfrac{1.6128}{0.3010} \approx 5.358$.

6.250 $5^{x+1} = 3.02$

▮ $\log 5^{x+1} = \log 3.02$, $(x + 1) \log 5 = \log 3.02$, $x + 1 = \dfrac{\log 3.02}{\log 5}$, or $x = \dfrac{\log 3.02}{\log 5} - 1 = \dfrac{0.4800}{0.6990} - 1 \approx -0.310$.

6.251 $195^x = 2.68$

▮ $\log 195^x = \log 2.68$, $x \log 195 = \log 2.68$, or $x = \dfrac{\log 2.68}{\log 195} = \dfrac{0.4281}{2.2900} \approx 0.187$.

6.252 $\log_5 (x - 1) + \log_5 (x + 3) = 1$

▮ Then $\log_5 (x-1)(x+3) = 1$, $5^{\log_5 (x-1)(x+3)} = 5^1$, $(x-1)(x+3) = 5$, $x^2 + 2x - 3 = 5$, $x^2 + 2x - 8 = 0$, $(x+4)(x-2) = 0$, or $x = -4, 2$. It is crucial to check these answers since the domain of $\log x$ is so restricted. If $x = -4$, $\log_5 (-5) + \log_5 (-1)$ is undefined, and $x = -4$ is extraneous. If $x = 2$, $\log_5 1 + \log_5 5 = 0 + 1 = 1$. Thus, the solution is $x = 2$.

6.253 $\log_5 (2x + 1) + \log_5 (3x - 1) = 2$

▮ Then $\log_5 (2x+1)(3x-1) = 2$, $(2x+1)(3x-1) = 5^2$, $6x^2 + x - 1 = 25$, $6x^2 + x - 26 = 0$, $(6x + 13)(x - 2) = 0$, $6x = -13$, or $x = -\frac{13}{6}, 2$. But $x = -\frac{13}{6}$ is extraneous, since $3x - 1 < 0$; thus $x = 2$ is the solution.

6.254 $\log_6 3 + \log_6 (x + 6) = 2$

▮ $\log_6 3(x+6) = 2$, $3x + 18 = 36$, $3x = 18$, or $x = 6$. Since $\log_6 3 + \log_6 12 = \log_6 36 = 2$, the answer checks.

6.255 $\log_3 (x + 2) + \log_3 (x + 4) = 1$

▮ $\log_3 (x+2)(x+4) = 1$, $(x+2)(x+4) = 3$, $x^2 + 6x + 8 = 3$, $x^2 + 6x + 5 = 0$, $(x+1)(x+5) = 0$, or $x = -1, -5$. But $x = -5$ is extraneous. If $x = -1$, $\log_3 1 + \log_3 3 = 0 + 1 = 1$. Thus, $x = -1$.

6.256 $\log_5 (2x + 4) - \log_5 (x - 1) = 1$

▮ $\log_5 \dfrac{2x+4}{x-1} = 1$, $\dfrac{2x+4}{x-1} = 5$, $2x + 4 = 5x - 5$, $3x = -9$, or $x = -3$. Since $x = -3$, $x - 1 < 0$; thus there are no solutions.

6.257 $\log_2 (3x + 1) - \log_2 (x - 3) = 3$

▮ $\log_2 \dfrac{3x+1}{x-3} = 3$, $3x + 1 = 8x - 24$ $(2^3 = 8)$, $5x = 25$, or $x = 5$. Since $\log_2 16 - \log_2 2 = 4 - 1 = 3$, the answer checks.

6.258 $\log_3 (x + 11) - \log_3 (x + 3) = 2$

▮ $\log_3 (x+11)(x+3) = 2$, $x^2 + 14x + 33 = 9$, $x^2 + 14x + 24 = 0$, $(x+12)(x+2) = 0$, or $x = -2, -12$. But $x = -12$ is extraneous. If $x = -2$, $\log_3 9 - \log_3 1 = 2 - 0 = 2$. Thus, $x = -2$.

6.259 $\log_5 (3x + 7) - \log_5 (x - 5) = 2$

▮ $\log_5 (3x+7)(x-5) = 2$, $3x^2 - 8x - 35 = 25$, $3x^2 - 8x - 60 = 0$, $(3x+10)(x-6) = 0$, or $x = -\frac{10}{3}, 6$. $x = 6$ is the solution.

6.260 $\log \dfrac{1}{2x} = -\log 6$

▮ $\log \dfrac{1}{2x} = \log 6^{-1}$, $\dfrac{1}{2x} = 6^{-1} = \frac{1}{6}$, $2x = 6$, or $x = 3$. Since $\log \frac{1}{6} = \log 6^{-1} = -\log 6$, the answer checks.

6.261 $\log (3x + 4) = \log (5x - 6)$

▮ Then $3x + 4 = 5x - 6$, $2x = 10$, or $x = 5$. Check this solution.

6.262 $2 \ln 5x = 3 \ln x$

▮ If $2 \ln 5x = 3 \ln x$, then $\ln(5x)^2 = \ln x^3$. Then $(5x)^2 = x^3$, $25x^2 = x^3$, $x^3 - 25x^2 = 0$, $x^2(x - 25) = 0$, or $x = 0, 25$. But $x = 0$ is extraneous, since $5x = 0$ and $\ln 0$ is undefined.

6.263 $\log x = \ln e$

▮ If $\log x = \ln e$, then $10^{\log x} = 10^{\ln e}$, or $x = 10^{\ln e}$. But $\ln e = 1$, so $x = 10$.

6.264 $\ln 25x - \ln 5x = 2 \ln 5x - \ln 25x$

▮ $\ln \dfrac{25x}{5x} = \ln \dfrac{(5x)^2}{25x}$, $\ln 5 = \ln \dfrac{25x^2}{25x}$, $\ln 5 = \ln x$, or $x = 5$.

6.265 $\log(\log x) = 1$

▮ Then $10^{\log(\log x)} = 10^1$. But $10^{\log A} = A$ for all A. Thus, $10^{\log(\log x)} = \log x$, $\log x = 10^1 = 10$, $10^{\log x} = 10^{10} = 100$, or $x = 100$.

6.266 $\underbrace{\log\{\log[\cdots(\log x)\cdots]\}}_{m \text{ logs}} = 1$

▮ $\log x = 1$ means $x = 10$, and $\log(\log x) = 1$ means $x = 10^2$. Thus, with m logs, $x = 10^m$.

6.267 $\log \dfrac{2}{x} - \log \dfrac{1}{x+1} = 1$

▮ $\log\left(\dfrac{2}{x} \div \dfrac{1}{x+1}\right) = 1$, $\log\left(\dfrac{2x+2}{x}\right) = 1$, $\dfrac{2x+2}{x} = 10$, $2x+2 = 10x$, $8x = 2$, or $x = \frac{1}{4}$.

6.268 $\log \dfrac{2}{x+1} + \log \dfrac{3}{x-1} = \log 4$

▮ $\log\left(\dfrac{2}{x+1} \cdot \dfrac{3}{x-1}\right) = \log 4$, $\dfrac{6}{(x+1)(x-1)} = 10^{\log 4} = 4$, $4(x^2 - 1) = 6$, $x^2 - 1 = \frac{6}{4}$, $x^2 = \frac{10}{4}$, or $x = \pm\dfrac{\sqrt{10}}{2}$. But $x = \dfrac{-\sqrt{10}}{2}$ is extraneous. Check it!

6.269 $\log_x 2 + \log_2 x = 2$

▮ $\log_x 2 = \dfrac{\log_2 2}{\log_2 x}$, but $\log_2 2 = 1$, so $\dfrac{1}{\log_2 x} + \log_2 x = 2$. Then $1 + (\log_2 x)^2 = 2$, $(\log_2 x)^2 = 1$, or $\log_2 x = 1, -1$. If $\log_2 x = 1$, $x = 2$; if $\log_2 x = -1$, $x = \frac{1}{2}$. But if $x = -\frac{1}{2}$, $\log_{1/2} 2 + \log_2\left(-\frac{1}{2}\right) \neq 2$. Thus, $x = 2$ is the only solution.

6.270 $\log_x 3 + \log_3 x = 2$

▮ $\log_x 3 = \dfrac{\log_3 3}{\log_3 x} = \dfrac{1}{\log_3 x}$, $\dfrac{1}{\log_3 x} + \log_3 x = 2$, $1 + (\log_3 x)^2 = 2$, $(\log_3 x)^2 = 1$, or $\log_3 x = \pm 1$. If $\log_3 x = 1$, $x = 3$. If $\log_3 x = -1$, $x = \frac{1}{3}$. But $x = \frac{1}{3}$ does not check, so $x = 3$ is the only solution.

6.271 $\log_4 x + \log_4 4x = 4$

▮ $\log_4 4x^2 = 4$, $4x^2 = 4^4 = 256$, $x^2 = 64$, or $x = \pm 8$. But $x = -8$ is extraneous because of the $\log_4 x$ domain. So $x = 9$ is the only solution.

6.272 $\log_2[\log_3(2x-1)] = 2$

▮ Then $\log_3(2x-1) = 2^2 = 4$ and $2x - 1 = 3^4 = 81$. Then $2x = 82$, or $x = 41$.

6.273 $\log_2 \sqrt{x^2 - 36} = 3$

▮ $\sqrt{x^2 - 36} = 2^3 = 8$, $x^2 - 36 = 64$, $x^2 = 100$, or $x = \pm 10$. Check these; they are both good.

6.274 $\sqrt{\log_2 \sqrt{x+6}} = 2$

▮ Square both sides: $\log_2 \sqrt{x+6} = 4$. Then $\sqrt{x+6} = 2^4$, $x + 6 = 2^8$, or $x = 2^8 - 6 = 250$.

6.275 $\log(x - 9) + \log 100x = 3$

▮ $\log(x-9)(100x) = 3$, $(x-9)(100x) = 1000$, $100x^2 - 900x - 1000 = 0$, $x^2 - 9x - 10 = 0$, $(x - 90)(x + 1) = 0$, or $x = 10, -1$. But $x = -1$ is extraneous, so $x = 10$ is the only solution.

6.276 $\log x - \log 5 = \log 2 - \log(x - 3)$

▮ $\log(x/5) = \log[2/(x-3)]$, $\dfrac{x}{5} = \dfrac{2}{x-3}$, $x^2 - 3x = 10$, $x^2 - 3x - 10 = 0$, $(x - 5)(x + 2) = 0$, or $x = 5, -2$. But $x = -2$ is extraneous, so $x = 5$ is the only solution.

6.277 $\log(6x + 5) - \log 3 = \log 2 - \log x$

▮ $\log \dfrac{6x+5}{3} = \log \dfrac{2}{x}$, $\dfrac{6x+5}{3} = \dfrac{2}{x}$, $6x^2 + 5x = 6$, $6x^2 + 5x - 6 = 0$, $(3x - 2)(2x + 3) = 0$, or $x = \frac{2}{3}, -\frac{3}{2}$. But $x = -\frac{3}{2}$ is extraneous, so $x = \frac{2}{3}$ is the only solution.

6.278 $(\ln x)^3 = \ln x^4$

▮ Let $u = \ln x$; then $u^3 = 4 \ln x = 4u$, $u^3 - 4u = 0$, $u(u^2 - 4) = 0$, or $u = 0, 2, -2$. If $u = 0$, $\ln x = 0$, or $x = 1$. If $u = 2$, $\ln x = 2$, or $x = e^2$. If $u = -2$, $\ln x = -2$, or $x = e^{-2}$. Thus, the solutions are $x = 1, e^2, e^{-2}$.

6.279 $x^{\log x} = 100x$

▮ $\log(x^{\log x}) = \log(100x)$, $\log x \log x = \log 100 + \log x$, $(\log x)^2 = 2 + \log x$, $(\log x)^2 - \log x - 2 = 0$, $(\log x - 2)(\log x + 1) = 0$, or $\log x = 2, -1$. If $\log x = 2$, $x = 100$. If $\log x = -1$, $x = \frac{1}{10}$.

For Probs. 6.280 to 6.284, solve for the indicated letter.

6.280 $I = I_0 e^{-kx}$ for x (x-ray absorption).

▮ $\ln I = \ln I_0 e^{-kx} = \ln I_0 + \ln e^{-kx}$, $\ln I - \ln I_0 = -kx$, $x = 1/k \ln(I/I_0) = \ln(I/I_0)^{1/k}$.

6.281 $A = P(1 + i)^n$ for n (compound interest).

▮ $\log A = \log P(1 + i)^n = \log P + \log(1 + i)^n$, $\log A - \log P = n \log(1 + i)$, $\dfrac{\log(A/P)}{\log(1 + i)} = n$.

6.282 $N = 10 \log(I/I_0)$ for I (sound intensity decibels).

▮ $N = 10 \log(I/I_0)$, $\dfrac{N}{10} = \log(I/I_0)$, $10^{N/10} = 10^{\log(I/I_0)}$, $10^{N/10} = I/I_0$, $I = I_0 10^{N/10}$.

6.283 $t = \dfrac{-1}{k}(\ln A - \ln A_0)$ for A (radioactive decay)

▮ $t = \dfrac{-1}{k}[\ln(A/A_0)]$, $\dfrac{t}{-1/k} = \ln(A/A_0)$, $-kt = \ln(A/A_0)$, $e^{-kt} = A/A_0$, or $A = A_0 e^{-kt}$.

6.284 $I = \dfrac{E}{R}(1 - e^{-Rt/L})$ for t (electric circuits)

▮ $I = \dfrac{E}{R} - \dfrac{E}{R} e^{-Rt/L}$, $I - \dfrac{E}{R} = -\dfrac{E}{R} e^{-Rt/L}$, $\dfrac{I - E/R}{-E/R} = e^{-Rt/L}$, $\dfrac{IR - E}{-E} = e^{-Rt/L}$, $\ln\left(\dfrac{IR - E}{-E}\right) = \dfrac{-Rt}{L}$, $-RT = L \ln \dfrac{IR - E}{-E}$, $t = \dfrac{-L}{R} \ln \dfrac{IR - E}{-E}$.

For Probs. 6.285 to 6.290, solve to three significant digits.

6.285 $10^{-x} = 0.0347$

▮ $\log 10^{-x} = \log 0.0347$, $-x = \log 0.0347 = -1.46$. (Use a calculator, or use a log table; be careful! The table shows 0.5403. Take $0.5403 - 2$, since this is $\log 0.0347$, *not* $\log 3.47$.)

6.286 $10^x = 14.3$

▮ $10^x = 14.3$, $\log 10^x = \log 14.3$, $x = \log 14.3 = 1.16$. Find 1.16 in the table, or use the calculator. The table shows 0.1553. Use 1.1553 and round to 1.16 since we want the answer to three significant digits.

6.287 $10^{3x+1} = 925$

▮ Then $3x + 1 = \log 925 = 2.9661$, $3x + 1 = 2.9661$, $3x = 1.9661$, or $x = 0.655$ (rounded).

6.288 $2 = 1.05^x$

▮ $\log 2 = \log 1.05^x = x \log 1.05$, or $x = \dfrac{\log 2}{\log 1.05} = \dfrac{0.3010}{0.0211} = 14.3$ (rounded).

6.289 $e^{-1.4x} = 13$

▮ $\ln e^{-1.4x} = \ln 13$, $-1.4x = \ln 13 = 2.565$, or $x = -1.83$

6.290 $438 = 200e^{0.25x}$

▮ $e^{0.25x} = \frac{438}{200} = 2.19$, $0.25x = \ln 2.19 = 0.7839$,   or $x = 3.14$.

For Probs. 6.291 and 6.292, find the fallacy.

6.291 $3 > 2$, $\log(\frac{1}{2})3 > \log(\frac{1}{2})2$, $3\log\frac{1}{2} > 2\log\frac{1}{2}$, $\log(\frac{1}{2})^3 > \log(\frac{1}{2})^2$, $(\frac{1}{2})^3 > (\frac{1}{2})^2$,   so $\frac{1}{8} > \frac{1}{4}$.

▮ The problem is that, although $3 > 2$, $3\log\frac{1}{2} \not> 2\log\frac{1}{2}$ (since $\log\frac{1}{2} < 0$ and the sign reverses).

6.292 $-2 < -1$, $\ln e^{-2} < \ln e^{-1}$, $2\ln e^{-1} < \ln e^{-1}$,   so $2 < 1$.

▮ $\ln e^{-1} = -1$; division by a negative reverses the signs.

For Probs. 6.293 to 6.300, solve the given equation.

6.293 $\log\dfrac{10}{x} = 0$

▮ Then $10^{\log(10/x)} = 10^0$, $\dfrac{10}{x} = 1$, or $x = 10$. See Prob. 6.294.

6.294 $\log\dfrac{10}{x} = 0$

▮ $\log 10 - \log x = 0$, $1 - \log x = 0$, $\log x = 1$, or $x = 10$.

6.295 $\ln ex = 0$

▮ $e^{\ln ex} = e^0$ $ex = 1$, or $x = 1/e = e^{-1}$.

6.296 $\ln\dfrac{x}{e} = -2$

▮ $\ln x - \ln e = -2$, $\ln x = -2 + \ln e = -2 + 1 = -1$, or $x = e^{-1} = 1/e$. Since $\ln\dfrac{x}{e} = \ln\dfrac{1/e}{e} = \ln\dfrac{1}{e^2} = -2$, the answer checks.

6.297 $\log_x\dfrac{x}{16} = 5$

▮ $\log_x x - \log_x 16 = 5$, $1 - \log_x 16 = 5$, $\log_x 16 = -4$, $x^{-4} = 16$, or $x = \frac{1}{2}$. [Why? $(\frac{1}{2})^4 = 16$.]

6.298 $1 + \ln x = 0$

▮ If $1 + \ln x = 0$, then $\ln x = -1$, and $x = e^{-1} = 1/e$.

6.299 $(1 + \ln x)\ln x = 0$

▮ If $(1 + \ln x)(\ln x) = 0$, then $1 + \ln x = 0$ or $\ln x = 0$. If $\ln x = 0$, $x = e^0 = 1$. If $1 + \ln x = 0$, $x = 1/e$ (see Prob. 6.298). Thus the solutions are $x = 1, 1/e$.

6.300 $5^{2x-1} = 10$

▮ Then $\log 5^{2x-1} = \log 10$, $(2x-1)\log 5 = \log 10 = 1$, $2x - 1 = \dfrac{1}{\log 5}$, $2x = (\log 5)^{-1} + 1$, or $x = \dfrac{(\log 5)^{-1} + 1}{2}$. (Note: you can substitute 1.4307 for $\dfrac{1}{\log 5}$ and simplify if you like.)

For Probs. 6.301 to 6.303, solve the given system.

6.301 $5^{x+2y} = 28$, $\quad 4x + 2y = 3$

▮ $2y = 3 - 4x$; thus, if $5^{x+2y} = 28$, then $5^{x+(3-4x)} = 28$ and $5^{-3x+3} = 28$. Then $(-3x + 3) \log 5 = \log 28$ and $-3x + 3 = \dfrac{\log 28}{\log 5}$. Thus, $x = 0.31$ (rounded) and $y = 0.88$ (rounded).

6.302 $10^{x+2y} = 2$, $\quad \log 3x - \log 2y = 1$

▮ If $10^{x+2y} = 2$, then $10^x 10^{2y} = 2$ and $10^x = 2 \cdot 10^{-2y}$. Thus, $x = \log(2 \cdot 10^{-2y}) = \log 2 + \log 10^{-2y}$ and $x = -2y + \log 2$. Since $\log 3x - \log 2y = \log \dfrac{3x}{2y} = \log \dfrac{3(-2y + \log 2)}{2y} = 1$, $\dfrac{-6y + 3\log 2}{2y} = 10$. Thus, $y = 0.035$ (rounded) and $x = 0.23$ (rounded).

6.303 $x - y = 6$, $\quad \log x + \log y = 1$

▮ $x - y = 6$ means $x = 6 + y$. Then $\log(6 + y) + \log y = 1$, $\log y(6 + y) = 1$, $y(6 + y) = 10^1 = 10$, and $y^2 + 6y - 10 = 0$. So $y = \dfrac{-6 \pm \sqrt{36 - 4(1)(-10)}}{2} = \dfrac{-6 \pm \sqrt{76}}{2} = \dfrac{-6 \pm 2\sqrt{19}}{2} = -3 \pm \sqrt{19}$. Thus, $y = -3 + \sqrt{19}$, $y = -3 - \sqrt{19}$. But $-3 - \sqrt{19}$ is an extraneous solution. So $x = 6 + (-3 + \sqrt{19}) = 3 + \sqrt{19}$.

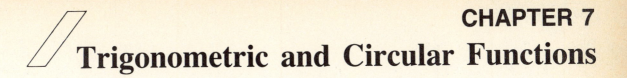

CHAPTER 7

Trigonometric and Circular Functions

7.1 ANGLE MEASUREMENT

For Probs. 7.1 to 7.8, determine the quadrant containing the given angle.   Do *not* sketch the angle to determine the quadrant.

7.1   115°

▮ Since $90° < 115° < 180°$,   115° is in quadrant II.

7.2   245.6°

▮ Since $180° < 245.6° < 270°$,   245.6° is in quadrant III.

7.3   −67°

▮ We note that $−90° < −67° < 0°$ and that, therefore, −67° is in quadrant IV.   We also might note that $−67° + 360° = 293°$.   Thus, −67° is coterminal with 293°.   Since $270° < 293° < 360°$,   293° (or −67°) is in quadrant IV.

7.4   −415°

▮ $−415° + 360° = −55°$ and $−55° + 360° = 305°$.   Thus, −415° and 305° are coterminal; 305° is in quadrant IV, and, therefore, so is −415°.

7.5   −1654°

▮ $360° \times 5 = 1800°$,   and $1800° + (−1654°) = 146°$.   Since 146° is in quadrant II and 146° and −1654° are coterminal, −1654° is in quadrant II.

7.6 $\pi/17$

▮ Since $0 < \pi/17 < \pi/2$, $\pi/17$ is in quadrant I.

7.7 $−\pi/10$

▮ Since $−\pi/2 < −\pi/10 < 0$, $−\pi/10$ is in quadrant IV.   Also, we might note that $2\pi + (−\pi/10) = 19\pi/10$ and that $19\pi/10$ is in quadrant IV.   Since $−\pi/10$ and $19\pi/10$ are coterminal, $−\pi/10$ is in quadrant IV.

7.8 $−46.5\pi$

▮ $2\pi \times 24 = 48\pi$,   and $48\pi + (−46.5\pi) = 1.5\pi = 3\pi/2$.   This is a quadrantal angle.

For Probs. 7.9 to 7.15, sketch the angle of each of the following measures.   In your sketch, place the angle in standard position.

7.9   225°

▮ See Fig. 7.1. $180° + 45° = 225°$.

7.10   420°

▮ See Fig. 7.2. $420° = 360° + 60°$.

7.11   2421°

▮ See Fig. 7.3.   When we divide 2421° by 360°, the quotient is 6. $360° \times 6 = 2160°$,   and $2421° − 2160° = 261° = 180° + 81°$.

7.12   −675°

▮ See Fig. 7.4. $−675° = −360° + (−315°) = −360° + (−270°) + (−45°)$.

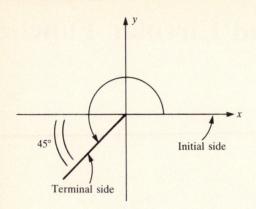

Fig. 7.1

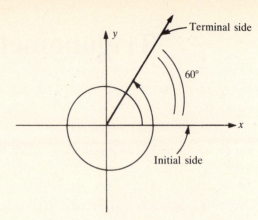

Fig. 7.2

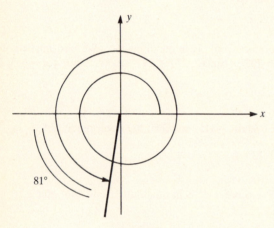

Fig. 7.3

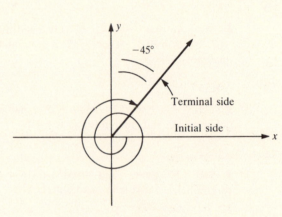

Fig. 7.4

7.13 $-\pi/3$

▮ See Fig. 7.5. $-\pi/2 < -\pi/3 < 0$.

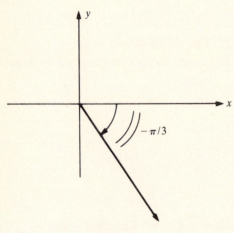

Fig. 7.5

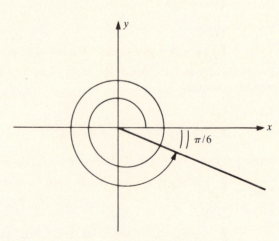

Fig. 7.6

7.14 $23\pi/6$

▮ See Fig. 7.6. $23\pi/6 = 2\pi + 11\pi/6 = 2\pi + (2\pi - \pi/6)$.

7.15 $-19\pi/4$

▮ See Fig. 7.7. $-19\pi/4 = -16\pi/4 + (-3\pi/4) = -4\pi + (-3\pi/4) = -4\pi + (-\pi/2) + (-\pi/4)$.

For Probs. 7.16 to 7.20, perform the indicated operation.

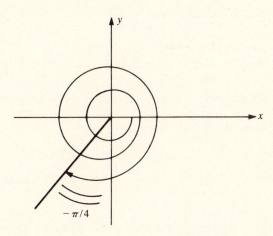

Fig. 7.7

7.16 $65°\,13'$
     $+71°\,26'$

▮ $65° + 71° = 136°$; $13' + 26' = 39'$. So the answer is $136°\,39'$.

7.17 $114°\,29'\,46''$
     $+\;81°\;\;4'\,11''$

▮ $114° - 81° = 33°$; $29' - 4' = 25'$; $46'' - 11'' = 35''$. So the answer is $33°\,25'\,35''$.

7.18 $127°\,10'\,14''$
     $-85°\;\;5'\,53''$

▮ $10'\,14'' = 9'\,74''$ $(1' = 60'')$. Then

$$127°\,9'\,74''$$
$$-85°\,5'\,53''$$
$$\overline{42°\,4'\,21''}$$

7.19 $11'\,11''$
     $+58'\,59''$

▮
$$11'\,11''$$
$$58'\,59''$$
$$\overline{69'\,70''}$$

Since $1' = 60''$, $69'\,70'' = 70'\,10''$. Then since $1° = 60'$, $70'\,10'' = 1°\,10'\,10''$.

7.20 $49°\;\;\;\;57''$
     $+61°\,59'\,53''$

▮
$$49°\;\;0'\;\;57''$$
$$+61°\,59'\;\;53''$$
$$\overline{110°\,59'\,110''} = 110°\,60'\,50'' = 111°\,0'\,50''$$

For Probs. 7.21 to 7.24, express the given angle in degrees, minutes, and seconds.

7.21   40.25°

▮ $40.25° = 40° + 0.25° = 40° + 0.25 \times$ (since $60' = 1°$) $= 40° + 15' = 40° 15' = 40° 15' 0''$.

7.22   75.2°

▮ $75.2° = 75° + 0.2° = 75° + 0.2 \times 60' = 75° + 12' = 75° 12' 0''$.

7.23   17.45°

▮ $17.45° = 17° \times 0.45 \times 60' = 17° + 27' = 17° 27'$.

7.24   14.458°

▮ $14.458° = 14° + 0.458 \times 60' = 14° + 27.48' = 14° + 27' + 0.48 \times 60'' = 14° + 27' + 28.8'' = 14° 27' 28.8''$.

For Probs. 7.25 to 7.27, rewrite the given angle in decimal form.

7.25   10° 10'

▮ $60' = 1°$;   thus, $10' = 1° \times \frac{10}{60} = (\frac{1}{6})° = 0.1\overline{6}°$.   Then $10° 10' = 10.1\overline{6}°$.

7.26   27° 15' 25''

▮ $27° 15' 25'' = 27° + (\frac{15}{60})° + [(25/(60)^2]° = 27° + 0.25° + 0.0069\overline{4}° = 27.2569\overline{4}°$.

7.27   45° 45' 45''

▮ $45° 45' 45'' = 45° + (\frac{45}{60})° + [45/(60)^2]° = 45° + 0.75° + 0.0125° = 45.7625°$.

For Probs. 7.28 to 7.37, convert the given angle to degrees if it is in radians and to radians if it is in degrees.

7.28 $\pi/4$

▮ (i) 2π rad $= 360°$.   Then $\pi/4$ rad $= 360° \times (\pi/4)/2\pi = 360° \times \frac{1}{8} = 45°$.   (ii) Alternatively, $\pi = 180°$ and $\pi/4 = 180°/4 = 45°$.

7.29 $6\pi/9$

▮ (i) $6\pi/9 = 180° \times \frac{6}{9} = 120°$;   or (ii) $6\pi/9 = 2\pi/3 = \frac{2}{3}(180°) = 120°$.

7.30 $-\pi/6$

▮ $-\pi/6 = 180° \times (-\frac{1}{6}) = -30°$.

7.31 -7π

▮ $-7\pi = -7(180°) = -1260°$.

7.32 $-17\pi/6$

▮ $-17\pi/6 = -\frac{17}{6}(180°) = -510°$.

7.33   2

▮ π rad $= 180°$.   Thus 1 rad $= (180/\pi)°$ and 2 rad $= 2(180/\pi)° = (360/\pi)°$.

7.34   45°

▮ $180° = \pi$ rad. $45° = \pi \times \frac{45}{180} = \pi \times \frac{1}{4} = \pi/4$ rad.

7.35   410°

▮ $410° = \pi \times \frac{410}{180} = \pi \times 2.2\overline{7} = 2.2\overline{7}\pi$ rad.

7.36 $-135°$

▮ $-135° = \pi \times (-\frac{135}{180}) = \pi \times (-0.75) = -0.75\pi = -3\pi/4$ rad.

7.37   7° 30′

▮ $7° 30′ = 7° + (\frac{30}{60})° = 7.5° = \pi \times 7.5/180 = \pi \times 0.041\overline{6} = 0.041\overline{6}$ rad.

For Probs. 7.38 and 7.39, find the radian measure of a central angle θ subtended by an arc s in a circle of radius R, where R and s are given in each of the following.

7.38 $R = 4$ cm, $s = 24$ cm.

▮ $\theta = \dfrac{s}{R}$ (in radians) $= \dfrac{24 \text{ cm}}{4 \text{ cm}} = 6$ rad.

7.39 $R = 10$ cm, $s = 10\pi$ cm.

▮ $\theta = \dfrac{s}{R} = \dfrac{10\pi \text{ cm}}{10 \text{ cm}} = \pi$.

7.40   Find five angles coterminal with $-155°$.

▮   See Fig. 7.8.   (*1*) $360° + (-155°) = 205°$.   (*2*) $2 \times 360° + (-155°) = 565°$.   (*3*) $3 \times 360° + (-155°) = 925°$.   (*4*) $15 \times 360° + (-155°) = 5245°$.   (*5*) $5245° - (35 \times 360°) = 5245° - 12\,600° = -7355°$.

For Probs. 7.41 to 7.43, answer true or false, and justify your answer.

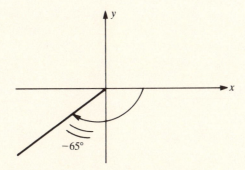

Fig. 7.8

7.41   If $\theta_1 = \theta_2$ in degree measure, then $\theta_1 = \theta_2$ in radian measure.

▮   True.   Radians and degrees are simply *units* of measure.

7.42   An angle measuring 4π rad is congruent to an angle measuring 2π rad.

▮   False.   Be very careful here.   These two angles are coterminal, but not congruent.   The same is true, for example, for $120°$ and $(360 + 120)°$.

7.43   If x is the supplement of y, then $x/2$ is the complement of y.

▮   False.   Here, $x + y = 180°$.   Then $(x + y)/2 = 90°$,   but $x/2 + y \neq 90°$.

7.2   THE WRAPPING FUNCTION

For Probs. 7.44 to 7.69, find the coordinates for each circular point.   Do not look them up in a table; do the calculation yourself!

7.44 $W(\pi)$

▮   See Fig. 7.9.   (We use the a, b instead of x, y axes so that we can use x later for the domain of the circular functions.)   To find the circular point associated with $x \in \mathcal{R}$,   begin at $(1, 0)$ and move along the circle $a^2 + b^2 = 1$ for $|x|$ units (if $x > 0$,   counterclockwise; if $x < 0$,   clockwise).   In this case, we move π units counterclockwise.   The circle was $C = 2\pi R = 2\pi(1) = 2\pi$.   Thus, moving π units counterclockwise puts us at $(-1, 0)$.   Thus, $W(\pi) = (-1, 0)$.

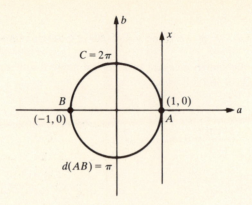

Fig. 7.9

7.45 $W(0)$

▮ We move 0 units from $(1, 0)$. $W(0) = (1, 0)$.

7.46 $W(4\pi)$

▮ See Fig. 7.10. We move 4π units around the circle from $(1, 0)$. This is $2(2\pi)$; thus, we are back at $(1, 0)$. So $W(4\pi) = (1, 0)$. Note that $W(6\pi) = W(8\pi)$ (etc.) $= (1, 0)$.

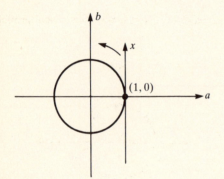

Fig. 7.10

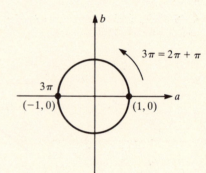

Fig. 7.11

7.47 $W(3\pi)$

▮ See Fig. 7.11. This is going to send us around the circle once, and then halfway around again. $W(3\pi) = (-1, 0)$.

7.48 $W(-\pi)$

▮ See Fig. 7.12. We will move π (one half circle) clockwise to $(-1, 0)$. $W(-\pi) = (-1, 0)$.

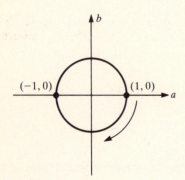

Fig. 7.12

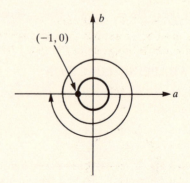

Fig. 7.13

7.49 $W(-3\pi)$

▮ See Fig. 7.13. $-3\pi = -2\pi + (-\pi)$. If we move -2π, we are back at $(1, 0)$. Next, move π units clockwise, and arrive at $(-1, 0)$. $W(-3\pi) = (-1, 0)$.

7.50 $W(3\pi/2)$

▮ See Fig. 7.14. $C = 2\pi$, and $W(\pi) = (-1, 0)$. Move $\pi/2$ more, and arrive at $(0, -1)$. $W(3\pi/2) = (0, -1)$.

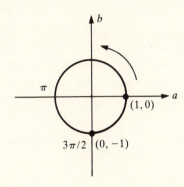

Fig. 7.14

7.51 $W(\pi/2)$

▮ We need to move one-fourth of the way around the circle, since $C = 2\pi$ and $(\pi/2)/2\pi = \frac{1}{4}$. Thus, $W(\pi/2) = (0, 1)$.

7.52 ▮ $W(-\pi/2)$.

▮ See Fig. 7.15. $C = 2\pi$ and $(-\pi/2)/2\pi = -\frac{1}{4}$. Move one-fourth of the way (clockwise) around the circle. $W(-\pi/2) = (0, -1)$.

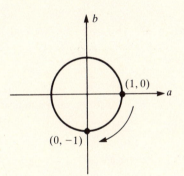

Fig. 7.15

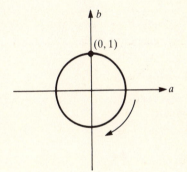

Fig. 7.16

7.53 $W(-3\pi/2)$

▮ See Fig. 7.16. $C = 2\pi$; $(-3\pi/2)/2\pi = -\frac{3}{4}$. Move three-fourths of the way (clockwise) around the circle. $W(-3\pi/2) = (0, 1)$.

7.54 $W(11\pi/2)$

▮ See Fig. 7.17. $(11\pi/2)/2\pi = \frac{11}{4}$. Move eleven-fourths of the way around the circle. Since $\frac{11}{4} = 2\frac{3}{4}$, we move twice around the circle which brings us to $(1, 0)$, and then three-fourths of the way around the circle which brings us to $(0, -1)$. $W(11\pi/2) = (0, -1)$.

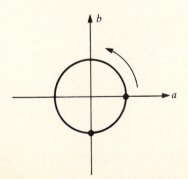

Fig. 7.17

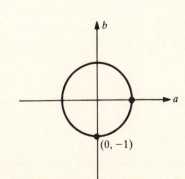

Fig. 7.18

7.55 $W(-15\pi/2)$

∎ See Fig. 7.18.   Since $\dfrac{-15\pi/2}{2\pi} = -\dfrac{15}{4}$,   we move fifteen-fourths of the way (counterclockwise) around the circle; $\frac{15}{4} = 3\frac{3}{4}$. $W(-15\pi/2) = (0, -1)$.

7.56 $W(\pi/4)$

∎ See Fig. 7.19. $\dfrac{\pi/4}{2\pi} = \dfrac{1}{8}$.   We move one-eighth of the way (clockwise) around the circle.   The point (s, t) is the intersection of the circle $a^2 + b^2 = 1$ and the line $a = b$,   which is $(\sqrt{2}/2, \sqrt{2}/2)$.
$W(\pi/4) = (\sqrt{2}/2, \sqrt{2}/2)$.

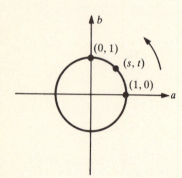

Fig. 7.19

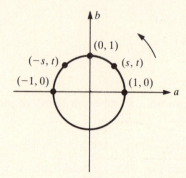

Fig. 7.20

7.57 $W(\pi/6)$

∎ See Fig. 7.20. $W(\pi/3) = (\frac{1}{2}, \sqrt{3}/2) = (s, t)$ above.   Then $W(2\pi/3) = (-\frac{1}{2}, \sqrt{3}/2)$.   We find $W(\pi/6)$ (using symmetry) to be $(\sqrt{3}/2, \frac{1}{2})$.

7.58 $W(-\pi/6)$

∎ See Fig. 7.21.   If $W(\pi/6) = (\sqrt{3}/2, \frac{1}{2})$,   then $W(-\pi/6) = (\sqrt{3}/2, -\frac{1}{2})$;   use the above indicated symmetry.

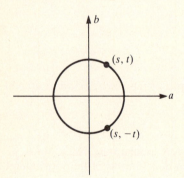

Fig. 7.21

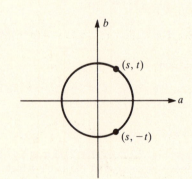

Fig. 7.22

7.59 $W(-\pi/3)$

∎ See Fig. 7.22. $W(\pi/3) = (\frac{1}{2}, \sqrt{3}/2)$.   Thus, $W(-\pi/3) = (\frac{1}{2}, -\sqrt{3}/2)$.

7.60 $W(-\pi/4)$

∎ See Fig. 7.23.   Again using symmetry, if $W(\pi/4) = (\sqrt{2}/2, \sqrt{2}/2)$,   then $W(-\pi/4) = (\sqrt{2}/2, -\sqrt{2}/2)$.

7.61 $W(2\pi/3)$

∎ See Fig. 7.24.   Again using symmetry, $W(\pi/3) = (\frac{1}{2}, \sqrt{3}/2)$,   and then $W(2\pi/3) = (-\frac{1}{2}, \sqrt{3}/2)$.

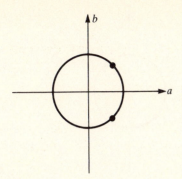

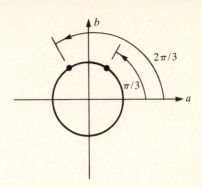

Fig. 7.23

Fig. 7.24

7.62 $W(11\pi/6)$

▮ See Fig. 7.25. $11\pi/6 = 2\pi - \pi/6$. But $W(\pi/6) = (\sqrt{3}/3, \frac{1}{2})$. Thus, $W(11\pi/6) = (\sqrt{3}/2, -\frac{1}{2})$.

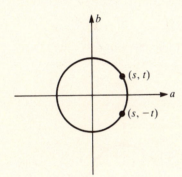

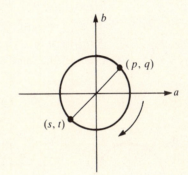

Fig. 7.25

Fig. 7.26

7.63 $W(-3\pi/4)$

▮ See Fig. 7.26. Going $\frac{3}{4}\pi$ clockwise, we arrive at (s, t) as indicated. Using symmetry, we notice that $s = -p$ and $t = -q$ where $(p, q) = W(\pi/4)$. Thus, $W(-3\pi/4) = (-\sqrt{2}/2, -\sqrt{2}/2)$.

7.64 $W(-7\pi/6)$

▮ See Fig. 7.27. If $(s, t) = W(\pi/6)$, then $(-s, t) = W(-7\pi/6) = (-\sqrt{3}/2, \frac{1}{2})$.

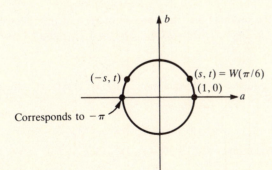

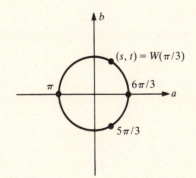

Fig. 7.27

Fig. 7.28

7.65 $W(5\pi/3)$

▮ See Fig. 7.28. If $W(\pi/3) = (s, t)$ then $W(5\pi/3) = (s, -t)$ (see above) $= (\frac{1}{2}, -\sqrt{3}/2)$.

7.66 $W(7\pi/4)$

▮ See Fig. 7.29. $\dfrac{7\pi}{4} = 2\pi - \dfrac{\pi}{4}$. If $W(\pi/4) = (s, t)$, then $W(7\pi/4) = (s, -t), = (\sqrt{2}/2, -\sqrt{2}/2)$.

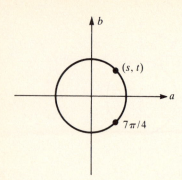

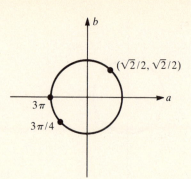

Fig. 7.29

Fig. 7.30

7.67 $W(13\pi/4)$

▮ See Fig. 7.30. $13\pi/4 = 12\pi/4 + \pi/4 = 3\pi + \pi/4$. Then, since $W(\pi/4) = (\sqrt{2}/2, \sqrt{2}/2)$,
$W(13\pi/4) = (-\sqrt{2}/2, -\sqrt{2}/2)$.

7.68 $W(-10\pi/3)$

▮ See Fig. 7.31. $-10\pi/3 = -9\pi/3 + (-\pi/3) = -3\pi + (-\pi/3)$. Then $W(-10\pi/3) = W(2\pi/3) = (-\frac{1}{2}, \sqrt{3}/2)$. (You can get the same result using other symmetries.)

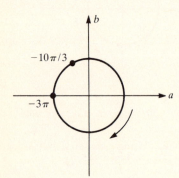

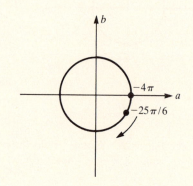

Fig. 7.31

Fig. 7.32

7.69 $W(-25\pi/6)$

▮ See Fig. 7.32. $-25\pi/6 = -24\pi/6 + (-\pi/6) = -4\pi + (-\pi/6)$. $W(-25\pi/6) = W(-\pi/6) = (\sqrt{3}/2, -\frac{1}{2})$.

For Probs. 7.70 to 7.78, determine whether the product of the ordinate and the abscissa of each of the following is >0 or <0. Do *not* calculate the actual coordinates! Use $\pi/2 \approx 1.57$, $\pi \approx 3.14$, $3\pi/2 \approx 4.71$, and $2\pi \approx 6.28$.

7.70 $W(2)$

▮ Note that $1.57 < 2 < 3.14$. Thus, $W(2)$ must be in the second quadrant. If $W(2) = (s, t)$, then $s < 0$, $t > 0$ (quadrant II), and $st < 0$.

7.71 $W(1)$

▮ $0 < 1 < 1.57$. Thus, $W(1)$ must be in the first quadrant. If $W(1) = (s, t)$, then $s > 0$, $t > 0$ (quadrant I), and $st > 0$.

7.72 $W(3)$

▮ Since $1.57 < 3 < 3.14$, $W(3)$ must be in quadrant II. Then, if $W(3) = (s, t)$, $s < 0$, $t > 0$ (quadrant II), and $st < 0$.

7.73 $W(4)$

▮ Since $3.14 < W(4) < 4.71$, $W(4)$ must be in quadrant III. Then, if $W(4) = (s, t)$, $s < 0$, $t < 0$ (quadrant III), and $st > 0$.

7.74 $W(5)$

▮ Since $4.71 < 5 < 6.28$, $W(5)$ must be in quadrant IV. Then, if $W(5) = (s, t)$, $s > 0$, $t < 0$ (quadrant IV), and $st < 0$.

7.75 $W(8)$

▮ See Fig. 7.33 and Prob. 7.70. Note that $7.85 < 8 < 9.42$ ($7.85 = 1.57 \times 5$; $9.42 = 1.57 \times 6$) and that $W(8)$ must be in quadrant II. Then $st < 0$.

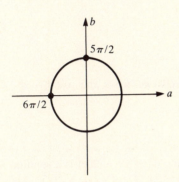

Fig. 7.33

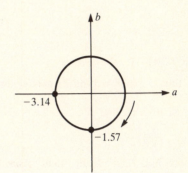

Fig. 7.34

7.76 $W(-2.5)$

▮ See Fig. 7.34 and Prob. 7.73. Note that $-3.14 < -2.5 < -1.57$. Thus, $W(-2.5)$ must be in quadrant III. Then $st > 0$.

7.77 $W(-6.1)$

▮ See Fig. 7.35 and Prob. 7.74. $-6.28 < -6.1 < -4.71$. Then $W(-6.1)$ is in quadrant IV and $st < 0$.

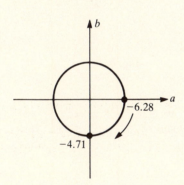

Fig. 7.35

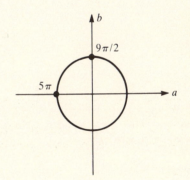

Fig. 7.36

7.78 $W(15)$

▮ See Fig. 7.36. $1.57 \times 10 = 15.7$, $1.57 \times 9 = 14.13$, and $14.13 < 15 < 15.7$. Since $\pi/2 \times 10 = 5\pi$, $\pi/2 \times 9 = 9\pi/2$, and $W(15)$ is in quadrant II, $st < 0$.

For Probs. 7.79 to 7.81, $W(x) = (s, t)$. Answer true or false, and justify your answer.

7.79 $W(x + \pi) = (-s, -t)$

▮ Suppose that $W(x) = (s, t)$ as in Fig. 7.37. Moving further along the circle brings us diametrically opposite the point (s, t). This point is $(-s, -t)$. True.

7.80 $W(-x) = (-s, t)$

▮ If $A = W(x)$, then $B = W(-x)$ in Fig. 7.38. But B is the point $(s, -t)$. False.

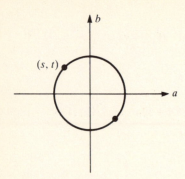

Fig. 7.37

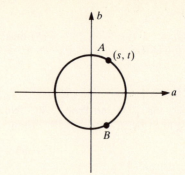

Fig. 7.38

7.81 $W(x + 2\pi) = (s, t)$

▮ True.   When we add 2π to our distance wrapped around the circle, we arrive at x again since $C = 2\pi$.

For Probs. 7.82 to 7.85, find all x, $-2\pi \le x \le 2\pi$,   for the given condition.

7.82 $W(x) = (\sqrt{2}/2, \sqrt{2}/2)$

▮ See Fig. 7.39.   Notice that $W(\pi/4) = (\sqrt{2}/2, \sqrt{2}/2)$.   But then $W(\pi/4 + n \cdot 2\pi)$ is a solution for any $n \in \mathscr{Z}$.   If $n = -1$, $\pi/4 + n \cdot 2\pi = \pi/4 - 2\pi = -7\pi/4$.   Then $W(-7\pi/4) = (\sqrt{2}/2, \sqrt{2}/2)$,   and $x = -7\pi/4, \pi/4$.   Note that any other choice of n places us outside our domain of $-2\pi \le x \le 2\pi$.

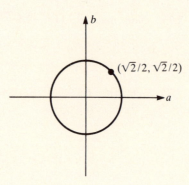

Fig. 7.39

7.83 $W(x) = (\sqrt{3}/2, \frac{1}{2})$

▮ We know that $W(\pi/6) = (\sqrt{3}/2, \frac{1}{2})$.   Then $x = \pi/6 + n \cdot 2\pi$, $\forall n \in \mathscr{Z}$,   will work.   If $n = -1$, then $x = \pi/6 - 2\pi = -11\pi/6$.   No other n will leave us in our domain of $-2\pi \le x \le 2\pi$.   Thus, $x = -11\pi/6, \pi/6$.

7.84 $W(x) = (-\sqrt{3}/2, -\frac{1}{2})$

▮ See Fig. 7.40.   Since $W(\pi/6) = (\sqrt{3}/2, \frac{1}{2})$, $(-\sqrt{3}/2, -\frac{1}{2}) = W(\pi + \pi/6) = W(7\pi/6)$.   Then, $7\pi/6 + n \cdot 2\pi$ $\forall n \in \mathscr{Z}$,   works as well; to be in the range $[-2\pi, 2\pi]$, we choose $n = -1$,   and get $7\pi/6 - 2\pi = -5\pi/6$.   Thus, $x = -5\pi/6, 77\pi/6$.

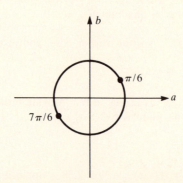

Fig. 7.40

7.85 $W(x) = (-\sqrt{2}/2, \sqrt{2}/2)$

▮ Since $W(\pi/4) = (\sqrt{2}/2, \sqrt{2}/2)$, $W(3\pi/4) = (-\sqrt{2}/2, \sqrt{2}/2)$. And for any $n \in \mathscr{Z}$, $W(3\pi/4 + n \cdot 2\pi) = (-\sqrt{2}/2, \sqrt{2}/2)$. To remain in $[-2\pi, 2\pi]$, we choose $n = -1$. Then $3\pi/4 = 2\pi = -5\pi/4$. Thus, $x = -5\pi/4, \ 3\pi/4$.

7.3 THE CIRCULAR FUNCTIONS

For Probs. 7.86 to 7.115, find each of the indicated circular function values. Do *not* use a calculator or a table of circular or trigonometric values.

7.86 $\cos 0$

▮ If $W(x) = (a, b)$, then $\cos x = a$. Here, $x = 0$. Then $W(0) = (1, 0)$ and $a = 1$. Thus, $\cos 0 = 1$.

7.87 $\sin 0$

▮ If $W(x) = (a, b)$, then $\sin x = b$. Here, $x = 0$. Then $W(0) = (1, 0)$ and $b = 0$. Thus, $\sin 0 = 0$.

7.88 $\sin \pi/2$

▮ $W(\pi/2) = (0, 1)$. Then $\sin \pi/2 = 1$ since $(a, b) = (0, 1)$; $b = 1$ and $\sin x = b$.

7.89 $\cos \pi/2$

▮ If $W(x) = (a, b)$, then $\cos x = a$. $W(\pi/2) = (0, 1)$; thus, $\cos \pi/2 = 0$.

7.90 $\sin \pi/6$

▮ Recall (or derive again) the fact that $W(\pi/6) = (\sqrt{3}/2, \frac{1}{2})$. Thus, $\sin \pi/6 = \frac{1}{2}$ (the ordinate is the sine value).

7.91 $\sin \pi/3$

$W(\pi/3) = (\frac{1}{2}, \sqrt{3}/2)$. Then $\sin \pi/3 = \frac{3}{2}$ (the ordinate of W is the sine value).

7.92 $\tan 0$

▮ If $W(x) = (s, t)$, then $\tan x = t/s$ (note that $s \neq 0$ here). $W(0) = (1, 0)$; thus $\tan 0 = \frac{0}{1} = 0$.

7.93 $\tan \pi/6$

▮ $\tan \pi/6 = t/s$ where $W(\pi/6) = (s, t)$. But $W(\pi/6) = (\sqrt{3}/2, \frac{1}{2})$. Thus, $\tan \dfrac{\pi}{6} = \dfrac{\frac{1}{2}}{\sqrt{3}/2} = \dfrac{1}{\sqrt{3}} = \dfrac{\sqrt{3}}{3}$.

7.94 $\tan \pi/3$

▮ $\tan \pi/3 = t/s$ where $W(\pi/3) = (s, t)$. Then $\tan \dfrac{\pi}{3} = \dfrac{\sqrt{3}/2}{\frac{1}{2}} = \sqrt{3}$.

7.95 $\cot \pi/3$

▮ (i) We can compute $\cot \pi/3$ directly from the definition: $\cot x = s/t \ (t \neq 0)$ where $W(x) = (s, t)$. Since $W(\pi/3) = (\frac{1}{2}, \sqrt{3}/2)$, $\cot \dfrac{\pi}{3} = \dfrac{\frac{1}{2}}{\sqrt{3}/2} = \dfrac{1}{\sqrt{3}} = \dfrac{\sqrt{3}}{3}$. (ii) We can compute $\cot \pi/3$ using the result in Prob. 7.94 and the identity $\cot x = \dfrac{1}{\tan x}$. Then, $\cot \dfrac{\pi}{3} = \dfrac{1}{\tan \pi/3} = \dfrac{1}{\sqrt{3}} = \dfrac{\sqrt{3}}{3}$.

7.96 $\csc \pi/2$

▮ $\csc \dfrac{\pi}{2} = \dfrac{1}{\sin \pi/2} \cdot \sin \dfrac{\pi}{2} = 1$ (see Prob. 7.88). Then $\csc \pi/2 = \frac{1}{1} = 1$.

7.97 $\sec 0$

▮ $\sec 0 = \dfrac{1}{\cos 0} = \dfrac{1}{1}$ (see Prob. 7.86) $= 1$.

7.98   sec 4 · cos 4

▮ $\sec x \cos x = 1$ (for all x in the domain of sec x) since $\cos x = 1/\sec x$.   Thus, $\sec 4 \cdot \cos 4 = 1$.

7.99 $(\sin \pi/12)^2 + (\cos \pi/12)^2$

▮ $(\sin x)^2 + (\cos x)^2 = 1$ for all values of x.   Thus, $(\sin \pi/12)^2 + (\cos \pi/12)^2 = 1$.

7.100   sec $\pi/2$

▮ If $W(x) = (q, r)$,   then $\sec x = 1/q \ (q \neq 0)$. $W(\pi/2) = (0, 1)$.   Thus, the abscissa is 0, and sec $\pi/2$ is undefined.   We can get the same result by noting that $\cos \pi/2 = 0$ and using the identity $\sec x = 1/\cos x$.

7.101   cot $\pi/2$

▮ $\cot \pi/2 = a/b$ where $W(\pi/2) = (a, b)$, $b \neq 0$.   Since, $W(\pi/2) = (0, 1)$, $\cot \pi/2 = \frac{0}{1} = 0$.

7.102   sin π

▮ $W(\pi) = (-1, 0)$.   Then $\sin \pi = 0$ since where $W(x) = (s, t)$, $t = \sin x$.

7.103   tan π

$W(\pi) = (-1, 0)$.   Then $\tan \pi = -\frac{0}{1} = 0$.

7.104   tan $3\pi/4$

See Fig. 7.41. $W\left(\dfrac{3\pi}{4}\right) = \left(\dfrac{-\sqrt{2}}{2}, \dfrac{\sqrt{2}}{2}\right)$.   Thus, $\tan \dfrac{3\pi}{4} = \dfrac{\sqrt{2}/2}{-\sqrt{2}/2} = -1$.

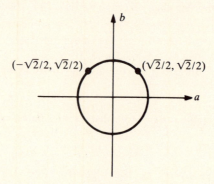

$(-\sqrt{2}/2, \sqrt{2}/2)$ $(\sqrt{2}/2, \sqrt{2}/2)$

Fig. 7.41

7.105   cot $3\pi/4$

▮ $W\left(\dfrac{3\pi}{4}\right) = \left(\dfrac{-\sqrt{2}}{2}, \dfrac{\sqrt{2}}{2}\right)$.   Then $\cot \dfrac{3\pi}{4} = \dfrac{-\sqrt{2}/2}{\sqrt{2}/2} = -1$.   (The result is the same as in Prob. 7.104, but the calculation is different!)

7.106   sin $2\pi/3$

▮ See Fig. 7.42. $W(2\pi/3) = (-\frac{1}{2}, \sqrt{3}/2)$.   Then $\sin 2\pi/3 = \sqrt{3}/2$ (the ordinate is the sine).

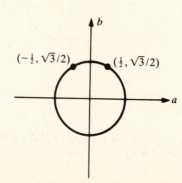

$(-\frac{1}{2}, \sqrt{3}/2)$ $(\frac{1}{2}, \sqrt{3}/2)$

Fig. 7.42

7.107 $\cos 2\pi/3$

▮ See Prob. 7.106. $W(2\pi/3) = (-\frac{1}{2}, \sqrt{3}/2)$. Then $\cos 2\pi/3 = -\frac{1}{2}$ (the abscissa is the cosine).

7.108 $\dfrac{\sin 2\pi/3}{\cos 2\pi/3}$

▮ (See Probs. 7.106 and 7.107.) $\sin 2\pi/3 = \sqrt{3}/2$ and $\cos 2\pi/3 = -\frac{1}{2}$. Then $\dfrac{\sin 2\pi/3}{\cos 2\pi/3} = \dfrac{\sqrt{3}/2}{-\frac{1}{2}} = \dfrac{-\sqrt{3}}{2} \cdot 2 = -\sqrt{3}$. Thus, $\tan 2\pi/3 = -\sqrt{3}$, since $\tan x = \dfrac{\sin x}{\cos x}$, $\cos x \neq 0$.

7.109 $\cot 2\pi/3$

▮ $\cot \dfrac{2\pi}{3} = \dfrac{1}{\tan 2\pi/3}$, and so $\tan 2\pi/3 = -\sqrt{3}$ (see Prob. 7.108). Thus, $\cot 2\pi/3 = 1/-\sqrt{3} = -\sqrt{3}/3$.

7.110 $\cot 5\pi/4$

▮ See Fig. 7.43. $W(5\pi/4) = (-\sqrt{2}/2, -\sqrt{2}/2)$. Then $\cot \dfrac{5\pi}{4} = \dfrac{-\sqrt{2}/2}{-\sqrt{2}/2} = 1$.

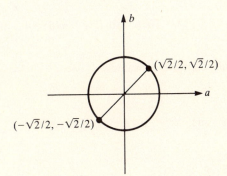

Fig. 7.43

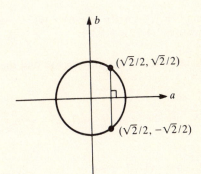

Fig. 7.44

7.111 $\cot 7\pi/4$

▮ See Fig. 7.44. $W(7\pi/4) = (\sqrt{2}/2, -\sqrt{2}/2)$. Then $\cot \dfrac{7\pi}{4} = \dfrac{\sqrt{2}/2}{-\sqrt{2}/2} = -1$.

7.112 $\sin 7\pi/6 + \cos 7\pi/6$

▮ See Fig. 7.45. $W(7\pi/6) = (-\sqrt{3}/2, -\frac{1}{2})$. Then, $\sin 7\pi/6 = -\frac{1}{2}$, $\cos 7\pi/6 = -\sqrt{3}/2$, and $\sin 7\pi/6 + \cos 7\pi/6 = (-\sqrt{3} - 1)/2$.

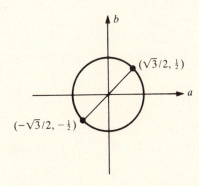

Fig. 7.45

7.113 $(\sin 7\pi/6)^2 + (\cos 7\pi/6)^2$

▮ (i) Since $(\sin x)^2 + (\cos x)^2 = 1$ for all x, $(\sin 7\pi/6)^2 + (\cos 7\pi/6)^2 = 1$. Calculating this sum directly, we get $\sin 7\pi/6 = -\frac{1}{2}$, so $(\sin 7\pi/6)^2 = \frac{1}{4}$, and $\cos 7\pi/6 = -\sqrt{3}/2$, so $(\cos 7\pi/6)^2 = \frac{3}{4}$. Then $\frac{1}{4} + \frac{3}{4} = 1$.

7.114 $\cot(-3\pi/4)$

▮ See Fig. 7.46. $W(-3\pi/4) = (-\sqrt{2}/2, -\sqrt{2}/2)$. Thus, $\cot\left(-3\dfrac{\pi}{4}\right) = \dfrac{-\sqrt{2}/2}{-\sqrt{2}/2} = 1$. Note also that $\cot 5\pi/4 = 1$.

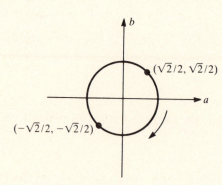

Fig. 7.46

7.115 $\sin 13\pi/6$

▮ $\sin 13\pi/6 = \sin(2\pi + \pi/6)$. $W(2\pi + \pi/6) = W(\pi/6) = (\sqrt{3}/2, \frac{1}{2})$. Then, $\sin(2\pi + \pi/6) = \frac{1}{2}$, the ordinate of $W(2\pi + \pi/6)$.

For Probs. 7.116 to 7.120, use identities to find the value of the circular function indicated.

7.116 Find $\sin(-x)$ if $\sin x = -\frac{1}{2}$.

▮ We use the identity $\sin(-x) = -\sin x$, for all x. Then $\sin(-x) = -\sin x = -(-\frac{1}{2})$; and $\sin(-x) = \frac{1}{2}$.

7.117 Find $\cos(-x)$ if $\cos x = -\frac{1}{2}$.

▮ $\cos(-x) = \cos x$ for all $x \in \mathscr{R}$. Then $\cos(-x) = \cos x = -\frac{1}{2}$.

7.118 Find $\cot(-x)$ if $\cot x = 5$.

▮ $\cot(-x) = -\cot x$ for all $x \in \mathscr{R}$ for which $\cot x$ is defined. Then $\cot(-x) = -\cot x = -5$.

7.119 Find $\csc(-x)$ if $\sin x = \frac{1}{3}$

▮ If $\sin x = \frac{1}{3}$, then $\csc x = 3$. Since $\csc(-x) = -\csc x$, $\csc(-x) = -\csc x = -3$.

7.120 Find $[\cot(-x)]^2$ if $\tan x = 2$.

▮ If $\tan x = 2$, then $\cot x = \frac{1}{2}$. Since $\cot(-x) = -\cot x$, $\cot(-x) = -2$, and $[\cot(-x)]^2 = (-2)^2 = 4$.

For Probs. 7.121 to 7.123, find $\sin x$ using identities.

7.121 $\tan x = 1$; $W(x)$ is in quadrant I.

▮ $\dfrac{\sin x}{\cos x} = \tan x = 1$, so $\sin x = \cos x$. Since $\sin^2 x + \cos^2 x = 1$, we have $2\sin^2 x = 1$, $2\sin^2 x = \frac{1}{2}$, and $\sin x = \pm\sqrt{\frac{1}{2}}$. Then $\sin x = \pm\sqrt{2}/2$, but $-\sqrt{2}/2$ is extraneous since $W(x)$ is in quadrant I, so the solution is $\sin x = \sqrt{2}/2$.

7.122 $\cos x = \sqrt{2}/2$; $\tan x < 0$

▮ If $\cos x > 0$ and $\tan x < 0$, then $W(x)$ is in quadrant IV. Since $\sin^2 x + \cos^2 x = 1$, $\sin^2 x + (\sqrt{2}/2)^2 = 1$. Then $\sin^2 x = 1 - \frac{1}{2} = \frac{1}{2}$, and $\sin x = \pm\sqrt{2}/2$; but $\sin x < 0$ is in quadrant IV. Thus, $\sin x = -\sqrt{2}/2$.

7.123 $\cot x = \frac{1}{4}$; $\csc x < 0$

▎ If $\cot x > 0$ and $\csc x < 0$, then $W(x)$ is in quadrant III. Since $\cot x = \frac{1}{4}$, $\tan x = \dfrac{\sin x}{\cos x} = 4$.
Then $4 \cos x = \sin x$ and $\cos x = \dfrac{\sin x}{4}$. Since $\sin^2 x + \cos^2 x = 1$, $\sin^2 x + \dfrac{\sin^2 x}{16} = 1$. Then
$\dfrac{17 \sin^2 x}{16} = 1$, $\sin^2 x = \frac{16}{17}$, and $\sin x = \pm 4 / \sqrt{17}$. We reject $+4/\sqrt{17}$. Thus, $\sin x = -4/\sqrt{17}$.

For Probs. 7.124 to 7.126, derive the given identity.

7.124 $\sin^2 x + \cos^2 x = 1$

▎ $\sin x = t$ and $\cos x = s$, where $W(x) = (s, t)$. Then $\sin^2 x + \cos^2 x = t^2 + s^2$; but (s, t) is on the
unit circle with center at $(0, 0)$. Thus, $s^2 + t^2 = 1$, and $\sin^2 x + \cos^2 x = t^2 + s^2 = 1$.

7.125 $1 + \cot^2 x = \csc^2 x$

▎ We know that $\sin^2 x + \cos^2 x = 1$. Multiply through by $1/\sin^2 x$ $(\sin x \neq 0)$: $\dfrac{\sin^2 x}{\sin^2 x} + \dfrac{\cos^2 x}{\sin^2 x} = \dfrac{1}{\sin^2 x}$,
or $1 + \cot^2 x = \csc^2 x$.

7.126 $\tan(-x) = -\tan x$

▎ Let $W(x) = (s, t)$. Then $W(-x) = (s, t)$ and $\tan(-x) = -t/s$. Then $\tan x = t/s$, and
$\tan(-x) = -\tan x$.

7.4 THE TRIGONOMETRIC FUNCTIONS

For Probs. 7.127 to 7.151, find the value of each of the following trigonometric functions at the given angle. Do *not*
use a calculator.

7.127 $\sin 90°$

▎ See Fig. 7.47. Here, the point $(0, 1)$ is chosen on the terminal ray of the angle. From the equation
$\sin \theta = b/R$ for any angle θ where b is the ordinate of (a, b) on the terminal ray, we have $\sin 90° = \frac{1}{1} = 1$.
[Clearly, $R = 1$, the distance from $(0, 0)$ to $(0, 1)$.]

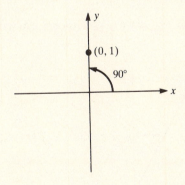

Fig. 7.47

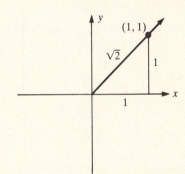

Fig. 7.48

7.128 $\sin 45°$

▎ See Fig. 7.48. For convenience, choose $(1, 1)$ on the terminal ray. Then $\sin \theta = b/R = 1/\sqrt{2} = \sqrt{2}/2$
[$\sqrt{2}$ = distance from $(0,0)$ to $(1, 1)$].

7.129 $\cos 60°$

▎ See Fig. 7.49. $\cos \theta = a/R = \frac{1}{2}$. We choose $(1, \sqrt{3})$ on the 60° ray for convenience.

7.130 $\sin 30°$

▎ See Fig. 7.50. Then $\sin 30° = b/R = \frac{1}{2}$.

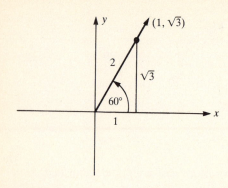

Fig. 7.49

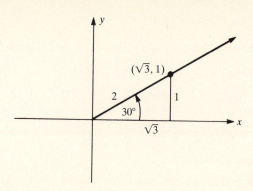

Fig. 7.50

7.131   cot 30°

▮ See Fig. 7.51. $\cot \theta = a/b$ for any angle θ. Then $\cot \theta = \sqrt{3}/1 = \sqrt{3}$ when $\theta = 30°$, so $\cot 30° = \sqrt{3}$.

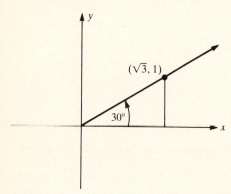

Fig. 7.51

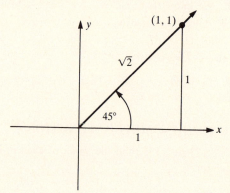

Fig. 7.52

7.132   csc 45°

▮ See Fig. 7.52. For any angle θ, $\csc \theta = 1/\sin \theta = R/b$. Thus, $\csc 45° = \sqrt{2}/1 = \sqrt{2}$.

7.133   tan 90°

▮ See Fig. 7.53. We choose $(0, 1)$ on the terminal ray. We know $\tan \theta = b/a \ (a \neq 0)$ for any θ, but here $a = 0$. Thus, $\tan 90°$ is *not* defined.

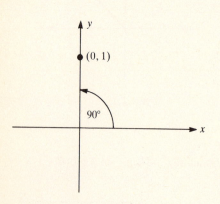

Fig. 7.53

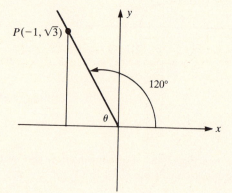

Fig. 7.54

7.134   sin 120°

▮ (i) We illustrate the 120° angle in Fig. 7.54. The reference angle is found by dropping a perpendicular line to the x axis. Here, $\theta = 180° - 120° = 60°$. Choose a convenient P on the terminal ray of 120°; $(-\sqrt{3}, 1)$ is convenient. Then $R = 2$, $b = \sqrt{3}$, and $\sin 120° = b/R = \sqrt{3}/2$.
(ii) Notice that *(1)* the reference angle for 120° is 60° (see Fig. 7.55). *(2)* $\sin \theta > 0$ in quadrant II, and *(3)* $\sin 60° = \sqrt{3}/2$. Thus, $\sin 120° = +(\sin 60°) = \sqrt{3}/2$.

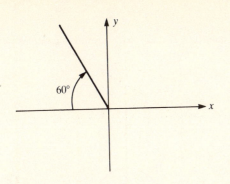

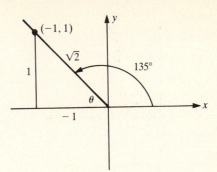

Fig. 7.55

Fig. 7.56

7.135 sin 135°

▮ See Fig. 7.56. (i) $\sin \theta = b/R = 1/\sqrt{2} = \sqrt{2}/2$.   (ii) $\sin 135° = +\sin 45°$ in quadrant II, where 45° is the reference angle.   Then $\sin \theta > 0 = \sqrt{2}/2$.

7.136 cos 135°

▮ See Fig. 7.57. (i) $\cos 135° = a/R = -1/\sqrt{2} = -\sqrt{2}/2$.
(ii) $\cos 135° = -\cos 45°$ ($\cos \theta < 0$ in quadrant II and   45° = reference angle) $= \frac{-\sqrt{2}}{2}$.

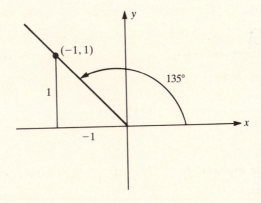

Fig. 7.57

7.137 tan 120°

▮ $\tan 120° = -\tan 60°$ (quadrant II) $= -\sqrt{3}$.   Note: If you forget that $\tan 60° = \sqrt{3}$,   then recall that $\sin 60° = \sqrt{3}/2$, $\cos 60° = \frac{1}{2}$,   and $\tan 60° = (\sqrt{3}/2) \cdot 2 = \sqrt{3}$.

7.138 cot 120°

▮ See Fig. 7.58. (i) $\cot 120° = -\cot 60° = -1/\tan 60° = -1/\sqrt{3} = -\sqrt{3}/3$.
(ii) Use the reference angle: $\cot 120° = a/b = -1/\sqrt{3} = -\sqrt{3}/3$.

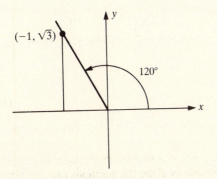

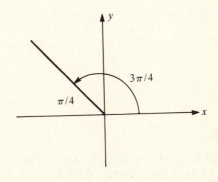

Fig. 7.58

Fig. 7.59

7.139 sin 3π/4

▮ See Fig. 7.59. $\sin 3\pi/4 = \sin \pi/4 = \sqrt{2}/2$.   (Recall: $\sin \pi/4 = b/R = 1/\sqrt{2} = \sqrt{2}/2$.)

7.140 $\cos 5\pi/6$

▮ See Fig. 7.60. (i) $\cos 5\pi/6 = -\cos \pi/6$ ($\cos \theta < 0$ in quadrant II and $\pi/6$ = reference angle) = $-\sqrt{3}/2$. (ii) $\cos 5\pi/6 = a/R = -\sqrt{3}/2$.

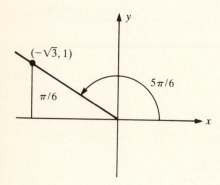

Fig. 7.60

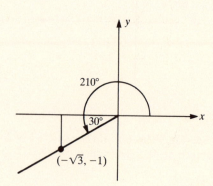

Fig. 7.61

7.141 $\csc 210°$

▮ See Fig. 7.61. (i) $\csc 210° = -\csc 30° = -\dfrac{1}{\sin 30°} = -\dfrac{1}{\frac{1}{2}} = -2$. (ii) $\csc 210° = R/b = \frac{2}{-1} = -2$.

7.142 $\sec 4\pi/3$

▮ See Fig. 7.62. (i) $\sec 4\pi/3 = -\sec \pi/3 = -\dfrac{1}{\cos \pi/3} = -\dfrac{1}{\frac{1}{2}} = -2$.
(ii) $\sec 4\pi/3 = R/a = \frac{2}{1} = -2$.

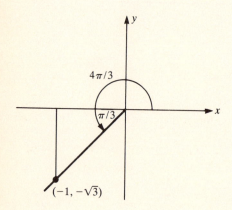

Fig. 7.62

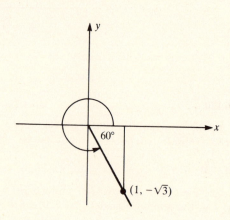

Fig. 7.63

7.143 $\cot 300°$

▮ See Fig. 7.63. (i) $\cot 300° = -\cot 60° = -\dfrac{\cos 60°}{\sin 60°} = \dfrac{-\frac{1}{2}}{\sqrt{3}/2} = -\frac{1}{2} \cdot (2/\sqrt{3}) = -1/\sqrt{3} = -\sqrt{3}/3$. (ii) $\cot 300° = a/b = 1/-\sqrt{3} = -\sqrt{3}/3$.

7.144 $\tan 330°$

▮ See Fig. 7.64. (i) $\tan 330° = -\tan 30° = -\dfrac{\sin 30°}{\cos 30°} = \dfrac{-\frac{1}{2}}{\sqrt{3}/2} = -\frac{1}{2} \cdot (2/\sqrt{3}) = -\sqrt{3}/3$. (ii) $\tan 330° = b/a = -1/\sqrt{3} = -\sqrt{3}/3$.

7.145 $\cos 315°$

▮ See Fig. 7.65 $\cos 315° = \cos 45° = \sqrt{2}/2$ ($\cos \theta > 0$ in quadrant IV).

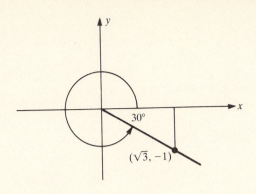

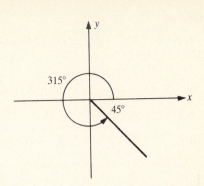

Fig. 7.64

Fig. 7.65

7.146 cos 270°

▮ See Fig. 7.66. $\cos 270° = a/R = \frac{0}{1} = 0.$

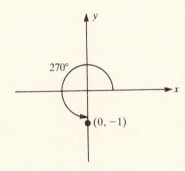

Fig. 7.66

7.147 sin 480°

▮ $\sin 480° = \sin (360° + 120°) = \sin 120° = \sin 60° = \sqrt{3}/2.$

7.148 cos 870°

▮ $\cos 870° = \cos (720° + 150°) = \cos 150° = -\cos 30° = -\sqrt{3}/2.$

7.149 cos (−870°)

▮ $\cos (-870°) = \cos 870°$ $[\cos (-\theta) = \cos \theta] = \cos (720° + 150°) = \cos 150° = -\cos 30° = -\sqrt{3}/2.$

7.150 sin (−960°)

▮ See Fig. 7.67. $\sin (-960°) = -\sin 960°$ $[\sin (-\theta) = -\sin \theta] = -\sin (720° + 240°) = -(-\sin 60°) = \sin 60° = \sqrt{3}/2.$

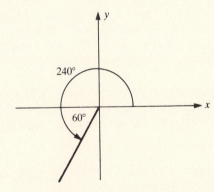

Fig. 7.67

7.151 cot (−1035°)

▮ $\cot (-1035°) = -\cot 1035° = -\cot (720° + 315°) = -\cot 315° = -\cot 45° = -1.$

For Probs. 7.152 to 7.160, evaluate the given expression.

7.152 $\sin^2 0° + \cos^2 0°$

▐ $\sin 0° = 0$, and $\cos 0° = 1$. Then $\sin^2 0° + \cos^2 0° = 0^2 + 1^2 = 1$.

7.153 $\sin^2 \pi/2 + \cos^2 \pi/2$

▐ $\sin \pi/2 = 1$, and $\cos \pi/2 = 0$. Then $\sin^2 \pi/2 + \cos^2 \pi/2 = 1^2 + 0^2 = 1$.

7.154 $\sin^2 315° + \cos^2 315°$

▐ $\sin 315° = -\sin 45° = -\sqrt{2}/2$, and $\cos 315° = \cos 45° = \sqrt{2}/2$. Then $\sin^2 315° + \cos^2 315° = (-\sqrt{2}/2)^2 + (\sqrt{2}/2)^2 = \frac{1}{2} + \frac{1}{2} = 1$.

7.155 $1 + \tan^2 \pi/2$

▐ $\tan \pi/2$ is undefined ($\tan \theta = b/a$, and $a = 0$). Thus, $1 + \tan^2 \pi/2$ is undefined.

7.156 $\cos \pi/4 \cos \pi/2 - \sin \pi/4 \sin \pi/2$

▐ $\cos \pi/4 = \sqrt{2}/2$, $\cos \pi/2 = 0$, $\sin \pi/4 = \sqrt{2}/2$, and $\sin \pi/2 = 1$. Thus, $\cos \pi/4 \cos \pi/2 - \sin \pi/4 \sin \pi/2 = \sqrt{2}/2 \cdot 0 - \sqrt{2}/2 \cdot 1 = -\sqrt{2}/2$.

7.157 $\sin 0 + \sin \pi/2 + \sin \pi$

▐ $\sin 0 = 0$, $\sin \pi/2 = 1$, and $\pi = 0$. Then $\sin 0 + \sin \pi/2 + \sin \pi = 0 + 1 + 0 = 1$.

7.158 $2 \sin \pi/2 \cos \pi/2$

▐ $\sin \pi/2 = 1$, and $\cos \pi/2 = 0$. Then $2 \sin \pi/2 \cos \pi/2 = 2 \cdot 1 \cdot 0 = 0$.

7.159 $\cos \pi/4 + \cos \pi/2 + \cos \pi$

▐ $\cos \pi/4 = \sqrt{2}/2$, $\cos \pi/2 = 0$, and $\cos \pi = -1$. Thus, $\cos \pi/4 + \cos \pi/2 + \cos \pi = \sqrt{2}/2 + 0 - 1 = \sqrt{2}/2 - 1$.

7.160 $\sin e^x$ where $x = \ln \pi$

▐ $\sin e^{\ln \pi} = \sin \pi = 0$.

For Probs. 7.161 to 7.166, find the value of the other five trigonometric functions for the indicated θ. Do *not* find θ.

7.161 $\sin \theta = \frac{3}{5}$, $\cos \theta < 0$

▐ $\sin \theta - \frac{3}{5} = b/R$. Then from $a^2 + b^2 = R^2$ we have $a^2 + 9 = 25$, or $a = \pm 4$. But $\cos \theta < 0$, so $a = -4$. Then $\cos \theta = a/R = -\frac{4}{5}$; $\tan \theta = b/a = \frac{3}{-4}$; $\cot \theta = -\frac{4}{3}$; $\cos \theta = 1/\sin \theta = \frac{5}{3}$; and $\sec \theta = 1/\cos \theta = \frac{5}{-4}$.

7.162 $\tan \theta = -\frac{4}{3}$, $\sin \theta < 0$

▐ $\tan \theta = -\frac{4}{3} = b/a$. Then $b = 4$, $a = -3$ or $b = -4$, $a = 3$. But $\sin \theta = b/R$ and $\sin \theta < 0$. Thus, $b < 0$. Then $b = -4$, $a = 3$, and $R^2 = a^2 + b^2 = 25$, or $R = 5$ ($R > 0$ always). Then $\cot \theta = -\frac{3}{4}$; $\sin \theta = b/R = -\frac{4}{5}$; $\csc \theta = 1/\sin \theta = \frac{5}{-4}$; $\cos \theta = a/R = -\frac{3}{5}$; and $\sec \theta = -\frac{5}{3}$.

7.163 $\cos \theta = -\sqrt{5}/3$, $\cot \theta > 0$

▐ If $\cos \theta = -\sqrt{5}/3$, then $R = 3$, $a = -\sqrt{5}$. From $a^2 + b^2 = R^2$, $b^2 = R^2 - a^2 = 9 - 5 - 4$. Then $b = \pm 2$. But $\cot \theta > 0$, so $a/b > 0$, and $a < 0$. Thus, $b < 0$, and $b = -2$. Then, $\sec \theta = -3/\sqrt{5}$; $\tan \theta = b/a = -2/-\sqrt{5} = 2/\sqrt{5}$; $\cot \theta = \sqrt{5}/2$; $\sin \theta = b/R = -\frac{2}{3}$; and $\csc \theta = -\frac{3}{2}$.

7.164 $\cos \theta = -\sqrt{5}/3$, $\tan \theta > 0$

▐ Note that $\tan \theta = b/a$ and $\cot \theta = a/b$. Then $\tan \theta > 0$ whenever $\cot \theta > 0$. Thus, this problem is equivalent to Prob. 7.163, and the answers are identical.

7.165 $\tan \theta = -\sqrt{2}$, $\sin \theta < 0$

▐ $\tan \theta = b/a = -\sqrt{2} = -\sqrt{2}/1$, so $b = -\sqrt{2}$, $a = 1$ or $b = \sqrt{2}$, $a = -1$. But $\sin \theta < 0$, so $b < 0$ and $b = -\sqrt{2}$, $a = 1$. Then $R = \sqrt{1^2 + (-2)^2} = \sqrt{3}$. Thus, $\cot \theta = -1/\sqrt{2}$; $\sin \theta = b/R = -2/\sqrt{3} = -\sqrt{2}/3$; $\csc \theta = -\sqrt{3}/2$; $\cos \theta = 1/\sqrt{3}$; and $\sec \theta = \sqrt{3}$.

7.166 $\tan\theta=-\sqrt2,\quad \cos\theta>0$

▮ $\tan\theta=-\sqrt2$, so $b=-\sqrt2$, $a=1$ or $b=\sqrt2$, $a=-1$. But $\cos\theta>0$, so $a>0$ and $b=-\sqrt2$, $a=1$. The results are identical to Prob. 7.167. Do you see why?

For Probs. 7.167 to 7.170, answer true or false, and justify your answer.

7.167 If x is an angle measured in radians, then $\cos(360+x)=\cos x$.

▮ False. In radian measure, x and $(360+x)$ are not coterminal. However, $\cos(2\pi+x)=\cos x$.

7.168 If $\sin x=\cos x$, then x must be $45°$.

▮ False $x=45°$ is one solution. Others are $45°+360°$, $45°+2(360°)$, etc.

7.169 $\sin(\sin\pi/2)>0$

▮ $\sin\pi/2=1$. Since $0<1<\pi/2$, the angle α measuring 1 rad is in quadrant I and $\sin\alpha>0$. True.

7.170 $\cos(\sin 3\pi/2)>0$

▮ $\sin 3\pi/2=-1$, but $-\pi/2<-1<0$; thus the angle β measuring -1 rad is in quadrant IV and $\cos\beta>0$. True.

7.5 THE INVERSE TRIGONOMETRIC FUNCTIONS

For Probs. 7.171 to 7.201, find the value of the given expression. Do *not* use a calculator or a table of values.

7.171 $\text{Sin}^{-1}\frac12$

▮ If $y=\text{Sin}^{-1}\frac12$, then $\sin y=\frac12$ (by definition of the $\text{Sin}^{-1}x$ function). Then $y=\pi/6$ ($30°$). Remember: If $y=\text{Sin}^{-1}x$, then $-\pi/2\le y\le\pi/2$. If $y=\text{Cos}^{-1}x$, then $0\le y\le\pi$. If $y=\text{Tan}^{-1}x$, then $-\pi/2<y<\pi/2$.

7.172 $\text{Sin}^{-1}\sqrt3/2$

▮ If $y=\text{Sin}^{-1}\sqrt3/2$, then $\sin y=\sqrt3/2$ ($-\pi/2\le y\le\pi/2$). Thus, $y=60°$ ($\pi/3$).

7.173 $\text{Arccos}\frac12$

▮ "Arc" notation is just another means of expressing inverse trigonometric functions. $\text{Arccos}\frac12=\text{Cos}^{-1}\frac12$. Then if $\text{Arccos}\frac12=y$, $\cos y=\frac12$ ($0\le y\le\pi$). Thus, $y=60°$.

7.174 $\text{Cos}^{-1}1/\sqrt2$

▮ If $y=\text{Cos}^{-1}1/\sqrt2$, then $\cos y=1/\sqrt2$ ($=\sqrt2/2$) ($0\le y\le\pi$); $y=45°$.

7.175 $\text{Sin}^{-1}1+\text{Sin}^{-1}0$

▮ If $y=\text{Sin}^{-1}1$, then $\sin y=1$ ($-\pi/2\le y\le\pi/2$), and $y=\pi/2$. $\text{Sin}^{-1}0=y$, so $\sin y=0$, and $y=0$. Thus, $\text{Sin}^{-1}1+\text{Sin}^{-1}0=\pi/2+0=\pi/2$.

7.176 $\text{Sin}^{-1}1+\text{Sin}^{-1}(-1)$

▮ $\text{Sin}^{-1}1=\pi/2$ (see Prob. 7.175). If $y=\text{Sin}^{-1}(-1)$, then $\sin y=-1$ ($-\pi/2\le y\le\pi/2$) and $y=-\pi/2$. (Note: $y=3\pi/2$ is incorrect. Look at the range!) Thus, $\text{Sin}^{-1}1+\text{Sin}^{-1}(-1)=\pi/2+(-\pi/2)=0$.

7.177 $\text{Arctan}\sqrt3$

▮ If $y=\text{Arctan}\sqrt3$, then $\tan y=\sqrt3$ ($-\pi/2<y<\pi/2$) and $y=60°$. $\Big($Note: Did you forget that $\tan 60°=\sqrt3$? If so, recall that $\sin 60°=\sqrt3/2$, $\cos 60°=1/2$, and $\tan 60°=\dfrac{\sin 60°}{\cos 60°}$.$\Big)$

7.178 $\text{Tan}^{-1}(-1/\sqrt3)$

▮ If $y=\text{Tan}^{-1}(-1/\sqrt3)$, then $\tan y=-1/\sqrt3$ ($-\pi/2<y<\pi/2$). Thus, $y=-30°$. (Note that $150°$ is wrong!)

7.179   Arccot $\sqrt{3}$

▮   If $y = \text{Arccot } \sqrt{3}$,   then $\cot y = \sqrt{3}$ $(0 < y < \pi)$ and $\tan y = 1/\sqrt{3}$.   Thus, $y = 30°$.

7.180   Arcsec $(-\sqrt{2})$

▮   See Fig. 7.68.   If $y = \text{Arcsec } (-\sqrt{2})$,   then $\sec y = -\sqrt{2}$ $(0 \le y \le \pi$,   but $y \ne \pi/2)$.   Then $\cos y = -1/\sqrt{2}$,   so $y = 135°$.

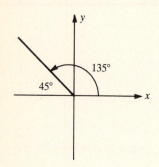

Fig. 7.68

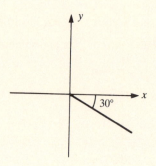

Fig. 7.69

7.181   Arccsc (-2)

▮   See Fig. 7.69.   If $y = \text{Arccsc } (-2)$,   then $\csc y = -2$ $(-\pi/2 \le y \le \pi/2$,   but $y \ne 0)$.   Then, $\sin y = -\frac{1}{2}$,   and $y = -30°$.

7.182 $\text{Tan}^{-1} (-1) + \text{Cot}^{-1} (-1)$

▮ $\text{Tan}^{-1} (-1) = y$,   so $\tan y = -1$ $(-\pi/2 < y < \pi/2)$ and $y = -\pi/4$. $\text{Cot}^{-1} (-1) = y$,   so $\cot y = -1$ $(0 < y < \pi)$ and $y = 3\pi/4$.   Then $\text{Tan}^{-1} (-1) + \text{Cot}^{-1} (-1) = -\pi/4 + 3\pi/4 = \pi/2$.

7.183 $\sin (\text{Arcsin } \frac{1}{2})$

▮   This really says "the sine of the angle whose sine is $\frac{1}{2}$."   Thus, $\sin (\text{Arcsin } \frac{1}{2}) = \frac{1}{2}$.

7.184 $\cos (\text{Cos}^{-1} 0.72134)$

▮   See Prob. 7.183.   This is simply 0.72134.   In fact, what is $\cos (\text{Cos}^{-1} l)$ where $-1 \le l \le 1$?   It must be l itself.

7.185   Arcsin $(\sin \pi/3)$

▮ $\sin \pi/3 = \sqrt{3}/2$. $\text{Arcsin } (\sin \pi/3) = \text{Arcsin } \sqrt{3}/2 =$ the angle whose sin is $\sqrt{3}/2$ (between $-\pi/2$ and $\pi/2$) $= \pi/3$.

7.186 $\cos (\text{Arccos } \sqrt{3}/2)$

▮   See Prob. 7.183. $\cos (\text{Arccos } \sqrt{3}/2) = \sqrt{3}/2$.

7.187 $\tan (\text{Arccos } 0.5)$

▮ $\text{Arccos } 0.5 = y$ means $\cos y = \frac{1}{2}$,   or $y = 60°$.   Then $\tan (\text{Arccos } 0.5) = \tan 60° = \dfrac{\sin 60°}{\cos 60°} = \dfrac{\sqrt{3}/2}{\frac{1}{2}} = \sqrt{3}$.

7.188   Arccos $(\cos \pi/2)$

▮ $\cos \pi/2 = 0$. $\text{Arccos } 0 = y$ means $\cos y = 0$ $(0 \le y \le \pi)$,   and $y = \pi/2$.   Thus,   Arccos $(\cos \pi/2) = \pi/2$.   Look at Prob. 7.185.   Do you see a pattern emerging?

7.189 $\sin [\text{Arccos } (-\sqrt{2}/2)]$

▮ $[\text{Arccos } (-\sqrt{2}/2) = y$ means $\cos y = -\sqrt{2}/2$ $(0 \le y \le \pi)$,   and $y = 135°$.   Thus, $\sin [\text{Arccos } (-\sqrt{2}/2)] = \sin 135° = \sqrt{2}/2$.

7.190 $\cos\left[\text{Arctan}\,(-1)\right]$

┃ Arctan $(-1) = -\pi/4$ $(-\pi/2 < y < \pi/2)$. Then $\cos\left[\text{Arctan}\,(-1)\right] = \cos\,(-\pi/4) = \sqrt{2}/2$.

7.191   Arccos $(\sin 3\pi/4)$

┃ See Fig. 7.70. $\sin 3\pi/4 = \sin \pi/4 = \sqrt{2}/2$. Then,   Arccos $\sqrt{2}/2 = \pi/4$ (since $\cos \pi/4 = \sqrt{2}/2$).

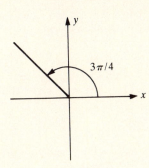

Fig. 7.70

7.192 $\sin\left[\text{Arccos}\,(-\sqrt{3}/2)\right]$

┃ Arccos $(-\sqrt{3}/2) = y$,   so $\cos y = -\sqrt{3}/2$ $(0 \le y \le \pi)$, and $y = 150°$. Then $\sin 150° = \sin 30° = 0.5$.

7.193 $\cos\,(2\,\text{Tan}^{-1}\,1)$

┃ $\text{Tan}^{-1}\,1 = \pi/4$. Then $2\,\text{Tan}^{-1}\,1 = \pi/2$ and $\cos \pi/2 = 0$. Thus, $\cos\,(2\,\text{Tan}^{-1}\,1) = 0$.

7.194   Sec $\left[\text{Arccos}\,(-\sqrt{3}/2)\right]$

┃ Arccos $(-\sqrt{3}/2) = 150°$ (since $0 \le y \le \pi$). Then $\sec 150° = \dfrac{1}{\cos 150°} = \dfrac{1}{-\sqrt{3}/2} = \dfrac{-2}{\sqrt{3}} = \dfrac{-2\sqrt{3}}{3}$.

7.195 $\csc\,(\text{Arcsin}\,\sqrt{2}/2)$

┃ Arcsin $\sqrt{2}/2 = \pi/4$. Then $\csc \dfrac{\pi}{4} = \dfrac{1}{\sin \pi/4} = \dfrac{1}{\sqrt{2}/2} = \dfrac{2}{\sqrt{2}} = \dfrac{2\sqrt{2}}{2} = \sqrt{2}$.

7.196 $\sec\,(\text{Arccos}\,\sqrt{3}/2)$

┃ Arccos $\sqrt{3}/2 = \pi/6$. Then $\sec \dfrac{\pi}{6} = \dfrac{1}{\cos \pi/6} = \dfrac{1}{\sqrt{3}/2} = \dfrac{2}{\sqrt{3}} = \dfrac{2\sqrt{3}}{3}$.

7.197 $\tan\,(\text{Arcsin}\,\sqrt{3}/2)$

┃ Arcsin $\sqrt{3}/2 = 60°$. Then $\tan 60° = \dfrac{\sin 60°}{\cos 60°} = \dfrac{\sqrt{3}/2}{\frac{1}{2}} = \sqrt{3}$.

7.198 $\cot\left[\text{Arccos}\,(-\sqrt{2}/2)\right]$

┃ Arccos $(-\sqrt{2}/2) = y$,   so $\cos y = -\sqrt{2}/2$ $(0 \le y \le \pi)$. Thus, $y = 135°$, and then $\cot 135° = \dfrac{\cos 135°}{\sin 135°} = \dfrac{-\sqrt{2}/2}{\sqrt{2}/2} = -1$.

7.199 $\csc\left[\text{Arctan}\,(-\sqrt{3})\right]$

┃ Arctan $(-\sqrt{3}) = y$ means that $\tan y = -\sqrt{3}$ $(-\pi/2 < y < \pi/2)$, so $y = -60°$. Then $\csc\,(-60°) = \dfrac{1}{\sin\,(-60°)} = \dfrac{1}{-\sin 60°} = \dfrac{1}{-\sqrt{3}/2} = \dfrac{-2}{\sqrt{3}}$.

7.200 $\text{Tan}^{-1}\left[\tan\,(\text{Cos}^{-1}\,\tfrac{1}{2})\right]$

┃ (i) $\text{Cos}^{-1}\,\tfrac{1}{2} = \pi/3$ $(0 \le y \le \pi)$. Then $\tan \pi/3 = \sqrt{3}$. Finally, $\text{Tan}^{-1}\sqrt{3} = \pi/3$.
(ii) Another method would be to note that $\text{Tan}^{-1}\,(\tan \theta) = \theta$ $(-\pi/2 < \theta < \pi/2)$. Then $\text{Tan}^{-1}\left[\tan\,(\text{Cos}^{-1}\,\tfrac{1}{2})\right] = \text{Cos}^{-1}\,\tfrac{1}{2} = \pi/3$.

7.201 Arccos (Arcsin 0)

▮ Arcsin $0 = 0$. Then Arccos (Arcsin 0) = Arccos $0 = \pi/2$ $(0 \le y \le \pi)$.

For Probs. 7.202 to 7.208, prove the given statement.

7.202 Arcsin $(-x) = -$Arcsin x.

▮ Let $y = $Arcsin $(-x)$. Then $\sin y = -x$, and $\sin(-y) = -(-x) = x$. Then $-y = $Arcsin x, and $y = -$Arcsin x. Thus, $-$Arcsin $x = y = $Arcsin $(-x)$.

7.203 Arccos $(-x) = \pi - $Arccos x.

▮ Let $y = \pi - $Arccos x. Then Arccos $x = \pi - y$, and $x = \cos(\pi - y) = \cos \pi \cos y + \sin \pi \sin y$ [using the cos of a sum formula (see Chap. 8)] $= -\cos y$. Thus, $y = $Arccos $(-x)$, and Arccos $(-x) = y = \pi - $Arccos x.

7.204 Arctan $(-x) = -$Arctan x.

▮ Let $y = $Arctan $(-x)$; then $\tan y = -x$, and $x = -\tan y = \tan(-y)$. Thus, $-y = $Arctan x, $y = -$Arctan x, and Arctan $(-x) = y = -$Arctan x.

7.205 Arctan $x + $Arctan $y = $Arctan $\dfrac{x + y}{1 - xy}$ where $|x| < 1$, $|y| < 1$.

▮ Let $a = $Arctan x; $b = $Arctan y. Then $x = \tan a$ and $y = \tan b$. Thus, Arctan $\dfrac{x + y}{1 - xy} = $Arctan $\dfrac{\tan a + \tan b}{1 - \tan a \tan b} = $Arctan $[\tan(a + b)]$ (for the tangent of a sum see Chap. 8) $= a + b = $Arctan $x + $Arctan y, and the two sides of the identity are equal.

7.206 Sec (Arccos x) $= 1/x$.

▮ Let $y = $Arccos x. Then $\cos y = x$, and $\sec y = \frac{1}{2}$; but $y = $Arccos x, so sec (Arccos x) $= 1/x$.

7.207 csc (Arcsin x) $= 1/x$.

▮ Let $y = $Arcsin x. Then $\sin y = x$, and $\csc y = 1/x$. Thus csc (Arcsin x) $= 1/x$.

7.208 2 Arctan $x = $Arctan $\dfrac{2x}{1 - x^2}$, $|x| < 1$.

▮ 2 Arctan $x = $Arctan $\dfrac{x + x}{1 - x \cdot x} = $Arctan $\dfrac{2x}{1 - x^2}$; see problem 7.205 and let $x = y$.

For Probs. 7.209 to 7.211, verify the given statement.

7.209 Arctan $\frac{1}{2} + $Arctan $\frac{1}{3} = \pi/4$

▮ Arctan $x + $Arctan $y = $Arctan $\dfrac{x + y}{1 - xy}$, so Arctan $\frac{1}{2} + $Arctan $\frac{1}{3} = $Arctan $\dfrac{\frac{1}{2} + \frac{1}{3}}{1 - \frac{1}{6}} = $Arctan $\dfrac{\frac{5}{6}}{\frac{5}{6}} = $Arctan $1 = \dfrac{\pi}{4}$.

7.210 2 Arctan $\frac{1}{3} + $Arctan $\frac{1}{7} = \pi/4$

▮ 2 Arctan $\frac{1}{3} = $Arctan $\dfrac{\frac{2}{3}}{1 - \frac{1}{9}} = $Arctan $\frac{3}{4}$; Then Arctan $\frac{3}{4} + $Arctan $\frac{1}{7} = $Arctan $\dfrac{\frac{3}{4} + \frac{1}{7}}{1 - \frac{3}{28}} = $Arctan $\dfrac{\frac{25}{28}}{\frac{25}{28}} = \dfrac{\pi}{4}$.

7.211 Arctan $\frac{1}{2} + $Arctan $\frac{1}{5} + $Arctan $\frac{1}{8} = \pi/4$

▮ Arctan $\frac{1}{2} + $Arctan $\frac{1}{5} = $Arctan $\dfrac{\frac{1}{2} + \frac{1}{5}}{1 - \frac{1}{10}} = $Arctan $\frac{7}{9}$. Then Arctan $\frac{7}{9} + $Arctan $\frac{1}{8} = $Arctan $\dfrac{\frac{7}{9} + \frac{1}{8}}{1 - \frac{7}{72}} = $Arctan $\dfrac{\frac{65}{72}}{\frac{65}{72}} = \dfrac{\pi}{4}$.

7.212 Write $\sin(\text{Cos}^{-1} x)$ as an algebraic expression in x free of all trigonometric functions.

▮ Let $\theta = \text{Cos}^{-1} x$. Then, $0 \le \theta \le \pi$ and if θ' is θ's reference angle, Fig. 7.71 describes θ. Then $\overline{AB} = \sqrt{1 - x^2}$ and $\sin \theta = \dfrac{\sqrt{1 - x^2}}{1} = \sqrt{1 - x^2}$. Thus, $\sin(\text{Cos}^{-1} x) = \sin \theta = \sqrt{1 - x^2}$. Note that whether θ is a quadrant I or II angle does not alter this.

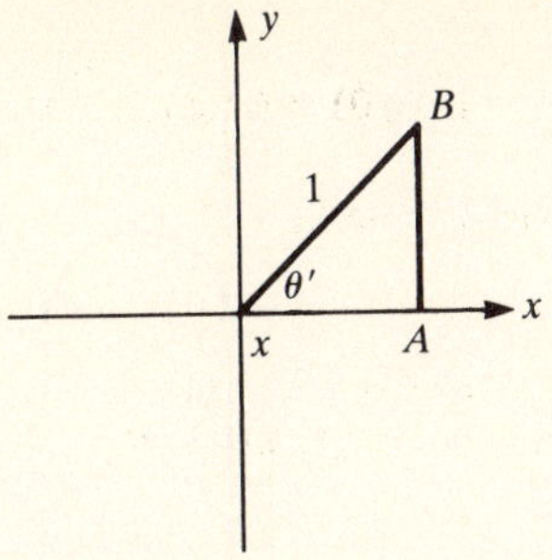

Fig. 7.71

7.6 GRAPHING THE TRIGONOMETRIC FUNCTIONS

For Probs. 7.213 to 7.222, find the period of the function.

7.213 $y = \csc x$

▮ We note that $\csc(x + 2\pi) + \csc x \; \forall x$ in the domain; also, no smaller number p then 2π satisfies the equation $\csc(x + p) = \csc x$ for all x.   Thus,   period $= 2\pi$.

7.214 $y = \cot x$

▮ $\cot(x + \pi) = \cot x$ for all x in the domain.   No smaller number p satisfies $\cot(x + p) = \cot x \forall x$.
period $= \pi$.

7.215 $f(x) = \sec x$

▮ $\sec(x + 2\pi) = \sec x \forall x \in$ domain;   thus, $2\pi =$ period since no smaller p satisfies $\sec(x + p) = \sec x \forall x$.

7.216 $f(x) = -\sec x$

▮ If $\sec(x + 2\pi) = \sec x \forall x$,   then $-\sec(x + 2\pi) = -\sec x$.   period $= 2\pi$.

7.217 $f(x) = 4 \sin x$

▮ The period is not affected by the number 4.   It is only affected by that which we take the sine of.
Here,   period $= 2\pi$.

7.218 $y = 7 \sin 3x$

▮ The period of $y = q \sin rx$ is $2\pi/|r|$.   Here, $r = 3$.   The period $= 2\pi/3$.

7.219 $y = -6 \cos 2x/7$

▮ The period of $y = q \cos rx$ is $2\pi/|r|$.   Here, $r = \frac{2}{7}$;   thus,   the period $= 2\pi/\frac{2}{7} = 2\pi \cdot \frac{7}{2} = 7\pi$.

7.220 $y = 4 \sin(2x + \pi)$

▮ $y = 4 \sin(2x + \pi)$ which is of the form $y = A \sin(Bx + C)$ where $A = 4$, $B = 2$.   Then   the
period $= 2\pi/B = 2\pi/2 = \pi$.

7.221 $f(x) = 3 \tan(2x + \pi)$

▮ Here, $f(x) = A \tan(Bx + C)$ where $B = 3$.   Then   the period $= \pi/B = \pi/3$.

7.222 $f(x) = |\sin x|$

▮ See Fig. 7.72.   The period is π (not 2π) since $|\sin x| \geq 0$ for all x; thus, between π and 2π, we get a
repeat of the graph from 0 to π.

For Probs. 7.223 to 7.225, determine the phase shift.

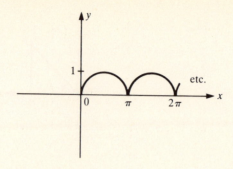

Fig. 7.72

7.223 $y = \sin(x + \pi)$

▮ If $y = A \sin(Bx + C)$, $B > 0$, then the phase shift $= |C/B|$ to the right if $C/B < 0$; C/B to the left if $C/B < 0$. Here, $A = 1$, $B = 1$, $C = \pi$, and $C/B = \pi/1 > 0$; thus, the phase shift $= \pi$ units to the left.

7.224 $f(x) = 2\cos(2x + \pi/3)$

▮ Here $B = 2$, and $C = \pi/3$. Then $C/B = (\pi/3)/2 = \pi/6 > 0$. Phase shift $= C/B = \pi/6$ units to the left.

7.225 $f(x) = \frac{1}{3}\sin(2x - \pi/7)$

▮ Here $B = 2$, and $C = -\pi/7$. Then $C/B = -\pi/14 < 0$. Phase shift $= |C/B| = \pi/14$ to the *right*.

For Probs. 7.226 to 7.228, use the trigonometric graphs to solve the given equation.

7.226 $\sin x = 0$

▮ See Fig. 7.73. Clearly, $\sin x = 0$, so $x = \pm n\pi$ for all whole numbers n.

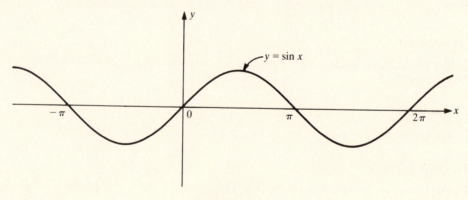

Fig. 7.73

7.227 $\tan x = 0$

▮ See Fig. 7.74. Clearly, $\tan x = 0$, so $x = n\pi \; \forall n \in \mathcal{Z}$.

7.228 $\csc x = 0$

▮ See Fig. 7.75. Clearly, $\csc x$ is never zero. Notice why! $\csc x = \dfrac{1}{\sin x}$ and $1 \neq 0$.

For Probs. 7.229 to 7.231, answer true or false and justify your answer.

7.229 The secant function is even.

▮ Does $f(-x) = f(x) \; \forall x$? $\sec(-x) = \dfrac{1}{\cos(-x)} = \dfrac{1}{\cos x} = \sec x$. Thus, $f(-x) = f(x)$ and $\sec x$ is even. True.

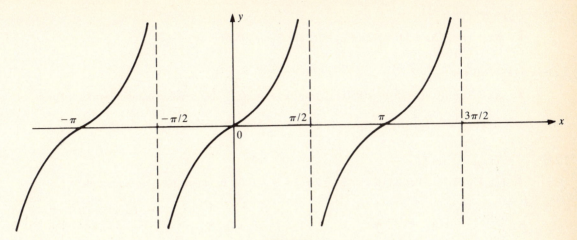

Fig. 7.74

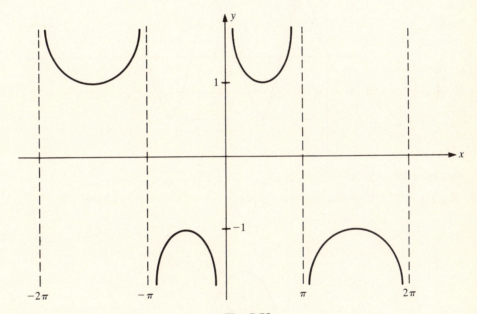

Fig. 7.75

7.230 The cosecant function is odd.

$\blacksquare$ $\csc(-x) = \dfrac{1}{\sin(-x)} = \dfrac{1}{-\sin x} = -\csc x$. Thus, $\csc(-x) = -\csc x$, and $\csc x$ is odd. True.

7.231 The tangent curve is symmetric about the x axis.

$\blacksquare$ Recall that if $y = f(x)$ and replacing y by $-y$ does not alter the function, then f exhibits x-axis symmetry. Let $y = \tan x$. Then $-y = -\tan x \neq \tan x \ \forall x$. The tangent curve is *not* x-axis symmetric.

For Probs. 7.232 to 7.235, let f be a function with period 2. Find the period of each function.

7.232 $g(x) = f(x - 1)$

$\blacksquare$ $g(x + 1) = f(x + 1 - 1) = f(x) \neq f(x - 1)$. $g(x + 2) = f(x + 2 - 1) = f(x + 1) = f(x - 1)$ since f has period 2. Thus, period of $g = 2$.

7.233 $g(x) = f(x) + 1$

$\blacksquare$ $g(x + 1) = f(x + 1) + 1 \neq f(x) + 1$. $g(x + 2) = f(x + 2) + 1 = f(x) + 1$ (period of $f = 2$) $= g(x)$. Period of $g = 2$.

7.234 $g(x) = f(2x)$

▮ $g(x + 1) = f(2(x + 1)) = f(2x + 2) = f(2x) = g(x)$.   Period of $g = 1$.

7.235 $g(x) = f(0.4x)$

▮ $g(x + 5) = f(0.4(x + 5)) = f(0.4x + 2) = f(0.4x) = g(x)$.   No smaller number works.   Period of $g = 5$.

For Probs. 7.236 to 7.241, sketch, on the same axes, one complete period of each function.

7.236 $y = \sin x$, $y = \sin 2x$

▮ See Fig. 7.76.   Period of $\sin 2x = 2\pi/2 = \pi$.   Both functions have an amplitude $= 1$.

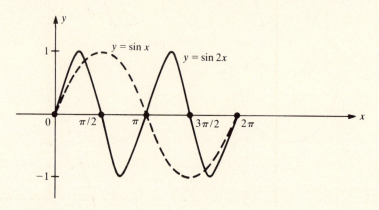

Fig. 7.76

7.237 $y = \sin x$, $y = 2 \sin x$

▮ See Fig. 7.77.   Both functions have a period $= 1$; $2 \sin x$ has amplitude $= 2$.

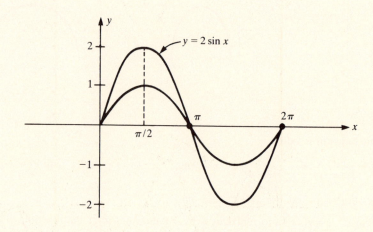

Fig. 7.77

7.238 $y = \cos x$, $y = \cos 2x$

▮ See Fig. 7.78.   Both functions have amplitude $= 1$;   period of $\cos 2x = 2\pi/2 = \pi$.

7.239 $y = \cos x$, $y = 2 \cos x$

▮ See Fig. 7.79.   Amplitude of $2 \cos x = 2$.

7.240 $y = \sin x$, $y = \sin 3x$

▮ See Fig. 7.80.   Period of $\sin 3x = 2\pi/3$.   Sketching hint: Sketch three copies of $\sin x$ for $\sin 3x$, and *then* label the points on the x axis.   Finally, sketch $\sin x$.

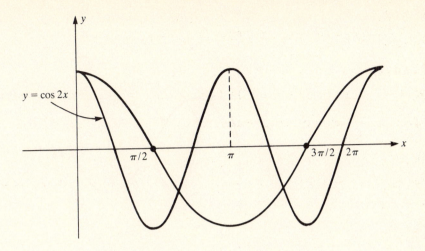

$y = \cos 2x$

Fig. 7.78

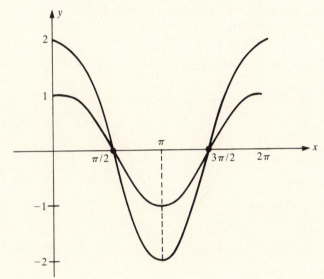

Fig. 7.79

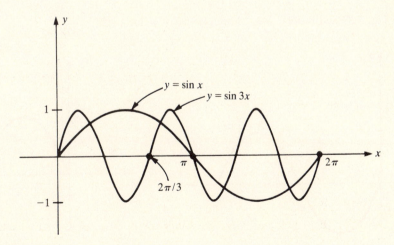

$y = \sin x$

$y = \sin 3x$

Fig. 7.80

7.241 $y = \sin x, \quad y = \sin \frac{1}{2} x$

∎ See Fig. 7.81. Period of $\sin x/2 = \dfrac{2\pi}{\frac{1}{2}} = 4\pi$. Sketching hint: Sketch $\sin x$ for $\sin x/2$, and then label the x axis from 0 to 4π.

For Probs. 7.242 to 7.253, sketch the graph of the given equation.

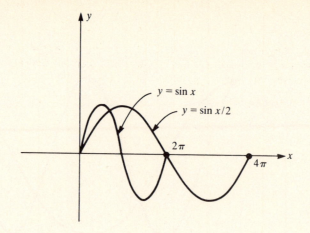

Fig. 7.81

7.242 $y = -\sin x$

▮ See Fig. 7.82. $\sin \pi/2 = -1$; $\sin 3\pi/2 = -(-1) = 1$.

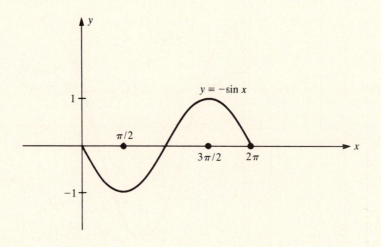

Fig. 7.82

7.243 $f(x) = -2 \sin x$

▮ See Fig. 7.83 and Prob. 7.242.   Multiply each y value by 2.

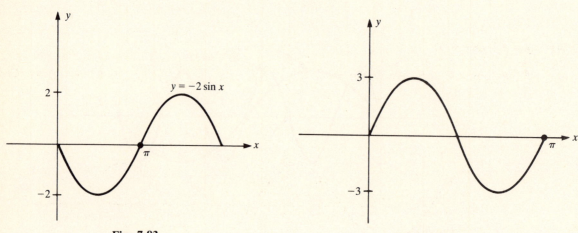

Fig. 7.83

Fig. 7.84

7.244 $y = -3 \sin 2x$

▮ See Fig. 7.84.   Amplitude $= 3$ $(= |-3|)$;   period $= 2\pi/2 = \pi$.   *Sketching hint*:   Sketch a copy of $\sin x$, but observe the period.   Label the x axis appropriately.   Observe the amplitude, and Label the y axis.

7.245 $y = \sin(x + \pi/2)$

▮ See Fig. 7.85. Amplitude $= 1$; $C = \pi/2$, $B = 1$.   Phase shift $= C/B = (\pi/2)/1 = \pi/2$ units to the left ($C/B > 0$).   Sketching hint: Sketch the sine curve and then put in the axes.   Notice the sine curve between the two slash marks on the graph.

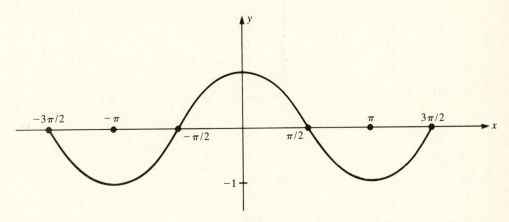

Fig. 7.85

7.246 $y = \sin(x + \pi/3)$

▮ See Fig. 7.86.   The phase shift is $C/B = (\pi/3)/1 = \pi/3$ units to the left   ($C/B > 0$).

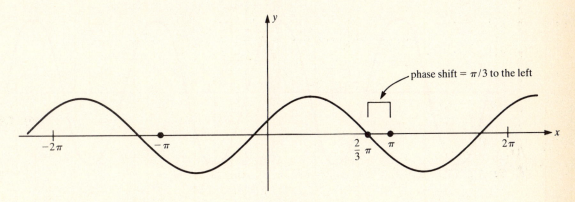

Fig. 7.86

7.247 $y = -\sin(x - \pi/4)$

▮ See Fig. 7.87.   Amplitude $= |-1| = 1$.   Phase shift $= \left|\dfrac{C}{B}\right| = \left|\dfrac{-\pi/4}{1}\right| = \dfrac{\pi}{4}$ units to the right   ($C/B < 0$).

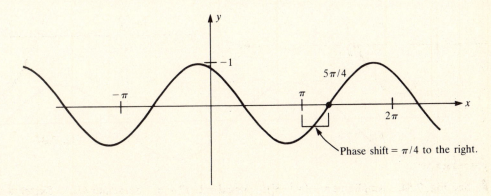

Fig. 7.87

7.248 $y = 2\cos(x + \pi/3)$

▮   See Fig. 7.88.   Amplitude = 2.   Phase shift $= C/B = \pi/3$ to the left.

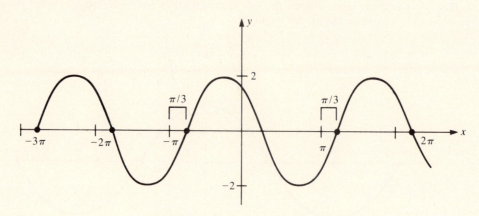

Fig. 7.88

7.249 $y = 2\cos(2x + \pi/2)$

▮   See Fig. 7.89.   Amplitude = 2.   Phase shift $= \dfrac{\pi/2}{2} = \dfrac{\pi}{4}$ to the left.   Period $= 2\pi/2 = \pi$.

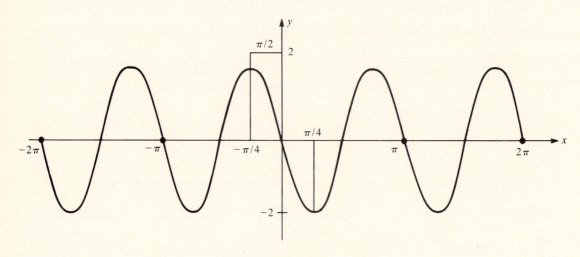

Fig. 7.89

7.250 $y = \frac{1}{2}\cos(x - \pi/4)$ $(-\pi \le x \le 3\pi)$

▮   See Fig. 7.90.   Amplitude $= \frac{1}{2}$.   Period $= 2\pi$ $(= 2\pi/1)$.   Phase shift $= |-\pi/4| = \pi/4$ units to the right.

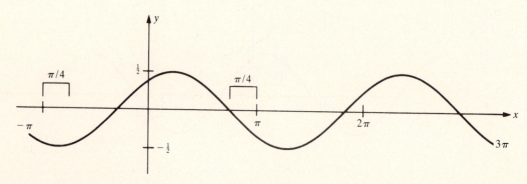

Fig. 7.90

7.251 $y = 3 \tan 2x$ $(-\pi \le x \le \pi)$

   ▮ See Fig. 7.91.   Period $= \pi/2$.

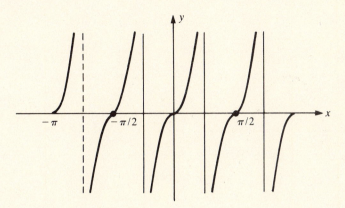

Fig. 7.91

7.252 $y = \text{Arcsin } x$

   ▮ See Fig. 7.92.   This is the inverse of $y = \sin x$ $(-\pi/2 \le x \le \pi/2)$.   Thus, if $y = \text{Arcsin } x$,   then $\sin y = x$ where $-\pi/2 \le y \le \pi/2$, $x \in [-1, 1]$.

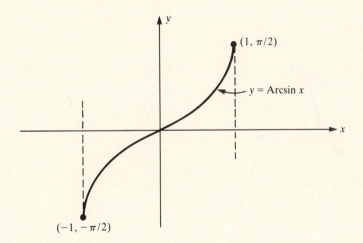

Fig. 7.92

7.253 $y = \text{Arccos } x$

   ▮ See Fig. 7.93 and Prob. 7.252.

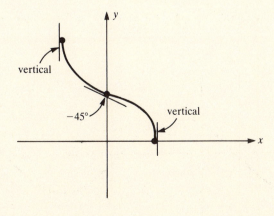

Fig. 7.93

CHAPTER 8
Trigonometric Equations and Inequalities

8.1 ELEMENTARY IDENTITIES

For Probs. 8.1 to 8.53, prove that the given equation is an identity.

8.1 $\sin x \, (\csc x - \sin x) = \cos^2 x$

▮ Ask yourself, "Which side of the equation is more complicated?" and begin working there. Continue to simplify that side of the equation until it looks exactly like the other side of the equation. $\sin x \, (\csc x - \sin x)$ $= \sin x \csc x - \sin^2 x = 1 - \sin^2 x$ (since $\sin x = 1/\csc x$) $= \cos^2 x$ (since $\sin^2 x + \cos^2 x = 1$),   and the identity is established.

8.2 $\csc x \, (\csc x - \sin x) = \cot^2 x$

▮ $\csc x \, (\csc x - \sin x) = \csc^2 x - \csc x \sin x = \csc^2 x - 1$ (since $\csc x = 1/\sin x$) $= \cot^2 x$ (since $\csc^2 x - 1 = \cot^2 x$).

8.3 $\cos x \, (\sec x - \cos x) = \sin^2 x$

▮ $\cos x \, (\sec x - \cos x) = \cos x \sec x - \cos^2 x = 1 - \cos^2 x$ (since $\sec x \cos x = 1$) $= \sin^2 x$.

8.4 $(\sin x + \cos x) \sin x = 1 - \cos x \, (\cos x - \sin x)$

▮ $(\sin x + \cos x) \sin x = \sin^2 x + \cos x \sin x$
$$= 1 - \cos^2 x + \cos x \sin x \quad (\text{since} \quad \sin^2 x + \cos^2 x = 1)$$
$$= 1 - (\cos^2 x - \cos x \sin x) = 1 - \cos x \, (\cos x - \sin x)$$

8.5 $\dfrac{\sec x}{\sin x} + \dfrac{\csc x}{\cos x} = \dfrac{2}{\sin x \cos x}$

▮ $\dfrac{\sec x}{\sin x} + \dfrac{\csc x}{\cos x} = \dfrac{1/\cos x}{\sin x} + \dfrac{1/\sin x}{\cos x} \quad \left(\text{since} \quad \sec x = \dfrac{1}{\cos x} \, , \quad \csc x - \dfrac{1}{\sin x} \right)$

$$= \dfrac{1}{\sin x \cos x} + \dfrac{1}{\sin x \cos x} = \dfrac{2}{\sin x \cos x}$$

8.6 $(\sin x + \cos x)^2 = 1 + 2 \sin x \cos x$

▮ $(\sin x + \cos x)^2 = \sin^2 x + 2 \sin x \cos x + \cos^2 x = (\sin^2 x + \cos^2 x) + 2 \sin x \cos x = 1 + 2 \sin x \cos x$.

8.7 $(\sin x - \cos x)^2 = 1 - 2 \sin x \cos x$

▮ $(\sin x - \cos x)^2 = \sin^2 x - 2 \sin x \cos x + \cos^2 x = 1 - 2 \sin x \cos x$.

8.8 $(\sin x - \cos x)^2 + (\sin x + \cos x)^2 = 2$

▮ $(\sin x - \cos x)^2 + (\sin x + \cos x)^2 = \sin^2 x + \cos^2 x - 2 \sin x \cos x + \sin^2 x + \cos^2 x + 2 \sin x \cos x$
$$= 1 - 2 \sin x \cos x + 1 + 2 \sin x \cos x = 2 + 0 = 2$$

8.9 $\dfrac{\sin x + \cos x}{\tan x} = \cos x + \dfrac{\cos^2 x}{\sin x}$

▮ $\dfrac{\sin x + \cos x}{\tan x} = \dfrac{\sin x}{\tan x} + \dfrac{\cos x}{\tan x} = \dfrac{\sin x}{\sin x/\cos x} + \dfrac{\cos x}{\sin x/\cos x} \quad \left(\text{since} \quad \tan x = \dfrac{\sin x}{\cos x} \right)$

$$= \dfrac{\sin x \cos x}{\sin x} + \dfrac{\cos x \cos x}{\sin x} = \cos x + \dfrac{\cos^2 x}{\sin x}$$

8.10 $\sin^4 x = 1 - (\cos^2 x + \sin^2 x \cos^2 x)$

▮ $\sin^4 x = \sin^2 x \sin^2 x = \sin^2 x \, (1 - \cos^2 x)$ (since $\sin^2 x + \cos^2 x = 1$)
$$= \sin^2 x - \sin^2 x \cos^2 x = 1 - \cos^2 x - \sin^2 x \cos^2 x = 1 - (\cos^2 x + \sin^2 x \cos^2 x)$$

8.11 $(\sin x + \tan x)^2 + 1 - \sec^2 x = \sin^2 x + \dfrac{2\sin^2 x}{\cos x}$

■ $(\sin x + \tan x)^2 + 1 - \sec^2 x = \sin^2 x + 2\sin x \tan x + \tan^2 x + 1 - \sec^2 x$

$$= \sin^2 x + 2\sin x \tan x + (\sec^2 x - 1) + 1 - \sec^2 x$$
$$(\text{since}\quad \tan^2 x = \sec^2 x - 1)$$
$$= \sin^2 x + 2\sin x \tan x = \sin^2 x + 2\sin x\, \frac{\sin x}{\cos x}$$
$$= \sin^2 x + \frac{2\sin^2 x}{\cos x}$$

8.12 $\dfrac{1}{\sin x + 1} - \dfrac{1}{\sin x - 1} = 2\sec^2 x$

■ $\dfrac{1}{\sin x + 1} - \dfrac{1}{\sin x - 1} = \dfrac{\sin x - 1 - (\sin x + 1)}{(\sin x + 1)(\sin x - 1)} = \dfrac{-2}{\sin^2 x - 1}$

$$= \frac{+2}{1 - \sin^2 x} = \frac{2}{\cos^2 x} = 2\sec^2 x$$

8.13 $\dfrac{\sin x}{1 - \sin^2 x} + \dfrac{\cos x}{1 - \cos^2 x} = \dfrac{\sin^3 x + \cos^3 x}{\sin^2 x \cos^2 x}$

■ $\dfrac{\sin x}{1 - \sin^2 x} + \dfrac{\cos x}{1 - \cos^2 x} = \dfrac{\sin x}{\cos^2 x} + \dfrac{\cos x}{\sin^2 x} \quad (\text{since}\quad \sin^2 x + \cos^2 x = 1)$

$$= \frac{\sin^3 x + \cos^3 x}{\cos^2 x \sin^2 x}$$

8.14 $\dfrac{\sin^2 x}{1 - \sin^2 x} + \dfrac{\cos^2 x}{1 - \cos^2 x} = \dfrac{\tan^4 x + 1}{\tan^2 x}$

■ $\dfrac{\sin^2 x}{1 - \sin^2 x} + \dfrac{\cos^2 x}{1 - \cos^2 x} = \dfrac{\sin^2 x}{\cos^2 x} + \dfrac{\cos^2 x}{\sin^2 x} = \tan^2 x + \cot^2 x \quad \left(\text{since}\quad \dfrac{\sin x}{\cos x} = \tan x\right)$

$$= \tan^2 x + \frac{1}{\tan^2 x} = \frac{\tan^4 x + 1}{\tan^2 x}$$

8.15 $(1 - \sin x)(1 + \sin x) + \sin^2 x = 1$

■ $(1 - \sin x)(1 + \sin x) + \sin^2 x = (1 - \sin^2 x) + \sin^2 x = \cos^2 x + \sin^2 x = 1,$ or, more simply, $(1 - \sin^2 x) + \sin^2 x = 1 + 0 = 1.$

8.16 $\dfrac{1 - \cos^2 x}{\sin^2 x \cos x} = \sec x$

■ $\dfrac{1 - \cos^2 x}{\sin^2 x \cos x} = \dfrac{\sin^2 x}{\sin^2 x \cos x} = \dfrac{1}{\cos x} = \sec x$

8.17 $(\sin x + \cot x)^2 = \dfrac{\sin^4 x + \cos^2 x}{\sin^2 x} + 2\cos x$

■ $(\sin x + \cot x)^2 = \sin^2 x + \cot^2 x + 2\sin x \cot x$

$$= \sin^2 x + \frac{\cos^2 x}{\sin^2 x} + 2\sin x\, \frac{\cos x}{\sin x} \quad \left(\text{since}\quad \cot x = \frac{\sin x}{\cos x}\right)$$
$$= \sin^2 x + \frac{\cos^2 x}{\sin^2 x} + 2\cos x = \frac{\sin^4 x + \cos^2 x}{\sin^2 x} + 2\cos x$$

8.18 $\dfrac{1}{\sin x + \cot x} - \dfrac{1}{\sin x - \cot x} = \dfrac{-2\sin x \cos x}{\sin^4 x - \cos^2 x}$

■ $\dfrac{1}{\sin x + \cos x} - \dfrac{1}{\sin x - \cot x} = \dfrac{\sin x - \cot x - (\sin x + \cot x)}{(\sin x + \cot x)(\sin x - \cot x)}$

$$= \frac{-2\cot x}{\sin^2 x - \cot^2 x} = \frac{(-2\cos x)/\sin x}{(\sin^2 x)/1 - (\cos^2 x)/\sin^2 x}$$
$$= \frac{-2\sin x \cos x}{\sin^4 x - \cos^2 x} \quad \left(\text{here we multiplied by}\quad \frac{\sin^2 x}{\sin^2 x}\right)$$

8.19 $(\sin x + \cos x)^4 = 1 + 4 \sin x \cos x + 4 (\sin x \cos x)^2$

▮ $(\sin x + \cos x)^4 = (\sin x + \cos x)^2 (\sin x + \cos x)^2$
$$= (\sin^2 x + 2 \sin x \cos x + \cos^2 x)(\sin^2 x + 2 \sin x \cos x + \cos^2 x)$$
$$= (1 + 2 \sin x \cos x)(1 + 2 \sin x \cos x) = 1 + 4 \sin x \cos x + 4 (\sin x \cos x)^2$$

8.20 $(\sin x - \cot x)(\sin x + \cot x) = 1 - \dfrac{\cos^2 x (\sin^2 x + 1)}{\sin^2 x}$

▮ $(\sin x - \cot x)(\sin x + \cot x) = \sin^2 x - \cot^2 x = 1 - \cos^2 x - \cot^2 x = 1 - (\cos^2 x + \cot^2 x)$
$$= 1 - \left(\frac{\cos^2 x}{1} + \frac{\cos^2 x}{\sin^2 x} \right) = 1 - \frac{\cos^2 x \sin^2 x + \cos^2 x}{\sin^2 x}$$
$$= 1 - \frac{\cos^2 x (\sin^2 x + 1)}{\sin^2 x}$$

8.21 $\dfrac{1 + \cot x}{\cot x} = \tan x + \csc^2 x - \cot^2 x$

▮ $\dfrac{1 + \cot x}{\cot x} = \dfrac{1}{\cot x} + \dfrac{\cot x}{\cot x} = \tan x + 1$
$$= \tan x + (\csc^2 x - \cot^2 x) \quad (\text{since} \quad \csc^2 x - \cot^2 x = 1; \quad \text{see Prob. 8.22}).$$

8.22 $\dfrac{1 + \cot x}{\cot x} = \tan x + \csc^2 x - \cot^2 x$

▮ This is the same as Prob. 8.21. Here, we begin with the right-hand side. Sometimes, both sides look fairly complicated, and it is hard to decide where to begin.
$$\tan x + \csc^2 x - \cot^2 x = \tan x + 1 = \frac{1}{\cot x} + \frac{\cot x}{\cot x} = \frac{1 + \cot x}{\cot x}$$

8.23 $\sec^4 x - \tan^4 x = \sec^2 x + \tan^2 x$

▮ $\sec^4 x - \tan^4 x = (\sec^2 x + \tan^2 x)(\sec^2 x - \tan^2 x) \quad (\text{difference of two perfect squares})$
$$= (\sec^2 x + \tan^2 x)(1) \quad (\text{since} \quad \sec^2 x - \tan^2 x = 1) \quad = \sec^2 x + \tan^2 x$$

8.24 $2 \sin^2 x - \cos^2 x + 1 = 3 \sin^2 x$

▮ $2 \sin^2 x - \cos^2 x + 1 = 2 \sin^2 x + (1 - \cos^2 x) = 2 \sin^2 x + \sin^2 x = 3 \sin^2 x$

8.25 $\sec^2 x \tan^2 x - \tan^2 x = \tan^4 x$

▮ $\sec^2 x \tan^2 x - \tan^2 x = \tan^2 x (\sec^2 x - 1)$
$$= \tan^2 x \tan^2 x \quad (\text{since} \quad \tan^2 x = \sec^2 x - 1) \quad = \tan^4 x$$

8.26 $\dfrac{\cos x}{\sin x} + \dfrac{\sin x}{\cos x} = \sec x \csc x$

▮ $\dfrac{\cos x}{\sin x} + \dfrac{\sin x}{\cos x} = \dfrac{\cos^2 x + \sin^2 x}{\sin x \cos x} = \dfrac{1}{\sin x \cos x} = \dfrac{1}{\sin x} \dfrac{1}{\cos x}$
$$= \csc x \sec x \quad (\text{since} \quad \sin x \csc x = 1 \quad \text{and} \quad \cos x \sec x = 1).$$

8.27 $\dfrac{1 + \sec x}{\tan x} + \dfrac{\tan x}{\sec x} = \dfrac{1 + \sec x}{\sec x \tan x}$

▮ $\dfrac{1 + \sec x}{\tan x} + \dfrac{\tan x}{\sec x} = \dfrac{(\sec x)(1 + \sec x) - \tan^2 x}{\sec x \tan x} = \dfrac{\sec x + \sec^2 x - \tan^2 x}{\sec x \tan x}$
$$= \frac{\sec x + 1}{\sec x \tan x} \quad (\text{since} \quad \sec^2 x - \tan^2 x = 1)$$

8.28 $\dfrac{1 - \cos x}{\sin x} + \dfrac{\sin x}{1 - \cos x} = 2 \csc x$

▮ $\dfrac{1 - \cos x}{\sin x} + \dfrac{\sin x}{1 - \cos x} = \dfrac{(1 - \cos x)^2 + \sin^2 x}{(\sin x)(1 - \cos x)} = \dfrac{1 - 2 \cos x + \cos^2 x + \sin^2 x}{(\sin x)(1 - \cos x)}$
$$= \frac{2 - 2 \cos x}{(\sin x)(1 - \cos x)} \quad (\text{since} \quad \sin^2 x + \cos^2 x = 1) \quad = \frac{2(1 - \cos x)}{\sin x (1 - \cos x)}$$
$$= \frac{2}{\sin x} \quad \left(\text{since} \quad \sin x = \frac{1}{\csc x} \right) \quad = 2 \csc x$$

8.29 $\dfrac{1-\cos x}{\csc x} = \dfrac{\sin^3 x}{1+\cos x}$

▮ $\dfrac{1-\cos x}{\csc x} = \dfrac{1-\cos x}{1/\sin x} = (\sin x)(1-\cos x)$

$= (\sin x)(1-\cos x)\cdot\dfrac{\sin^2 x}{\sin^2 x} = \dfrac{(\sin^3 x)(1-\cos x)}{\sin^2 x} = \dfrac{(\sin^3 x)(1-\cos x)}{1-\cos^2 x}$

$= \dfrac{(\sin^3 x)(1-\cos x)}{(1-\cos x)(1+\cos x)} = \dfrac{\sin^3 x}{1+\cos x}$ (see Prob. 8.30)

8.30 $\dfrac{1-\cos x}{\csc x} = \dfrac{\sin^3 x}{1+\cos x}$

▮ $\dfrac{\sin^3 x}{1+\cos x} = \dfrac{\sin x \sin^2 x}{1+\cos x} = \dfrac{(\sin x)(1-\cos^2 x)}{1+\cos x}$

$= \dfrac{(\sin x)(1-\cos x)(1+\cos x)}{1+\cos x} = \sin x(1-\cos x) = \dfrac{1-\cos x}{\csc x}$.   Which method do you

think is preferable, that of Prob. 8.29 or Prob. 8.30?   Why?

8.31 $\dfrac{\cos^2\theta = \sin^2\theta}{\sin\theta\cos\theta} = \cot\theta - \tan\theta$

▮ $\dfrac{\cos^2\theta - \sin^2\theta}{\sin\theta\cos\theta} = \dfrac{\cos^2\theta}{\sin\theta\cos\theta} - \dfrac{\sin^2\theta}{\sin\theta\cos\theta}$

$= \dfrac{\cos\theta\cos\theta}{\sin\theta\cos\theta} - \dfrac{\sin\theta\sin\theta}{\sin\theta\cos\theta} = \cot\theta - \tan\theta$

Note the first line of equations.   There were many algebraic techniques that seemed applicable here.   The minus sign on both sides convinced us to use this technique.   Do not be afraid to abort an attempt that appears futile!

8.32 $\dfrac{\cos\theta}{\sec\theta} + \dfrac{\sin\theta}{\csc\theta} = 1$

▮ $\dfrac{\cos\theta}{\sec\theta} + \dfrac{\sin\theta}{\csc\theta} = \dfrac{\cos\theta}{1/\cos\theta} + \dfrac{\sin\theta}{1/\cos\theta} = \cos^2\theta + \sin^2\theta = 1$

8.33 $\dfrac{\csc\theta}{\sec\theta} + \dfrac{\cos\theta}{\sin\theta} = 2\cot\theta$

▮ $\dfrac{\csc\theta}{\sec\theta} + \dfrac{\cos\theta}{\sin\theta} = \dfrac{1/\sin\theta}{1/\cos\theta} + \dfrac{\cos\theta}{\sin\theta} = \dfrac{\cos\theta}{\sin\theta} + \dfrac{\cos\theta}{\sin\theta}$

$= \cot\theta + \cot\theta = 2\cot\theta$

8.34 $(1-\cos\theta)(1+\sec\theta) = \sec\theta - \cos\theta$

▮ $(1-\cos\theta)(1+\sec\theta) = 1 - \cos\theta + \sec\theta - \sec\theta\cos\theta$

$= 1 - \cos\theta + \sec\theta - 1 = \sec\theta - \cos\theta$

8.35 $\sec^2\theta + \csc^2\theta = \sec^2\theta\csc^2\theta$

▮ $\sec^2\theta + \csc^2\theta = \dfrac{1}{\cos^2\theta} + \dfrac{1}{\sin^2\theta} = \dfrac{\sin^2\theta + \cos^2\theta}{\sin^2\theta\cos^2\theta}$

$= \dfrac{1}{\sin^2\theta\cos^2\theta} = \dfrac{1}{\sin^2\theta}\dfrac{1}{\cos^2\theta} = \sec^2\theta\csc^2\theta$

Note that at the beginning we went from $\sec\theta$ and $\csc\theta$ to $\sin\theta$ and $\cos\theta$, and back, then at the end went to $\sec\theta$ and $\csc\theta$.   This is a common technique: We tend to feel more comfortable with the sine and cosine functions.

8.36 $\tan^2\theta - \sin^2\theta = \tan^2\theta\sin^2\theta$

$$\tan^2 \theta - \sin^2 \theta = \frac{\sin^2 \theta}{\cos^2 \theta} - \sin^2 \theta = \frac{\sin^2 \theta - \sin^2 \theta \cos^2 \theta}{\cos^2 \theta}$$

$$= \frac{\sin^2 \theta(1 - \cos^2 \theta)}{\cos^2 \theta} = \frac{\sin^2 \theta}{\cos^2 \theta} \cdot \frac{1 - \cos^2 \theta}{1} = \tan^2 \theta \sin^2 \theta$$

8.37 $\csc \theta - \cot \theta \cos \theta = \sin \theta$

▮ (See the discussion for Prob. 8.35.) $\csc \theta - \cot \theta \cos \theta = \dfrac{1}{\sin \theta} - \dfrac{\cos \theta}{\sin \theta} \cos \theta = \dfrac{1 - \cos^2 \theta}{\sin \theta}$

$$= \frac{\sin^2 \theta}{\sin \theta} = \sin \theta$$

8.38 $\dfrac{\tan^2 \theta - \sin^2 \theta}{\sin^2 \theta} = \tan^2 \theta$

▮ $\dfrac{\tan^2 \theta - \sin^2 \theta}{\sin^2 \theta} = \dfrac{\tan^2 \theta}{\sin^2 \theta} - \dfrac{\sin^2 \theta}{\sin^2 \theta} = \dfrac{(\sin^2 \theta)/\cos^2 \theta}{\sin^2 \theta} - 1 \quad \left(\text{since} \quad \tan^2 \theta = \dfrac{\sin^2 \theta}{\cos^2 \theta}\right)$

$$= \frac{1}{\cos^2 \theta} - 1 = \sec^2 \theta - 1 = \tan^2 \theta \quad (\text{since} \quad \sec^2 \theta - 1 = \tan^2 \theta)$$

8.39 $\dfrac{\cot \theta - \tan \theta}{\sin \theta \cos \theta} = \csc^2 \theta - \sec^2 \theta$

▮ $\dfrac{\cot \theta - \tan \theta}{\sin \theta \cos \theta} = \dfrac{\cot \theta}{\sin \theta \cos \theta} - \dfrac{\tan \theta}{\sin \theta \cos \theta} = \dfrac{(\cos \theta)/\sin \theta}{\sin \theta \cos \theta} - \dfrac{(\sin \theta)/\cos \theta}{\sin \theta \cos \theta}$

$$= \frac{\cos \theta}{\sin^2 \theta \cos \theta} - \frac{\sin \theta}{\sin \theta \cos^2 \theta} = \frac{1}{\sin^2 \theta} - \frac{1}{\cos^2 \theta}$$

$$= \csc^2 \theta - \sec^2 \theta \quad \left(\text{since} \quad \csc \theta = \frac{1}{\sin \theta} \quad \text{and} \quad \sec \theta = \frac{1}{\cos \theta}\right)$$

8.40 $\dfrac{\sin \theta}{\csc \theta} + \dfrac{\cos \theta}{\sec \theta} = \sin \theta \csc \theta$

▮ $\dfrac{\sin \theta}{\csc \theta} + \dfrac{\cos \theta}{\sec \theta} = \dfrac{\sin \theta}{1/\sin \theta} + \dfrac{\cos \theta}{1/\cos \theta} = \sin^2 \theta + \cos^2 \theta$

$$= 1 = \sin \theta \csc \theta \quad \left(\text{since} \quad \sin \theta = \frac{1}{\csc \theta}\right)$$

8.41 $\dfrac{\sin \theta}{\csc \theta} + \dfrac{\cos \theta}{\sec \theta} = \sin \theta \csc \theta$

▮ See Prob. 8.40. We can work on both sides of the equation as long as we *do not* make use of the equal sign.

$$\frac{\sin \theta}{\csc \theta} + \frac{\cos \theta}{\sec \theta} = \sin \theta \csc \theta$$

$$\frac{\sin \theta}{1/\sin \theta} + \frac{\cos \theta}{1/\cos \theta} = \sin \theta \csc \theta$$

$$\sin^2 \theta + \cos^2 \theta = 1$$

$$1 = 1$$

8.42 $\sec^2 \theta + \cot^2 \theta = \csc^2 \theta + \tan^2 \theta$

▮

$$\sec^2 \theta + \cot^2 \theta = \csc^2 \theta + \tan^2 \theta$$

$$\frac{1}{\cos^2 \theta} + \frac{\cos^2 \theta}{\sin^2 \theta} = \frac{1}{\sin^2 \theta} + \frac{\sin^2 \theta}{\cos^2 \theta}$$

$$\frac{\sin^2 \theta + \cos^4 \theta}{\cos^2 \theta \sin^2 \theta} = \frac{\cos^2 \theta + \sin^4 \theta}{\sin^2 \theta \cos^2 \theta}$$

This seems to be futile. Try this instead:

$$\sec^2 \theta + \cot^2 \theta = \csc^2 \theta + \tan^2 \theta$$

$$(\tan^2 \theta + 1) + \cot^2 \theta = (1 - \cot^2 \theta) + \tan^2 \theta$$

$$\tan^2 \theta + 1 + \cot^2 \theta = \tan^2 \theta + 1 + \cot^2 \theta$$

8.43 $\dfrac{\sec^2 \theta}{\sec^2 \theta - 1} = \csc^2 \theta$

▮ $\dfrac{\sec^2 \theta}{\sec^2 \theta - 1} = \dfrac{\sec^2 \theta}{\tan^2 \theta} = \dfrac{1/\cos^2 \theta}{(\sin^2 \theta)/\cos^2 \theta} = \dfrac{1}{\sin^2 \theta} = \csc^2 \theta$

8.44 $\dfrac{\sin^4 \theta - \cos^4 \theta}{\sin^2 \theta - \cos^2 \theta} = 1$

▮ $\dfrac{\sin^4 \theta - \cos^4 \theta}{\sin^2 \theta - \cos^2 \theta} = \dfrac{\cancel{(\sin^2 \theta - \cos^2 \theta)}(\sin^2 \theta + \cos^2 \theta)}{\cancel{\sin^2 \theta - \cos^2 \theta}} = \sin^2 \theta + \cos^2 \theta = 1$

8.45 $\dfrac{\cos \theta - \sin \theta}{\cos \theta + \sin \theta} = \dfrac{\cot \theta - 1}{\cot \theta + 1}$

▮ Think! What will change $\cos \theta$ into $\cot \theta$? $\quad \dfrac{\cos \theta - \sin \theta}{\cos \theta + \sin \theta} = \dfrac{-\sin \theta}{\cos \theta + \sin \theta} \cdot \dfrac{1/\sin \theta}{1/\sin \theta} = \dfrac{\cot \theta - 1}{\cot \theta + 1}$

8.46 $\tan^2 x \csc^2 x \cot^2 x \sin^2 x = 1$

▮ $\tan^2 x \csc^2 x \cot^2 x \sin^2 x = \dfrac{\sin^2 x}{\cos^2 x} \dfrac{1}{\sin^2 x} \dfrac{\cos^2 x}{\sin^2 x} \sin^2 x = \dfrac{1}{1} = 1$

8.47 $\dfrac{1}{\sec \theta + \tan \theta} = \sec \theta - \tan \theta$

▮ We begin by multiplying by the conjugate.

$$\dfrac{1}{\sec \theta + \tan \theta} = \dfrac{1}{\sec \theta + \tan \theta} \dfrac{\sec \theta - \tan \theta}{\sec \theta - \tan \theta} = \dfrac{\sec \theta - \tan \theta}{(\sec \theta + \tan \theta)(\sec \theta - \tan \theta)}$$

$$= \dfrac{\sec \theta - \tan \theta}{\sec^2 \theta - \tan^2 \theta} = \dfrac{\sec \theta - \tan \theta}{1} = \sec \theta - \tan \theta$$

8.48 $\dfrac{\sin x + \tan x}{\cot x + \csc x} = \sin x \tan x$

▮ $\dfrac{\sin x + \tan x}{\cot x + \csc x} = \dfrac{\sin x + (\sin x)/\cos x}{(\cos x)/\sin x + 1/\sin x} = \dfrac{(\sin x \cos x + \sin x)/\cos x}{(\cos x + 1)/\sin x}$

$= \dfrac{(\sin x \cos x + \sin x)\sin x}{\cos x(\cos x + 1)} = \dfrac{\sin^2 x \cos x + \sin^2 x}{\cos^2 x + \cos x}$

$= \dfrac{(\sin^2 x)(\cos x + 1)}{(\cos x)(\cos x + 1)} = \dfrac{\sin^2 x}{\cos x} = \sin x \cdot \dfrac{\sin x}{\cos x}$

$= \sin x \tan x$

8.49 $\cot \theta + \dfrac{\sin \theta}{1 + \cos \theta} = \csc \theta$

▮ $\cot \theta + \dfrac{\sin \theta}{1 + \cos \theta} = \dfrac{\cos \theta}{\sin \theta} + \dfrac{\sin \theta}{1 + \cos \theta} = \dfrac{\cos \theta(1 + \cos \theta) + \sin^2 \theta}{\sin \theta(1 + \cos \theta)}$

$= \dfrac{\cos \theta + \cos^2 \theta + \sin^2 \theta}{\sin \theta(1 + \cos \theta)} = \dfrac{\cancel{\cos \theta + 1}}{\sin \theta\cancel{(1 + \cos \theta)}} = \dfrac{1}{\sin \theta} = \csc \theta$

8.50 $\dfrac{\sin \theta \cos \theta}{\cos^2 \theta - \sin^2 \theta} = \dfrac{\tan \theta}{1 - \tan^2 \theta}$

▮ $\dfrac{\tan \theta}{1 - \tan^2 \theta} = \dfrac{(\sin \theta)/\cos \theta}{1 - (\sin^2 \theta)/\cos \theta} \cdot \dfrac{\cos \theta}{\cos \theta}$

$= \dfrac{\sin \theta}{\cos \theta - (\sin^2 \theta)/\cos \theta} \cdot \dfrac{\cos \theta}{\cos \theta} = \dfrac{\sin \theta \cos \theta}{\cos^2 \theta - \sin^2 \theta}$

8.51 $(x \sin \theta - y \cos \theta)^2 + (x \cos \theta + y \sin \theta)^2 = x^2 + y^2$

▮ $(x \sin \theta - y \cos \theta)^2 + (x \cos \theta + y \sin \theta)^2 = x^2 \sin^2 \theta - 2xy \sin \theta \cos \theta + y^2 \cos^2 \theta$
$$+ x^2 \cos^2 \theta + 2xy \sin \theta \cos \theta + y^2 \sin^2 \theta$$
$$= x^2 \sin^2 \theta + x^2 \cos^2 \theta + y^2 \sin^2 \theta + y^2 \cos^2 \theta$$
$$= x^2 (\sin^2 \theta + \cos^2 \theta) + y^2 (\sin^2 \theta + \cos^2 \theta)$$
$$= x^2 \cdot 1 + y^2 \cdot 1 = x^2 + y^2$$

8.52 $\ln \tan x = \ln \sin x - \ln \cos x$

▮ $\ln \tan x = \ln \dfrac{\sin x}{\cos x} \left(\text{since} \quad \tan x = \dfrac{\sin x}{\cos x} \right) = \ln \sin x - \ln \cos x$ (property of logs)

8.53 $\dfrac{\sin x}{1 - \cos x} = \dfrac{1 + \cos x}{\sin x}$

▮ $\dfrac{\sin x}{1 - \cos x} = \dfrac{\sin x}{1 - \cos x} \cdot \dfrac{1 + \cos x}{1 + \cos x} = \dfrac{(\sin x)(1 + \cos x)}{1 - \cos^2 x}$

$$= \dfrac{(\sin x)(1 + \cos x)}{\sin^2 x} = \dfrac{1 + \cos x}{\sin x}$$

8.2 ADDITION AND SUBTRACTION IDENTITIES

For Probs. 8.54 to 8.63, find the value of the given expression. Do *not* use tables or a calculator.

8.54 $\sin 15°$

▮ Here we will use the identities $\sin(x + y) = \sin x \cos y + \cos x \sin y$ and $\sin(x - y) = \sin x \cos y - \cos x \sin y$. $15° = 45° - 30°$. Thus, $\sin 15° = \sin(45° - 30°) = \sin 45° \cos 30° - \cos 45° \sin 30° = \dfrac{\sqrt{2}}{2} \cdot \dfrac{\sqrt{3}}{2} - \dfrac{\sqrt{2}}{2} \cdot \dfrac{1}{2} = \dfrac{\sqrt{6}}{4} - \dfrac{\sqrt{2}}{4} = \dfrac{\sqrt{6} - \sqrt{2}}{4}$.

8.55 $\cos 15°$

▮ Compare this to Prob. 8.54. Here we will use the identities $\cos(x + y) = \cos x \cos y - \sin x \sin y$ and $\cos(x - y) = \cos x \cos y + \sin x \sin y$. Thus, $\cos 15° = \cos(60° - 45°) = \cos 60° \cos 45° + \sin 60° \sin 45° = \dfrac{1}{2} \cdot \dfrac{\sqrt{2}}{2} + \dfrac{\sqrt{3}}{2} \cdot \dfrac{\sqrt{2}}{2} = \dfrac{\sqrt{2} + \sqrt{6}}{4}$.

8.56 $\tan 75°$

▮ Here we will use the identity $\tan(x \pm y) = \dfrac{\tan x \pm \tan y}{1 \mp \tan x \tan y}$. Thus,

$$\tan 75° = \tan(45° + 30°) = \dfrac{\tan 45° + \tan 30°}{1 - \tan 45° \tan 30°} = \dfrac{1 + 1/\sqrt{3}}{1 - 1(1/\sqrt{3})} = \dfrac{\sqrt{3} + 1)/\sqrt{3}}{(\sqrt{3} - 1)/\sqrt{3}} = \dfrac{\sqrt{3} + 1}{\sqrt{3} - 1}.$$

If you forgot what $\tan 30°$ is, use $\sin 30° / \cos 30°$.

8.57 $\cot 75°$

▮ $\cot 75° = \dfrac{1}{\tan 75°} = \dfrac{1}{(\sqrt{3} + 1)/(\sqrt{3} - 1)}$ (See Prob. 8.56) $= \dfrac{\sqrt{3} - 1}{\sqrt{3} + 1}$.

8.58 $\csc \pi/12$

▮ $\pi/12 = \pi/3 - \pi/4$ ($=4\pi/12 - 3\pi/12$). Thus,

$$\csc \dfrac{\pi}{12} = \csc\left(\dfrac{\pi}{3} - \dfrac{\pi}{4} \right) = \dfrac{1}{\sin(\pi/3 - \pi/4)} = \sin\left(\dfrac{\pi}{3} - \dfrac{\pi}{4} \right)$$

$$= \sin \dfrac{\pi}{3} \cos \dfrac{\pi}{4} - \cos \dfrac{\pi}{3} \sin \dfrac{\pi}{4} = \dfrac{\sqrt{3}}{2} \cdot \dfrac{\sqrt{2}}{2} - \dfrac{1}{2} \cdot \dfrac{\sqrt{2}}{2} = \dfrac{\sqrt{6} - \sqrt{2}}{4}.$$

Thus, $\csc \dfrac{\pi}{12} = \dfrac{4}{\sqrt{6} - \sqrt{2}}$.

8.59 $\tan \dfrac{7\pi}{12}$

▮ $\tan \dfrac{7\pi}{12} = \tan \left(\dfrac{\pi}{3} + \dfrac{\pi}{4} \right) = \dfrac{\tan \pi/3 + \tan \pi/4}{1 - \tan \pi/3 \tan \pi/4} = \dfrac{\sqrt{3}+1}{1 - \sqrt{3} \cdot 1} = \dfrac{\sqrt{3}+1}{1 - \sqrt{3}}.$

8.60 $\sin 22° \cos 38° + \cos 22° \sin 38°$

▮ $22° + 38° = 60°;$ thus, $\sin 22° \cos 38° + \cos 22° \sin 38° = \sin (22° + 38°) = \sin 60° = \sqrt{3}/2.$

8.61 $\cos 14° \cos 76° - \sin 14° \sin 76°$

▮ $14° + 76° = 90°;$ thus, $\cos 14° \cos 76° - \sin 14° \sin 76° = \cos (14° + 76°) = \cos 90° = 0.$

8.62 $\dfrac{\tan 80° + \tan 40°}{1 - \tan 80° \tan 40°}$

▮ $\dfrac{\tan 80° + \tan 40°}{1 - \tan 80° \tan 40°} = \tan (80° + 40°) = \tan 120° = -\tan 60°$ (quadrant II) $= -3.$

8.63 $\cos 260° \cos 70° - \sin 260° \sin 70°$

▮ $\cos 260° \cos 70° - \sin 260° \sin 70° = \cos (260° + 70°) = \cos 330° = \cos 30°$ $(\cos x > 0$ in quadrant IV$) = \frac{3}{2}.$

For Probs. 8.64 to 8.82, prove that the given equation is an identity.

8.64 $\dfrac{\tan 4\theta - \tan 3\theta}{1 + \tan 4\theta \tan 3\theta} = \tan \theta$

▮ $\dfrac{\tan 4\theta - \tan 3\theta}{1 + \tan 4\theta \tan 3\theta}$ is the formula for $\tan (x - y),$ and therefore is equal to $\tan (4\theta - 3\theta) = \tan \theta.$

8.65 $\dfrac{\tan (45° - x)}{\tan (45° + x)} = \dfrac{(1 - \tan x)^2}{(1 + \tan x)^2}$

▮ $$\tan (45 - x) = \dfrac{\tan 45° - \tan x}{1 + \tan 45° \tan x} = \dfrac{1 - \tan x}{1 + \tan x}$$

$$\tan (45 + x) = \dfrac{\tan 45° + \tan x}{1 - \tan 45° \tan x} = \dfrac{1 + \tan x}{1 - \tan x}$$

Thus, $\dfrac{\tan (45° - x)}{\tan (45° + x)} = \dfrac{\dfrac{1 - \tan x}{1 + \tan x}}{\dfrac{1 + \tan x}{1 - \tan x}} = \dfrac{(1 - \tan x)^2}{(1 + \tan x)^2}.$

8.66 $\sec (\pi/2 - x) = \csc x$

▮ $\sec (\pi/2 - x) = \dfrac{1}{\cos (\pi/2 - x)} = \dfrac{1}{\cos \pi/2 \cos x + \sin \pi/2 \sin x}$

$= [\text{from the identity for}\ \cos (x - y)]$

$= \dfrac{1}{0 \cos x + \sin x} = \dfrac{1}{\sin x} = \csc x$ (since $\sin x \csc x = 1$)

8.67 $\cot (\pi/2 - x) = \tan x$

▮ $\cot (\pi/2 - x) = \dfrac{1}{\tan (\pi/2 - x)} = \dfrac{1}{\dfrac{\tan \pi/2 - \tan x}{1 + \tan \pi/2 \tan x}}.$ But $\tan \pi/2$ is undefined.   Abort this attempt.

(This happens frequently.)   Try using $\tan x = \dfrac{\sin x}{\cos x}$ and $\cot x = \dfrac{\cos x}{\sin x}$ instead: $\cot (\pi/2 - x) = \dfrac{\cos (\pi/2 - x)}{\sin (\pi/2 - x)}.$ But $\cos (\pi/2 - x) = \sin x$ and $\sin (\pi/2 - x) = \cos x.$ Thus, $\cot (\pi/2 - x) = \dfrac{\cos (\pi/2 - x)}{\sin (\pi/2 - x)} = \dfrac{\sin x}{\cos x} = \tan x.$

8.68 $\sin(\pi - x) = \sin x$

▮ $\sin(\pi - x) = \sin\pi\cos x - \cos\pi\sin x = 0\cos x - (-1)\sin x = 0 + \sin x = \sin x.$

8.69 $\dfrac{\sin(\pi - x)}{\sin(\pi + x)} = -1 \quad (\sin \neq 0)$

▮ $\sin(\pi - x) = \sin x$ (see Prob. 8.68).   Then $\sin(\pi + x) = \sin\pi\cos x + \cos\pi\sin x = 0 + (-\sin x) = -\sin x.$
Thus, $\dfrac{\sin(\pi - x)}{\sin(\pi + x)} = \dfrac{\sin x}{-\sin x} = -1 \quad (\sin x \neq 0).$

8.70 $\cos 2x = \cos^2 x - \sin^2 x$

▮ $\cos 2x = \cos(x + x) = \cos x\cos x - \sin x\sin x = \cos^2 x - \sin^2 x.$

8.71 $\cos 2x = 2\cos^2 x - 1 = 1 - 2\sin^2 x$

▮ (1) $\cos 2x = \cos^2 x - \sin^2 x$ (see Prob. 8.70) $= \cos^2 x - (1 - \cos^2 x) = 2\cos^2 x - 1.$ (2) $\cos 2x = \cos^2 x - \sin^2 x = (1 - \sin^2 x) - \sin^2 x = 1 - 2\sin^2 x.$ Thus, $\cos 2x = \cos^2 x - \sin^2 x = 1 - 2\sin^2 x = 2\cos^2 x - 1.$

8.72 $\tan(2x) = \dfrac{2\tan x}{1 - \tan^2 x}$

▮ $\tan 2x = \tan(x + x) = \dfrac{\tan x + \tan x}{1 - \tan x\tan x} = \dfrac{2\tan x}{1 - \tan^2 x}.$

8.73 $\sin 2x = 2\sin x\cos x$

▮ $\sin 2x = \sin(x + x) = \sin x\cos x + \cos x\sin x = 2\sin x\cos x.$

8.74 $\sin 3x = \sin 2x\cos x + \cos 2x\sin x$

▮ $3x = 2x + x.$ Then $\sin 3x = \sin(2x + x) = \sin 2x\cos x + \cos 2x\sin x.$

8.75 $\sin 3x = 2\sin x\cos^2 x + \cos^2 x\sin x - \sin^3 x$

▮ $\sin 3x = \sin 2x\cos x + \cos 2x\sin x$ (see Prob. 8.74)
$= (2\sin x\cos x)\cos x + (\cos^2 x - \sin^2 x)\sin x$
$= 2\sin x\cos^2 x + \cos^2 x\sin x - \sin^3 x$

8.76 $2\sin\left(\dfrac{\pi}{6} + x\right) = \sqrt{3}\sin x + \cos x$

▮ $2\sin\left(\dfrac{\pi}{6} + x\right) = 2\left(\sin\dfrac{\pi}{6}\cos x + \cos\dfrac{\pi}{6}\sin x\right) = 2\left(\tfrac{1}{2}\cos x + \dfrac{\sqrt{3}}{2}\sin x\right) = \cos x + \dfrac{\sqrt{3}}{2}\cdot 2\sin x$
$= \sqrt{3}\sin x + \cos x$

8.77 $\sqrt{2}\cos\left(\dfrac{\pi}{4} - x\right) = \sin x + \cos x$

▮ $\sqrt{2}\cos\left(\dfrac{\pi}{4} - x\right) = \sqrt{2}\left(\cos\dfrac{\pi}{4}\cos x + \sin\dfrac{\pi}{4}\sin x\right) = \sqrt{2}\left(\dfrac{\sqrt{2}}{2}\cos x + \dfrac{\sqrt{2}}{2}\sin x\right)$
$= \tfrac{2}{2}\cos x + \tfrac{2}{2}\sin x = \cos x + \sin x$

8.78 $\cot(x + y) = \dfrac{\cot x\cot y - 1}{\cot x + \cot y}$

▮ $\cot(x + y) = \dfrac{1}{\tan(x + y)} = \dfrac{1}{\dfrac{\tan x + \tan y}{1 - \tan x\tan y}} = \dfrac{1 - \tan x\tan y}{\tan x + \tan y}$

$= \dfrac{1 - \dfrac{1}{\cot x}\dfrac{1}{\cot y}}{\dfrac{1}{\cot x} + \dfrac{1}{\cot y}}\cdot\dfrac{\cot x\cot y}{\cot x\cot y} = \dfrac{\cot x\cot y - 1}{\cot y + \cot x}$

8.79 $\dfrac{\sin(\theta+h)-\sin\theta}{h}=(\cos\theta)\dfrac{\sin h}{h}-(\sin\theta)\dfrac{1-\cos h}{h}$

▮ $\dfrac{\sin(\theta+h)-\sin\theta}{h}=\dfrac{\sin\theta\cos h+\cos\theta\sin h-\sin\theta}{h}$

$\qquad\qquad = \dfrac{\sin\theta\cos h}{h}+\dfrac{\cos\theta\sin h}{h}-\dfrac{\sin\theta}{h}$

$\qquad\qquad = \dfrac{\sin\theta\cos h}{h}-\dfrac{\sin\theta}{h}+\dfrac{\cos\theta\sin h}{h}$

$\qquad\qquad = \dfrac{\sin\theta(\cos h-1)}{h}+\dfrac{\cos\theta\sin h}{h}$

$\qquad\qquad = \dfrac{-\sin\theta(1-\cos h)}{h}+\dfrac{\cos\theta\sin h}{h}$

8.80 $\tan x-\tan y=\dfrac{\sin(x-y)}{\cos x\cos y}$

▮ $\dfrac{\sin(x-y)}{\cos x\cos y}=\dfrac{\sin x\cos y-\cos x\sin y}{\cos x\cos y}=\dfrac{\sin x\,\cancel{\cos y}}{\cos x\,\cancel{\cos y}}-\dfrac{\cancel{\cos x}\sin\cdot y}{\cancel{\cos x}\cos y}$

$\qquad\qquad =\dfrac{\sin x}{\cos x}-\dfrac{\sin y}{\cos y}=\tan x-\tan y$

8.81 $\cot x-\tan y=\dfrac{\cos(x+y)}{\sin x\cos y}$

▮ $\dfrac{\cos(x+y)}{\sin x\cos y}=\dfrac{\cos x\cos y-\sin x\sin y}{\sin x\cos y}=\dfrac{\cos x\,\cancel{\cos y}}{\sin x\,\cancel{\cos y}}-\dfrac{\cancel{\sin x}\sin y}{\cancel{\sin x}\cos y}$

$\qquad\qquad =\dfrac{\cos x}{\sin x}-\dfrac{\sin y}{\cos y}=\cot x-\tan y$

8.82 $\cot 2x=\dfrac{\cot^2 x-1}{2\cot x}$

▮ $\cot 2x=\dfrac{1}{\tan 2x}=\dfrac{1}{(2\tan x)/(1-\tan^2 x)}$ (see Prob. 8.72)

$\qquad =\dfrac{1-(1/\cot^2 x)}{2/\cot x}\cdot\dfrac{\cot^2 x}{\cot^2 x}=\dfrac{\cot^2 x-1}{2\cot x}$

8.83 Evaluate $\sin[\operatorname{Cos}^{-1}(-\tfrac{4}{5})+\operatorname{Sin}^{-1}(-\tfrac{3}{5})]$

▮ See Fig. 8.1. $\sin(x+y)=\sin x\cos y+\cos x\sin y$. Let $x=\operatorname{Cos}^{-1}(-\tfrac{4}{5})$, $y=\operatorname{Sin}^{-1}(-\tfrac{3}{5})$. We get

$\sin\underbrace{\operatorname{Cos}^{-1}(-\tfrac{4}{5})}_{+\frac{3}{5}(\text{quad. II})}\underbrace{\cos\operatorname{Sin}^{-1}(-\tfrac{3}{5})}_{+\frac{4}{5}(\text{quad. IV})}+\cos\underbrace{\operatorname{Cos}^{-1}(-\tfrac{4}{5})}_{\text{quad. II}}\sin\underbrace{\operatorname{Sin}^{-1}(-\tfrac{3}{5})}_{\text{quad. IV}}=(+\tfrac{3}{5})\cdot(+\tfrac{4}{5})+(-\tfrac{4}{5})(-\tfrac{3}{5})=+\tfrac{12}{25}+\tfrac{12}{25}=\tfrac{24}{25}$

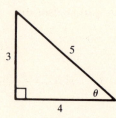

Fig. 8.1

8.84 Evaluate $\sin(x+y+\pi/2)$

▮ $x+y+\pi/2=(x+y)+\pi/2$. Then $\sin[(x+y)+\pi/2]=\sin(x+y)\cos\pi/2+\cos(x+y)\sin\pi/2$

$\qquad =\sin(x+y)\cdot 0+\cos(x+y)\cdot 1=\cos(x+y)$

8.85 Simplify $\sin(x+y)\dfrac{\sqrt{3}}{2}+\cos(x+y)\cdot\frac{1}{2}$

▮ This is of the form $\sin A\cos B+\cos A\sin B$ where $A=(x+y)$, $B=\pi/6$. But $\sin A\cos B+\cos A\sin B=\sin(A+B)$. Thus, the solution is $\sin(x+y+\pi/6)$.

8.3 DOUBLE- AND HALF-ANGLE IDENTITIES

For Probs. 8.86 to 8.91, prove that the given formula is true for all x.

8.86 $\sin 2x = 2\sin x\cos x$

▮ $\sin(x+y)=\sin x\cos y+\cos x\sin y$. Thus, letting $x=y$ in the above equation, $\sin(x+x)=\sin x\cos x+\cos x\sin x=2\sin x\cos x$, and the proof is complete.

8.87 $\cos 2x=\cos^2 x-\sin^2 x=1-2\sin^2 x=2\cos^2 x-1$

▮ Letting $x=y$, $\cos(x+y)=\cos x\cos y-\sin x\sin y=\cos x\cos x-\sin x\sin x=\cos^2 x-\sin^2 x=(1-\sin^2 x)-\sin^2 x=1-2\sin^2 x$. From $\cos^2 x-\sin^2 x$ we also have $=\cos^2 x-(1-\cos^2 x)=2\cos^2 x-1$, and the proof is complete.

8.88 $\tan 2x=\dfrac{2}{\cot x-\tan x}$

▮ $\tan(x+y)=\dfrac{\tan x+\tan y}{1-\tan x\tan y}$. Thus, $\tan 2x=\tan(x+x)=\dfrac{2\tan x}{1-\tan^2 x}=\dfrac{2/\cot x}{\frac{1}{1}-1/\cot^2 x}\cdot\dfrac{\cot^2 x}{\cot^2 x}=\dfrac{2\cot x}{\cot^2 x-1}\cdot\dfrac{1/\cot x}{1/\cot x}=\dfrac{2}{\cot x-\tan x}$, and the proof is complete.

8.89 $\sin\dfrac{x}{2}=\pm\sqrt{\dfrac{1-\cos x}{2}}$

▮ $\cos 2t=1-2\sin^2 t$. Let $t=\dfrac{x}{2}$. Then $\cos x=1-2\sin^2\dfrac{x}{2}$, $\sin^2\dfrac{x}{2}=\dfrac{1-\cos x}{2}$, $\sin^2\dfrac{x}{2}=\pm\sqrt{\dfrac{1-\cos x}{2}}$, and the proof is complete.

8.90 $\cos\dfrac{x}{2}=\pm\sqrt{\dfrac{1+\cos x}{2}}$

▮ $\cos 2t=2\cos^2 t-1$. Let $t=x/2$. Then $\cos x=2\cos^2 x/2-1$, $2\cos^2 x/2=\cos x+1$, $\cos\dfrac{x}{2}=\pm\sqrt{\dfrac{1+\cos x}{2}}$, and the proof is complete.

8.91 $\tan\dfrac{x}{2}=\dfrac{\sin x}{1+\cos x}=\dfrac{1-\cos x}{\sin x}$

▮ $\tan\dfrac{x}{2}=\dfrac{\sin x/2}{\cos x/2}=\dfrac{\pm\sqrt{(1-\cos x)/2}}{\pm\sqrt{(1+\cos x)/2}}=\pm\sqrt{\dfrac{1-\cos x}{1+\cos x}}=\pm\sqrt{\dfrac{1-\cos}{1+\cos}\cdot\dfrac{1+\cos x}{1+\cos x}}$ (multiplying by 1) $=$

$\pm\sqrt{\dfrac{1-\cos^2 x}{(1+\cos x)^2}}=\pm\sqrt{\dfrac{\sin^2 x}{(1+\cos x)^2}}$. Thus, $\left|\tan\dfrac{x}{2}\right|=\dfrac{|\sin x|}{|1+\cos x|}=\dfrac{|\sin x|}{1+\cos x}$ since $|\cos x|\le 1$ for all x. Actually, $\tan x/2$ and $\sin x$ always agree in sign (check this!), so $\tan\dfrac{x}{2}=\dfrac{\sin x}{1+\cos x}=\dfrac{\sin x}{1+\cos x}$. $\dfrac{1-\cos x}{1-\cos x}=\dfrac{(\sin x)(1-\cos x)}{1-\cos^2 x}=\dfrac{(\sin x)(1-\cos x)}{\sin^2 x}=\dfrac{1-\cos x}{\sin x}$. Thus, $\tan\dfrac{x}{2}=\dfrac{\sin x}{1+\cos x}=\dfrac{1-\cos x}{\sin x}$, and the proof is complete.

For Probs. 8.92 to 8.99, find the exact value of the given expression without using a table or calculator.

8.92 $\sin(22.5°)$

▮ $\sin(22.5°)=\sin 45°/2$. Since $\sin\dfrac{x}{2}=\pm\sqrt{\dfrac{1-\cos x}{2}}$, $\sin\dfrac{45°}{2}=\pm\sqrt{\dfrac{1-\cos 45°}{2}}$. We reject the minus answer since this is a quadrant I angle. Thus, the answer is $=\sqrt{\dfrac{1-\sqrt{2}}{2}}$.

8.93   tan 22.5°

▮ $\tan \dfrac{x}{2} = \dfrac{\sin x}{1 + \cos x}$.   Then $\tan 22.5° = \dfrac{\sin 45°}{1 + \cos 45°} = \dfrac{\sqrt{2}/2}{1 + \sqrt{2}/2} \cdot \dfrac{2}{2} = \dfrac{\sqrt{2}}{2 + \sqrt{2}}$.

8.94   sin 165°

▮ $\sin 165° = \sin \dfrac{330°}{2} = \pm\sqrt{\dfrac{1 - \cos 330°}{2}} = \sqrt{\dfrac{1 - \cos 330°}{2}}$ (sin x > 0   in quadrant II) $= \sqrt{\dfrac{1 - \sqrt{3}/2}{2} \cdot \dfrac{2}{2}} =$
$\sqrt{\dfrac{2 - \sqrt{3}}{2}}$.

8.95   cos 165°

▮ $\cos 165° = -\sqrt{\dfrac{1 + \cos 300°}{2}}$ (Be careful!   We have a minus sign here because   cos x < 0   in quadrant II.)
$= -\sqrt{\dfrac{1 + \sqrt{3}/2}{2} \cdot \dfrac{2}{2}} = -\dfrac{\sqrt{2 + \sqrt{3}}}{2}$.

8.96   tan 165°

▮ $\tan \dfrac{x}{2} = \dfrac{1 - \cos x}{\sin x}$.   Then $\tan 165° = \dfrac{1 - \cos 330°}{\sin 330°} = \dfrac{1 - \sqrt{3}/2}{-\frac{1}{2}} \cdot \dfrac{2}{2} = \dfrac{2 - \sqrt{3}}{-1} = \sqrt{3} - 2$.

8.97   sin 75°

$75° = \dfrac{150°}{2}$.   Then $\sin 75° = \sqrt{\dfrac{1 - \cos 150°}{2}} = \sqrt{\dfrac{1 - (-\sqrt{3}/2)}{2}} = \sqrt{\dfrac{1 + \sqrt{3}/2}{2} \cdot \dfrac{2}{2}} = \sqrt{\dfrac{2 + \sqrt{3}}{4}} = \dfrac{\sqrt{2 + \sqrt{3}}}{2}$.

8.98   cos 15°

▮ $\cos 15° = \cos \dfrac{30°}{2} = \sqrt{\dfrac{1 + \cos 30°}{2}} = \sqrt{\dfrac{1 + \sqrt{3}/2}{2}} = \dfrac{\sqrt{2 + \sqrt{3}}}{2}$.   Look at Prob. 8.97; does the fact that these
two results are the same surprise you?   It shouldn't!

8.99   cos 375°

▮ $\cos 375° = \cos \dfrac{750°}{2} = \dfrac{\sqrt{1 + \cos 750°}}{2}$ (Here we have a positive answer since 375° is in quadrant I, and
thus,   cos x > 0.) $= \sqrt{\dfrac{1 + \cos (750° - 720°)}{2}} = \sqrt{\dfrac{1 + \cos 30°}{2}} = \sqrt{\dfrac{1 + \sqrt{3}/2}{2}} = \dfrac{\sqrt{2 + \sqrt{3}}}{2}$.

For Probs. 8.100 to 8.114, establish that the given statement is an identity.

8.100 $\sec 2\theta = \dfrac{1}{2 \cos^2 \theta - 1}$

▮ $\sec 2\theta = \dfrac{1}{\cos 2\theta} = \dfrac{1}{2 \cos^2 \theta - 1}$ (since $\cos 2\theta = 2 \cos^2 \theta - 1$).

8.101 $\csc 2\theta = \dfrac{\sec \theta \csc \theta}{2}$

▮ $\csc 2\theta = \dfrac{1}{\sin 2\theta} = \dfrac{1}{2 \sin \theta \cos \theta}$ (since $\sin 2\theta = 2 \sin \theta \cos \theta$) $= \dfrac{1}{2} \dfrac{1}{\sin \theta} \dfrac{1}{\cos \theta} = \dfrac{\sec \theta \csc \theta}{2}$ (since
$\sin \theta \csc \theta = 1$ and $\cos \theta \sec \theta = 1$).

8.102 $\sec 2\theta = \dfrac{\sec^2 \theta}{2 - \sec^2 \theta}$

▮ See Prob. 8.100. $\sec 2\theta = \dfrac{1}{2 \cos^2 \theta - 1} = \dfrac{1}{2/(\sec^2 \theta) - 1} \cdot \dfrac{\sec^2 \theta}{\sec^2 \theta}$ (multiplying by 1) $= \dfrac{\sec^2 \theta}{2 - \sec^2 \theta}$.

8.103 $\cos 2\theta = \dfrac{\cot^2 \theta - 1}{2 \cot \theta}$

$\blacksquare$ $\tan 2\theta = \dfrac{2 \tan \theta}{1 - \tan^2 \theta}$.   Thus, $\cot 2\theta = \dfrac{1}{\tan 2\theta} = \dfrac{1}{(2 \tan \theta)/(1 - \tan^2 \theta)} = \dfrac{1 - \tan^2 \theta}{2 \tan \theta}$

$= \dfrac{1 - 1/\cot^2 \theta}{2/\cot \theta} \cdot \dfrac{\cot^2 \theta}{\cot^2 \theta}$ $(\tan \theta \cot \theta = 1) = \dfrac{\cot^2 \theta - 1}{2 \cot \theta}$.

8.104 $\dfrac{\sin 2x}{1 + \cos 2x} = \tan x$

$\blacksquare$ We want the 1 to drop out, so we let $\cos 2x = 2 \cos^2 x - 1$.   Then $\dfrac{\sin 2x}{1 + \cos 2x} = \dfrac{2 \sin x \cos x}{1 + (2 \cos^2 x - 1)} =$
$\dfrac{2 \sin x \cos x}{2 \cos^2 x} = \dfrac{\sin x}{\cos x} = \tan x$.

8.105 $\cot 4x = \dfrac{1 - \tan^2 2x}{2 \tan 2x}$

$\blacksquare$ $\tan 4x = \tan 2(2x)$,   so $\cot 4x = \dfrac{1}{\tan 2(2x)} = \dfrac{1}{(2 \tan 2x)/(1 - \tan^2 2x)} = \dfrac{1 - \tan^2 2x}{2 \tan 2x}$.

8.106 $4 \sin^2 \theta \cos^2 \theta = 1 - \cos^2 2\theta$

$\blacksquare$ $\sin 2\theta = 2 \sin \theta \cos \theta$.   Thus, $(2 \sin \theta \cos \theta)^2 = (\sin 2\theta)^2$,   or $4 \sin^2 \theta \cos^2 \theta = \sin^2 2\theta = 1 - \cos^2 2\theta$.

8.107 $\cos 2x + 2 \sin^2 x = 1$

$\blacksquare$ $2 \sin^2 x = 2(1 - \cos^2 x) = 2 - 2 \cos^2 x$.   Thus, $\cos 2x + 2 \sin^2 x = (2 \cos^2 x - 1) + (2 - 2 \cos^2 x) = -1 + 2 = 1$.

8.108 $\tan(\theta - 45°) + \tan(\theta + 45°) = 2 \tan 2\theta$

$\blacksquare$ $\tan(\theta - 45°) + \tan(\theta + 45°) = \dfrac{\tan \theta - \tan 45°}{1 + \tan \theta \tan 45°} + \dfrac{\tan \theta + \tan 45°}{1 - \tan \theta \tan 45°}$

$= \dfrac{\tan \theta - 1}{1 + \tan \theta} + \dfrac{\tan \theta + 1}{1 - \tan \theta}$

$= \dfrac{(\tan \theta - 1)(1 - \tan \theta) + (\tan \theta + 1)^2}{1 - \tan^2 \theta}$

$= \dfrac{\tan \theta - 1 - \tan^2 \theta + \tan \theta + \tan^2 \theta + 2 \tan \theta + 1}{1 - \tan^2 \theta}$

$= \dfrac{4 \tan \theta}{1 - \tan^2 \theta} = 2 \dfrac{2 \tan \theta}{1 - \tan^2 \theta} = 2 \tan 2\theta$

8.109 $\cot \dfrac{x}{2} = \dfrac{1 + \cos x}{\sin x}$

$\blacksquare$ $\cot \dfrac{x}{2} = \dfrac{1}{\tan x/2} = \dfrac{1}{(1 - \cos x)/\sin x} = \dfrac{\sin x}{1 - \cos x} \cdot \dfrac{1 + \cos x}{1 + \cos x} = \dfrac{\sin x (1 + \cos x)}{1 - \cos^2 x} = \dfrac{\sin x (1 + \cos x)}{\sin^2 x}$

$= \dfrac{1 + \cos x}{\sin x}$.

8.110 $\sec^2 \dfrac{\theta}{2} = \dfrac{2}{1 + \cos \theta}$

$\blacksquare$ $\cos^2 \dfrac{\theta}{2} = \dfrac{1 + \cos \theta}{2}$.   Then $\sec^2 \dfrac{\theta}{2} = \dfrac{1}{\cos^2 \theta/2} = \dfrac{1}{(1 + \cos \theta)/2} = \dfrac{2}{1 + \cos \theta}$.

8.111 $\csc^2 \dfrac{\theta}{2} = \dfrac{2}{1 - \cos \theta}$

▮ $\csc^2 \dfrac{\theta}{2} = \dfrac{1}{\sin^2 \theta/2} = \dfrac{1}{(1 - \cos \theta)/2} = \dfrac{2}{1 - \cos \theta}$.

8.112 $\sin^2 \dfrac{\theta}{2} = \dfrac{\sec \theta - 1}{2 \sec \theta}$

▮ $\sin^2 \dfrac{\theta}{2} = \dfrac{1 - \cos \theta}{2} = \dfrac{1 - 1/\sec \theta}{2} \cdot \dfrac{\sec \theta}{\sec \theta}$ $\left(\text{since} \quad \cos \theta = \dfrac{1}{\sec \theta}\right) = \dfrac{\sec \theta - 1}{2 \sec \theta}$.
(Question: When would this formula be useful?)

8.113 $\cot^2 \dfrac{\theta}{2} = \dfrac{1 + \cos \theta}{1 - \cos \theta}$

▮ $\cot^2 \dfrac{\theta}{2} = \dfrac{\cos^2 \theta/2}{\sin^2 \theta/2} = \dfrac{(1 + \cos \theta)/2}{(1 - \cos \theta)/2} = \dfrac{1 + \cos \theta}{1 - \cos \theta}$.

8.114 $\tan \dfrac{x}{2} \sin x = \dfrac{\tan x - \sin x}{\sin x \sec x}$

▮ Here we will simplify both sides. (1) $\tan \dfrac{x}{2} \sin x = \left(\dfrac{1 - \cos x}{\sin x}\right) \sin x = 1 - \cos x$ (LHS).
(2) $\dfrac{\tan x - \sin x}{\sin x \sec x} = \dfrac{(\sin x)/(\cos x) - \sin x}{(\sin x)/(\cos x)}$. Then multiplying by $\dfrac{\cos x}{\cos x}$ we have $\dfrac{\sin x - \sin x \cos x}{\sin x} =$
$\dfrac{(\sin x)(1 - \cos x)}{\sin x} = 1 - \cos x$ (RHS). Since LHS = RHS, the identity is established.

For Probs. 8.115 to 8.120, find the exact value without using a table or calculator.

8.115 $\cos (2 \operatorname{Cos}^{-1} \tfrac{3}{5})$

▮ Let $x = \operatorname{Cos}^{-1} \tfrac{3}{5}$. Then $\cos 2x = 2 \cos^2 x - 1$. Therefore, $\cos (2 \operatorname{Cos}^{-1} \tfrac{3}{5}) = 2 \cos^2 (\operatorname{Cos}^{-1} \tfrac{3}{5}) - 1 =$
$2(\tfrac{3}{5})^2 - 1 = 2 \cdot \tfrac{9}{25} - 1 = \tfrac{18}{25} - 1 = -\tfrac{7}{25}$.

8.116 $\sin (2 \operatorname{Cos}^{-1} \tfrac{3}{5})$

▮ See Fig. 8.2 $\sin 2x = 2 \sin x \cos x$. Then $\sin (2 \operatorname{Cos}^{-1} \tfrac{3}{5}) = 2 \sin (\operatorname{Cos}^{-1} \tfrac{3}{5}) \cos (\operatorname{Cos}^{-1} \tfrac{3}{5}) = 2 \cdot \tfrac{4}{5} \cdot \tfrac{3}{5} = \tfrac{24}{25}$
(since $\sin \operatorname{Cos}^{-1} \tfrac{3}{5} = \tfrac{4}{5}$).

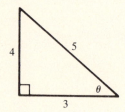

Fig. 8.2

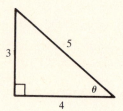

Fig. 8.3

8.117 $\tan [2 \operatorname{Cos}^{-1} (-\tfrac{4}{5})]$

▮ See Fig. 8.3. $\tan 2x = \dfrac{2 \tan x}{1 - \tan^2 x}$. Then $\tan [2 \operatorname{Cos}^{-1} (-\tfrac{4}{5})] = \dfrac{2 \tan \operatorname{Cos}^{-1} (-\tfrac{4}{5})}{1 - \tan^2 [\operatorname{Cos}^{-1} (-\tfrac{4}{5})]}$. Since
$\tan \operatorname{Cos}^{-1} (\tfrac{4}{5}) = \tfrac{3}{4}$ and $\tan \operatorname{Cos}^{-1} (-\tfrac{4}{5}) = -\tfrac{3}{4}$ ($\tan x > 0$ in quadrant II), $\dfrac{2 \tan \operatorname{Cos}^{-1} (-\tfrac{4}{5})}{1 - \tan^2 [\operatorname{Cos}^{-1} (-\tfrac{4}{5})]} =$
$\dfrac{2(-\tfrac{3}{4})}{1 - (-\tfrac{3}{4})^2} = \dfrac{-\tfrac{6}{4}}{1 - \tfrac{9}{16}} = -\tfrac{6}{4} \cdot \tfrac{16}{7} = -\tfrac{24}{7}$.

8.118 $\sin x/2$ if $\cos x = \tfrac{1}{3}$, $0° < x < 90°$.

▮ $\sin x/2 = \sqrt{\dfrac{1 - \cos x}{2}}$ (quadrant I) $= \sqrt{\dfrac{1 - \tfrac{1}{3}}{2}} = \sqrt{\dfrac{\tfrac{2}{3}}{2}} = \sqrt{\dfrac{1}{3}}$.

8.119 $\cos x/2$ if $\sin x = -\frac{1}{3}$, $\pi < x < 3\pi/2$.

▮ See Fig. 8.4. $\cos \dfrac{x}{2} = -\sqrt{\dfrac{1 = \cos x}{2}}$ (quadrant III) $= -\sqrt{\dfrac{1 + \sqrt{8}/3}{2}} = -\sqrt{\dfrac{1 + 2\sqrt{2}/3}{2}} = -\sqrt{\dfrac{3 + 2\sqrt{2}}{6}}$.

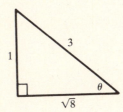

Fig. 8.4

8.120 $\tan 2x$ if $\cot x = -\frac{5}{12}$, $-\pi/2 < x < 0$.

▮ $\tan 2x = \dfrac{2 \tan x}{1 - \tan^2 x} = \dfrac{2(1/\cot x)}{1 - (1/\cot x)^2} = \dfrac{2(-\frac{5}{12})}{1 - (-\frac{5}{12})^2} = \dfrac{-\frac{10}{12}}{1 - \frac{25}{144}} = -\dfrac{10}{12} \cdot \dfrac{144}{119} = -\dfrac{120}{119}$.

8.4 PRODUCT AND SUM IDENTITIES

8.121 State the four basic product identities.

▮ $2 \sin A \cos B = \sin (A + B) + \sin (A - B)$ $2 \cos A \sin B = \sin (A + B) - \sin (A - B)$
$2 \cos A \cos B = \cos (A + B) + \cos (A - B)$ $2 \sin A \sin B = -\cos (A + B) + \cos (A - B)$

8.122 State the four basic sum identities.

▮ $\sin A + \sin B = 2 \sin \dfrac{A + B}{2} \cos \dfrac{A - B}{2}$ $\sin A - \sin B = 2 \cos \dfrac{A + B}{2} \sin \dfrac{A - B}{2}$

$\cos A + \cos B = 2 \cos \dfrac{A + B}{2} \cos \dfrac{A - B}{2}$ $\cos B - \cos A = 2 \sin \dfrac{A + B}{2} \sin \dfrac{A - B}{2}$

Be careful with the $\cos B - \cos A$ identity. Look carefully at A and B.

8.123 Derive the identity $2 \sin A \cos B = \sin (A + B) + \sin (A - B)$.

▮ $\sin (A + B) = \sin A \cos B + \cos A \sin B$ and $\sin (A - B) = \sin A \cos B - \cos A \sin B$. Adding these together we get $\sin (A + B) + \sin (A - B) = 2 \sin A \cos B$. (You should try to establish the other three!)

8.124 Derive the identity $\sin X + \sin Y = 2 \sin \dfrac{X + Y}{2} \cos \dfrac{X - Y}{2}$.

▮ Let $X = A + B$ and $Y = A - B$. Then, solving for A and B, $X + Y = 2A$, and $X - Y = 2B$, so $A = \dfrac{X + Y}{2}$ and $B = \dfrac{X - Y}{2}$. Thus, $2 \sin \dfrac{X + Y}{2} \cos \dfrac{X - Y}{2} = \sin X + \sin Y$ by substituting directly into the identity in Prob. 8.123 above.

For Probs. 8.125 to 8.129 write each expression as a sum or difference.

8.125 $2 \sin 45° \cos 73°$

▮ $2 \sin A \cos B = \sin (A + B) + \sin (A - B)$. Then letting $A = 45°$ and $B = 73°$, $\sin (45° + 73°) + \sin (45° - 73°) = \sin (118°) + \sin (-28°) = \sin 118° - \sin 28°$. See Prob. 8.126.

8.126 $2 \sin 45° \cos 73°$

▮ This is the same as Prob. 8.125, but here we find the answer a different way. $2 \cos A \sin B = \sin (A + B) - \sin (A - B)$. Then letting $A = 45°$ and $B = 73°$, we have $\sin 118° - \sin 28°$. Notice that either formula could be used here. Obviously, the answers will be the same either way.

8.127 $2 \cos 28° \cos 16°$

▮ $2 \cos A \cos B = \cos (A + B) + \cos (A - B)$. Then letting $A = 28°$ and $B = 16°$, $\cos (28° + 16°) + \cos (28° - 16°) = \cos 44° + \cos 12°$.

8.128 $2 \cos 15° \cos 17°$

▮ $2 \cos 15° \cos 17° = \cos (A + B) + \cos (A - B)$.   Letting $A = 15°$ and $B = 17°$, $\cos (15° + 17°) + \cos (15° - 17°) = \cos 32° + \cos (-2°) = \cos 32° + \cos 2°$ ($\cos x > 0$ in quadrant IV).

8.129 $\sin 5p \sin 6p$

▮ $2 \sin A \sin B = -\cos (A + B) + \cos (A - B)$,   and $\sin A \sin B = \dfrac{-\cos (A + B) + \cos (A - B)}{2}$.   Letting $A = 5p$ and $B = 6p$, $\dfrac{-\cos (5p + 6p) + \cos (5p - 6p)}{2} = \dfrac{-\cos 11p + \cos (-p)}{2} = \dfrac{-\cos 11p}{2} + \dfrac{\cos (-p)}{2}$.

For Probs. 8.130 to 8.134, rewrite each expression as a product.

8.130 $\sin 20° + \sin 15°$

▮ $\sin A + \sin B = 2 \sin \dfrac{A + B}{2} \cos \dfrac{A - B}{2}$.   Letting $A = 20°$ and $B = 15°$, $\sin 20° + \sin 15° = 2 \sin \dfrac{20° + 15°}{2} \cos \dfrac{20° - 15°}{2} = 2 \sin 17.5° \cos 2.5°$.

8.131 $\sin 51° - \sin 61°$

▮ $\sin A - \sin B = 2 \cos \dfrac{A + B}{2} \sin \dfrac{A - B}{2}$.   Letting $A = 51°$ and $B = 61°$, $\sin 51° - \sin 61° = 2 \cos \dfrac{51° + 61°}{2} \sin \dfrac{51° - 61°}{2} = 2 \cos 56° \sin (-5°) = -2 \cos 56° \sin 5°$ ($\sin x < 0$ in quadrant IV).

8.132 $\cos \dfrac{\pi}{12} - \cos \dfrac{\pi}{6}$

▮ $\cos \dfrac{\pi}{12} - \cos \dfrac{\pi}{6} = 2 \sin \dfrac{\pi/6 + \pi/12}{2} \sin \dfrac{\pi/6 - \pi/12}{2} = 2 \sin \dfrac{3\pi/12}{2} \sin \dfrac{\pi/12}{2} = 2 \sin \dfrac{\pi}{8} \sin \dfrac{\pi}{24}$.

8.133 $\cos 5x + \cos 7x$

▮ $\cos 5x + \cos 7x = 2 \cos \dfrac{5x + 7x}{2} \cos \dfrac{5x - 7x}{2} = 2 \cos \dfrac{12x}{2} \cos \dfrac{-2x}{2} = 2 \cos 6x \cos (-x) = 2 \cos 6x \cos x$.

8.134 $\sin 620° - \sin 720°$

▮ $\sin 620° - \sin 720° = 2 \cos \dfrac{620° + 720°}{2} \sin \dfrac{620° - 720°}{2} = 2 \cos \dfrac{1340°}{2} \sin \dfrac{-100°}{2} = 2 \cos 670° \sin (-50°)$
$= -2 \cos 310° \sin 50°$.

For Probs. 8.135 to 8.145, use a product or sum identity to evaluate the given expression.

8.135 $2 \sin 75° \cos 15°$

▮ $2 \sin 75° \cos 15° = \sin (75° + 15°) + \sin (75° - 15°) = \sin 90° + \sin 60° = 1 + \dfrac{\sqrt{3}}{2} = \dfrac{2 + 2\sqrt{3}}{2}$.

8.136 $\cos 75° \cos 15°$

▮ $\cos 75° \cos 15° = \dfrac{\cos (75° + 15°) + \cos (75° - 15°)}{2} = \dfrac{\cos 90° + \cos 60°}{2} = \dfrac{0 + \frac{1}{2}}{2} = \frac{1}{4}$.

8.137 $\sin 15° \sin 45°$

▮ $\sin 15° \sin 45° = \dfrac{-\cos (15° + 45°) + \cos (15° - 45°)}{2} = \dfrac{-\cos 60° + \cos (-30°)}{2} = \dfrac{-\frac{1}{2} + \sqrt{3}/2}{2} = \dfrac{-1 + \sqrt{3}}{4}$.

8.138   ▮ $\sin 45° \sin 15°$ (See Prob. 8.137.)

▮ $\sin 45° \sin 15° = \dfrac{-\cos (45° + 15°) + \cos (45° - 15°)}{2} = \dfrac{-\cos 60° + \cos 30°}{2} = \dfrac{-\frac{1}{2} + \sqrt{3}/2}{2} = \dfrac{-1 + \sqrt{3}}{2}$.   You should not be surprised that these answers are the same.   What law governs this situation?

8.139 $\cos 52\frac{1}{2}° \cos 7\frac{1}{2}°$

▮ $\cos 52\frac{1}{2}° \cos 7\frac{1}{2}° = \dfrac{\cos (52\frac{1}{2}° + 7\frac{1}{2}°) + \cos (52\frac{1}{2}° - 7\frac{1}{2}°)}{2} = \dfrac{\cos 60° + \cos 45°}{2} = \dfrac{\frac{1}{2} + \sqrt{2}/2}{2} = \dfrac{\sqrt{2} + 1}{4}$.

8.140 $\sin 75° + \sin 15°$

▮ $\sin 75° + \sin 15° = 2 \sin \dfrac{75° + 15°}{2} \cos \dfrac{75° - 15°}{2} = 2 \sin 45° \cos 30° = 2 \dfrac{\sqrt{2}}{2} \cdot \dfrac{\sqrt{3}}{2} = \dfrac{2\sqrt{6}}{4} = \dfrac{\sqrt{6}}{2}$.

8.141 $\cos 75° - \cos 15°$

▮ $\cos 75° - \cos 15° = 2 \sin \dfrac{15° + 75°}{2} \sin \dfrac{15° - 75°}{2} = 2 \sin 45° \sin (-30°) = (2\sqrt{2}/2) \cdot -\frac{1}{2} = -\dfrac{\sqrt{2}}{2}$. Note that $\cos 75° - \cos 15° = -\sqrt{2}/2 = -\cos 45°$.

8.142 $\sin 75° - \sin 15°$

▮ $\sin 75° - \sin 15° = 2 \cos \dfrac{75° + 15°}{2} \sin \dfrac{75° - 15°}{2} = 2 \cos 45° \sin 30° = (2\sqrt{2}/2) \cdot \frac{1}{2} = \sqrt{2}/2$. Note that $\cos 75° - \cos 15° = -(\sin 75° - \sin 15°)$.

8.143 $\cos 300° + \cos 120°$

▮ $\cos 300° + \cos 120° = 2 \cos \dfrac{300° + 120°}{2} \cos \dfrac{300° - 120°}{2} = 2 \cos 210° \cos 90° = 0$. Note: You can check this answer by noting that: $\cos 300° = \cos 60° = \frac{1}{2}$, $\cos 120° = -\cos 60° = -\frac{1}{2}$, and $\cos 300° + \cos 120° = 0$.

8.144 $\cos 75° \sin 45°$

▮ $\cos 75° \sin 45° = \dfrac{\sin (75° + 45°) - \sin (75° - 45°)}{2} = \dfrac{\sin 120° - \sin 30°}{2} = \dfrac{\sin 60° - \sin 30°}{2} = \dfrac{\sqrt{3}/2 - \frac{1}{2}}{2}$

$= \dfrac{\sqrt{3} - 1}{4}$.

8.145 $\cos 75° \cos 45°$

▮ $\cos 75° \cos 45° = \dfrac{\cos (75° + 45°) + \cos (75° - 45°)}{2} = \dfrac{\cos 120° + \cos 30°}{2} = \dfrac{-\cos 60° + \cos 30°}{2} = \dfrac{-\frac{1}{2} + \sqrt{3}/2}{2}$

$= \dfrac{\sqrt{3} - 1}{4}$. Note that the answer here is the same as in Prob. 8.144.

For Probs. 8.146 to 8.156, establish that the given equation is an identity.

8.146 $\dfrac{\cos 3x + \cos x}{\sin 3x + \sin x} = \cot 2x$

▮ Using the sum formulas, $A = 3x$, and $B = x$, $\dfrac{\cos 3x + \cos x}{\sin 3x + \sin x} = \dfrac{2 \cos \dfrac{3x + x}{2} \cos \dfrac{3x - x}{2}}{2 \sin \dfrac{3x + x}{2} \cos \dfrac{3x - x}{2}} = \dfrac{2 \cos 2x}{2 \sin 2x} =$

$\cot 2x$.

8.147 $\dfrac{\sin x + \sin 3x}{\cos x - \cos 3x} = 1$

▮ $\dfrac{\sin x + \sin 3x}{\cos x - \cos 3x} = \dfrac{2 \sin \dfrac{x + 3x}{2} \cos \dfrac{x - 3x}{2}}{2 \sin \dfrac{3x + x}{2} \cos \dfrac{3x - x}{2}} = \dfrac{\cos (-x)}{\cos x} = \dfrac{\cos x}{\cos x} = 1$.

8.148 $\dfrac{\cos 5x - \cos 7x}{\sin 7x - \sin 5x} = \tan 6x$

▮ $\dfrac{\cos 5x - \cos 7x}{\sin 7x - \sin 5x} = \dfrac{2 \sin \dfrac{7x + 5x}{2} \sin \dfrac{7x - 5x}{2}}{2 \cos \dfrac{7x + 5x}{2} \sin \dfrac{7x - 5x}{2}} = \dfrac{2 \sin 6x \sin x}{2 \cos 6x \sin x} = \dfrac{\sin 6x}{\cos 6x} = \tan 6x$.

8.149 $\dfrac{\sin 5x - \sin 3x}{\cos 5x + \cos 3x} = \tan x$

▮ $\dfrac{\sin 5x - \sin 3x}{\cos 5x + \cos 3x} = \dfrac{\cancel{2\cos 4x}\sin x}{\cancel{2\cos 4x}\cos x} = \dfrac{\sin x}{\cos x} = \tan x$.

8.150 $\dfrac{\sin B + \sin A}{\cos B - \cos A} = \cot \dfrac{A - B}{2}$

▮ $\dfrac{\sin B + \sin A}{\cos B - \cos A} = \dfrac{\cancel{2\sin} \dfrac{B + A}{2} \cos \dfrac{B - A}{2}}{\cancel{2\sin} \dfrac{A + B}{2} \sin \dfrac{A - B}{2}}$

$= \dfrac{\cos \dfrac{B - A}{2}}{\sin \dfrac{A - B}{2}} = \dfrac{\cos\left[-\left(\dfrac{A - B}{2}\right)\right]}{\sin \dfrac{A - B}{2}}$

$= \dfrac{\cos \dfrac{A - B}{2}}{\sin \dfrac{A - B}{2}}$ [since $\cos(-x) = \cos x$] $= \cot \dfrac{A - B}{2}$

8.151 $\dfrac{\sin A + \sin B}{\cos A + \cos B} = \tan \dfrac{A + B}{2}$

▮ $\dfrac{\sin A + \sin B}{\cos A + \cos B} = \dfrac{2\sin \dfrac{A + B}{2} \cancel{\cos \dfrac{A - B}{2}}}{2\cos \dfrac{A + B}{2} \cancel{\cos \dfrac{A - B}{2}}} = \dfrac{\sin \dfrac{A + B}{2}}{\cos \dfrac{A + B}{2}} = \tan \dfrac{A + B}{2}$.

8.152 $\dfrac{\sin 5x - \sin 9x}{\cos 5x - \cos 9x} = -\cot 7x$

▮ $\dfrac{\sin 5x - \sin 9x}{\cos 5x - \cos 9x} = \dfrac{2\cos \dfrac{5x + 9x}{2} \sin \dfrac{5x - 9x}{2}}{2\sin \dfrac{9x + 5x}{2} \sin \dfrac{9x - 5x}{2}} = \dfrac{\cancel{2\cos 7x}\sin(-2x)}{\cancel{2\sin 7x}\sin 2x}$

$= \dfrac{-\cos 7x \cancel{\sin 2x}}{\sin 7x \cancel{\sin 2x}}$ [since $\sin(-x) = -\sin x$] $= -\cot 7x$

8.153 $(\cos x - \cos 5x)(\cos x + \cos 5x) = \sin 4x \sin 6x$

▮ $\cos x - \cos 5x = 2\sin \dfrac{5x + x}{2} \sin \dfrac{5x - x}{2}$ and $\cos x + \cos 5x = 2\cos \dfrac{5x + x}{2} \cos \dfrac{5x - x}{2}$. Thus, $(\cos 5x - \cos 5x)(\cos x + \cos 5x) = 4\sin 3x \sin 2x \cos 3x \cos 2x = 4\sin 3x \cos 3x \sin 2x \cos 2x = 2\sin 3x \cos 3x)(2\sin 2x \cos 2x)$. Letting $A = 3x$, $2\sin 3x \cos 3x = 2\sin A \cos A = \sin 2A = \sin 6x$. Letting $B = 2x$, $2\sin 2x \cos 2x = 2\sin B \cos B = \sin 2B = \sin 4x$. Therefore, $(\cos x - \cos 5x)(\cos x + \cos 5x) = \sin 4x \sin 6x$.

8.154 $\dfrac{\cos 2\theta - \sin 2\theta}{\sin \theta \cos \theta} = \cot \theta - \tan \theta - 2$

▮ $\dfrac{\cos 2\theta - \sin 2\theta}{\sin \theta \cos \theta} = \dfrac{\cos^2 \theta - \sin^2 \theta - 2\sin \theta \cos \theta}{\sin \theta \cos \theta} = \dfrac{\cos^2 \theta}{\sin \theta \cancel{\cos \theta}} - \dfrac{\sin^2 \theta}{\cancel{\sin \theta} \cos \theta} - \dfrac{2\sin \theta \cos \theta}{\sin \theta \cos \theta} =$ $\cot \theta - \tan \theta - 2$. Note that we did not use one of the four sum formulas here. Why?

8.155 $\dfrac{4}{\cos \theta - \cos 3\theta} = 2\csc 2\theta \csc \theta$

▮ $\dfrac{4}{\cos \theta - \cos 3\theta} = \dfrac{4}{2\sin \dfrac{3\theta + \theta}{2} \sin \dfrac{3\theta - \theta}{2}} = \dfrac{4}{2\sin 2\theta \sin \theta} = \dfrac{2}{\sin 2\theta \sin \theta} = 2\csc 2\theta \csc \theta$.

8.156 Assume that $A + B + C = \pi$, and prove that $\sin(A + B) = \sin C$ is an identity.

▮ $\sin(A + B) = \sin(\pi - C)$ (since $A + B = \pi - C$) $= \sin \pi \cos C - \cos \pi \sin C$ [from the $\sin(A + B)$ formula)] $= 0 + \sin C = \sin C$.

8.5 MISCELLANEOUS IDENTITIES

For Probs. 8.157 to 8.176, verify the given identity. Problems in this section combine the techniques of the first four sections.

8.157 $\tan(\pi - x) = -\tan x$

▮ $\tan(\pi - x) = \dfrac{\tan \pi - \tan x}{1 + \tan \pi \tan x} = \dfrac{0 - \tan x}{1 + 0} = -\tan x.$

8.158 $\csc(\pi - u) = \csc u$

▮ $\csc(\pi - u) = \dfrac{1}{\sin(\pi - u)} = \dfrac{1}{\sin \pi \cos u - \cos \pi \sin u} = \dfrac{1}{0 - (-\sin u)} = \dfrac{1}{\sin u} = \csc u.$

8.159 $\sec(\pi - x) = -\sec x$

▮ $\sec(\pi - x) = \dfrac{1}{\cos(\pi - x)} = \dfrac{1}{\cos \pi \cos x + \sin \pi \sin x} = \dfrac{1}{-\cos x + 0} = -\dfrac{1}{\cos x} = -\sec x.$

8.160 $\cot(\pi - x) = -\cot x$

▮ $\cot(\pi - x) = \dfrac{1}{\tan(\pi - x)} = \dfrac{1}{-\tan x}$ (see Prob. 8.157) $= -\cot x.$

8.161 $\tan\left(\dfrac{\pi}{2} + u\right) = -\cot u$

▮ $\tan\left(\dfrac{\pi}{2} + u\right) = \dfrac{\sin(\pi/2 + u)}{\cos(\pi/2 + u)} = \dfrac{\sin \pi/2 \cos u + \cos \pi/2 \sin u}{\cos \pi/2 \cos u - \sin \pi/2 \sin u} = \dfrac{\cos u + 0}{0 - \sin u} = -\dfrac{\cos u}{\sin u} = -\cot u.$

8.162 $\tan(\pi + x) = \tan x$

▮ $\tan(\pi + x) = \dfrac{\tan \pi + \tan x}{1 - \tan \pi \tan x} = \dfrac{0 + \tan x}{1 - 0} = \tan x.$ (Note: You may also obtain this result by using the technique of Prob. 8.161.)

8.163 $\sec(\pi + x) = -\sec x$

▮ $\sec(\pi + x) = \dfrac{1}{\cos(\pi + x)} = \dfrac{1}{\cos \pi \cos x - \sin \pi \sin x} = \dfrac{1}{-\cos x - 0} = -\dfrac{1}{\cos x} = -\sec x.$

8.164 $\tan(2\pi - x) = \tan(-x)$

▮ $\tan(2\pi - x) = \dfrac{\tan 2\pi - \tan x}{1 + \tan 2\pi \tan x} = \dfrac{0 - \tan x}{1 + 0} = -\tan x = \tan(-x).$

8.165 $\cot(2\pi - u) = \cot(-u)$

▮ $\cot(2\pi - u) = \dfrac{1}{\tan(2\pi - u)} = \dfrac{1}{\tan(-u)}$ (see Prob. 8.164) $= \cot(-u).$

8.166 $\dfrac{\cot x}{\csc x - 1} = \dfrac{\csc x + 1}{\cot x}$

▮ $\dfrac{\cot x}{\csc x - 1} = \dfrac{\cot x}{\csc x - 1} \cdot \dfrac{\csc x + 1}{\csc x + 1} = \dfrac{(\cot x)(\csc x - 1)}{\csc^2 x - 1} = \dfrac{(\cot x)(\csc x + 1)}{\cot^2 x} = \dfrac{\csc x + 1}{\cot x}.$

8.167 $\dfrac{\tan x}{\sec x - 1} - \dfrac{\tan x}{\sec x + 1} = 2 \cot x$

▮ $\dfrac{\tan x}{\sec x - 1} - \dfrac{\tan x}{\sec x + 1} = \dfrac{(\tan x)(\sec x + 1) - (\tan x)(\sec x - 1)}{\sec^2 x - 1} = \dfrac{2 \tan x}{\tan^2 x} = \dfrac{2}{\tan x} = 2 \cot x$.

8.168 $\ln |\cos x| + \ln |\sec x| = 0$

▮ $\ln |\cos x| = \ln \left| \dfrac{1}{\sec x} \right|$.   Thus, $\ln |\cos x| + \ln |\sec x| = \ln \left| \dfrac{1}{\sec x} \right| + \ln |\sec x| = \ln \left| \dfrac{1}{\sec x} \sec x \right| = \ln 1 = 0$.

8.169 $\tan \theta \sin 2\theta = 2 \sin^2 \theta$

▮ $\tan \theta \sin 2\theta = (\tan \theta)(2 \sin \theta \cos \theta) = \dfrac{\sin \theta}{\cos \theta} 2 \sin \theta \cos \theta = 2 \sin^2 \theta$.

8.170 $\cot \theta \sin 2\theta = 1 + \cos 2\theta$

▮ $\cot \theta \sin 2\theta = \dfrac{\cos \theta}{\sin \theta} \cdot 2 \sin \theta \cos \theta = 2 \cos^2 \theta = 1 + (2 \cos^2 \theta - 1) = 1 + \cos 2\theta$.

8.171 $\dfrac{1 + \cos 2\theta}{\sin 2\theta} = \cot \theta$

▮ $\dfrac{1 + \cos 2\theta}{\sin 2\theta} = \dfrac{1 + (2 \cos^2 \theta - 1)}{2 \sin \theta \cos \theta} = \dfrac{2 \cos^2 \theta}{2 \sin \theta \cos \theta} = \dfrac{\cos \theta}{\sin \theta} = \cot \theta$.

8.172 $\sin x = 2 \sin \frac{1}{2} x \cos \frac{1}{2} x$

▮ $\sin 2\theta = 2 \sin \theta \cos \theta$.   Let $\theta = \frac{1}{2} x$.   Then, $\sin (2 \cdot \frac{1}{2} x) = 2 \sin \frac{1}{2} x \cos \frac{1}{2} x$,   or $\sin x = 2 \sin \frac{1}{2} x \cos \frac{1}{2} x$.

8.173 $\cos (x - 3\pi/2) = -\sin x$

▮ $\cos (x - 3\pi/2) = \cos x \cos 3\pi/2 + \sin x \sin 3\pi/2 = 0 + (\sin x)(-1) = -\sin x$.

8.174 $(1 - \cos x)(\csc x + \cot x) = \sin x$

▮ $(1 - \cos x)(\csc x + \cot x) = \csc x - \cos x \csc x + \cot x - \cos x \cot x$

$$= \csc x - \cos x \cdot \dfrac{1}{\sin x} + \dfrac{\cos x}{\sin x} - \cos x \cdot \dfrac{\cos x}{\sin x}$$

$$= \csc x - \dfrac{\cos^2 x}{\sin x} = \dfrac{1}{\sin x} - \dfrac{\cos^2 x}{\sin x}$$

$$= \dfrac{1 - \cos^2 x}{\sin x} = \dfrac{\sin^2 x}{\sin x} = \sin x$$

8.175 $\left(\dfrac{1 - \cot x}{\csc x} \right)^2 = 1 - \sin 2x$

▮ $\left(\dfrac{1 - \cot x}{\csc x} \right)^2 = \dfrac{1 - 2 \cot x + \cot^2 x}{\csc^2 x} = \dfrac{\csc^2 x - 2 \cot x}{\csc^2 x}$ (since $\csc^2 x = 1 + \cot^2 x$) $= 1 - \dfrac{2 \cot x}{\csc^2 x} =$
$1 - \dfrac{2 \cos x}{\sin x} \cdot \sin^2 x = 1 - 2 \cos x \sin x = 1 - \sin 2x$.

8.176 $\sin (x + 9\pi/2) = \cos x$

▮ $\sin (x + 9\pi/2) = \sin x \cos 9\pi/2 + \cos x \sin 9\pi/2 = \sin x \cos \pi/2 + \cos x \sin \pi/2 = 0 + \cos x = \cos x$.

8.177 A and B are acute; find $A + B$ given that $\tan A = \frac{1}{4}$ and $\tan B = \frac{3}{5}$.

▮ $\tan (A + B) = \dfrac{\tan A + \tan B}{1 - \tan A \tan B} = \dfrac{\frac{1}{4} + \frac{3}{5}}{1 - \frac{1}{4} \cdot \frac{3}{5}} = \dfrac{(5 + 12)/20}{1 - \frac{3}{20}} = \dfrac{\frac{17}{20}}{\frac{17}{20}} = 1$.   If $\tan (A + B) = 1$ and A, B are
acute, then $A + B = 45°$.

8.178 If $\tan(x+y) = 33$ and $\tan x = 3$, show that $\tan y = 0.3$.

▮ $\tan(x+y) = 33 = \dfrac{\tan x + \tan y}{1 - \tan x \tan y} = \dfrac{3 + \tan y}{1 - 3\tan y}$. Then $3 + \tan y = 33 - 99\tan y$, $99\tan y = 30$, and $\tan y = \frac{30}{99} = 0.\overline{30} \approx 0.3$.

8.179 Prove that $\tan 50° - \tan 40° = 2\tan 10°$.

▮ $\tan(50° - 40°) = \dfrac{\tan 50° - \tan 40°}{1 + \tan 50° \tan 40°}$. Then $\tan 50° - \tan 40° = (1 + \tan 50° \tan 40°)\tan 10° = 2\tan 10°$.

8.6 TRIGONOMETRIC EQUATIONS

For Probs. 8.180 to 8.190, determine whether the given number is a solution of the given equation.

8.180 $\pi/3$, $2\sin x = \sqrt{3}$

▮ Letting $x = \pi/3$, $2\sin \pi/3 = 2 \cdot \sqrt{3}/2 = \sqrt{3}$. Yes, it is a solution.

8.181 $3\pi/4$, $1 + \tan x = 0$

▮ Letting $x = 3\pi/4$, $1 + \tan 3\pi/4 = 1 + (-\tan \pi/4) = 1 - 1 = 0$. Yes, it is a solution.

8.182 $\pi/2$, $1 + \tan x = 0$

▮ $\tan \pi/2$ is undefined. No, it is not a solution.

8.183 π, $\sin x = \cos x$

▮ $\sin \pi = 0$, and $\cos \pi = -1$, but $0 \neq -1$. No, it is not a solution.

8.184 $\pi/4$, $2\tan^2 x + \tan x - 3 = 0$

▮ $2\tan^2 \pi/4 + \tan \pi/4 - 3 = 2 \cdot 1^2 + 1 - 3 = 3 - 3 = 0$. Yes, it is a solution.

8.185 $\pi/3$, $2\sin 2x = 1$

▮ $2\sin 2 \cdot \pi/3 = 2\sin 2\pi/3 = 2\sin \pi/3 = 2\sqrt{3}/2 = \sqrt{3} \neq 1$. No, it is not a solution.

8.186 $\pi/6$, $2\sec x = \tan x + \cot x$

▮ $2\sec \pi/6 = \dfrac{2}{\cos \pi/6} = \dfrac{2}{\sqrt{3}/2} = \dfrac{4}{\sqrt{3}}$, and $\tan \pi/6 + \cot \pi/6 = \dfrac{1}{\sqrt{3}} + \sqrt{3} = \dfrac{1 + (\sqrt{3})^2}{\sqrt{3}} = \dfrac{4}{\sqrt{3}}$. Yes, it is a solution.

8.187 $\pi/3$, $\csc x + \cot x = \sqrt{3}$

▮ $\csc \pi/3 + \cot \pi/3 = \dfrac{1}{\sin \pi/3} + \dfrac{\cos \pi/3}{\sin \pi/3} = \dfrac{1}{\sqrt{3}/2} + \dfrac{\frac{1}{2}}{\sqrt{3}/2} = (\frac{3}{2})(2/\sqrt{3}) = (3/\sqrt{3})(\sqrt{3}/\sqrt{3}) = \dfrac{3\sqrt{3}}{3} = \sqrt{3}$. Yes, it is a solution.

8.188 $5\pi/4$, $\tan x + 3\cot x = 4$

▮ $\tan 5\pi/4 + 3\cot 5\pi/4 = \tan \pi/4 + 3\cot \pi/4 = 1 + 3 = 4$. Yes, it is a solution.

8.189 $11\pi/6$, $2\sin x - \csc x = 1$

▮ $2\sin 11\pi/6 - \csc 11\pi/6 = -2\sin \pi/6 - (-1/\sin \pi/6) = -2 \cdot \frac{1}{2} + 1/\frac{1}{2} = -1 + 2 = 1$. Yes, it is a solution.

8.190 $2\pi/3$, $\cos x - \sqrt{3}\sin x = 1$

▮ $\cos 2\pi/3 - \sqrt{3}\sin 2\pi/3 = -\cos \pi/3 - \sqrt{3}\sin \pi/3 = -\frac{1}{2} - \sqrt{3} \cdot \sqrt{3}/2 = -\frac{1}{2} - \frac{3}{2} = -\frac{4}{2} \neq 1$. No, it is not a solution.

For Probs. 8.191 to 8.215, solve the given equation, finding all solutions which lie in the interval $[0, 2\pi)$.

8.191 $1 + \cos x = 0$

▮ If $1 + \cos x = 0$, then $\cos x = -1$, or $x = \pi$. There are no other angles in $[0, 2\pi)$ whose cosines are -1.

8.192 $1 - \sin x = 0$

▮ If $1 - \sin x = 0$, then $\sin x = 1$, or $x = \pi/2$. There are no others in $[0, 2\pi)$; for example, $\sin 3\pi/2 = -1$.

8.193 $1 + \sqrt{2} \sin \theta = 0$

▮ If $1 + \sqrt{2} \sin \theta = 0$, then $\sqrt{2} \sin \theta = -1$, or $\sin \theta = -1/\sqrt{2}\ (=-\sqrt{2}/2)$. Then $\theta =$ the angle in quadrants II and III with a $\pi/4$ reference angle ($\sin \theta < 0$), so $\theta = 5\pi/4$ or $\theta = 7\pi/4$.

8.194 $1 - \sqrt{2} \cos \theta = 0$

▮ If $1 - \sqrt{2} \cos \theta = 0$, then $-\sqrt{2} \cos \theta = -1$, $\sqrt{2} \cos \theta = 1$, or $\cos \theta = 1/\sqrt{2} = \sqrt{2}/2$. We know $\cos \theta > 0$ in quadrant I and IV. Since $\cos \theta = \sqrt{2}/2$, the reference angle is $\pi/4$, so $\theta = \pi/4$ or $\theta = 7\pi/4$.

8.195 $4 \cos^2 x - 3 = 0$

▮ $4 \cos^2 x = 3$, $\cos^2 x = \frac{3}{4}$, and $\cos x = \pm\sqrt{3/4} = \sqrt{3}/2$ or $-\sqrt{3}/2$. If $\cos x \sqrt{3}/2$, then x has a $\pi/6$ reference angle. x must be a quadrant I or IV angle since $\cos x > 0$. $x = \pi/6$ or $x = 11\pi/6$. If $\cos x = \sqrt{3}/2$, x is in quadrant II or III. $x = 5\pi/6$ or $7\pi/6$. *Check*: $4 \cos^2 7\pi/6 \overset{?}{=} 3$, $4(-\sqrt{3}/2)^2 \overset{?}{=} 3$, $3 = 3$. The other three solutions check as well.

8.196 $2 \sin^2 x - 1 = 0$

▮ $2 \sin^2 x = 1$, $\sin^2 x = \frac{1}{2}$, or $\sin x = \pm\sqrt{\frac{1}{2}} = \pm 1/\sqrt{2} = \pm\sqrt{2}/2$. Then $\sin x = \sqrt{2}/2$ or $\sin x = -\sqrt{2}/2$. If $\sin x = \sqrt{2}/2$, x has a $\pi/4$ reference angle and is in quadrant I or II. If $\sin x = -\sqrt{2}/2$, x has the same reference angle but is in quadrant III or IV. Thus, $x = \pi/4, 3\pi/4, 7\pi/4, 5\pi/4$. *Check*: $2 \sin^2 (3\pi/4) \overset{?}{=} 1$, $2(\sqrt{2}/2)^2 \overset{?}{=} 1$, $1 = 1$. The others check as well.

8.197 $\sin^2 \theta = \sin \theta$

▮ $\sin^2 \theta - \sin \theta = 0$, $(\sin \theta)(\sin \theta - 1) = 0$, and $\sin \theta = 0$ or $\sin \theta = 1$. If $\sin \theta = 0$, then $\theta = 0$ or $\theta = \pi$. If $\sin \theta = 1$, then $\theta = \pi/2$ (*not* $3\pi/2$). Thus, $\theta = 0$, $\pi/2$, π. *Check*: $\sin^2 \pi/2 - \sin \pi/2 = 1 - 1 = 0$. Check the others.

8.198 $\cos^2 \theta = \cos \theta$

▮ $\cos^2 \theta - \cos \theta = 0$, $(\cos \theta)(\cos \theta - 1) = 0$, and $\cos \theta = 0$ or $\cos \theta = 1$. If $\cos \theta = 0$, then $\theta = \pi/2$ or $\theta = 3\pi/2$. If $\cos \theta = 0$, then $\theta = 0$. Thus, $\theta = 0, \pi/2, 3\pi/2$.

8.199 $2 \sin x \cos x - \cos x$

▮ $2 \sin x \cos x - \cos x = 0$, and $\cos x (2 \sin x - 1) = 0$. Then, $\cos x = 0$, and $x = \pi/2$ or $3\pi/2$; *or* $2 \sin x = 1$, $\sin x = \frac{1}{2}$, and $x = \pi/6$ or $5\pi/6$. Thus, $x = \pi/6, \pi/2, 3\pi/2, 5\pi/6$. Check these in the original equation.

8.200 $2 \sin x \cos x = \sin x$

▮ $2 \sin x \cos x - \sin x = 0$, $\sin x (2 \cos x - 1) = 0$, and $\sin x = 0$ or $2 \cos x - 1 = 0$. If $\sin x = 0$, then $x = 0$. If $\cos x = \frac{1}{2}$, $x = \pi/3$ or $5\pi/3$. Check these answers.

8.201 $1 + \cos x/2 = 0$

▮ $\cos x/2 = -1$. Then $x/2 = \pi$, so $x = 2\pi$. *Check*: $1 + \cos (2\pi/2) \overset{?}{=} 0$, $1 + \cos \pi \overset{?}{=} 0$, $1 + (-1) = 0$.

8.202 $1 - \sin x/2 = 0$

▮ $\sin x/2 = 1$, $x/2 = \pi/2$, and $x = \pi$.

8.203 $1 - \sin x/8 = 0$

▮ $\sin x/8 = 1$, $x/8 = \pi/2$, $2x = 8\pi$, and $x = 4\pi$. There are *no* solutions in the interval $[0, 2\pi)$.

8.204 $\sin 2x = 1$

▮ $\sin 2x = 1$, $2x = \pi/2$,   or $x = \pi/4$.

8.205 $\sin x \cos x = 0$

▮ $\sin x \cos x = 0$,   so $\sin x = 0$ or $\cos x = 0$. If $\sin x = 0$,   then $x = 0$ or π. If $\cos x = 0$,   then $x = \pi/2$ or $3\pi/2$. See Prob. 8.206.

8.206 $8 \sin x \cos x = 0$
▮ $2 \sin x \cos x = \sin 2x$.   Then multiplying by 4, $8 \sin x \cos x = 4 \sin 2x$.   Thus, $4 \sin 2x = 0$, $\sin 2x = 0$, and $2x = 0, \pi, 2\pi$, or 3π. If $2x = 0$, $x = 0$. If $2x = \pi$, $x = \pi/2$. If $2x = 2\pi$, $x = \pi$. If $2x = 3\pi$, $x = 3\pi/2$. See Prob. 8.205.

8.207 $\tan x/2 = \sqrt{3}$

▮   If $\tan x/2 = \sqrt{3}$,   then $x/2 = \pi/3$ or $4\pi/3$. If $x/2 = \pi/3$, $3x = 2\pi$ and $x = 2\pi/3$. If $x/2 = 4\pi/3$, $3x = 8\pi$ and $x = 8\pi/3$.   But this is extraneous since it is not in the interval $[0, 2\pi)$.   Hence, the solution is $x = 2\pi/3$.

8.208 $1 + \sin x = 1 - \sin x$

▮   Then $2 \sin x = 0$, $\sin x = 0$,   and $x = 0$ or π.

8.209 $\tan x + 1 = 0$

▮ $\tan x = -1$.   Then $x =$ the quadrant II and IV angles with a $\pi/4$ reference angle.   Thus, $x = 3\pi/4$ or $7\pi/4$.

8.210 $\cos^2 x + \cos x = 0$

▮ $\cos x (\cos x + 1) = 0$. If $\cos x = 0$,   then $x = \pi/2$ or $3\pi/2$. If $\cos x = -1$,   then $x = \pi$.

8.211 $\sin x \tan x = \sin x$

▮ $\sin x \tan x - \sin x = 0$; $\sin x (\tan x - 1) = 0$. If $\sin x = 0$,   then $x = 0$ or π. If $\tan x = 1$,   then $x = \pi/4$ or $5\pi/4$.

8.212 $(2 \sin x + 1)(\cos x + 1) = 0$

▮   If $2 \sin x + 1 = 0$,   then $2 \sin x = -1$ or $\sin x = -\frac{1}{2}$.   In this case $x = 7\pi/6$ or $11\pi/6$. If $\cos x + 1 = 0$,   then $\cos x = -1$,   and in this case $x = \pi$. Then $2 \sin x = -1$,   and $\sin x = -\frac{1}{2}$. Thus, $x = 7\pi/6$ or $11\pi/6$ or $\cos x + 1 = 0$, $\cos x = -1$, $x = \pi$.

8.213 $\sin x + \cos x = 0$

▮   If $\sin x + \cos x = 0$,   then $\sin x = -\cos x$, $\dfrac{\sin x}{\cos x} = -1$ $(\cos x \neq 0)$,   and $\tan x = -1$.   Thus, $x = 3\pi/4$ or $7\pi/4$. (Note that $3\pi/4 \neq 0$, $\cos 7\pi/4 \neq 0$,   and thus division by $\cos x$ was okay; check the results.)

8.214 $1 - \tan^2 x = 0$

▮ $1 - \tan^2 x = 0$ is the difference of two squares, so we have $(1 - \tan x)(1 + \tan x) = 0$. If $\tan x = 1$, then $x = \pi/4$ or $5\pi/4$. If $\tan x = -1$,   then $x = 3\pi/4$ or $7\pi/4$.

8.215 $2 \cos^2 x = 1 + \sin x$

▮ $2 \cos^2 x - 1 = \sin x$,   or $\cos 2x = \sin x$. (1) x has a $\pi/6$ reference angle $(\cos \pi/3 = \frac{1}{2} = \sin \pi/6)$.   But if the signs of $\cos 2x$ and $\sin x$ must agree, then $x = \pi/6, 5\pi/6$. (2) Otherwise x is $3\pi/2$, since $\sin 3\pi/2 = -1$ and $\cos 6\pi/2 = 0$.   The solutions are therefore $\pi/6, 5\pi/6, 3\pi/2$.

For Probs. 8.216 to 8.224, solve for the indicated variable over the given interval

8.216 $2 \sin^2 x = 3 \sin x - 1$, $[0, 2\pi)$

▮ $2 \sin^2 x - 3 \sin x + 1 = 0$,   so $(2 \sin x - 1)(\sin x - 1) = 0$. If $2 \sin x = 1$,   then $\sin x = \frac{1}{2}$ and $x = \pi/6$ or $5\pi/6$. If $\sin x = 1$,   then $x = \pi/2$.

8.217 $2\cos^2 x + \cos x = 1$, $\quad [0, 2\pi)$

▐ $2\cos^2 x + \cos x - 1 = 0$, so $(2\cos x - 1)(\cos x + 1) = 0$. If $2\cos x = 1$, then $\cos x = \frac{1}{2}$, and $x = \pi/3$, or $5\pi/3$. If $\cos x = -1$, then $x = \pi$.

8.218 $2\cos 2x = 1$, $\quad [0, 2\pi)$

▐ If $2\cos 2x = 1$, then $2(2\cos^2 x - 1) = 1$, $4\cos^2 x - 2 = 1$, $4\cos^2 x = 3$, $\cos^2 x = \frac{3}{4}$, or $\cos x = \pm\sqrt{3}/2$. If $\cos x = \sqrt{3}/2$, then $x = \pi/6$ or $11\pi/6$. If $\cos x = -\sqrt{3}/2$, then $x = 5\pi/6$ or $7\pi/6$.

8.219 $\sqrt{3}\sin x - \cos x = 0$, $\quad (-\infty, \infty)$

▐ $\sqrt{3}\sin x = \cos x$, or $\tan x = \sqrt{3}/3 (=1/\sqrt{3})$. Then x has a reference angle of $\pi/6$. The solution is $x = \pi/6 + K\pi$ $\quad \forall K \in \mathbb{Z}$.

8.220 $4\cos^2 2x - 4\cos 2x + 1 = 0$, $\quad [0, \pi]$

▐ $(2\cos 2x - 1)(2\cos 2x - 1) = 0$. If $2\cos 2x - 1 = 0$, then $\cos 2x = \frac{1}{2}$. Thus, $x = \pi/6$ or $5\pi/6$ (note the domain); no other angles in $[0, \pi]$ have the property that the cosine of their double is $\frac{1}{2}$.

8.221 $\sin^2 \theta + 2\cos \theta = -2$, $\quad [0°, 360°)$

▐ $\sin^2 \theta = 1 - \cos^2 \theta$. Then $1 - \cos^2 \theta + 2\cos \theta + 2 = 0$, $\cos^2 \theta - 2\cos \theta - 3 = 0$, or $(\cos \theta - 3)(\cos \theta + 1) = 0$. If $\cos \theta = 3$, there is no solution. If $\cos \theta = -1$, then $\theta = \pi = 180°$. Thus, the solution is $\theta = 180°$

8.222 $2\sin^2 x + 3\cos x = 3$, $\quad (-\infty, \infty)$

▐ $2(1 - \cos^2 x) + 3\cos x = 3$, $2 - 2\cos^2 x + 3\cos x = 3$, $2\cos^2 x - 3\cos x + 1 = 0$, or $(2\cos x - 1)(\cos x - 1) = 0$. If $2\cos x = 1$, then $\cos x = \frac{1}{2}$, and $x = \pi/3 + K(2\pi)$ $\quad \forall K \in \mathbb{Z}$. If $\cos x = 1$, then $x = K(2\pi)$ $\quad \forall K \in \mathbb{Z}$.

8.223 $\sec x + \tan x = 1$, $\quad (0, 2\pi)$

▐ The idea here is to put the equation in a form so that the $\sec^2 \theta = 1 + \tan^2 \theta$ identity can be used. Here is one way: $\sec x = 1 - \tan x$. Squaring both sides of the equation we get, $\sec^2 x = (1 - \tan x)^2 = 1 - 2\tan x + \tan^2 x$ or $\tan^2 x + 1 = 1 - 2\tan x + \tan^2 x$. Then $2\tan x = 0$, $\tan x = 0$, and $x = 0$ or π. It is crucial to check these answers since we squared and may have picked up extraneous solutions. If $x = 0$, then $\sec 0 + \tan 0 \overset{?}{=} 1$, and $1 = 1$. If $x = \pi$, then $\sec \pi + \tan \pi \overset{?}{=} 0$, and $-1 \neq 1$ No! Thus, the only solution is $x = 0$.

8.224 $\sin 3x + \sin x = 0$, $\quad [0, \pi]$

▐ $\sin 3x + \sin x = 2\sin \dfrac{3x + x}{2} \cos \dfrac{3x - x}{2} = 0$. (This is the formula for a sum! See section 4.) Then $2\sin 2x \cos x = 0$. If $\sin 2x = 0$, then $2x = 0, \pi, 2\pi$, and $x = 0, \pi/2, \pi$. If $\cos x = 0$, then $x = \pi/2$. Thus the solutions are $x = 0, \pi/2, \pi$.

For Probs. 8.225 to 8.230, solve for all real x using a calculator. Give answers to four significant digits.

8.225 $\sin x = 0.2977$

▐ Recall that if $\sin x = r (r \in [-1, 1])$ then $x = 2k\pi + \text{Sin}^{-1} r$ or $x = 2k + (\pi - \text{Sin}^{-1} r) = (2k + 1)\pi - \text{Sin}^{-1} r$ where $k \in \mathbb{Z}$. Here, using a calculator, we find $\text{Sin}^{-1} 0.2977 \approx 0.3023$. Thus, $x = 2k\pi + 0.3023$ or $x = (2k + 1)\pi - 0.3023$ $\quad \forall k \in \mathbb{Z}$.

8.226 $\cos x = -0.8861$

▐ See Prob. 8.225. If $\cos x = r$, then $r = 2k\pi + \text{Cos}^{-1} r$ or $2k\pi - \text{Cos}^{-1} r$ $\quad \forall k \in \mathbb{Z}$, where $r \in (-1, 1)$. Here, using a calculator: $\text{Cos}^{-1} (-0.8861) \approx 2.660$ and $x = 2k\pi + 2.660$ or $2k\pi - 2.660$ $\quad \forall k \in \mathbb{Z}$.

8.227 $\tan x = 13.08$

    ▮   See Probs. 8.225 and 8.226. If $\tan x = r$, $r \in \mathbb{R}$, then $x = k\pi + \text{Arctan } r$, $\forall k \in \mathbb{Z}$. Here, Arctan $13.08 \approx 1.494$, and $x = k\pi + 1.494$ $\forall k \in \mathbb{Z}$.

8.228 $\sec x = -8.613$

    ▮ $\cos x \approx -0.1161$ and $\text{Cos}^{-1}(-0.1161) \approx 1.687$. Then $x = 2k\pi + 1.687$ or $2k\pi - 1.687$ $\forall k \in \mathbb{Z}$.

8.229 $\cot x = -3.478$

    ▮ $\tan x \approx -0.2875$ and $\text{Tan}^{-1}(-0.2875) \cong -2.800$. Then $x = k\pi + (-2.800)$ $\forall k \in \mathbb{Z}$.

8.230 $\csc x = 42.29$

    ▮ $\sin x = 0.0236$ and $\text{Sin}^{-1}(0.0236) = 0.0236$. Then $x = 2k\pi + 0.0236$ or $(2k + 1)\pi - 0.0236$ $\forall k \in \mathbb{Z}$.

CHAPTER 9
Additional Topics in Trigonometry

9.1 RIGHT TRIANGLES

For Probs. 9.1 to 9.9, find the remaining parts of $\triangle ABC$, where $\angle C = 90°$. Round off all angles to the nearest minute, and all lengths to the nearest hundredth.

9.1. $b = 12$, $c = 13$

▮ See Fig. 9.1. Then $12^2 + a^2 = 13^2$ (by the pythagorean theorem) and $a^2 = 169 - 144 = 25$, or $a = 5$. Using a calculator or a table, $\sin A = \text{opposite/hypotenuse} = \frac{5}{13}$, $\angle A = \sin^{-1} \frac{5}{13} \approx 22°37'$; and using a calculator or a table, $\sin B = \frac{12}{13}$, $\angle B = \sin^{-1} \frac{12}{13} \approx 67°23'$.

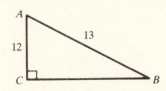

Fig. 9.1

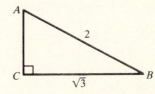

Fig. 9.2

9.2 $a = \sqrt{3}$, $c = 2$

▮ See Fig. 9.2. Then $b^2 + (\sqrt{3})^2 = 4$, $b^2 + 3 = 4$, $b^2 = 1$, or $b = 1$. Then $\cos B = \sqrt{3}/2$, $B = \cos^{-1} \sqrt{3}/2 = 30°$. Thus, $\angle A = 60°$ ($90° = 60° + 30°$).

9.3 $a = 1$, $b = 1$

▮ See Fig. 9.3. If $a = b = 1$, $\triangle ABC$ is an isosceles right triangle, and $\angle A = \angle B = 45°$. Then $c^2 = a^2 + b^2 = 1^2 + 1^2 = 2$, $c^2 = 2$, or $c = \sqrt{2}$.

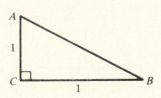

Fig. 9.3

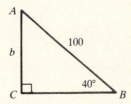

Fig. 9.4

9.4 $c = 100$, $B = 40°$

▮ See Fig. 9.4. $\sin B = b/100$, $b = 100 \sin B \approx 64.28$ (use a calculator or table). Since $\angle B = 40°$, $\angle A = 50°$. Then $\sin 50° = a/100$, and $a = 100 \sin 50° \approx 76.60$.

9.5 $b = 14.2$, $c = 23.7$

▮ See Fig. 9.5. $a^2 + (14.2)^2 = (23.7)^2$, or $a \approx 18.98$. Then $\cos A = 14.2/23.7$, $A = \cos^{-1} (14.2/23.7) \approx 53°11'$. Thus, $B \approx 90° - (53°11') = 36°49'$.

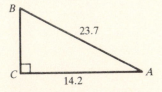

Fig. 9.5

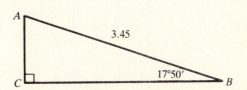

Fig. 9.6

9.6 $B = 17°50'$, $c = 3.45$

▮ See Fig. 9.6. $\angle A = 90° - 17°50' = 72°10'$. Then $\sin B = \sin 17°50' = b/3.45$, or $b = 3.45 \sin 17°50' \approx 1.06$. Thus, $\sin A = a/3.45$, $a = 3.45 \sin A \approx 3.28$.

9.7 $a = 6$, $b = 8.46$

     ▮   See Fig. 9.7. $c^2 = (8.46)^2 + 6^2 = 107.57$,   so $c \approx 10.37$.   Then $\tan B = 8.46/6$, $\measuredangle B =$ Tan$^{-1}\,(8.46/6) \approx 54°40'$.   Thus, $\measuredangle A = 35°20'$.

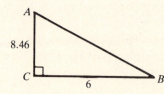

Fig. 9.7

9.8 $b = 10$, $c = 12.6$

     ▮ $a^2 + b^2 = c^2$, $a^2 + 10^2 = (12.6)^2$, $a^2 = 58.76$,   or $a \approx 7.67$.   Then, $\sin B = 10/12.6$,   so $\measuredangle B =$ Sin$^{-1}\,(10/12.6) \approx 52°30'$.   Thus, $\measuredangle A = 90° - \measuredangle B = 37°30'$.   Finally, $\tan B = 10/a$, $a \tan B = 10$, $a = 10/\tan B = 10/(\tan 52°30')$,   or $a \approx 7.67$ (using a calculator).

9.9 $a = 2.42$, $c = 3.22$

     ▮ $a^2 + b^2 = c^2$, $b^2 = c^2 - a^2 = (3.22)^2 - (2.42)^2 \approx 16.22$,   or $b \approx 4.03$.   Then $\cos B = 2.42/3.22$, $\measuredangle B =$ Cos$^{-1}\,(2.42/3.22) \approx 41°20'$.   Thus, $\measuredangle A = 90° - 41°20' = 48°40'$.

For Probs. 9.10 to 9.12 find the perimeter of the described regular polygon.

9.10   A hexagon inscribed in a circle of radius 5 m.

     ▮   See Fig. 9.8.   Then $AOB = 360°/6 = 60°$,   so $\measuredangle AOD = 60°/2 = 30°$.   Since $\sin 30° = \sin(\measuredangle AOD) = AD/5$, $AD = 5 \sin 30° = 5 \cdot \frac{1}{2} = 2.5$ m.   Thus, $AB = 5$ m,   and   the perimeter $= P = 6 \cdot 5 = 30$ m.

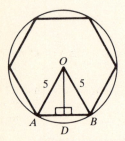

Fig. 9.8

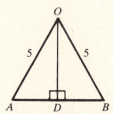

Fig. 9.9

9.11   An octagon inscribed in a circle of radius 5 m.

     ▮   See Fig. 9.9 and Prob. 9.10.   Here, the central angle, $\measuredangle AOB$, $= 360°/8 = 45°$.   Then $\frac{1}{2}$(central angle) $= 45°/2 = 22.5°$,   and $\measuredangle AOD = 22.5°$.   Since $\sin 22.5° = AD/5$, $AD = 5 \sin 22.5° \approx 1.91$ m.   Thus, $AB \approx 3.82$ m,   and $P = 30.56$ m.

9.12   A pentagon circumscribed about a circle of radius 5 m.

     ▮   See Fig. 9.10. $\measuredangle AOG = \frac{1}{2} \cdot (360°/5) = 36°$.   Then $\measuredangle AGO = 90°$ (it is formed by a tangent and radius),   and $AO = 5$ m   (radius of the circle).   Thus, $\sin 36° = AG/5$,   and $AG = 5 \sin 36° \approx 2.94$ m, $AE \approx 5.88$ m, $P \approx 29.4$ m.

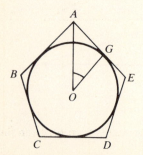

Fig. 9.10

Fig. 9.11

9.13 If a train climbs at a constant angle of 1°23′, how many vertical feet has it climbed after going 1 m?

▮ See Fig. 9.11. Let x = number of feet the train climbs. Then $\tan 1°23′ = x/5280$, and $x = 5280 \tan 1°23′ \approx 127.5$ ft.

9.14 Find the diameter of the moon (to the nearest rule) if at 239 000 mi from earth, it subtends an angle of 32′ relative to an observer on the earth.

▮ See Fig. 9.12 where $\angle ACB = 32′$ and $\angle ACD = 16′$. The diameter of the moon $= 2r$. Then $\tan 16′ = r/239\,000$, and $r = (\tan 16′) \cdot 239\,000$. Thus, the moon's diameter $= 2 \cdot (\tan 16′) \cdot 239\,000 \approx 2225$ mi.

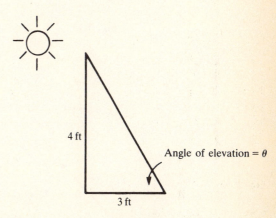

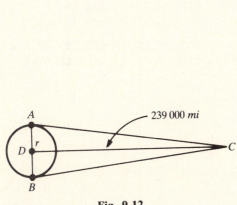

Fig. 9.12   Fig. 9.13

9.15 An object 4 ft tall casts a 3 ft shadow when the angle of elevation of the sun is $\theta°$. Find θ to the nearest degree.

▮ See Fig. 9.13. Then $\tan \theta = \frac{4}{3}$, and $\theta = \operatorname{Arctan} \frac{4}{3} \approx 59°$.

9.16 Find the angle formed by the intersection of a diagonal of the face of a cube with a diagonal of the cube drawn from the same vertex.

▮ See Fig. 9.14a. Let a = the edge of the cube, b = face diagonal, and d = cube's diagonal. Then $a^2 + a^2 = b^2$, $b^2 = 2a^2$, or $b = \sqrt{2}a$. Also, $a^2 + b^2 = d^2$, $a^2 + (\sqrt{2}a)^2 = d^2$, $a^2 + 2a^2 = d^2$, $d^2 = 3a^2$, or $d = \sqrt{3}a$. See Fig. 9.14b. Then $\sin \theta = a/d = a/\sqrt{3}a = 1/\sqrt{3}$, and $\theta = \operatorname{Sin} 1/\sqrt{3} \approx 35°$ (to the nearest degree).

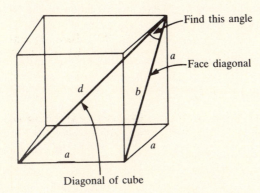

Diagonal of cube   Fig. 9.14a

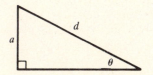

Fig. 9.14b

9.2 POLAR COORDINATES AND GRAPHS

For Probs. 9.17 to 9.26, plot the given points in a polar coordinate system.

9.17 $A(4, 0°)$, $B(7, 180°)$, $C(9, 45°)$

▮ See Fig. 9.15. Recall that to plot (r, θ) in polar coordinates, we go out r units from $(0, 0)$ on the θ ray, where θ is measured from the positive x axis.

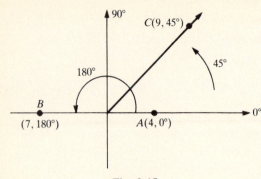

Fig. 9.15

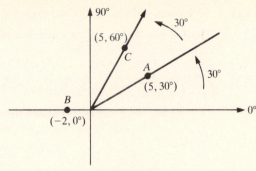

Fig. 9.16

9.18 $A(5, 30°)$, $B(-2, 0°)$, $C(5, 60°)$

▮ See Fig. 9.16.

9.19 $A(3, 45°)$, $B(3, -45°)$, $C(3, 90°)$

▮ See Fig. 9.17.

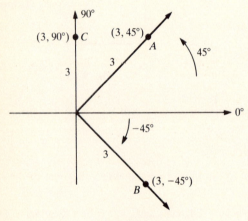

Fig. 9.17

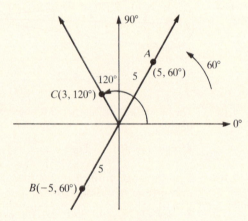

Fig. 9.18

9.20 $A(5, 60°)$, $B(-5, 60°)$, $C(3, 120°)$

▮ See Fig. 9.18. Note that to plot point B, we move five units in the negative direction on the 60° ray.

9.21 $A(5, 45°)$, $B(7, 45°)$, $C(-7, -45°)$

▮ See Fig. 9.19. Note that to plot point C, we move seven units in the negative direction on the $-45°$ ray.

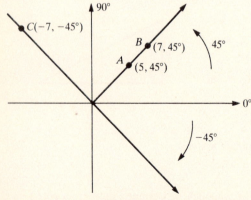

Fig. 9.19

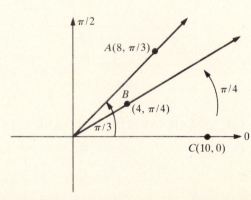

Fig. 9.20

9.22 $A(8, \pi/3)$, $B(4, \pi/4)$, $C(10, 0)$

$\blacksquare$ See Fig. 9.20.   In this case the angles are measured in radians.

9.23 $A(0, 3\pi/2)$, $B(3, \pi/10)$, $C(-3, \pi/10)$

$\blacksquare$ See Fig. 9.21. $A(0, 3\pi/2)$ is located at $(0, 0)$.   Do you see why?   We move zero units on a ray!

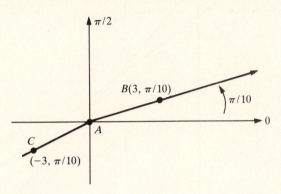

Fig. 9.21

9.24 $A(-6, \pi/6)$, $B(-5, \pi/2)$, $C(-8, \pi/4)$

$\blacksquare$ See Fig. 9.22.

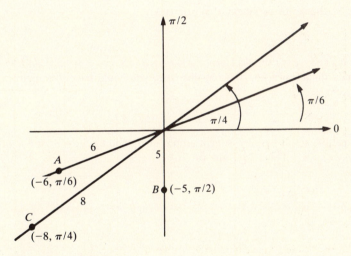

Fig. 9.22

9.25 $A(-3, -45°)$, $B(-5, -60°)$, $C(-7, -310°)$

$\blacksquare$ See Fig. 9.23.

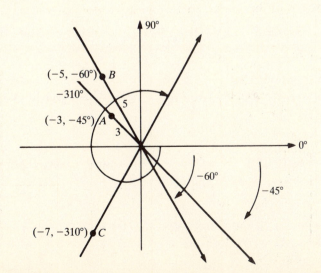

Fig. 9.23

9.26 $A(-6, -\pi/6)$, $B(-5, -\pi/2)$, $C(-8, -\pi/4)$, $D(5, -\pi/2)$

▮   See Fig. 9.24.

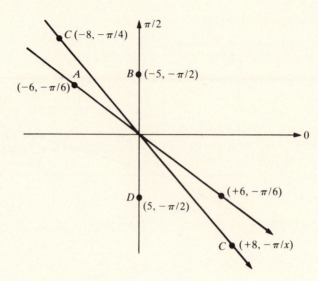

Fig. 9.24

For Probs. 9.27 to 9.39, change to rectangular coordinates.

9.27 $(5, \pi)$

▮ $x = r \cos \theta$ and $y = r \sin \theta$.   Here, $x = r \cos \theta = 5 \cdot \cos \pi = -5$,   and $y = r \sin \theta = 5 \cdot \sin \pi = 0$.   The rectangular coordinates are $(-5, 0)$.   In this case, you could have obtained the result by looking at $(5, \pi)$ in the polar system.

9.28 $(4, \pi/2)$

▮   You can get the result here by using a graph, or by finding $x = r \cos \theta = 4 \cos \pi/2 = 0$ and $y = r \sin \theta = 4 \sin \pi/2 = 4$.   Thus, $(x, y) = (0, 4)$.

9.29 $(-3, \pi/3)$

▮ $x = r \cos \theta = -3 \cos \pi/3 = -3 \cdot \frac{1}{2} = -\frac{3}{2}$. $y = r \sin \theta = -3 \sin \pi/3 = -3 \cdot (\sqrt{3}/2) = -3\sqrt{3}/2$.   Thus, $(x, y) = (-\frac{3}{2}, -3\sqrt{3}/2)$.

9.30 $(4, \pi/4)$

▮ $x = r \cos \theta = 4 \cos \pi/4 = 4\sqrt{2}/2 = 2\sqrt{2}$. $y = r \sin \theta = 4 \sin \pi/4 = 2\sqrt{2}$.   Thus, $(x, y) = (2\sqrt{2}, 2\sqrt{2})$.

9.31 $(-3, \pi/6)$

▮ $x = r \cos \theta = -3 \cos \pi/6 = -3(\sqrt{3}/2) = -3\sqrt{3}/2$. $y = r \sin \theta = -3 \sin \pi/6 = -\frac{3}{2}$.   Thus, $(x, y) = (-3\sqrt{3}/2, -\frac{3}{2})$.

9.32 $(-3, -\pi/6)$

▮ $x = r \cos \theta = -3 \cos (-\pi/6) = -3 \cos \pi/6 = -3\sqrt{3}/2$. $y = r \sin \theta = -3 \sin (-\pi/6) = 3 \sin \pi/6 = \frac{3}{2}$.   Thus, $(x, y) = (-3\sqrt{3}/2, \frac{3}{2})$.

9.33 $(5, -120°$

▮ $x = r \cos \theta = r \cos (-120°) = 5 \cos 120° = -\frac{5}{2}$. $y = r \sin \theta = 5 \sin (-120°) = -5 \sin 120° = -5 \sin 60° = -5\sqrt{3}/2$.   Thus, $(x, y) = (-\frac{5}{2}, -5\sqrt{3}/2)$.

9.34 $(4, -210°)$

▮ $x = 4 \cos (-210°) = 4 \cos 210° = -4 \cos 30° = -4\sqrt{3}/2 = -2\sqrt{3}$. $y = 4 \sin (-210°) = -4 \sin 210° = 4 \sin 30° = 2$.   Thus, $(x, y) = (-2\sqrt{3}, 2)$.

9.35 $(-3, -210°)$

■ $x = -3 \cos(-210°) = -3 \cos 210° = 3 \cos 30° = 3\sqrt{3}/2.$ $y = -3 \sin(-210°) = 3 \sin 210° = -3 \sin 30° = -\frac{3}{2}.$
Thus, $(x, y) = (3\sqrt{3}/2, -\frac{3}{2}).$

9.36 $(5, 315°)$

■ $x = 5 \cos 315° = 5 \cos 45° = 5\sqrt{2}/2.$ $y = 5 \sin 315° = -5 \sin 45° = -5\sqrt{2}/2.$ Thus, $(x, y) = (5\sqrt{2}/2, -5\sqrt{2}/2).$

9.37 $(-3, 330°)$

■ $x = -3 \cos 330° = -3 \cos 30° = -3\sqrt{3}/2.$ $y = -3 \sin 330° = 3 \sin 30° = \frac{3}{2}.$ Thus, $(x, y) = (-3\sqrt{3}/2, \frac{3}{2}).$

9.38 $(-3, -330°)$

■ $x = -3 \cos(-330°) = -3 \cos 330° = -3 \cos 30° = -3\sqrt{3}/2.$ $y = -3 \sin(-330°) = 3 \sin 330° = -3 \sin 30° = -\frac{3}{2}.$ $(x, y) = (-3\sqrt{3}/2, -\frac{3}{2}).$

9.39 $(-6, -420°)$

■ $x = -6 \cos(-420°) = -6 \cos 420° = -6 \cos 60° = -\frac{6}{2} = -3.$ $y = -6 \sin(-420°) = 6 \sin 420° = 6 \sin 60° = 6\sqrt{3}/2 = 3\sqrt{3}.$ Thus, $(x, y) = (-3, 3\sqrt{3}).$

For Probs. 9.40 to 9.49, change the given cartesian coordinates to polar coordinates, with $r \geq 0$ and $\theta \in [0, 2\pi]$ or $\theta \in [0°, 360°]$.

9.40 $(1, \sqrt{3})$

■ $r^2 = x^2 + y^2 = 1^2 + (\sqrt{3})^2 = 4,$ or $r = 2.$ Then $\cos \theta = x/r = \frac{1}{2},$ and $\sin \theta = y/r = \sqrt{3}/2.$ Thus, $\theta = \pi/3,$ and the answer is $(2, \pi/3).$

9.41 $(0, -5)$

■ $x^2 + y^2 = r^2 = 0 + 25 = 25,$ or $r = 5.$ Then $\cos \theta = \frac{0}{5} = 0,$ and $\sin \theta = -\frac{5}{5} = -1.$ Thus $\theta = 3\pi/2,$ and the answer is $(5, 3\pi/2).$

9.42 $(1, -1)$

■ $r^2 = x^2 + y^2 = 1^2 + (-1)^2 = 2,$ or $r = \sqrt{2}.$ Then $\cos \theta = x/r = 1/\sqrt{2},$ and $\sin \theta = y/r = -1\sqrt{2}.$
Thus, $\theta = 225°,$ and the answer is $(\sqrt{2}, 225°).$

9.43 $(-12, 5)$

■ $r^2 = (-12)^2 + 5^2 = 169,$ or $r = 13.$ Then $\cos \theta = -\frac{12}{13},$ and $\sin \theta = \frac{5}{13}.$ Thus, $\theta = \pi - \text{Arctan } \frac{5}{12}$ (quadrant II), and the answer is $(13, \pi - \text{Arctan } \frac{5}{12}).$

9.44 $(-6\sqrt{3}, 6)$

■ $r^2 = (-6\sqrt{3})^2 + 6^2 = 144,$ or $r = 12.$ Then $\cos \theta = -6\sqrt{3}/12 = -\sqrt{3}/2,$ and $\sin \theta = \frac{6}{12} = \frac{1}{2}.$ Thus, $\theta = 150°,$ and the answer is $(12, 150°).$

9.45 $(-4\sqrt{2}, 4\sqrt{2})$

■ $r^2 = (-4\sqrt{2})^2 + (4\sqrt{2})^2 = 32 + 32 = 64,$ or $r = 8.$ Then $\cos \theta = -4\sqrt{2}/8,$ and $\sin \theta = 4\sqrt{2}/8.$
Thus, $\theta = 3\pi/4,$ and the answer is $(8, 3\pi/4).$

9.46 $(-10, 0)$

■ $r^2 = 10^2 + 0^2,$ or $r = 10.$ Then $\cos \theta = -\frac{10}{10} = -1,$ and $\sin \theta = \frac{0}{10} = 0.$ Thus $\theta = \pi,$ and the answer is $(10, \pi).$

9.47 $(s, 0),$ where $s \in \mathscr{R}^+$

■ $r^2 = s^2 + 0^2,$ or $r = s.$ Then $\cos \theta = x/r = s/r = s/s = 1,$ and $\sin \theta = y/r = 0.$ Thus, $\theta = 0,$ and the answer is $(s, 0).$

9.48 $(s, 0)$,   where $s \in \mathbb{R}^-$

    ▮ $r^2 = s^2$, or $r = -s$. (Do you see why? $s < 0$, so $\sqrt{s^2} = -s$). Then $\cos \theta = x/r = s/-s = -1$, and $\sin \theta = y/r = 0/r = 0$. Thus, $\theta = \pi$, and the answer is $(-s, \pi)$.

9.49 $(0, -7)$

    ▮ $r^2 = x^2 + y^2 = 0^2 + 49 = 49$, or $r = 7$. Then $\cos \theta = \frac{0}{7} = 0$, and $\sin \theta = -\frac{7}{7} = -1$. Thus, $\theta = 3\pi/2$, and the answer is $(7, 3\pi/2)$.

For Probs. 9.50 to 9.64, transform the given equation into polar form.

9.50 $x = 3$

    ▮ $x = r \cos \theta$. Thus, $x = 3 = r \cos \theta$. Then $r = 3/\cos \theta = 3 \sec \theta$. (Note: Division by $\cos \theta$ is legal. Although $\cos \theta$ can be 0, $\cos \theta = 0$ occurs only when $\sec \theta$ is not defined.)

9.51 $x + 2y = 1$

    ▮ $x = r \cos \theta$ and $y = r \sin \theta$. Thus, $r \cos \theta + 2r \sin \theta = 1$, or $r(\cos \theta + 2 \sin \theta) = 1$.

9.52 $3x - 5y = 6$

    ▮ Then $3(r \cos \theta) - 5(r \sin \theta) = 6$, or $r(3 \cos \theta - 5 \sin \theta) = 6$.

9.53 $2y = \pi$

    ▮ If $2y = \pi$, then $2(r \sin \theta) = \pi$. Thus, $r = \pi/(2 \sin \theta) = (\pi/2) \csc \theta = (\pi \csc \theta)/2$. (See the note in Prob. 9.50.)

9.54 $x^2 + y^2 = 9$

    ▮ $(r \cos \theta)^2 + (r \sin \theta)^2 = 9$, $r^2 \cos^2 \theta + r^2 \sin^2 \theta = 9$, $r^2(\cos^2 \theta + \sin^2 \theta) = 9$, $r^2(1) = 9$, or $r^2 = 9$. (Note the advantage here of using polar coordinates!)

9.55 $x^2 - y^2 = 9$

    ▮ $r^2 \cos^2 \theta - r^2 \sin^2 \theta = 9$, $r^2(\cos^2 \theta - \sin^2 \theta) = 9$, $r^2 \cos 2\theta = 9$ (since $\cos 2\theta = \cos^2 \theta - \sin^2 \theta$), $r^2 = 9/(\cos 2\theta) = 9 \sec 2\theta$.

9.56 $x^2 + y^2 = 25$

    ▮ $r^2 \cos^2 \theta + r^2 \sin^2 \theta = 25$, $r^2(\cos^2 \theta + \sin^2 \theta) = 25$, $r^2(1) = 25$, or $r^2 = 25$. (Compare this to Prob. 9.54.)

9.57 $x^2 - 3y^2 = 9$

    ▮ $r^2 \cos^2 \theta - 3r^2 \sin^2 \theta = 9$, or $r^2(\cos^2 \theta - 3 \sin^2 \theta) = 9$. (Compare this to Prob. 9.55.)

9.58 $y = \sqrt{3}x$

    ▮ $r \sin \theta = \sqrt{3}r \cos \theta$, $\sin \theta = \sqrt{3} \cos \theta$, $\tan \theta = \sqrt{3}$, or $\theta = \operatorname{Tan}^{-1} \sqrt{3} = \pi/3$. [Note: Division by r is legal since, when $r = 0$ we have the point $(0, 0)$. But we haven't lost that point, since it occurs on the ray $\theta = \pi/3$.]

9.59 $xy = 12$

    ▮ $(r \cos \theta)(r \sin \theta) = 12$, $r^2 \sin \theta \cos \theta = 12$, but $\sin \theta \cos \theta = (\sin 2\theta)/2$, so $r^2((\sin 2\theta)/2) = 12$, and $r^2 \sin 2\theta = 24$.

9.60 $(x^2 + y^2)x = 4y^2$

    ▮

$$(r^2 \cos^2 \theta + r^2 \sin^2 \theta)r \cos \theta = 4r^2 \sin^2 \theta$$
$$r^3 \cos^3 \theta + r^3 \cos \theta \sin^2 \theta = 4r^2 \sin^2 \theta$$
$$r \cos^3 \theta + r \cos \theta \sin^2 \theta = 4 \sin^2 \theta$$
$$r \cos^3 \theta + r \cos \theta \,(1 - \cos^2 \theta) = 4 \sin^2 \theta$$
$$r \cos^3 \theta + r \cos \theta - r \cos^3 \theta = 4 \sin^2 \theta$$
$$r \cos \theta = 4 \sin^2 \theta$$

Thus, $r = 4(\sin^2 \theta)/\cos \theta = 4(\sin \theta/\cos \theta) \cdot \sin \theta = 4 \tan \theta \sin \theta$.

9.61 $x^2 = 4y$

▮ $r^2 \cos^2 \theta = 4r \sin \theta$, $r \cos^2 \theta = 4 \sin \theta$,   or $r = (4 \sin \theta)/\cos^2 \theta = 4[(\sin \theta)/\cos \theta] \cdot 1/\cos \theta = 4 \tan \theta \sec \theta$.

9.62 $y^2 = x^3$

▮ $r^2 \sin^2 \theta = r^3 \cos^3 \theta$, $\sin^2 \theta = r \cos^3 \theta$,   or $r = (\sin^2 \theta)/\cos^3 \theta = [(\sin^2 \theta)/\cos^2 \theta] \cdot 1/\cos \theta = \tan^2 \theta \sec \theta$.

9.63 $x^4 - y^4 = 2xy$

▮

$$r^4 \cos^4 \theta - r^4 \sin^4 \theta = 2r \cos \theta \, r \sin \theta$$
$$r^4(\cos^4 \theta - \sin^4 \theta) = 2r^2 \sin \theta \cos \theta$$
$$\underbrace{r^4(\cos^2 \theta + \sin^2 \theta)}_{1}\underbrace{(\cos^2 \theta - \sin^2 \theta)}_{\cos 2\theta} = r^2 \cdot \underbrace{2 \sin \theta \cos \theta}_{\sin 2\theta}$$

Divide by r^2: $r^2(1) \cos 2\theta = \sin 2\theta$, $r^2 = (\sin 2\theta)/\cos 2\theta$, $r^2 = \tan 2\theta$.

9.64 $(x^2 + y^2)^{3/2} = x^2 - y^2 - 2xy$

▮ $(r^2 \cos^2 \theta + r^2 \sin^2 \theta)^{3/2} = r^2 \cos^2 \theta - r^2 \sin^2 \theta - 2r \cos \theta \, r \sin \theta$
$[r^2(\cos^2 \theta + \sin^2 \theta)]^{3/2} = r^2(\cos^2 \theta - \sin^2 \theta) - r^2 \cdot 2 \sin \theta \cos \theta$.

Then $(r^2 \cdot 1)^{3/2} = r^2 \cos 2\theta - r^2 \sin 2\theta$, $r^3 = r^2 \cos 2\theta - r^2 \sin 2\theta$,   or $r = \cos 2\theta - \sin 2\theta$.

For Probs. 9.65 to 9.75, transform the given equation into rectangular form.

9.65 $r = \sin \theta$

▮   Multiply both sides by r: $r^2 = r \sin \theta$.   But $x^2 + y^2 = r^2$,   so $y = r \sin \theta$.   Thus, $x^2 + y^2 = y$,   or $x^2 + y^2 - y = 0$.

9.66 $r = 2 \sec \theta$

▮ $\sec \theta = 1/\cos \theta$.   Thus, if $r = 2 \sec \theta$, $r = 2/\cos \theta$, $r \cos \theta = 2$,   or $x = 2$.

9.67 $r = -5 \sec \theta$

▮ $r = -5 \sec \theta$ becomes $r = -5/\cos \theta$.   Thus, $r \cos \theta = -5$,   or $x = -5$.   (See Prob. 9.66.   What generalization can you make?)

9.68 $r = 4 \csc \theta$

▮ $\csc \theta = 1/\sin \theta$.   Thus, if $r = 4 \csc \theta$, $r = 4/\sin \theta$.   Thus, $r \sin \theta = 4$,   or $y = 4$.

9.69 $r = -\pi \csc \theta$

▮   Then $r = -\pi/\sin \theta$, $r \sin \theta = -\pi$,   or $y = -\pi$.   (See Prob. 9.68 and generalize: What would $r = r_1 \csc \theta$ be in a rectangular system with $r_1 \in \mathbb{R}$?)

9.70 $r^2 = 16$

▮ $r^2 = x^2 + y^2$.   Thus, if $r^2 = 16$, $x^2 + y^2 = 16$.

9.71 $r = 5$

▮ $r^2 = x^2 + y^2$.   Then $r = \sqrt{x^2 + y^2}$, $\sqrt{x^2 + y^2} = 5$,   or $x^2 + y^2 = 25$.   (Note: We did not pick up extraneous points by squaring.)

9.72 $r = 14$

▮   See Prob. 9.71 and generalize: $\sqrt{x^2 + y^2} = 14$,   or $x^2 + y^2 = 196$.

9.73 $r = \sin 2\theta$

▮ $r = 2 \sin \theta \cos \theta = 2 \cdot (y/r) \cdot (x/r) = 2yx/r^2$.   Then $2yx = r^3$, $= (\sqrt{x^2 + y^2})^3$, $= (x^2 + y^2)^{3/2}$,   or $4y^2x^2 = (x^2 + y^2)^3$.

9.74 $r = \dfrac{7}{3\cos\theta - 4\sin\theta}$

▮ $r = \dfrac{7}{3x/r - 4y/r} = \dfrac{7r}{3x - 4y}$. Then $1 = 7/(3x - 4y)$, or $3x - 4y = 7$. (Division by r is fine here. Why?)

9.75 $r^2\cos 2\theta = -16$

▮ $x^2 + y^2 = r^2$. $\cos 2\theta = \cos^2\theta - \sin^2\theta$. Thus, $r^2\cos 2\theta = -16$ becomes $(x^2 + y^2)(\cos^2\theta - \sin^2\theta) = -16$, $(x^2 + y^2)(x^2/r^2 - y^2/r^2) = -16$, $(x^2 + y^2)[(x^2 - y^2)/r^2] = -16$, $(x^2 + y^2)[(x^2 - y^2)/(x^2 + y^2)] = -16$, or $x^2 - y^2 = -16$.

For Probs. 9.76 to 9.78, find the eccentricity and directrix. Identify the conic.

9.76 $r = \dfrac{4}{1 + 2\cos\theta}$

▮ Recall that the equation of a conic (in polar form) with focus at $(0, 0)$ and eccentricity $= e$ is

$$r = \begin{cases} \dfrac{ep}{1 \pm e\cos\theta} & \text{if directrix is } x = \pm p, \quad p > 0 \\[2mm] \dfrac{ep}{1 \pm e\sin\theta} & \text{if directrix is } y = \pm p, \quad p > 0 \end{cases}$$

Here $r = \dfrac{4}{1 + 2\cos\theta}$. Thus, $e = 2$. Then if $ep = 4$ and $e = 2$, we have $2p = 4$, or $p = 2$. Thus, the eccentricity $= 2$. The directrix is $x = \pm p$ (see the general forms above) so that $x = 2$. Since $e = 2 > 1$, the conic is a hyperbola.

9.77 $r = \dfrac{3}{1 - \sin\theta}$

▮ Here, $r = \dfrac{ep}{1 - e\sin\theta}$. Thus, $e = 1$ and $ep = 3$, so $p = 3$. Then the directrix is $y = -3$, and the eccentricity $= 1$. The conic is a parabola.

9.78 $r = \dfrac{5}{2 - 2\cos\theta}$

▮ We must put this polar equation in the form $r = \dfrac{ep}{1 - e\cos\theta}$. Then $r = \dfrac{5}{2(1 - \cos\theta)} = \dfrac{\frac{5}{2}}{1 - \cos\theta}$ (here we multiplied the numerator and denominator by $\frac{1}{2}$). Then $e = 1$, and since $1 \cdot p = 5$, $p = 5$. Thus, eccentricity $= 1$ (parabola), and the directrix is $x = -5$.

For Probs. 9.79 to 9.81, write the polar equation of each given item.

9.79 A straight line bisecting the second and fourth quadrants.

▮ See Fig. 9.25. All points on l are such that $\theta = 3\pi/4$. Thus, the equation of l is $\theta = 3\pi/4$.

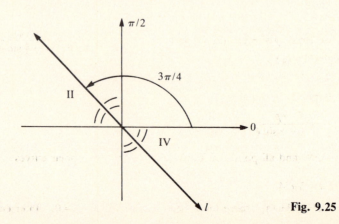

Fig. 9.25

9.80   A straight wire through the point $(4, 2\pi/3)$ and perpendicular to the polar axis (x axis, $x > 0$).

▮ See Fig. 9.26. We are looking for the equation of line l above. Since the line l is vertical, we only need to find x. See Fig. 9.26b. Then $x = 4\cos \pi/3 = 2$,   and the line is $x = -2$ or $r\cos \theta = -2$.

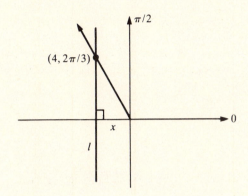

Fig. 9.26a

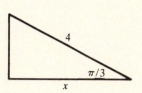

Fig. 9.26b

9.81   Circle with center at $(0,0)$ and   radius $= \sqrt{17}$.

▮ See Fig. 9.27. For every A on the circle, $r = \sqrt{17}$. Thus, the circle has equation $r = \sqrt{17}$.

For Probs. 9.82 to 9.85, write the polar equation of the conic with focus at the pole and with the given eccentricity and directrix.

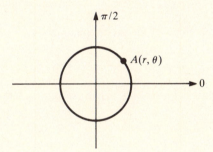

Fig. 9.27

9.82 $e = 1$, $x = 4$

▮ Since the directrix is $x = 4$, $r = \dfrac{ep}{1 + e\cos \theta} = \dfrac{4}{1 + \cos \theta}$.

9.83 $e = \frac{2}{3}$, $x = -6$

▮ With directrix $x = -6$, $r = \dfrac{ep}{1 - e\cos \theta} = \dfrac{\frac{2}{3}(6)}{1 - \frac{2}{3}\cos \theta} = \dfrac{4}{1 - \frac{2}{3}\cos \theta}$.

9.84 $e = \frac{2}{5}$, $y = -\frac{5}{6}$

▮ With directrix $y = -\frac{5}{6}$,   the equation is $r = \dfrac{ep}{1 - e\sin \theta} = \dfrac{\frac{2}{5}(\frac{5}{6})}{1 - \frac{2}{5}\sin \theta} = \dfrac{\frac{1}{3}}{1 - \frac{2}{5}\sin \theta} = \dfrac{5}{15 - 6\sin \theta}$ (here we multiplied by $\frac{15}{15}$).

9.85 $e = 1$, $y = -2$

▮ Then $r = \dfrac{ep}{1 - e\sin \theta} = \dfrac{1(2)}{1 - (1)\sin \theta} = \dfrac{2}{1 - \sin \theta}$.

For Probs. 9.86 to 9.89, find all points of intersection of the given polar curves.

9.86 $\theta = \pi/4$, $\theta = 3\pi/4$

▮ Since $\pi/4 \neq 3\pi/4$,   these lines intersect only when $r = 0$. Intersection point: $(0, \pi/4) = (0, 3\pi/4)$.

9.87 $r = 4$, $\theta = \pi/4$

▎ See Fig. 9.28. $r = 4$ is the circle with center at the pole and radius $= 4$. $\theta = \pi/4$ is the line making a $\pi/4$ angle with the polar axis.   Then, for (r, θ) at the left, $r = 4$, $\theta = \pi/4$.   At (r', θ'), $r' = -4$ and $\theta' = \pi/4$.   Intersection points: $(4, \pi 4)$ and $(-4, \pi/4)$.

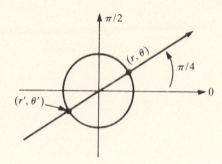

Fig. 9.28

9.88 $r = 3$, $r = \sin \theta$

▎ If $r = 3$ and $r = \sin \theta$,   then $3 = \sin \theta$;   but $0 \le \sin \theta \le 1$ $\forall \theta$,   so there are no intersection points.

9.89 $r = a \cos 3\theta$, $r = a$

▎ If $r = a \cos 3\theta$, $r = a$.   Then $a = a \cos 3\theta$, $1 = \cos 3\theta$,   or $3\theta = 0, 2\pi, 4\pi$.   If $3\theta = 0$, $\theta = 0$.   If $3\theta = 2\pi$, $\theta = 2\pi/3$.   If $3\theta = 4\pi$, $\theta = 4\pi/3$.   Intersection points: $(a, 0)$, $(a, 2\pi/3)$,   and $(a, 4\pi/3)$.

For Probs. 9.90 to 9.105, sketch the graph of the given equation.

9.90 $r = 3 \sin \theta$

▎ See Fig. 9.29.   We construct a table of values as follows: If $\theta = 0$, $r = 3 \sin 0 = 0$;   if $\theta = \pi/6$, $r = 3 \sin \pi/6 = \frac{3}{2}$,   etc.

θ	0	$\pi/6$	$\pi/2$	$5\pi/6$	π	$7\pi/6$	$3\pi/2$	$11\pi/6$
$r = 3 \sin \theta$	0	$\frac{3}{2}$	3	$\frac{3}{2}$	0	$-\frac{3}{2}$	-3	$-\frac{3}{2}$

Notice that, since $(\frac{3}{2}, \pi/6)$ and $(-\frac{3}{2}, 7\pi/6)$ are the same point, we get no new points for $\theta > \pi$ in the above table.

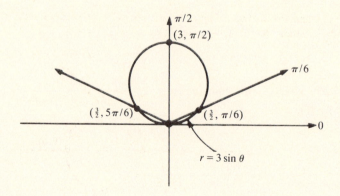

Fig. 9.29

9.91 $r = 7 \sin \theta$

▎ See Fig. 9.30 and Prob. 9.90.   For each θ in the table of values, we find $7 \sin \theta$ instead of $3 \sin \theta$.

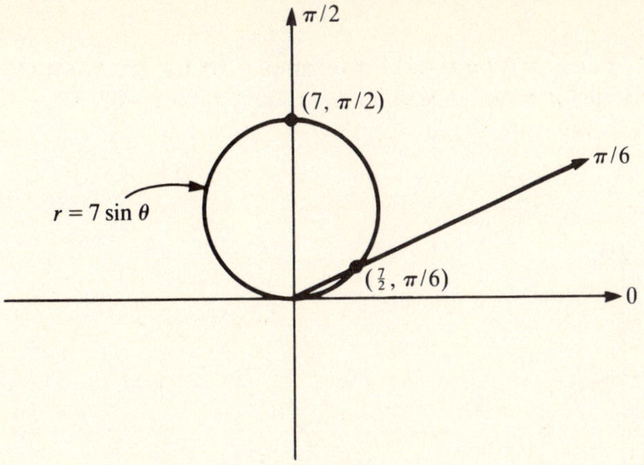

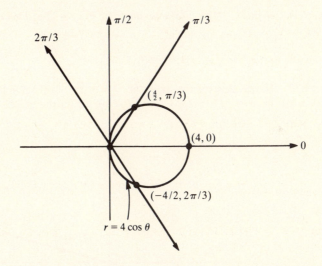

Fig. 9.30

9.92 $r = 4 \cos \theta$

▮ See Fig. 9.31. Construct a table of values like the one in Prob. 9.90.

θ	0	$\pi/3$	$\pi/2$	$2\pi/3$	π	$4\pi/3$	$3\pi/2$	$5\pi/3$	2π
$r = 4 \cos \theta$	4	$\frac{4}{2}$	0	$-\frac{4}{2}$	-4	$-\frac{4}{2}$	0	$\frac{4}{2}$	4

Observe that for $\theta > \pi$,   we got no new points. $(4, 0)$ and $(-4, \pi)$ are the same point.

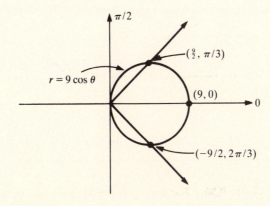

Fig. 9.31

9.93 $r = 9 \cos \theta$

▮ See Fig. 9.32 and Prob. 9.92. We construct the table of values in the exact same way, but when $\theta = 0$,   we get 9, when $\theta = \pi/3$,   we get $\frac{9}{2}$, etc.

Fig. 9.32

9.94 $r = 6$

▮ See Fig. 9.33. For any θ, $r = 6$. For example, $(6, 0)$, $(6, \pi/8)$, $(6, \pi)$, etc., are all on this graph. Clearly, this is a circle with center at the origin and radius = 6.

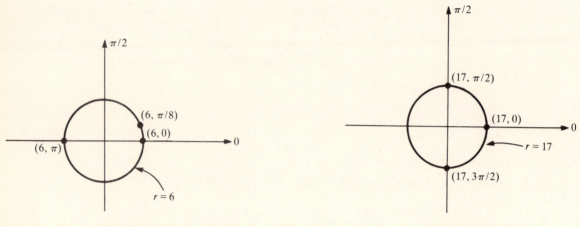

Fig. 9.33   Fig. 9.34

9.95 $r = 17$

▮ See Fig. 9.34 and Prob. 9.94. For any choice of θ, $r = 17$. This is the circle with center at the origin and radius = 17.

9.96 $\theta = \pi/6$

▮ See Fig. 9.35. Notice that for the graph of $\theta = \pi/6$, every point must satisfy $\theta = \pi/6$. For example, $(0, \pi/6)$, $(1, \pi/6)$, $(7, \pi/6)$, $(-11, \pi/6)$ are all on the graph.

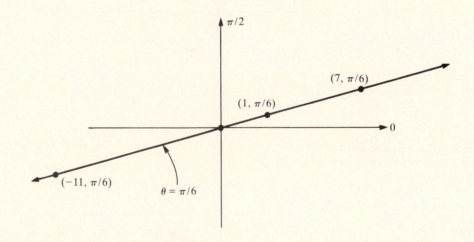

Fig. 9.35

9.97 $r = 4 + 4 \sin \theta$

▮ See Fig. 9.36.

θ	0	$\pi/6$	$\pi/2$	$5\pi/6$	π	$7\pi/6$	$3\pi/2$	$11\pi/6$	2π
r	4	6	8	6	4	2	0	2	4

Notice that we needed to use angles from 0 to 2π to get a complete sketch.

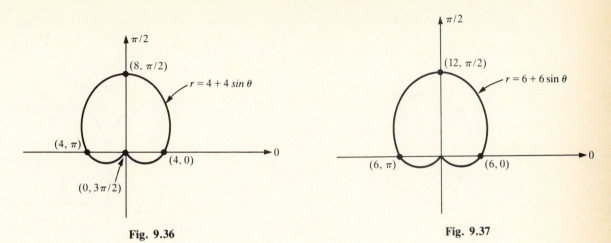

Fig. 9.36   Fig. 9.37

9.98 $r = 6 + 6 \sin \theta$

▮ See Fig. 9.37 and Prob. 9.97.   This will be the same, except we will have $(6, 0)$ instead of $(4, 0)$; $(12, \pi/2)$ instead of $(8, \pi/2)$, etc.

9.99 $r = 3 + 3 \cos \theta$

▮ See Fig. 9.38.

θ	0°	60°	90°	120°	180°	270°	300°	360°
r	6	$4\frac{1}{2}$	3	$1\frac{1}{2}$	0	3	$4\frac{1}{2}$	6

Notice that we did need to use the interval $[0, 360°]$ to sketch the entire curve here.

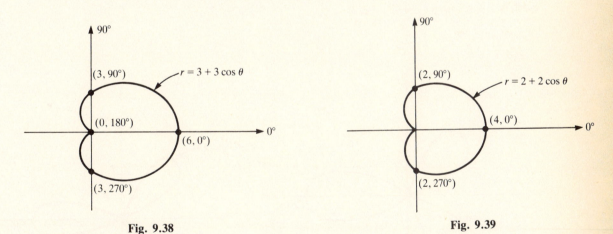

Fig. 9.38   Fig. 9.39

9.100 $r = 2 + 2 \cos \theta$

▮ See Fig. 9.39.   This will be the same as the curve in Prob. 9.99, except we will have $(4, 0°)$ instead of $(6, 0°)$; $(3, 60°)$ instead of $(4\frac{1}{2}, 60°)$, etc.

9.101 $r = \theta$

▮ See Fig. 9.40.

θ	0	$\pi/6$	$\pi/2$	π	2π	5π	$\cdots$
r	0	$\pi/6$	$\pi/2$	π	2π	5π	$\cdots$

Notice that, as θ increases, the value of r increases.   The curve here is spiral-shaped.

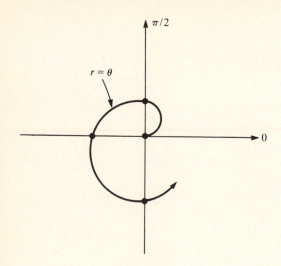

Fig. 9.40

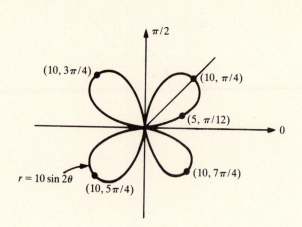

Fig. 9.41

9.102 $r = 10 \sin 2\theta$

▮ See Fig. 9.41.

θ	0	$\pi/12$	$\pi/4$	$\pi/3$	$\pi/2$
r	0	5	10	$5\sqrt{3}$	0

9.103 $r = 5 \cos 3\theta$

▮ See Fig. 9.42.

θ	0	$\pi/9$	$\pi/6$	$2\pi/3$	$4\pi/3$
r	5	$\frac{5}{2}$	0	5	5

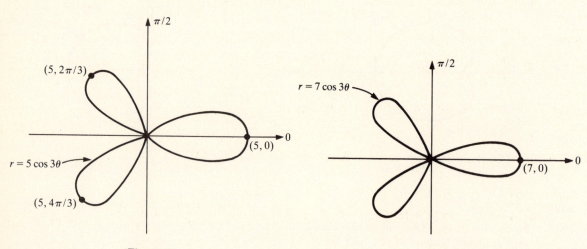

Fig. 9.42

Fig. 9.43

9.104 $r = 7 \cos 3\theta$

▮ See Fig. 9.43 and Prob. 9.103.   Do you see why these are almost the same?

9.105 $r = \sin 3\theta$

▮ See Fig. 9.44 and Prob. 9.102.

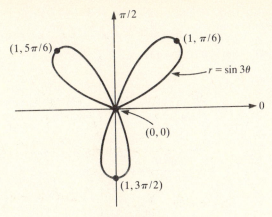

Fig. 9.44

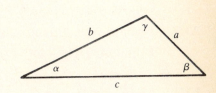

Fig. 9.45

9.3 THE LAW OF SINES

For Probs. 9.106 to 9.117, find the measure of the indicated angle or the length of the indicated side. Refer to Fig. 9.45 and use a calculator.

9.106 $\beta = 12°40'$, $\gamma = 100°$, $b = 13.1$; find a.

▮ $\alpha = 180° - 12°40' - 100° = 67°20'$. The law of sines is $\dfrac{\sin \alpha}{a} = \dfrac{\sin \beta}{b}$. Then $a = \dfrac{b \sin \alpha}{\sin \beta} = \dfrac{13.1 \sin 67°20'}{\sin 12°40'} = 55.1$.

9.107 Find c in Prob. 9.106.

▮ $\dfrac{\sin \gamma}{c} = \dfrac{\sin \beta}{b}$, so $c = \dfrac{b \sin \gamma}{\sin \beta} = \dfrac{13.1 \sin 100°}{\sin 12°40'} = 58.8$.

9.108 $\beta = 27°30'$, $\gamma = 54°30'$, $a = 9.27$; find b.

▮ $\alpha = 180° - 54°30' - 27°30' = 98°$. Then from $\dfrac{\sin \beta}{b} = \dfrac{\sin \alpha}{a}$, $b = \dfrac{a \sin \beta}{\sin \alpha} = \dfrac{9.27 \sin 27°30'}{\sin 98°} = 4.32$.

9.109 Find c in Prob. 9.108.

▮ $\dfrac{\sin \gamma}{c} = \dfrac{\sin \alpha}{a}$, so $c = \dfrac{a \sin \gamma}{\sin \alpha} = \dfrac{9.27 \sin 54°30'}{\sin 98°} = 7.62$.

9.110 $\alpha = 25°50'$, $a = 65$, $b = 105$, β obtuse; find β.

▮ $\dfrac{\sin \beta}{b} = \dfrac{\sin \alpha}{a}$, so $\sin \beta = \dfrac{b \sin \alpha}{a} = \dfrac{105 \sin 25°50'}{65} = 0.7039$. We choose $180 - \operatorname{Sin}^{-1} 0.7039$, not $\operatorname{Sin}^{-1} 0.7039$, since β is obtuse. Then $180 - \operatorname{Sin}^{-1} 0.7039 = 135°15'$ $(=135.25°)$.

9.111 Find c in Prob. 9.110.

▮ $\gamma = 180° - (25°50' + 135°15') = 18°55'$. Then from $\dfrac{\sin \alpha}{a} = \dfrac{\sin \gamma}{c}$, $c = \dfrac{a \sin \gamma}{\sin \alpha} = \dfrac{65 \sin 18°55'}{\sin 25°50'} = 48.36$.

9.112 $\beta = 31°40'$, $a = 12$, $b = 8$, α acute; find α.

▮ $\dfrac{\sin \alpha}{a} = \dfrac{\sin \beta}{b}$, so $\sin \alpha = \dfrac{a \sin \beta}{b} = \dfrac{12 \sin 31°40'}{8} = 0.7875$. We choose $\alpha = \operatorname{Sin}^{-1} 0.7875$, *not* $\alpha = 180 - \operatorname{Sin}^{-1} 0.7875$ since α is acute. Then $\alpha = \operatorname{Sin}^{-1} 0.7875 = 52°$.

9.113 Find c in Prob. 9.112.

▮ $\gamma = 180° - (52° + 31°40') = 96°20'$. Then from $\dfrac{\sin \beta}{b} = \dfrac{\sin \gamma}{c}$, $c = \dfrac{b \sin \gamma}{\sin \beta} = \dfrac{\gamma \sin 96°20'}{\sin 31°40'} = 15.1$.

9.114 $\alpha = 50$, $c = 40$, $\gamma = 30'$; find a.

    / α can be obtuse or acute here.   Then from $\dfrac{\sin \alpha}{a} = \dfrac{\sin \gamma}{c}$, $\sin \alpha = \dfrac{a \sin \gamma}{c} = \dfrac{50 \sin 30°}{40} = 0.6250$.
Thus,   (1) $\alpha = \text{Sin}^{-1} 0.6250 = 39°$,   or   (2) $\alpha = 180° - \text{Sin}^{-1} 0.6250 = 141°$.

9.115   Find b in Prob. 9.114 if $\cdot \alpha < 90°$.

    $\beta = 180° - 39° - 30° = 111°$.   Then from $\dfrac{\sin \beta}{b} = \dfrac{\sin \gamma}{c}$, $b = \dfrac{40 \sin 111°}{\sin 30°} = 75$.

9.116   Find b in Prob. 9.114 if $\alpha > 90°$.

    / $\beta = 180° - 141° - 30° = 9°$.   Then from $\dfrac{\sin \beta}{b} = \dfrac{\sin \gamma}{c}$, $b = \dfrac{40 \sin 9°}{\sin 30°} = 13$.

9.117 $a = 14$, $b = 23$, $\alpha = 41°$; find β.

    /   Then from $\dfrac{\sin \alpha}{a} = \dfrac{\sin \beta}{b}$, $\sin \beta = \dfrac{23 \sin 41°}{14} = 1.078$.   Thus $\beta = \text{Sin}^{-1} 1.078$;   there is no solution.
Draw the triangle; do you see why there is no solution?

9.4   THE LAW OF COSINES

For Probs. 9.118 to 9.128, find the indicated piece of information concerning the triangle in Fig. 9.45.   Use a calculator.

9.118 $\alpha = 50°40'$, $b = 7.03$, $c = 7.00$;   find a.

    /   From the law of cosines $a^2 = b^2 + c^2 - 2bc \cos \alpha$,   we have $a^2 = (7.03)^2 + (7)^2 - 2(7.03)(7) \cos 50°40' = 36.03925 \cdots$,   or $a = 6.00$.

9.119   Find β in Prob. 9.118.

    /   From the law of sines $\dfrac{a}{\sin \alpha} = \dfrac{b}{\sin \beta}$, $\sin \beta = \dfrac{7.03 \sin 50°40'}{6} = 0.9063$.   Then $\beta = \text{Sin}^{-1} 0.9063 = 65°0'$.
(Do *not* choose $180° - \text{Sin}^{-1} 0.9063$ since there cannot be two obtuse angles in a triangle.)

9.120 $\gamma = 120°20'$, $a = 5.73$, $b = 10.2$;   find c.

    / $c^2 = a^2 + b^2 - 2ab \cos \gamma = (5.73)^2 + (10.2)^2 - 2(5.73)(10.2) \cos (120°20') = 195.90686 \cdots$.   Thus, $c = 14.0$.

9.121   Find β in Prob. 9.120.

    /   From $\dfrac{\sin \beta}{b} = \dfrac{\sin \gamma}{c}$, $\sin \beta = \dfrac{b \sin \gamma}{\sin c}$.   Then $\beta = \text{Sin}^{-1} 0.6288 = 39°0'$.   (Do *not* use $180° - \text{Sin}^{-1} 0.6288$ since γ is obtuse.)

9.122   Find α in Prob. 9.120.

    /   We do not need the law of sines or cosines here; $\alpha = 180° - (\beta + \gamma) = 180° - 39°0' - 120°20' = 20°40'$.

9.123 $a = 4.00$, $b = 10.0$, $c = 9.00$;   find α.

    / $a^2 = b^2 + c^2 - 2bc \cos \alpha$.   Then $\cos \alpha = (10.0)^2 + (9.00)^2 - (4.00)^2 = 0.9167$.   Thus, $\alpha = \text{Cos}^{-1} 0.9167 = 23°30'$.   (Note that, given the three lengths we have, α and γ are both less than $90°$.)

9.124   Find γ in Prob. 9.123.

    /   From $\dfrac{\sin \gamma}{c} = \dfrac{\sin \alpha}{a}$, $\sin \gamma = \dfrac{(9.00) \sin 23°30'}{4.00} = 0.8972$.   Thus, $\gamma = \text{Sin}^{-1} 0.8972$ ($\gamma < 90°$) = $63°50'$.

9.125   Find γ in Prob. 9.123, using the law of cosines.

    / $\cos \gamma = \dfrac{(4.00)^2 + (10.0)^2 - (9.00)^2}{2(4)(10)} = 0.4375$.   Thus $\gamma = \text{Cos}^{-1} 0.4375 = 64°0'$.   (Do not be concerned by the small difference in results; rounding causes the discrepancy.)

9.126 $a = 10.5$, $b = 20.7$, $c = 12.2$;   find α.

▮ $a^2 = b^2 + c^2 - 2bc \cos \alpha$.   Then $\cos \alpha = \dfrac{b^2 + c^2 - a^2}{2bc} = \dfrac{(20.7)^2 + (12.2)^2 - (10.5)^2}{2(20.7)(12.2)} = 0.9248$.   Thus, $\alpha = \text{Cos}^{-1} (0.9248) \doteq 22°20'$.   (Here, again, α and γ are acute.   Draw a picture of the triangle whose sides are the lengths given in this problem.)

9.127   Find γ in Prob. 9.126.

▮   From $\dfrac{\sin \gamma}{c} = \dfrac{\sin \alpha}{a}$, $\gamma = \text{Sin}^{-1} \left[\dfrac{12.2 \sin (22°20')}{10.5} \right] = \text{Sin}^{-1} 0.4415 = 26°10' \ (\gamma < 90°!)$.

9.128   Find β in Prob. 9.126.

▮ $\beta = 180° - (\alpha + \gamma) = 180° - 22°20' - 26°10' = 131°30'$.

9.129   Prove that $c^2 = a^2 + b^2$,   where $\gamma = 90°$.

▮   From the law of cosines $c^2 = a^2 + b^2 - 2ab \cos \gamma$,   we have $c^2 = a^2 + b^2 - 2ab \cos 90°$ $(\gamma = 90°) = a^2 + b^2 - 2ab \cdot 0$.   Thus, $c^2 = a^2 + b^2$.   This is a proof of the pythagorean theorem.

9.130   Two adjacent sides of a parallelogram meet at an angle of $35°10'$ and have lengths of 3 ft and 8 ft.   What is the length of the shortest diagonal of the parallelogram (to three digits)?

▮   See Fig. 9.46.   We need to find d.   Since $d^2 = 3^2 + 8^2 - 2 \cdot 3 \cdot 8 \cos 35°10' = 33.76096\cdots$, $d = 5.81$ ft.

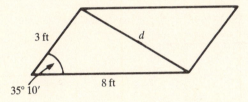

3 ft d

$35°\,10'$ 8 ft   **Fig. 9.46**

9.5   VECTORS

For Probs. 9.131 to 9.136, the magnitude and direction of a vector are given.   Represent them geometrically.

9.131   3, N 40°E

▮   See Fig. 9.47.   The "length" of the vector is 3 (magnitude), and we move 40° in the easterly direction from north (y axis).

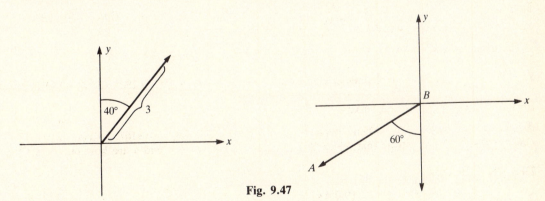

Fig. 9.47   **Fig. 9.48**

9.132   5, S 60° W

▮   See Fig. 9.48. $AB = 5$,   and we move 60° west from the "south axis."

9.133   2, S 45° E

▮   See Fig. 9.49. $AB = 2$.

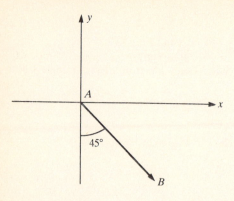

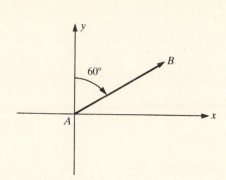

Fig. 9.49   Fig. 9.50

9.134 3, course of 60°

▮ See Fig. 9.50. $AB = 3$. "Course of 60°" means move 60° clockwise from north (y axis).

9.135 4, course of 160°.

▮ See Fig. 9.51. $AB = 4$, and there is a 160° angle between the resulting vector and the positive y axis.

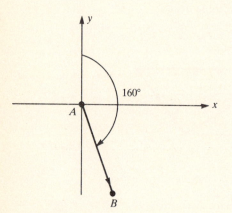

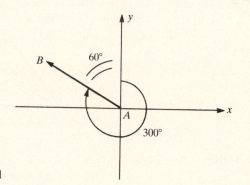

Fig. 9.51   Fig. 9.52

9.136 3, course of 300°

▮ See Fig. 9.52. $AB = 3$.

For Probs. 9.137 to 9.139, find the magnitude of the given vector.

9.137 $(4, 5)$

▮ The magnitude of $\mathbf{v} = (4, 5)$, written $|\mathbf{v}|$ or $|(4, 5)| = \sqrt{4^2 + 5^2} = \sqrt{16 + 25} = \sqrt{41}$.

9.138 $(6, -14)$

▮ $|(6, -14)| = \sqrt{6^2 + (-14)^2} = \sqrt{36 + 196} = \sqrt{232} = 2\sqrt{58}$.

9.139 $\mathbf{v}$ where $\mathbf{v} = (a, d)$

▮ $|\mathbf{v}| = \sqrt{a^2 + d^2}$. No further simplification is possible.

For Probs. 9.140 to 9.142, determine the direction of the given vector.

9.140 $(4, -3)$

▮ If $\mathbf{v} = (x, y)$, then θ is the direction of $\mathbf{v}$, where $\tan \theta = y/x$. Here, $\tan \theta = -\frac{3}{4}$ and $\theta = 323°8'$ (use a calculator).

9.141 $\mathbf{v} = (-2, -1)$

▮ See Prob. 9.140. $\tan \theta = -1/-2 = \frac{1}{2}$. Then $\theta = 206°34'$.

9.142 $(3, 0)$

$\blacksquare$ $\tan \theta = \frac{0}{3} = 0;$ $\theta = 0°$.

For Probs. 9.143 to 9.149, let $\mathbf{a} = (1, 0)$, $\mathbf{b} = (3, 0)$, $\mathbf{c} = (4, 6)$, $\mathbf{d} = (4, 9)$, $\mathbf{e} = (1, 6)$.   Find each of the following.

9.143 $2\mathbf{a}$

$\blacksquare$ Recall that $k(x, y) = (kx, ky)$.   Thus $2\mathbf{a} = 2(1, 0) = (2, 0)$.

9.144 $\frac{2}{3}\mathbf{b}$

$\blacksquare$ $\frac{2}{3}\mathbf{b} = \frac{2}{3}(3, 0) = (\frac{2}{3} \cdot 3, \; \frac{2}{3} \cdot 0) = (2, 0) = 2\mathbf{a}$ (see Prob. 9.143).

9.145 $\mathbf{a} + \mathbf{c}$

$\blacksquare$ Recall that $(x_1, y_1) + (x_2, y_2) = (x_1 + x_2, \; y_1 + y_2)$.   Thus, $\mathbf{a} + \mathbf{c} = (1, 0) + (4, 6) = (1 + 4, 0 + 6) = (5, 6)$.

9.146 $\mathbf{a} - \mathbf{d}$

$\blacksquare$ Recall that $\mathbf{a} - \mathbf{b} = \mathbf{a} + (-\mathbf{b})$.   Here, $\mathbf{d} = (4, 9)$,   so $-\mathbf{d} = -1\mathbf{d} = (-4, -9)$ and $\mathbf{a} - \mathbf{d} = \mathbf{a} + (-\mathbf{d}) = (1, 0) + (-4, -9) = (1 - 4, 0 - 9) = (-3, -9)$.

9.147 $\mathbf{e} - 2\mathbf{c}$

$\blacksquare$ $\mathbf{e} - 2\mathbf{c} = (1, 6) + (-2)(4, 6) = (1, 6) + (-8, -12) = (-7, -6)$.

9.148 $-6(\mathbf{b} + 2\mathbf{d})$

$\blacksquare$ $-6(\mathbf{b} + 2\mathbf{d}) = -6\mathbf{b} - 12\mathbf{d}$ (distributive law) $= -6(3, 0) - 12(4, 9) = (-18, 0) + (-48, -108) = (-66, -108)$.

9.149 $\mathbf{a} \cdot \mathbf{c}$

$\blacksquare$ If $\mathbf{x} = (x_1 y_1)$ and $\mathbf{y} = (x_2, y_2)$,   then $\mathbf{x} \cdot \mathbf{y} = x_1 x_2 + y_1 y_2 =$ the dot product.   Here, $\mathbf{a} \cdot \mathbf{c} = (1, 0) \cdot (4, 6) = 1 \cdot 4 + 0 \cdot 6 = 4 + 0 = 4$.

For Probs. 9.150 to 9.155, prove the given statement.

9.150 $\mathbf{u} \cdot \mathbf{u} = |\mathbf{u}|^2$

$\blacksquare$ Let $\mathbf{u} = (a, b)$.   Then $\mathbf{u} \cdot \mathbf{u} = (a, b) \cdot (a, b) = a \cdot a + b \cdot b = a^2 + b^2$ or $|\mathbf{u}|^2 = (\sqrt{a^2 + b^2})^2 = a^2 + b^2$, and $\mathbf{u} \cdot \mathbf{u} = |\mathbf{u}|^2$.

9.151 $\mathbf{u} \cdot \mathbf{v} = \mathbf{v} \cdot \mathbf{u}$

$\blacksquare$ Let $\mathbf{u} = (a, b)$, $\mathbf{v} = (c, d)$.   Then $\mathbf{u} \cdot \mathbf{v} = ac + bd = ca + db$ (multiplication is commutative) $= (c, d) \cdot (a, b) = \mathbf{v} \cdot \mathbf{u}$.   Thus, $\mathbf{u} \cdot \mathbf{v} = \mathbf{v} \cdot \mathbf{u}$.

9.152 $\mathbf{u} \cdot (\mathbf{v} + \mathbf{w}) = \mathbf{u} \cdot \mathbf{v} + \mathbf{u} \cdot \mathbf{w}$

$\blacksquare$ Let $\mathbf{u} = (a, b)$, $\mathbf{v} = (c, d)$, $\mathbf{w} = (e, f)$.   Then $\mathbf{u} \cdot (\mathbf{v} + \mathbf{w}) = (a, b) \cdot (c + e, d + f) = a(c + e) + b(d + f) = ac + ae + bd + bf = ac + bd + ae + bf = \mathbf{u} \cdot \mathbf{v} + \mathbf{u} \cdot \mathbf{w}$.

9.153 $(k\mathbf{u}) \cdot \mathbf{v} = k(\mathbf{u} \cdot \mathbf{v})$

$\blacksquare$ $k\mathbf{u} = (ka, kb)$ where $\mathbf{u} = (a, b)$.   Let $\mathbf{v} = (c, d)$.   Then $(k\mathbf{u}) \cdot \mathbf{v} = (ka, kb) \cdot (c, d) = (ka)c + (kb)d$. Since $\mathbf{u} \cdot \mathbf{v} = ac + bd$,   we have $k(\mathbf{u} \cdot \mathbf{v}) = k(ac + bd) = k(ac) + k(bd) = (ka)c + (kb)d$ (associative law for multiplication), and $(k\mathbf{u}) \cdot \mathbf{v} = k(\mathbf{u} \cdot \mathbf{v})$.

9.154   If $\mathbf{u} \perp \mathbf{v}$,   then $\mathbf{u} \cdot \mathbf{v} = 0$ $(\mathbf{u}, \mathbf{v} \neq \mathbf{0})$

$\blacksquare$ If $\mathbf{u} \perp \mathbf{v}$,   then $\mathbf{u} \cdot \mathbf{v} = |\mathbf{u}||\mathbf{v}| \cos \theta$ (θ = angle between $\mathbf{u}, \mathbf{v}$) $= |\mathbf{u}||\mathbf{v}| \cos \pi/2$ $(\mathbf{u} \perp \mathbf{v}) = |\mathbf{u}||\mathbf{v}| \cdot 0 = 0$.

9.155   If $\mathbf{u} \cdot \mathbf{v} = 0$,   then $\mathbf{u} \perp \mathbf{v}$ $(\mathbf{u}, \mathbf{v} \neq \mathbf{0})$

$\blacksquare$ Of $\mathbf{u} \cdot \mathbf{v} = 0$,   then $|\mathbf{u}||\mathbf{v}| \cos \theta = 0$, $\cos \theta = 0$ $(\mathbf{u}, \mathbf{v} \neq \mathbf{0})$, $\theta = \pi/2$,   and $\mathbf{u} \perp \mathbf{v}$.

For Probs. 9.156 to 9.159, prove the given statement.

9.156 $\mathbf{a} + \mathbf{c} = \mathbf{c} + \mathbf{a}$

▮ Let $\mathbf{a} = (x_1 \; y_1)$ and $\mathbf{c} = (x_2, y_2)$. Then $\mathbf{a} + \mathbf{c} = (x_1, y_1) + (x_2, y_2) = (x_1 + x_2, \; y_1 + y_2) = (x_2 + x_1, \; y_2 + y_1)$ (by commutative law for addition) $= \mathbf{c} + \mathbf{a}$.

9.157 $\mathbf{a} - \mathbf{c} = -(\mathbf{c} - \mathbf{a})$

▮ Let $\mathbf{a} = (x_1, y_1)$ and $\mathbf{c} = (x_2, y_2)$. Then $\mathbf{a} - \mathbf{c} = (x_1 - x_2, \; y_1 - y_2)$ and $-(\mathbf{c} - \mathbf{a}) = -1(x_2 - x_1, \; y_2 - y_1) = (x_1 - x_2, y_1 - y_2) = \mathbf{a} - \mathbf{c}$.

9.158 $\mathbf{u} + \mathbf{0} = \mathbf{u}$

▮ Let $\mathbf{u} = (x, y)$. Then $\mathbf{u} + \mathbf{0} = (x, y) + (0, 0) = (x + 0, \; y + 0) = (x, y) = \mathbf{u}$.

9.159 $\mathbf{u} + (-\mathbf{u}) = \mathbf{0}$

▮ Let $\mathbf{u} = (a, b)$. Then $\mathbf{u} + (-\mathbf{u}) = (a, b) + (-a, -b) = (a - a, b - b) = (0, 0) = \mathbf{0}$.

For Probs. 9.160 and 9.161, rewrite the given vector in terms of $\mathbf{i}$ and $\mathbf{j}$.

9.160 $(4, 7)$

▮ $\mathbf{i} = (1, 0)$ and $\mathbf{j} = (0, 1)$. Then $(4, 7) = 4(1, 0) + 7(0, 1) = 4\mathbf{i} + 7\mathbf{j}$.

9.161   5, 530° E

▮ See Fig. 9.53. The length of line $AB = |AB| = 5$. Then B has coordinates $(\frac{5}{2}, -\frac{5}{2}\sqrt{3})$, and so the vector is $(\frac{5}{2}, -\frac{5}{2}\sqrt{3}) = \frac{5}{2}(1, 0) - \frac{5}{2}\sqrt{3}(0, 1) = \frac{5}{2}\mathbf{i} - \frac{5}{2}\sqrt{3}\mathbf{j}$.

For Probs. 9.162 and 9.163, find the unit vector in the direction of the given vector.

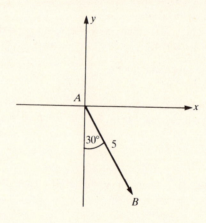

Fig. 9.53

9.162 $(6, 4)$

▮ $|(6, 4)| = \sqrt{36 + 16} = \sqrt{52}$. Then $|\mathbf{A}| = 1$ where $\mathbf{A} = (6/\sqrt{52}, 4/\sqrt{52})$ and the direction of $\mathbf{A} =$ the direction of $(6, 4)$.

9.163 (e, f)

▮ $|(e, f)| = \sqrt{e^2 + f^2}$. Let $\mathbf{A} = \left(\dfrac{e}{\sqrt{e^2 + f^2}}, \; \dfrac{f}{\sqrt{e^2 + f^2}} \right)$. Then $|\mathbf{A}| = 1$, and the direction of $\mathbf{A} =$ the direction of (e, f).

9.164   Find the angle between the vectors $(1, 0)$ and $(\sqrt{2}, \sqrt{2})$.

▮ $(1, 0) \cdot (\sqrt{2}, \sqrt{2}) = 1 \cdot \sqrt{2} + 0 \cdot \sqrt{2} = \sqrt{2}$. Also, $(1, 0) \cdot (\sqrt{2}, \sqrt{2}) = |(1, 0)||(\sqrt{2}, \sqrt{2})| \cos \theta$. Thus $\sqrt{2} = 1 \cdot 2 \cos \theta$, $\cos \theta = \sqrt{2}/2$, and $\theta = \pi/4$.

9.165   An automobile weighing 4000 lbs is standing on a smooth driveway that is inclined 5.0° with the horizontal. Find the force parallel to the driveway necessary to keep the car from rolling down the hill. Neglect all friction.

▮ See Fig. 9.54. We need the component of ray xy parallel to the driveway; i.e., we need $|TX|$. Since $|TX|/4000 = \sin 5°$, $|TX| = 4000 \sin 5° = 349$ lbs.

For Probs. 9.166 to 9.168, answer true or false, and justify your answer.

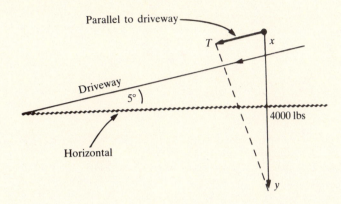

Fig. 9.54

9.166   If $\mathbf{u} = \mathbf{v}$,   then $|\mathbf{u}| = |\mathbf{v}|$.

▮ True. Two vectors are equal if and only if their magnitude and direction are the same. Thus, if $\mathbf{u} = \mathbf{v}$, $|\mathbf{u}|$ must equal $|\mathbf{v}|$.

9.167   Vector $(1, 1) =$ ray $\mathbf{AB}$ where the coordinate of $A = (1, 0)$ and the coordinate of $B = (2, 1)$.

▮ $|AB| = \sqrt{(2 - 1)^2 + (1 - 0)^2} = \sqrt{1 + 1} = \sqrt{2}$. Thus, $|AB| = |(1, 1)|$. Are they in the same direction? Since the slope of the line connecting the points $(0, 0)$ and $(1, 1) =$ the slope of line AB, the directions are the same, and the vector $(1, 1) = AB$. True.

9.168   If $\mathbf{a} \cdot \mathbf{b} = \mathbf{c} \cdot \mathbf{d}$,   then $\mathbf{a} = \mathbf{c}$ and $\mathbf{b} = \mathbf{d}$ or $\mathbf{a} = \mathbf{d}$ and $\mathbf{b} = \mathbf{c}$.

▮ False. Let $\mathbf{a} = (1, 0)$, $\mathbf{b} = (2, 0)$. Then $\mathbf{a} \cdot \mathbf{b} = 2 + 0 = 2$. Let $\mathbf{c} = (1, 1) = \mathbf{d}$. Then $\mathbf{c} \cdot \mathbf{d} = (1, 1) \cdot (1, 1) = 1 + 1 = 2$. But $\mathbf{a} \neq \mathbf{d}$ and $\mathbf{a} \neq \mathbf{c}$.

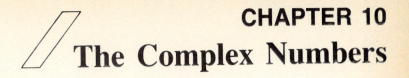

10.1 POLAR FORM

For Probs. 10.1 to 10.5, plot the given complex number.

10.1 $1 + i$

▮ Remember that, given $a + bi$ where $a, b \in \mathcal{R}$, $a + bi$ corresponds to the point (a, b) in the cartesian plane. Thus, $1 + i$ corresponds to $(1, 1)$. See Fig. 10.1.

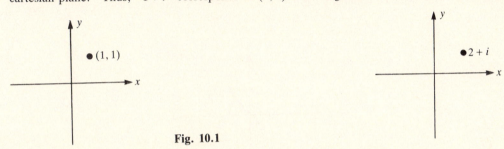

Fig. 10.1   Fig. 10.2

10.2 $2 + i$

▮ See Prob. 10.1. $2 + i$ corresponds to the point $(2, 1)$. See Fig. 10.2.

10.3 $3 - i$

▮ $3 - i$ corresponds to $(3, -1)$. See Fig. 10.3.

Fig. 10.3   Fig. 10.4

10.4 i

▮ Since $i = 0 + 1i$, i corresponds to $(0, 1)$. See Fig. 10.4.

10.5 7

▮ $7 = 7 + 0i$, and corresponds to the point $(7, 0)$. See Fig. 10.5.

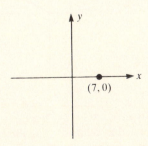

Fig. 10.5

For Probs. 10.6 to 10.26, give the polar (or trigonometric) form for the given complex number.

10.6 $-1-i$

∎

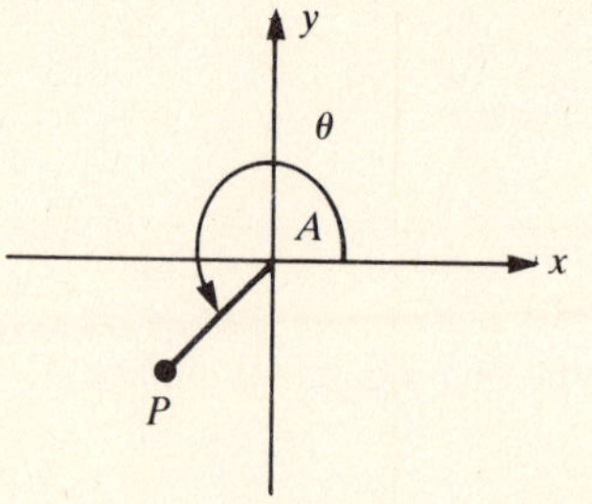

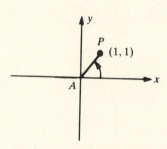

Fig. 10.6

Let $P = (-1, -1)$ which corresponds to $-1 - i$. $d(A, P) = r = \sqrt{1^2 + 1^2} = \sqrt{2}$. $\theta = $ angle from the positive x axis to the distance segment $= 225^0 = 5\pi/4$. $-1 - i = \sqrt{2}(\cos 5\pi/4 + i \sin 5\pi/4)$. See Fig. 10.6.

10.7 $1 + i$

∎

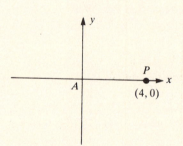

Fig. 10.7

$d(A, P) = \sqrt{2}$ and $\theta = \pi/4$, so $1 + i = \sqrt{2}(\cos \pi/4 + i \sin \pi/4)$. See Fig. 10.7.

10.8 $4 + 0i$

∎

Fig. 10.8

$d(A, P) = 4$ and $\theta = 0$ (θ is the angle between AP and the positive x axis), so $4 + 0i = 4(\cos 0 + i \sin 0)$. See Fig. 10.8.

10.9 $\sqrt{3} + i$

∎

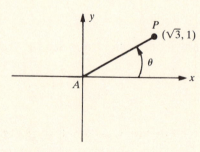

Fig. 10.9

$d(A, P) = \sqrt{(\sqrt{3})^2 + 1^2} = \sqrt{3 + 1} = 2$. To find θ: $\tan \theta = b/a$ for $a + bi$. In this case, $\tan \theta = 1/\sqrt{3} = \sqrt{3}/3$. Thus, $\theta = \text{Tan}^{-1} \sqrt{3}/3 = \pi/6$. $\sqrt{3} + i = 2(\cos \pi/6 + i \sin \pi/6)$. *Note*: The formulas we are using in these problems can be summarized as follows. For the number $a + bi$, with corresponding point (a, b), $r^2 = a^2 + b^2$, $\tan \theta = b/a$. $a = r \cos \theta$, $b = r \sin \theta$, and $a + bi = r(\cos \theta + i \sin \theta)$. See Fig. 10.9.

10.10 $2\sqrt{3} - 2i$

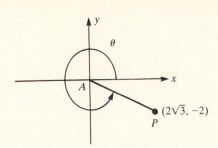

Fig. 10.10

$d(A, P) = \sqrt{(2\sqrt{3})^2 + 4} = \sqrt{12 + 4} = 4 = r.$ $\tan\theta = \dfrac{b}{a} = \dfrac{-2}{2\sqrt{3}} = \dfrac{-1}{\sqrt{3}} = \dfrac{-\sqrt{3}}{3}$, so $\theta = \dfrac{11\pi}{6}$. $2\sqrt{3} - 2i = 4(\cos 11\pi/6 + i\sin 11\pi/6)$. See Fig. 10.10.

10.11 $1 - i\sqrt{3}$

$r = \sqrt{a^2 + b^2} = \sqrt{1^2 + (-\sqrt{3})^2} = \sqrt{1 + 3} = 2.$ $\tan\theta = b/a = -\sqrt{3}/1 = -\sqrt{3}$, so $\theta = 5\pi/3$. $1 - i\sqrt{3} = 2(\cos 5\pi/3 + i\sin 5\pi/3)$. See Fig. 10.11.

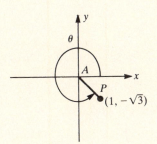

Fig. 10.11

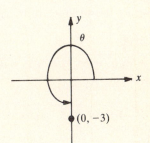

Fig. 10.12

10.12 $0 - 3i$

$r = \sqrt{a^2 + b^2} = \sqrt{0 + 9} = 3.$ $\tan\theta = b/a = -3/0$ which is undefined, so $\theta = 3\pi/2$. $0 - 3i = 3(\cos 3\pi/2 + i\sin 3\pi/2)$. See Fig. 10.12.

10.13 $-4 + 0i$

$r = \sqrt{16 + 0} = 4.$ $\tan\theta = 0/-4 = 0$, so $\theta = \pi$. $-4 + 0i = 4(\cos\pi + i\sin\pi)$. See Fig. 10.13.

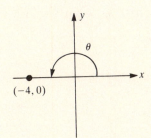

Fig. 10.13

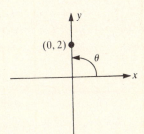

Fig. 10.14

10.14 $0 + 2i$

$r = \sqrt{0 + 4} = 2.$ $\tan\theta = 2/0$, which is undefined, so $\theta = \pi/2$. $0 + 2i = 2(\cos\pi/2 + i\sin\pi/2)$. See Fig. 10.14.

10.15 $\sqrt{2} + i\sqrt{2}$

$r = \sqrt{(\sqrt{2})^2 + (\sqrt{2})^2} = 2.$ $\tan\theta = \sqrt{2}/\sqrt{2} = 1$, so $\theta = \pi/4$. $\sqrt{2} + i\sqrt{2} = 2(\cos\pi/4 + i\sin\pi/4)$. See Fig. 10.15.

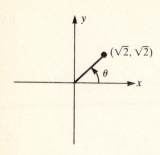

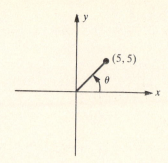

Fig. 10.15

Fig. 10.16

10.16 $5 + 5i$

▮ $r = \sqrt{5^2 + 5^2} = \sqrt{50} = 5\sqrt{2}$. $\tan \theta = 5/5 = 1$, so $\theta = \pi/4$. $5 + 5i = 5\sqrt{2}(\cos \pi/4 + i \sin \pi/4)$. See Fig. 10.16.

10.17 $7 + 7i$

▮ $r = \sqrt{7^2 + 7^2} = \sqrt{98} = 7\sqrt{2}$. $\tan \theta = 7/7 = 1$, so $\theta = \pi/4$. $7 + 7i = 7\sqrt{2}(\cos \pi/4 + i \sin \pi/4)$. Compare this to Prob. 10.16 above. Do you see a pattern developing?

10.18 $6 - 6i\sqrt{3}$

▮ $r = \sqrt{6^2 + (6\sqrt{3})^2} = \sqrt{36 + 108} = 12$. $\tan \theta = \dfrac{-6\sqrt{3}}{6} = -\sqrt{3}$, so $\theta = 5\pi/3$. $6 - 6i\sqrt{3} = 12(\cos 5\pi/3 + i \sin \pi/3)$. See Fig. 10.17.

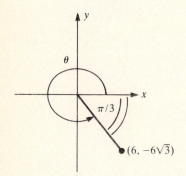

Fig. 10.17

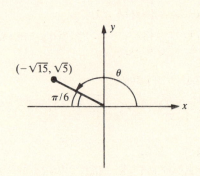

Fig. 10.18

10.19 $-\sqrt{15} + i\sqrt{5}$

▮ $r = \sqrt{(-\sqrt{15})^2 + (\sqrt{5})^2} = \sqrt{20} = 2\sqrt{5}$. $\tan \theta = \dfrac{\sqrt{5}}{-\sqrt{15}} = \dfrac{\sqrt{5}}{-\sqrt{5}\sqrt{3}} = \dfrac{-1}{\sqrt{3}}$, so $\theta = \dfrac{5\pi}{6}$. $-\sqrt{15} + i\sqrt{5} = 2\sqrt{5}(\cos 5\pi/6 + i \sin 5\pi/6)$. See Fig. 10.18.

10.20 $-\sqrt{7} - i\sqrt{21}$

▮ $r = \sqrt{(-7)^2 + (-\sqrt{21})^2} = \sqrt{49 + 21} = \sqrt{70}$. $\tan \theta = -\sqrt{21}/-\sqrt{7} = \sqrt{3}$, so $\theta = \dfrac{4\pi}{3}$. $-7 - i\sqrt{21} = \sqrt{70}\left(\cos \dfrac{4\pi}{3} + i \sin \dfrac{4\pi}{3}\right)$. See Fig. 10.19.

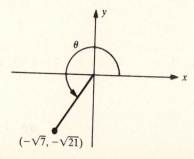

Fig. 10.19

10.21 $3 + 4i$

$\blacksquare$ $r = \sqrt{3^2 + 4^2} = 5$. $\tan \theta = \frac{4}{3}$,   so $\theta \approx 0.93$ (quadrant I, in radians). $3 + 4i \approx 5(\cos 0.93 + i \sin 0.93)$.

10.22 $2 + i$

$\blacksquare$ $r = \sqrt{2^2 + 1^2} = \sqrt{5}$. $\tan \theta = b/a = \frac{1}{2} = 0.5000$,   so $\theta \approx 0.46$ (quadrant I, in radians). $2 + i \approx$
$5(\cos 0.46 + i \sin 0.46)$.

10.23 $1 + \sqrt{3}i$

$\blacksquare$ $r = \sqrt{1 + 3} = 2$. $\tan \theta = \sqrt{3}/1 = \sqrt{3}$ (quadrant I),   so $\theta = \pi/3$. $1 + \sqrt{3}i = 2(\cos \pi/3 + i \sin \pi/3)$.

10.24 $-2\sqrt{3} - 2i$

$\blacksquare$ $r = \sqrt{12 + 4} = 4$. $\tan \theta = -2/-2\sqrt{3} = 1/\sqrt{3}$ in quadrant III, so $\theta = 7\pi/6$. $-2\sqrt{3} - 2i =$
$4(\cos 7\pi/6 + i \sin 7\pi/6)$.

10.25 $\sqrt{2} - i\sqrt{2}$

$\blacksquare$ $r = \sqrt{2 + 2} = 2$. $\tan \theta = -\sqrt{2}/\sqrt{2} = -1$ in quadrant IV, so $\theta = 7\pi/4$. $\sqrt{2} - i\sqrt{2} =$
$2(\cos 7\pi/4 + i \sin 7\pi/4)$.

10.26 -8

$\blacksquare$ $-8 = -8 + 0i$. $r = \sqrt{64} = 8$. $\tan \theta = 0/-8 = 0$ in quadrant II, so $\theta = \pi$. $-8 = 8(\cos \pi + i \sin \pi)$.

For Probs. 10.27 to 10.35, change the given complex number to polar form.

10.27 $\sqrt{2}(\cos 5\pi/4 + i \sin 5\pi/4)$

$\blacksquare$ $\sqrt{2} \cos 5\pi/4 = \sqrt{2} \cdot (-\sqrt{2}/2) = -1$,   and $\sqrt{2}(i \sin 5\pi/4) = \sqrt{2} \cdot i \cdot -\sqrt{2}/2 = -i$.   So the answer is
$-1 - i$.   See Prob. 10.6.

10.28 $\sqrt{2}(\cos \pi/4 + i \sin \pi/4)$

$\blacksquare$ $\sqrt{2} \cos \pi/4 = \sqrt{2} \cdot (\sqrt{2}/2) = 1$,   and $i\sqrt{2} \sin \pi/4 = i \cdot \sqrt{2} \cdot (\sqrt{2}/2) = i$.   So the answer is $1 + i$.   See
Prob. 10.7.

10.29 $4(\cos 0 + i \sin 0)$

$\blacksquare$ $4 \cos 0 = 4 \cdot 1 = 4$,   and $i \cdot 4 \sin 0 = i \cdot 0 = 0$.   So the answer is $4 + 0i = 4$.   See Prob. 10.8.

10.30 $2(\cos \pi/6 + i \sin \pi/6)$

$\blacksquare$ $2 \cos \pi/6 = 2 \cdot (\sqrt{3}/2) = \sqrt{3}$,   and $i \cdot 2 \sin \pi/6 = i \cdot 2 \cdot \frac{1}{2} = i$.   So the answer is $3 + i$.   See Prob. 10.9.

10.31 $4(\cos 11\pi/6 + i \sin 11\pi/6)$

$\blacksquare$ $4 \cos 11\pi/6 = 4 \cdot (\sqrt{3}/2) = 2\sqrt{3}$,   and $i \cdot 4 \sin 11\pi/6 = -2i$.   So the answer is $2\sqrt{3} - 2i$.   See Prob.
10.10.

10.32 $2(\cos 5\pi/3 + i \sin 5\pi/3)$

$\blacksquare$ $2 \cos 5\pi/3 = 1$,   and $i \cdot 2 \sin 5\pi/3 = i \cdot -2(\sqrt{3}/2) = -i\sqrt{3}$.   So the answer is $1 - i\sqrt{3}$.   See Prob.
10.11.

10.33 $3(\cos 3\pi/2 + i \sin 3\pi/2)$

$\blacksquare$ $\cos 3\pi/2 = 0$,   and $\sin 3\pi/2 = -1$.   Then, $3 \cos 3\pi/2 = 0$ and $i \cdot 3 \sin 3\pi/2 = -3i$.   So the answer is
$0 - 3i$.   See Prob. 10.12.

10.34 $4(\cos \pi + i \sin \pi)$

$\blacksquare$ $\cos \pi = -1$,   and $\sin \pi = 0$.   Then, $4 \cos \pi = -4$,   and $i \cdot 4 \sin \pi = 0i$.   So the answer is $-4 + 0i$.
See Prob. 10.13.

10.35 $2(\cos \pi/4 + i \sin \pi/4)$

▮ $\cos \pi/4 = \sin \pi/4 = \sqrt{2}/2$. $2 \cos \pi/4 = 2 \cdot \sqrt{2}/2 = \sqrt{2}$,   and $i \cdot 2 \sin \pi/4 = i \cdot 2\sqrt{2}/2 = \sqrt{2}i$.   So the answer is $\sqrt{2} + i\sqrt{2}$.   See Prob. 10.15.

For Probs. 10.36 to 10.42, perform the indicated operations.

10.36 $\sqrt{2}(\cos \pi/4 + i \sin \pi/4) \cdot 3(\cos \pi/2 + i \sin \pi/2)$

▮ Recall that the product is found by multiplying moduli and adding corresponding arguments.   In this case, $\sqrt{2} \cdot 3 = 3\sqrt{2}$,   and $\pi/4 + \pi/2 = 3\pi/4$.   So the answer is $3\sqrt{2}(\cos 3\pi/4 + i \sin 3\pi/4)$.

10.37 $(\cos \pi/2 + i \sin \pi/2) \cdot 2(\cos \pi/3 + i \sin \pi/3)$.

▮ $2 \cdot 1 = 2$,   and $\pi/2 + \pi/3 = 5\pi/6$.   So the answer is $2(\cos 5\pi/6 + i \sin 5\pi/6)$.

10.38 $\sqrt{3}(\cos \pi/6 + i \sin \pi/6) \cdot \sqrt{3}(\cos \pi/2 + i \sin \pi/2)$

▮ $3(\cos 4\pi/6 + i \sin 4\pi/6)$.

10.39 $4(\cos \pi/8 + i \sin \pi/8) \cdot 2(\cos \pi/7 + i \sin \pi/7)$

▮ $8(\cos 15\pi/56 + i \sin 15\pi/56)$.

10.40 $2(\cos 0 + i \sin 0) \cdot 4(\cos \pi + i \sin \pi)$

▮ $8(\cos \pi + i \sin \pi)$.

10.41 $\sqrt{3}(\cos \pi + i \sin \pi) \div 2(\cos \pi/2 + i \sin \pi/2)$

▮ We divide moduli and subtract corresponding arguments. $\sqrt{3} \div 2 = \sqrt{3}/2$,   and $\pi - \pi/2 = \pi/2$.   So the answer is $(\sqrt{3}/2)(\cos \pi/2 + i \sin \pi/2)$.

10.42 $2(\cos \pi/3 + i \sin \pi/3) \div 2(\cos \pi/6 + i \sin \pi/6)$

▮ $2/2 = 1$ and $\pi/3 - \pi/6 = \pi/6$.   So the answer is $\cos \pi/6 + i \sin \pi/6$.

For Probs. 10.43 to 10.47, perform the operation indicated, and write the result in $a + bi$ form.

10.43 $2(\cos \pi/6 + i \sin \pi/6) \cdot 3(\cos \pi/3 + i \sin \pi/3)$

▮ The product is $6(\cos 4\pi/6 + i \sin 4\pi/6) = 6(-\frac{1}{2}) + 6(i \cdot \sqrt{3}/2) = -3 + 3i\sqrt{3}$.

10.44 $3(\cos 25° + i \sin 25°) \cdot 8(\cos 200° + i \sin 200°)$

▮ The product is $24(\cos 225° + i \sin 225°) = -12\sqrt{2} - 12\sqrt{2}i$.

10.45 $4(\cos 50° + i \sin 50°) \cdot 2(\cos 100° + i \sin 100°)$

▮ The product is $8(\cos 150° + i \sin 150°) = -4\sqrt{3} + 4i$.

10.46 $\dfrac{4(\cos 190° + i \sin 190°)}{2(\cos 70° + i \sin 70°)}$

▮ The quotient is $2(\cos 120° + i \sin 120°) = -1 + i\sqrt{3}$.

10.47 $\dfrac{12(\cos 200° + i \sin 200°)}{3(\cos 350° + i \sin 350°)}$

▮ The quotient is $4[\cos (-150°) + i \sin (-150°)] = -2\sqrt{3} - 2i$.

For Probs. 10.48 to 10.51, use polar form to perform the given calculation.

10.48 $(1 + i)(\sqrt{2} - i\sqrt{2})$

▮ $1 + i = \sqrt{2}(\cos \pi/4 + i \sin \pi/4)$,   and $\sqrt{2} - i\sqrt{2} = 2(\cos 7\pi/4 + i \sin 7\pi/4)$.   The product is $2\sqrt{2}(\cos 2\pi + i \sin 2\pi) = 2\sqrt{2} \cdot 1 = 2\sqrt{2}$.

10.49 $\dfrac{1+i}{1+i}$

$\blacksquare$ $1+i=\sqrt{2}(\cos \pi/4 + i \sin \pi/4)$. Then, $\dfrac{1+i}{1+i}=(\sqrt{2}/\sqrt{2})(\cos 0 + i \sin 0)=(\sqrt{2}/\sqrt{2})\cdot 1 = 1$ (but you knew that prior to doing the calculation!)

10.50 $(1+i)^2$

$\blacksquare$ $(1+i)^2=(1+i)(1+i)=\sqrt{2}(\cos \pi/4 + i \sin \pi/4)\sqrt{2}(\cos \pi/4 + i \sin \pi/4)=2(\cos \pi/2 + i \sin \pi/2)$
$=2(0+i)=2i$.

10.51 $(1+i)^3$

$\blacksquare$ $(1+i)^3=(1+i)^2(1+i)=2(\cos \pi/2 + i \sin \pi/2)\ \sqrt{2}(\cos \pi/4 + i \sin \pi/4)$ (see Prob. 10.50) $=$
$2\sqrt{2}(\cos 3\pi/4 + i \sin 3\pi/4)=2\sqrt{2}\cos 3\pi/4 + i\cdot 2\sqrt{2}\sin 3\pi/4=2\sqrt{2}\cdot(-\sqrt{2}/2)+i\cdot 2\sqrt{2}\cdot(\sqrt{2}/2)=-2+2i$.

10.2 ROOTS AND DE MOIVRE'S THEOREM

10.52 State De Moivre's theorem.

$\blacksquare$ Let $n\in \mathcal{Z}$. Then $[r(\cos \theta + i \sin \theta)]^n = r^n(\cos n\theta + i \sin n\theta)$.

For Probs. 10.53 to 10.56, find the indicated powers.

10.53 $[2(\cos 35° + i \sin 35°)]^2$

$\blacksquare$ Since $2^2 = 4$, and $2\cdot 35° = 70°$, the answer is $4(\cos 70° + i \sin 70°)$.

10.54 $[3(\cos 10° + i \sin 10°)]^3$

$\blacksquare$ Since $3^3 = 27$, and $10°\cdot 3 = 30°$, the answer is $27(\cos 30° + i \sin 30°)$.

10.55 $(\cos 15° + i \sin 15°)^6$

$\blacksquare$ Since $1^6 = 1$ and $15°\cdot 6 = 90°$, the answer is $\cos 90° + i \sin 90°$. (Do you notice that this is just 1? That is a surprising result of the original problem!)

10.56 $[4(\cos 310° + i \sin 310°)]^3$

$\blacksquare$ $64(\cos 930° + i \sin 930°)$.

For Probs. 10.57 to 10.63, use De Moivre's theorem to rewrite each of the following in standard form.

10.57 $(\cos 40° + i \sin 40°)^3$

$\blacksquare$ $(\cos 40° + i \sin 40°)^3 = \cos 120° + i \sin 120° = \dfrac{-1}{2}+\dfrac{\sqrt{3}}{2}\,i$.

10.58 $(\cos \pi/3 + i \sin \pi/3)^5$

$\blacksquare$ $(\cos \pi/3 + i \sin \pi/3)^5 = \cos 5\pi/3 + i \sin 5\pi/3 = \dfrac{1}{2}-\dfrac{\sqrt{3}}{2}\,i$.

10.59 $8(1-i)^6$

$\blacksquare$ $1-i=\sqrt{2}(\cos 7\pi/4 + i \sin 7\pi/4)$. Then, $8(1-i)^6 = 8[\sqrt{2}(\cos 7\pi/4 + i \sin 7\pi/4)]^6$
$= 8\cdot(\sqrt{2})^6(\cos 42\pi/4 + i \sin 42\pi/4)=64(0+i)=64i$. See Fig. 10.20.

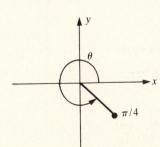

Fig. 10.20

10.60 $(\sqrt{3} + i)^{12}$

▮

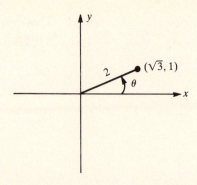

Fig. 10.21

$\tan \theta = b/a = 1/\sqrt{3}$. $\sqrt{3} + i = 2(\cos \pi/6 + i \sin \pi/6)$. Then $(\sqrt{3} + i)^{12} = (2)^{12}(\cos 2\pi + i \sin 2\pi) = 4096 + 0i$. See Fig. 10.21.

10.61 $(1 + i)^{-6}$

▮ $1 + i = \sqrt{2}(\cos \pi/4 + i \sin \pi/4)$. Then $(1 + i)^{-6} = (\sqrt{2})^{-6}(\cos -6\pi/4 + i \sin -6\pi/4) = \dfrac{1}{(\sqrt{2})^6}(0 + i) = \dfrac{1}{8} i$.

10.62 $(1 + i)^{16}(1 - i)^{10}$

▮ $(1 + i)^{16} = [\sqrt{2}(\cos \pi/4 + i \sin \pi/4)]^{16} = (\sqrt{2})^{16}(\cos 4\pi + i \sin 4\pi) = 256(1 + 0) = 256$, and $(1 - i)^{10} = (\sqrt{2})^{10}(\cos 70\pi/4 + i \sin 70\pi/4) = 32(0 - i) = -32i$. Therefore, the answer is $(32i)(256) = -8192i$.

10.63 $\dfrac{(1 - i)^6}{(1 + i)^{10}}$

▮ $(1 - i)^6 = 8i$. (See Prob. 10.59.) $(1 + i)^{10} = (\sqrt{2})^{10}(\cos 10\pi/4 + i \sin 10\pi/4) = 32(0 + i) = 32i$. Therefore, the answer is $\dfrac{8i}{32i} = \dfrac{1}{4}$.

For Probs. 10.64 to 10.68, find the indicated roots.

10.64 The square roots of i.

▮ See Fig. 10.22. Recall that if $z(\neq 0) = r(\cos \theta + i \sin \theta)$, then z has n, n^{th} roots given by the formula

$$r^{1/n}\left(\cos \frac{\theta + 2\pi k}{n} + i \sin \frac{\theta + 2\pi k}{n}\right) \quad \text{for} \quad k = 0, 1, \ldots, n - 1$$

Since $i = 0 + i = 1(\cos \pi/2 + i \sin \pi/2)$, the square roots of i are $1^{1/2}\left(\cos \dfrac{\theta + 2\pi k}{n} + i \sin \dfrac{\theta + 2\pi k}{n}\right)$ for $k = 0, 1$ where $\theta = \pi/2$, $n = 2$. The two roots are

$$w_1 = \cos \frac{\pi/2}{2} + i \sin \frac{\pi/2}{2} = \cos \frac{\pi}{4} + i \sin \frac{\pi}{4} = \frac{\sqrt{2}}{2} + i \frac{\sqrt{2}}{2}$$

and

$$w_2 = \cos\left(\frac{\pi/2 + 2\pi}{2}\right) + i \sin\left(\frac{\pi/2 + 2\pi}{2}\right) = \frac{-\sqrt{2}}{2} - i \frac{\sqrt{2}}{2}$$

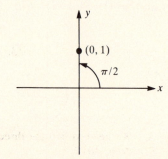

Fig. 10.22

10.65 The square roots of $1 + i\sqrt{3}$.

$\blacksquare$ See Fig. 10.23. $1 + i\sqrt{3} = 2(\cos \pi/3 + i \sin \pi/3)$. Then $w_1 = 2^{1/2}\left(\cos \dfrac{\pi/3}{2} + i \sin \dfrac{\pi/3}{2}\right) =$
$\sqrt{6}/2 + i\sqrt{2}/2$ and $w_2 = 2^{1/2}\left(\cos \dfrac{2\pi + \pi/3}{2} + i \sin \dfrac{2\pi + \pi/3}{2}\right) = -\sqrt{6}/2 - i\sqrt{2}/2$.

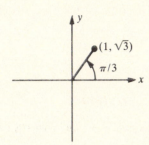
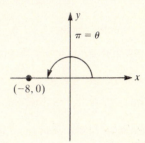

Fig. 10.23   **Fig. 10.24**

10.66 The cube roots of -8.

$\blacksquare$ See Fig. 10.24. $-8 = -8 + 0i = 8(\cos \pi + i \sin \pi)$. The cube roots are: $w_1 = 8^{1/3}\left(\cos \dfrac{\pi}{3} + i \sin \dfrac{\pi}{3}\right) =$
$1 + \sqrt{3}i$, $w_2 = 8^{1/3}\left(\cos \dfrac{2\pi + \pi}{3} + i \sin \dfrac{2\pi + \pi}{3}\right) = -2$, and $w_3 = 8^{1/3}\left(\cos \dfrac{4\pi + \pi}{3} + i \sin \dfrac{4\pi + \pi}{3}\right) =$
$1 - \sqrt{3}i$.

10.67 The cube roots of -27.

$\blacksquare$ $-27 = -27 + 0i = 27(\cos \pi + i \sin \pi)$. Then, the cube roots are $w_1 = 27^{1/3}\left(\cos \dfrac{\pi}{3} + i \sin \dfrac{\pi}{3}\right) = \dfrac{3}{2} +$
$\dfrac{3\sqrt{3}}{2} i$, $w_2 = 27^{1/3}\left(\cos \dfrac{2\pi + \pi}{3} + i \sin \dfrac{2\pi + \pi}{3}\right) = -3$, and $w_3 = 27^{1/3}\left(\cos \dfrac{4\pi + \pi}{3} + i \sin \dfrac{4\pi + \pi}{3}\right) =$
$\dfrac{3}{2} - \dfrac{3\sqrt{3}}{2} i$.

10.68 The fifth roots of $1 + i$.

$\blacksquare$ $1 + i = \sqrt{2}(\cos 45° + i \sin 45°)$. Then the fifth roots of $1 + i$ are $(\sqrt{2})^{1/5}\left(\cos \dfrac{45° + k \cdot 360°}{5} +\right.$
$\left. i \sin \dfrac{45° + k \cdot 360°}{5}\right)$ where $k = 0, 1, 2, 3, 4$. Then $w_1 = 2^{1/10}(\cos 9° + i \sin 9°)$, and $w_2 = 2^{1/10}(\cos 81° + i \sin 81°)$, etc.

For Probs. 10.69 to 10.73, solve the given equation.

10.69 $x^3 + 1 = 0$

$\blacksquare$ $x^3 + 1 = 0$; then $x^3 = -1$ and x can be any one of the three cube roots of -1. $-1 = -1 + 0i =$
$1(\cos \pi + i \sin \pi)$. Then $w_1 = -1$, $w_2 = \dfrac{1 + \sqrt{3}i}{2}$, and $w_3 = \dfrac{1 - \sqrt{3}i}{2}$.

10.70 $x^2 - i = 0$

$\blacksquare$ If $x^2 - i = 0$, then $x^2 = i$. The two square roots of i are $w_1 = \dfrac{\sqrt{2}}{2} + \dfrac{i\sqrt{2}}{2}$, and $w_2 = \dfrac{-\sqrt{2}}{2} - \dfrac{i\sqrt{2}}{2}$. (See Prob. 10.63.)

10.71 $x^2 - (1 + i\sqrt{3}) = 0$

$\blacksquare$ If $x^2 - (1 + i\sqrt{3}) = 0$, $x =$ the square roots of $1 + i\sqrt{3}$. Then, $w_1 = \dfrac{\sqrt{6}}{2} + \dfrac{i\sqrt{2}}{2}$, and $w_2 = \dfrac{-\sqrt{6}}{2} - \dfrac{i\sqrt{2}}{2}$. (See Prob. 10.65.)

10.72 $x^3 + 8 = 0$

$\blacksquare$ If $x^3 + 8 = 0$, then $x^3 = -8$. We need to find the three cube roots of -8. Then $w_1 = 1 + \sqrt{3}i$, $w_2 = -2$, and $w_3 = 1 - \sqrt{3}i$. See Prob. 10.66.

10.73 $x^5 - (1 + i) = 0$

▮ See Prob. 10.68 for the fifth roots of $1 + i$. If $x^5 - (1 + i) = 0$, then $x^5 = 1 + i$ and x is the 5 fifth roots of $1 + i$.

For Probs. 10.74 to 10.77, tell whether the given statement is true or false and why.

10.74   Every complex number $a + bi$ has n distinct n^{th} roots.

▮ True; this is the major theorem concerning roots of complex numbers.

10.75   Only one of the fifth roots of 32 is real.

▮ True; $2^5 = 32$, and no other real has that property.

10.76   The modulus of the product of two complex numbers is the sum of their moduli.

▮ False; If $z_1 = r_1(\cos \theta + i \sin \theta)$ and $z_2 = r_2(\cos \theta_2 + i \sin \theta_2)$, then modulus $(z_1, z_2) = r_1 \cdot r_2$.

10.77   De Moivre's theorem is true for $n = -7$.

▮ True; it is true for all integers, not just the natural or whole numbers.

10.78   Prove: $(\cos \theta + i \sin \theta)^{-n} = \cos n\theta - i \sin n\theta$.

▮ $(\cos \theta + i \sin \theta)^{-n} = \cos(-n\theta) + i \sin(-n\theta)$, but $\cos(-n\theta) = \cos n\theta$ and $\sin(-n\theta) = -\sin n\theta$. Thus, $(\cos \theta + i \sin \theta)^{-n} = \cos n\theta - i \sin n\theta$.

10.79   Prove: De Moivre's theorem when $n = 2$.

▮ Let $Z = r(\cos \theta + i \sin \theta)$. Then

$$r(\cos \theta + i \sin \theta)^2 = r(\cos \theta + i \sin \theta) \cdot r(\cos \theta + i \sin \theta)$$
$$= r^2(\cos \theta + i \sin \theta)(\cos \theta = i \sin \theta)$$
$$= r^2 \cos^2 \theta + i^2 \sin^2 \theta + i(\sin \theta \cos \theta + \sin \theta \cos \theta)$$
$$= r^2 \cos^2 \theta - \sin^2 \theta + i(\sin 2\theta)$$
$$= r^2(\cos 2\theta + i \sin 2\theta)$$

10.80   Find the reciprocal of i using De Moivre's theorem.

▮ $1/i = (i)^{-1}$, and $i = 0 + 1i = 1(\cos \pi/2 + i \sin \pi/2)$. Thus, $i^{-1} = 1^{-1}[\cos(-\pi/2) + i \sin(-\pi/2)] = \cos(-\pi/2) + i \sin(-\pi/2) = \cos \pi/2 - i \sin \pi/2 = 0 - i = -i$.

10.81   Prove that your result in Prob. 10.80 is correct.

▮ $i \cdot (-i) = -(i^2) = -(-1) = 1$. Thus, $-i$ is the reciprocal of i.

10.3   MISCELLANEOUS PROBLEMS

For Probs. 10.82 to 10.86, find the distance between the two given complex numbers.

10.82 $z_1 = 1 + i$, $z_2 = 2 + 2i$

▮ If $z = a + bi$, and $w = c + di$, then $d(z, w) = \sqrt{(a - c)^2 + (b - d)^2}$. Notice that this formula comes from the ordinary distance formula in the cartesian plane. Thus, $d(z_1, z_2) = \sqrt{(1 - 2)^2 + (1 - 2)^2} = \sqrt{2}$.

10.83 $z_1 = 3i$, $z_2 = 5 + 6i$

▮ $d(z_1, z_2) = \sqrt{(0 - 5)^2 + (3 - 6)^2} = \sqrt{25 + 9} = \sqrt{34}$.

10.84 $z_1 = 4$, $z_2 = 6 - zi$

$\quad\blacksquare\quad d(z_1, z_2) = \sqrt{(4 - 6)^2 + (0 + 2)^2} = \sqrt{4 + 4} = 2\sqrt{2}.$

10.85 $z_1 = -7$, $z_2 = 7i$

$\quad\blacksquare\quad d(z_1, z_2) = \sqrt{(-7 - 0)^2 + (0 - 7)^2} = \sqrt{98} = 7\sqrt{2}.$

10.86 $z_1 = a - bi$, $z_2 = b - ai$

$\quad\blacksquare\quad d(z_1, z_2) = \sqrt{(a - b)^2 + (-b + a)^2} = \sqrt{2(a - b)^2} = |a - b|\sqrt{2}.$

For Probs. 10.87 to 10.90, perform the given operations graphically.

10.87 $(1 + i) + (2 + 3i)$

$\quad\blacksquare\quad$ See Fig. 10.25.   We represent $1 + i$ and $2 + 3i$ as vectors in the plane and find the diagonal of the resulting parallelogram.   The point $A = 3 + 4i$ represents the sum.

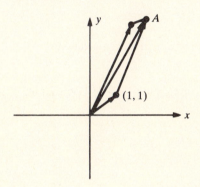

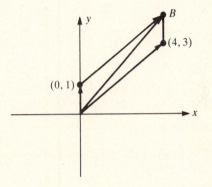

Fig. 10.25

Fig. 10.26

10.88 $i + (4 + 3i)$

$\quad\blacksquare\quad$ See Fig. 10.26.   We proceed as in Prob. 10.87 above.   The point $B = 4 + 4i$ represents the sum.

10.89 $(2 + i) - (4 - i)$

$\quad\blacksquare\quad$ See Fig. 10.27. $(2 + i) - (4 - i) = (2 + i) + [-(4 - i)] = (2 + i) + (-4 + i)$.   The point $C = -2 + 2i$ represents the difference.

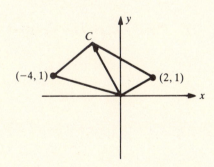

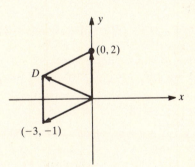

Fig. 10.27

Fig. 10.28

10.90 $2i - (3 + i)$

$\quad\blacksquare\quad$ See Fig. 10.28.   We find $2i + (-3 - i)$ graphically.   The point $D = (-3 + i)$ represents the difference.

For Probs. 10.91 and 10.92, solve the given equation.

10.91 $ix + 3i = 5i$

$\quad\blacksquare\quad$ Let $x = a + bi$;   then $5i = i(a + bi) + 3i = ai + 3i + bi^2 = i(a + 3) - b$.   Thus, $a + 3 = 5$,   and $-b = 0$,   so $a = 2$,   and $b = 0$.   Thus, $x = 2 + 0i$.   In this problem, you could see that $x = 2$ without going through this procedure by looking carefully at the equation.

10.92 $(1-i)x + 3i = 2i$

▮   Let $x = a + bi$;   then $(1-i)(a + bi) + 3i = 2i$.   Then $a - ai + bi - bi^2 + 3i = 2i$,   and $i(-a + b + 3) + (b + a) = 0 + 2i$.   If $b + a = 0$,   then $a = -b$.   If $-a + b + 3 = 2$,   then $-a + (-a) + 3 = 2$,   so $-2a = -1$,   or $a = \frac{1}{2}$.   Then $b = -\frac{1}{2}$.   Thus, $x = \frac{1}{2} - \frac{1}{2}i$.

For Probs. 10.93 to 10.95, the given number is a solution of a quadratic equation with real coefficients.   Find the other solution.

10.93 $1 + 2i$

▮   If $1 + 2i = Z$ is a solution of a quadratic equation with real coefficients, then so is $\bar{Z} = 1 - 2i$.

10.94 $2 - 5i$

▮   If $Z = 2 - 5i$ is a solution, so is $\bar{Z} = 2 + 5i$.

10.95 $4i$

▮   If $Z = 4i$,   then $Z = 0 + 4i$ and $\bar{Z} = 0 - 4i = -4i$.

For Probs. 10.96 to 10.100, find a quadratic equation with real coefficients having the given solution(s)

10.96 $1 \pm 3i$

▮ $[x - (1 + 3i)][x - (1 - 3i)] = 0$.   Then $[(x - 1) - 3i][(x - 1) + 3i] = 0$.   Thus, $0 = (x - 1)^2 - 9i^2 = (x - 1)^2 + 9 = (x^2 - 2x + 1) + 9 = x^2 - 2x + 10$.

10.97 $1 + i$

▮ $1 + i$ and $1 - i$ must be solutions.   Then $0 = [x - (1 + i)][x - (1 - i)] = [(x - 1) - i][(x - 1) + i] = (x - 1)^2 - i^2 = x^2 - 2x + 1 + 1 = x^2 - 2x + 2$.

10.98 i

▮ i and $-i$ are solutions.   Then $0 = (x - i)(x + i) = x^2 - ix + ix - i^2 = x^2 + 1$.

10.99 $-2i$

▮ $0 = (x - 2i)(x + 2i) = x^2 - 4i^2 = x^2 + 4$.

10.100 $\frac{1}{2} - i$

▮ $0 = [x - (\frac{1}{2} - i)][x - (\frac{1}{2} + i)] = [(x - \frac{1}{2}) + i][(x - \frac{1}{2}) - i] = (x - \frac{1}{2})^2 - i^2 = x^2 - x + \frac{1}{4} + 1 = x^2 - x + \frac{5}{4}$.

10.101   Prove: $d(Z, \bar{Z}) = 2|b|$,   where $Z = a + bi$.

▮   If $Z = a + bi$,   then $\bar{Z} = a - bi$. $d(Z, \bar{Z}) = \sqrt{(a - a)^2 + [b - (-b)]^2} = \sqrt{(2b)^2} = \sqrt{4b^2} = 2|b|$.

10.102   Prove: $|Z| = |\bar{Z}|$ for all Z.

▮   Let $Z = a + bi$;   then $\bar{Z} = a - bi$. $|Z| = d(Z, (0, 0)) = \sqrt{a^2 + b^2}$,   and $|\bar{Z}| = d(\bar{Z}, (0, 0)) = \sqrt{a^2 + (-b)^2} = \sqrt{a^2 + b^2}$.   Thus, $|Z| = |\bar{Z}|$.

10.103   Prove: $\overline{(\bar{z})} = Z$ for all Z.

▮   Let $Z = a + bi$.   Then, $\bar{Z} = a - bi = a + (-bi)$,   and $\overline{(\bar{z})} = a + [-(-bi)] = a + bi = Z$.   (Question: What is Z's fifteenth conjugate?   Sixteenth?)

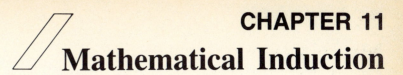

CHAPTER 11
Mathematical Induction

Prove each of the following using mathematical induction. In each case, the variable is a positive integer.

11.1 $\quad 1 + 2 + 3 + \cdots + n = \dfrac{n(n+1)}{2}$

▮ When $n = 1$, the above is written as $1 \overset{?}{=} \dfrac{1(1+1)}{2}$. Since $1 = \dfrac{1(1+1)}{2}$, the statement is verified for $n = 1$. Assume the statement is true for $n = k$. Then, $1 + 2 + 3 + \cdots + k = k\,\dfrac{(k+1)}{2}$. Add $k+1$ to both sides. Then

$$1 + 2 + 3 + \cdots + k + k + 1 = \frac{k(k+1)}{2} + k + 1 \qquad (1)$$

What is it that we want to prove? We want to show that the above statement is true for $n = k+1$ (assuming it is true for $n = k$). When $n = k+1$,

$$1 + 2 + 3 + \cdots + k + k + 1 \overset{?}{=} \frac{(k+1)(k+1)}{2} \qquad (2)$$

We assume statement (1) to be true; statement (2) is what we need to prove. Note that the LHS of both are the same. Working on the RHS of (1), we get:

$$\frac{k(k+1)}{2} + \frac{k+1}{1} = \frac{k(k+1) + 2(k+1)}{2} = \frac{(k+1)(k+2)}{2} = \text{RHS of } (2)$$

Thus,

$$\underbrace{1 + 2 + 3 + \cdots + k + k + 1}_{\text{LHS of } (1)} = \frac{(k+1)(k+2)}{2}$$
$$(2)$$

Statement (2) is verified, and the proof is complete.

11.2 $\quad 1 + 4 + 7 + \cdots + (3n - 2) = \dfrac{n(3n - 1)}{2}$

▮ When $n = 1$, the above is $3(1) - 2 \overset{?}{=} \dfrac{1(3 - 1)}{2}$, or $1 = 1$, and the statement is true when $n = 1$. If the statement is true for $n = k$, then, $1 + 4 + 7 + \cdots + (3k - 2) = \dfrac{k(3k - 1)}{2}$. Add the $(k+1)$st term for the LHS to both sides; then, $1 + 4 + 7 + \cdots + (3k - 2) + (3k + 1) - 2 = \dfrac{k(3k - 1)}{2} + 3(k + 1) - 2$. Thus, $1 + 4 + 7 + \cdots + (3k - 2) + 3(k + 1) - 2 = \dfrac{3k^2 - k}{2} + 3k + 3 - 2 = \dfrac{3k^2 - k}{2} + 3k + 1 = \dfrac{3k^2 - k + 6k + 2}{2} = \dfrac{3k^2 + 5k + 2}{2} = \dfrac{(k + 1)(3k + 2)}{2}$. But, $\dfrac{(k + 1)(3k + 2)}{2}$ is the RHS of our given statement when $n = k + 1$. The given statement is now proved true since we have proved it to be true for $n = k + 1$ assuming it is true when $n = k$.

11.3 $\quad 5 + 9 + 13 + \cdots + (4n + 1) = n(2n + 3)$

▮ When $n = 1$, $5 \overset{?}{=} 1(2 + 3)$, or $5 = 5$. Assume the statement is true when $n = k$ (write that as S_k). Then, $5 + 9 + 13 + \cdots + 4k + 1 = k(2k + 3)$. Add $4(k + 1)$ to the LHS and RHS; then, $5 + 9 + 13 + \cdots + (4k + 1) + [4(k + 1) + 1] = k(2k + 3) + [4(k + 1) + 1] = 2k^2 + 3k + 4k + 5 = 2k^2 + 7k + 5 = (k + 1)(2k + 5)$. [This is $n(2n + 3)$ when $n = k + 1$.] = RHS of required statement. Thus, S_k true implies S_{k+1} true, and the proof is complete.

11.4 $\quad 3 + 7 + 11 + \cdots + (4n - 1) = n(2n + 1)$

When $n = 1$, $3 \overset{?}{=} 1(2 + 1)$, or $3 = 3$. Assume S_k is true; then, $3 + 7 + 11 + \cdots + 4k - 1 = k(2k + 1)$. Add $4(k + 1) - 1 (= 4k + 3)$ to both sides. Then, $3 + 7 + 11 + \cdots + 4k - 1 + 4(k + 1) - 1 = k(2k + 1) + 4k + 3 = 2k^2 + k + 4k + 3 = 2k^2 + 5k + 3 = (k + 1)(2k + 3)$. Thus, S_{k+1} is true, and the proof is complete.

11.5 $3 + 5 + 7 + \cdots + (2n + 1) = n(n + 2)$

▮ When $n = 1$, $3 \overset{?}{=} 1(1 + 2)$, or $3 = 3$. Assume S_k is true. Then, $3 + 5 + 7 + \cdots + 2k + 1 = k(k + 2)$. Add $2(k + 1) + 1$ $(=2k + 3)$ to both sides. Then, $3 + 5 + 7 + \cdots + 2k + 1 + 2(k + 1) + 1 = k(k + 2)$ $+ 2k + 3 = k^2 + 2k + 2k + 3 = k^2 + 4k + 3 = (k + 1)(k + 3)$. Thus, S_{k+1} is true, and the proof is complete.

11.6 $2 + 7 + 12 + \cdots + (5n - 3) = \dfrac{n(5n - 1)}{2}$

▮ When $n = 1$, $2 \overset{?}{=} \dfrac{1(5 - 1)}{2}$, or $2 = 2$. Assume S_k is true; then $2 + 7 + 12 + \cdots + 5k - 3 = \dfrac{k(5k - 1)}{2}$. Add $5(k + 1) - 3(=5k + 2)$ to both sides. Then, $2 + 7 + 12 + \cdots + 5k - 3 + 5(k + 1) - 3 = \dfrac{k(5k - 1)}{2} + 5k + 2 = \dfrac{5k^2 - k}{2} + \dfrac{10k + 4}{2} = \dfrac{5k^2 + 9k + 4}{2} = \dfrac{(k + 1)(5k + 4)}{2}$. Thus, S_{k+1} is true, and the proof is complete.

11.7 $2 + 5 + 8 + \cdots + (3n - 1) = \dfrac{n(3n + 1)}{2}$

▮ When $x = 1$, $2 \overset{?}{=} \dfrac{1(3 + 1)}{2}$, or $2 = 2$. Assume S_k is true. Then, $2 + 5 + 8 + \cdots + 3(k - 1) = \dfrac{k(3k + 1)}{2}$. Add $3(k + 1) - 1(=3k + 2)$ to both sides. Then, $2 + 5 + 8 + \cdots + (3k - 1) + 3(k + 1) - 1 = \dfrac{k(3k + 1)}{2} + 3k + 2 = \dfrac{3k^2 + k}{2} + \dfrac{6k + 4}{2} = \dfrac{3k^2 + 7k + 4}{2} = \dfrac{(k + 1)(3k + 4)}{2}$. Thus, S_{k+1} is true, and the proof is complete.

11.8 $6 + 11 + 16 + \cdots + (5n + 1) = \dfrac{n(5n + 7)}{2}$

▮ When $n = 1$, $6 \overset{?}{=} \dfrac{1(5 + 7)}{2}$, or $6 = 6$. Assume S_k is true. Then $6 + 11 + 16 + \cdots + 5k + 1 = \dfrac{k(5k + 7)}{2}$. Add $5(k + 1) + 1(=5k + 6)$ to both sides. Then, $6 + 11 + 16 + \cdots + 5k + 1 + 5(k + 1) + 1 = \dfrac{k(5k + 7)}{2} + 5k + 6 = \dfrac{5k^2 + 7k}{2} + \dfrac{10k + 12}{2} = \dfrac{5k^2 + 17k + 12}{2} = \dfrac{(k + 1)(5k + 12)}{2}$. Thus, S_{k+1} is true, and the proof is complete.

11.9 $2 + 6 + 12 + \cdots + n(n + 1) = \dfrac{n(n + 1)(n + 2)}{3}$

▮ When $n = 1, 2 \overset{?}{=} \dfrac{1(1 + 1)(1 + 2)}{3}$, or $2 = 2$. Assume S_k is true. Then, $2 + 6 + 12 + \cdots + k(k + 1) = \dfrac{k(k + 1)(k + 2)}{3}$. Add $(k + 1)(k + 2)$ to both sides. Then, $2 + 6 + 12 + \cdots + k(k + 1) + (k + 1)(k + 2) = \dfrac{k(k + 1)(k + 2)}{3} + (k + 1)(k + 2) = \dfrac{k(k + 1)(k + 2)}{3} + \dfrac{3(k + 1)(k + 2)}{3} = \dfrac{(k + 1)(k + 2)(k + 3)}{3}$. Thus, S_{k+1} is true, and the proof is complete.

11.10 $4 + 10 + 18 + \cdots + n(n + 3) = \dfrac{n(n + 1)(n + 5)}{3}$

▮ When $n = 1$, $4 \overset{?}{=} \dfrac{1(2)(6)}{3}$, or $4 = 4$. Assume S_k is true. Then, $4 + 10 + 18 + \cdots + k(k + 3) = \dfrac{k(k + 1)(k + 5)}{3}$. Add $(k + 1)(k + 4)$ to both sides. Then, $4 + 10 + 18 + \cdots + k(k + 3) + (k + 1)(k + 4) = \dfrac{k(k + 1)(k + 5)}{3} + \dfrac{3(k + 1)(k + 4)}{3} = \dfrac{(k + 1)(k + 2)(k + 6)}{3}$. Thus S_{k+1} is true, and the proof is complete.

11.11 $n^2 + n$ divisible by 2.

▮ When $n = 1$, is $1^2 + 1$ divisible by 2? Yes, because 2 is divisible by 2. Assume S_k is true. Show S_{k+1} is true. Then, $k^2 + k$ is divisible by 2. But $(k + 1)^2 + (k + 1) = k^2 + 2k + 1 + k + 1 = (k^2 + k) + (2k + 2)$. Both $k^2 + k$ and $2k + 2$ are divisible by 2 since S_k is true. Thus, S_{k+1} is true, and the proof is complete.

11.12 $a + b$ is a factor of $a^{2n-1} + b^{2n-1}$

▎ When $n=1$, $a+b$ is a factor of a^1+b^1. Assume S_k is true. Then, $a+b$ is a factor of $a^{2k-1}+b^{2k-1}$. Show S_{k+1} is true, i.e., show $a+b$ is a factor of $a^{2(k+1)-1}+b^{2(k+1)-1}(=a^{2k+1}+b^{2k+1})$. But $a^{2k+1}+b^{2k+1}=(a+b)(a^{2k}-a^{2k-1}b+\cdots+b^{2k})$ which is divisible by $a+b$. Thus, S_{k+1} is true, and the proof is complete.

11.13 $(ab)^n=a^nb^n$

▎ When $n=1$, $(ab)^1\overset{?}{=}a^1b^1$, or $ab=ab$. Assume S_k is true. Then, $(ab)^k=a^kb^k$. But $(ab)^{k+1}=(ab)^k(ab)=(a^kb^k)ab$ $[(ab)^k=a^kb^k$ since S_k is true!] $=a^{k+1}b^{k+1}$. Thus, S_{k+1} is true, and the proof is complete.

11.14 $\log(a_1a_2\cdots a_n)=\log a_1+\log a_2+\cdots+\log a_n$

▎ When $n=1$, $\log a_1\overset{?}{=}\log a_1$, or $\log a_1=\log a_1$. Assume S_k is true. Then, $\log(a_1a_2\cdots a_k)=\log a_1+\log a_2+\cdots+\log a_k$. $\log(a_1a_2\cdots a_{k+1})=\log[(a_1a_2\cdots a_k)(a_{k+1})]=\log[(a_1a_2)a_3\cdots a_{k+1}]$ $[a_1a_2$ is 1 term; $a_3\cdots a_{k+1}$ is $k-1$ terms. We are assuming s_k is true.] $=\log a_1a_2+\log a_3+\log a_4+\cdots+\log a_{k+1}=\log a_1+\log a_2+\cdots+\log a_{k+1}$. Thus S_{k+1} is true, and the proof is complete.

11.15 $\sin(2n\pi+\theta)=\sin\theta$

▎ When $n=1$, $\sin(2\pi+\theta)=\sin\theta$, or $\sin\theta=\sin\theta$. Assume S_k is true. Then, $\sin(2\pi k+\theta)=\sin\theta$. $\sin[2\pi(k+1)+\theta]=\sin[(2\pi k+2\pi)+\theta]=\sin[2\pi+(2\pi k+\theta)]=\sin(2\pi k+\theta)$ (since $n=1$) $=\sin\theta$ (since S_k is true). Thus, S_{k+1} is true, and the proof is complete.

11.16 $\cos(2n\pi+\theta)=\cos\theta$

▎ When $n=1$, $\cos(2\pi+\theta)=\cos\theta$, or $\cos\theta=\cos\theta$. When $n=k$, $\cos(2k\pi+\theta)=\cos\theta$. Assume that S_k is true. Then, $\cos[2(k+1)\pi+\theta]=\cos(2k\pi+2\pi+\theta)=\cos[2\pi+(2k\pi+\theta)]=\cos(2k\pi+\theta)=\cos\theta$. S_{k+1} is true, and the proof is complete.

11.17 For any $x\in\mathscr{R}$, where $x\geq-1$, $(1+x)^n\geq1+nx$.

▎ When $n=1$, $(1+x)^1\overset{?}{\geq}1+x$, or $1+x=1+x$. Assume $(1+x)^k\geq1+kx$. Then, $(1+x)^{k+1}=(1+x)(1+x)^k\geq(1+x)(1+kx)$ (since S_k is assumed to be true) $=1+x+kx+kx^2\geq1+x+kx=1+(k+1)x$, and $(1+x)^{k+1}\geq1+(k+1)x$. Thus, S_{k+1} is true, and the proof is complete.

11.18 $2^n\geq2n$

▎ When $n=1$, $2^1\overset{?}{\geq}2(1)$, $2\overset{?}{\geq}2$, or $2=2$. Assume $2^k\geq2k$. $2^{k+1}=2^k\cdot2\geq2k\cdot2=4k=2k+2k\geq2k+2=2(k+1)$. Thus, $2^{k+1}\geq2(k+1)$, and S_{k+1} is true. The proof is complete.

11.19 $3^n\geq2n+1$

▎ When $n=1$, $3^1\overset{?}{\geq}2(1+1)$, or $3=3$. Assume $3^k\geq2k+1$. Then, $3^{k+1}=3^k\cdot3\geq(2k+1)^3=6k+3\geq2k+3=2(k+1)+1$. Thus, $3^{k+1}\geq2(k+1)+1$, and S_{k+1} is true.

11.20 Let $f(x+y)=f(x)+f(y)$. Prove $f(n)=nf(1)$.

▎ When $n=1$, $f(1)\overset{?}{=}1f(1)$, or $f(1)=f(1)$. Then $f(k+1)=f(k)+f(1)$. But $f(k)=kf(1)$, so $f(k+1)=kf(1)+f(1)=(k+1)f(1)$. So, S_{k+1} is true, and the proof is complete.

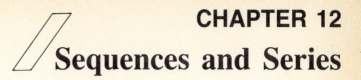

CHAPTER 12
Sequences and Series

12.1 SEQUENCES

For Probs. 12.1 to 12.10, write the first four terms of the sequence whose general term is given.

12.1 $n - 2$

▮ Let $n = 1, 2, 3$, and 4 to generate the first four terms: $1 - 2$, $2 - 2$, $3 - 2$, $4 - 2$; or $-1, 0, 1, 2$.

12.2 $n^2 + 1$

▮ $1^2 + 1$, $2^2 + 1$, $3^2 + 1$, $4^2 + 1$; or, $2, 5, 10, 17$.

12.3 $1/n$

▮ $\frac{1}{1}, \frac{1}{2}, \frac{1}{3}, \frac{1}{4}$; or $1, \frac{1}{2}, \frac{1}{3}, \frac{1}{4}$.

12.4 $1/2^n$

▮ $1/2^1, 1/2^2, 1/2^3, 1/2^4$; or $\frac{1}{2}, \frac{1}{4}, \frac{1}{8}, \frac{1}{16}$.

12.5 $\dfrac{1}{4^{n-2} + 1}$

▮ $\dfrac{1}{4^{1-2} + 1}$, $\dfrac{1}{4^{2-2} + 1}$, $\dfrac{1}{4^{3-2} + 1}$, $\dfrac{1}{4^{4-2} + 1}$; or $\frac{4}{5}, 1, \frac{1}{5}, \frac{1}{17}$.

12.6 $\dfrac{n+1}{n^2+1}$

▮ $\dfrac{1+1}{1-2}$, $\dfrac{2+1}{4-2}$, $\dfrac{3+1}{9-2}$, $\dfrac{4+1}{16-2}$; or $-2, \frac{3}{2}, \frac{4}{7}, \frac{5}{14}$.

12.7 $(1 + 1/n)^n$

▮ $(1 + \frac{1}{1})^1$, $(1 + \frac{1}{2})^2$, $(1 + \frac{1}{3})^3$, $(1 + \frac{1}{4})^4$; or 2, $\frac{9}{4}, \frac{64}{27}, \frac{625}{256}$.

12.8 $\dfrac{(-1)^n}{n+2}$

▮ When $n = 1$, $(-1)^n = -1^1 = -1$. Then the sign of the $n = 1$ term is -1 and the signs alternate -1, $+1$, -1, etc. Thus, the sequence begins: $-\frac{1}{3}, \frac{1}{4}, -\frac{1}{5}, \frac{1}{6}$.

12.9 $\sin(n+1)x$

▮ $\sin(1+1)x$, $\sin(2+1)x$, $\sin(3+1)x$, $\sin(4+1)x$; or $\sin 2x, \sin 3x, \sin 4x, \sin 5x$.

12.10 $\log nx$

▮ $\log 1x, \log 2x, \log 3x, \log 4x$, which can be simplified as follows: $\log x$, $\log 2 + \log x$, $\log 3 + \log x$, $\log 4 + \log x$.

For Probs. 12.11 to 12.20, find the general term of the given sequence.

12.11 $2, 3, 4, 5, \ldots$

▮ Notice that the terms of this sequence begin $a + 2$ (which is $n + 1$ when $n = 1$) and progress consecutively. The general term, then, is $1/(n + 1)$. (Check this by writing out the terms.)

12.12 $-1, -2, -3, -4, \ldots$

▮ When $n = 1$, $-n = -1$. The sequence is $-n$. (Check this by writing out the terms.)

12.13 $-4, -3, -2, -1, 0, 1, \ldots$

   ▮ When $n = 1$, $-4 = n - 5$. The general term of this sequence is $n - 5$. (Check this by writing out the terms.)

12.14 $2, 4, 6, 8, 10, \ldots$

   ▮ When $n = 1$, $2 = 2n$. The general term is $2n$. Notice that you could *not* use $n + 1$ since we must generate *only* evens.

12.15 $2, -4, 6, -8, 10, \ldots$

   ▮ This is similar to Prob. 12.14 above, but we want the terms to alternate in sign beginning with a positive. Using $(-1)^{n+1}$ to alternate signs (-1^{n+1} is $+1$ when $n = 1$), the general term is $(-1)^{n+1} \cdot 2n$.

12.16 $-2, 4, -6, 8, -10, \ldots$

   ▮ See Prob. 12.15 above. This is very similar, except we want the sign of the $n = 1$ term to be negative. The sequence is $(-1)^n \cdot 2n$.

12.17 $x, x^2/2, x^3/3, x^4/4, \ldots$

   ▮ Each term is of the form x^n/n (beginning with $n = 1$).

12.18 $\frac{1}{2}, \frac{3}{4}, \frac{5}{6}, \frac{7}{8}, \ldots$

   ▮ Notice that each term's numerator is 1 less than its denominator. Also, the numerator of the first term is 1. We try $n/(n + 1)$. Checking, we notice this is not the general term! Check it. However, when we notice that the numerators are all odd and the denominators even, we find the general term: $(2n - 1)/2n$. Check this and note that it is correct.

12.19 $x, -x^3, x^5, -x^7, \ldots$

   ▮ Alternate signs beginning with positive; moreover, we need to generate odd powers of x: $(-1)^{n+1} \cdot x^{2n-1}$.

12.20 $-x, x^3/2, -x^5/3, x^7/4, \ldots$

   ▮ See Prob. 12.19. We reverse the alternation in sign, and divide each term by n: $(-1)^n (x^{2n-1}/n)$.

For Probs. 12.21 to 12.25, find the first four terms of the sequence defined recursively by the given equations.

12.21 $A_1 = 1$, $A_{n+1} = A_n + 1$

   ▮ $A_1 = 1$; $A_2 = A_{1+1} = A_1 + 1 = 1 + 1 = 2$; $A_3 = A_{2+1} = A_2 + 1 = 2 + 1 = 3$, etc. The first four terms are 1, 2, 3, 4.

12.22 $A_1 = 2$, $A_{n+1} = 2A_n + 1$

   ▮ $A_2 = A_{1+1} = 2A_1 + 1 = 2(2) + 1 = 5$; $A_3 = A_{2+1} = 2A_2 + 1 = 11$; $A_4 = A_{3+1} = 2A_3 + 1 = 23$. The terms are 2, 5, 11, 23.

12.23 $A_1 = 1$, $A_{n+1} = -A_n$

   ▮ $A_1 = 1$; $A_2 = -A_1$; $A_3 = -A_2$; etc. The terms are 1, -1, 1, -1.

12.24 $A_1 = 1$, $A_2 = 1$, $A_{n+2} = A_n + A_{n+1}$

   ▮ This is the famous Fibonacci sequence. $A_1 = 1$; $A_2 = 1$; $A_3 = A_2 + A_1 = 1 + 1 = 2$; $A_4 = A_3 + A_2 = 2 + 1 = 3$. The terms are 1, 1, 2, 3.

12.25 $A_1 = 5$, $A_2 = 3$, $A_{n+2} = \dfrac{A_{n+1}}{A_n}$

   ▮ $A_3 = A_{1+2} = \dfrac{A_2}{A_1} = \dfrac{3}{5}$; $A_4 = A_{2+2} = \dfrac{A_3}{A_2} = \dfrac{\frac{3}{5}}{3} = \dfrac{1}{5}$. The terms are 5, 3, $\frac{3}{5}$, $\frac{1}{5}$.

For Probs. 12.26 and 12.27, define the given sequence using a recursive formula.

12.26 $1, 2, 3, 5, 8, \ldots$

▮ $A_1 = 1$; $A_2 = 2$; $A_{n+2} = A_{n+1} + A_n$.   Do you see how this resembles the Fibonacci sequence?

12.27 $1, 1, 5, 17, 61, \ldots$

▮ How can we get 5 from 1 and 1?   How about $3(1) + 2(1)$.   Note that $17 = 3(5) + 2(1)$,   etc. $A_1 = 1$; $A_2 = 1$; $A_{n+2} = 3A_{n+1} + 2A_n$.

12.2   SERIES

For Probs. 12.28 to 12.38, write the series without summation (sigma) notation.

12.28 $\displaystyle\sum_{i=1}^{5} i$

▮ Remember that to expand a given series with sigma notation, we replace the variable (in this case i) in the series expression one number at a time and add successively.   In this case: $1 + 2 + 3 + 4 + 5$.

12.29 $\displaystyle\sum_{i=1}^{4} 2i$

▮ $2(1) + 2(2) + 2(3) + 2(4) = 2 + 4 + 6 + 8$

12.30 $\displaystyle\sum_{i=2}^{5} i + 3$

▮ $(2+3) + (3+3) + (4+3) + (5+3) = 5 + 6 + 7 + 8$.

12.31 $\displaystyle\sum_{i=0}^{5} i^2$

▮ $0^2 + 1^2 + 2^2 + 3^2 + 4^2 + 5^2 = 0 + 1 + 4 + 9 + 16 + 25$.

12.32 $\displaystyle\sum_{i=1}^{4} i^i$

▮ $1^1 + 2^2 + 3^3 + 4^4 = 1 + 4 + 27 + 256$.

12.33 $\displaystyle\sum_{i=1}^{4} (i-1)$

▮ $1(1-1) + 2(2-1) + 3(3-1) + 4(4-1) = 0 + 2 + 6 + 12$.

12.34 $\displaystyle\sum_{i=1}^{4} (-1)^i i(i-1)$

▮ This is the same as Prob. 12.33, but with an alternating sign beginning with a negative.   Thus, we have, $-0 + 2 - 6 + 12$.

12.35 $\displaystyle\sum_{i=1}^{3} (-1)^{i+1}(x^i - 1)$

▮ When $i = 1$, $(-1)^2(x^1 - 1) = x - 1$.   When $i = 2$, $(-1)^3(x^2 - 1) = -(x^2 - 1) = 1 - x^2$.   When $i = 3$, $(-1)^4(x^3 - 1) = x^3 - 1$.   Thus, we have $(x - 1) + (1 - x^2) + (x^3 - 1)$.

12.36 $\displaystyle\sum_{i=1}^{\infty} 1/i$

▮ $1 + \frac{1}{2} + \frac{1}{3} + \frac{1}{4} + \cdots + 1/n + \cdots$.   (This is the harmonic series.   It is an important series as you will see later on.)

12.37 $\displaystyle\sum_{i=1}^{\infty} (-1)^i \frac{1}{i+1}$

▮ When $i = 1$,   we get $-\frac{1}{2}$; after that, the signs alternate and the denominators increase consecutively: $-\frac{1}{2} + \frac{1}{3} - \frac{1}{4} + \frac{1}{5} \cdots + \cdots (-1)^n [1/(n+1)] + \cdots$.

12.38 $\displaystyle\sum_{i=1}^{\infty} (-1)^{i-1} \frac{1}{i^2}$

▮ When $i = 1$, $(-1)^{i-1}(1/i^2) = \frac{1}{1} = 1$. When $i = 2$, $(-1)^{i-1}(1/i^2) = -\frac{1}{4}$; and then the signs alternate. $1 - \frac{1}{4} + \frac{1}{9} - \frac{1}{16} + \cdots - \cdots + (-1)^{n-1}(1/n^2) + \cdots$.

For Probs. 12.39 to 12.49, write the given series using summation (sigma) notation:

12.39 $1 + 2 + 3 + 4 + 5$

▮ The terms of this series are increasing consecutively, beginning at $i = 1$: $\displaystyle\sum_{i=1}^{5} i$.

12.40 $2 + 3 + 4 + 5 + 6$

▮ This is the same as the series in Prob. 12.39, except i's range is from 2 to 6: $\displaystyle\sum_{i=2}^{6} i$. See Prob. 12.41.

12.41 $2 + 3 + 4 + 5 + 6$

▮ We could have written this as $\displaystyle\sum_{i=1}^{5} (i + 1)$. Notice that you can change the general term if you want (or need) to if you make an appropriate change in the range of the variable i.

12.42 $2 + 4 + 6 + 8 + 10 + 12$

▮ These are the consecutive evens $2i$ beginning with $i = 1$: $\displaystyle\sum_{i=1}^{6} 2i$.

12.43 $1 - 3 + 5 - 7 + 9 - 11$

▮ These are the consecutive odds $2i - 1$ beginning with $i = 1$, alternating with a positive lead term: $\displaystyle\sum_{i=1}^{6} (-1)^{i+1}(2i - 1)$.

12.44 $1 - 3 + 5 - 7 + 9 - 11 + \cdots - \cdots$

▮ This is the series in Prob. 12.43, but is the infinite version of it: $\displaystyle\sum_{i=1}^{\infty} (-1)^{i+1}(2i - 1)$.

12.45 $1 - \frac{1}{2} + \frac{1}{4} - \frac{1}{8}$

▮ The denominators are the powers of 2, beginning with 2, which is 2^{i-1} when $i = 1$. $\displaystyle\sum_{i=1}^{4} (-1)^{i+1} \frac{1}{2^{i-1}}$.

12.46 $-1 + \frac{1}{4} - \frac{1}{9} + \frac{1}{16} + \cdots$

▮ The denominators are the squares of the consecutive integers beginning at 1: $\displaystyle\sum_{i=1}^{\infty} (-1)^{i} \frac{1}{i^2}$.

12.47 $1 - 1 + 1 - 1 + 1 - 1 + \cdots$

▮ This is the infinite series of alternating signs of 1: $\displaystyle\sum_{i=1}^{\infty} (-1)^{i+1}$.

12.48 $1 + 2 + \frac{3}{2} + \frac{4}{6} + \frac{5}{24} + \cdots$

▮ The denominators are the $i!$ beginning with 0! (which is 1): $\displaystyle\sum_{i=0}^{\infty} \frac{i+1}{i!}$.

12.49 $1 + x + x^2/2 + x^3/3! + \cdots$

▮ Remember that $0! = 1! = 1$: $\displaystyle\sum_{i=0}^{\infty} \frac{x^i}{i!}$.

12.50 Show that $\displaystyle\sum_{k=1}^{n} ca_k = c \sum_{k=1}^{n} a_k$.

▮ $\displaystyle\sum_{k=1}^{n} ca_k = ca_1 + ca_2 + \cdots + ca_n = c(a_1 + a_2 + \cdots + a_n) = c \sum_{k=1}^{n} a_k$.

12.51 Show that $\sum\limits_{k=1}^{n} (a_k + b_k) = \sum\limits_{k=1}^{n} a_k + \sum\limits_{k=1}^{n} b_k$.

▮ $\sum\limits_{k=1}^{n} (a_k + b_k) = (a_1 + b_1) + (a_2 + b_2) + \cdots + (a_n + b_n) = (a_1 + a_2 + \cdots + a_n) + (b_1 + b_2 + \cdots + b_n) = \sum\limits_{k=1}^{n} a_k + \sum\limits_{k=1}^{n} b_k$.

12.52 Solve for x: $\sum\limits_{i=1}^{2} ix = 5$.

▮ The LHS $= 1x + 2x = 3x$. If $3x = 5$, $x = \frac{5}{3}$.

12.53 Solve for x: $\sum\limits_{i=0}^{3} (i+1)x^2 = 10$.

▮ Then $1x^2 + 2x^2 + 3x^2 + 4x^2 = 10$, $10x^2 = 10$, $x^2 = 1$, and $x = 1$ or -1.

12.3 ARITHMETIC AND GEOMETRIC SEQUENCES

For Probs. 12.54 to 12.60 tell whether the given sequence is arithmetic or geometric or neither.

12.54 $1, 3, 5, 7, 9, \ldots$

▮ Arithmetic; the common difference is 2.

12.55 $2, 6, 10, 11, 15, 19, \ldots$

▮ Neither; there is neither a common difference nor a common ratio.

12.56 $1, \frac{1}{2}, \frac{1}{3}, \frac{1}{4}, \ldots$

▮ Neither; there is neither a common difference nor a common ratio.

12.57 $1, \frac{1}{2}, \frac{1}{4}, \frac{1}{8}, \ldots$

▮ Geometric; ratio $= \frac{1}{2}$.

12.58 $1, -\frac{1}{2}, \frac{1}{4}, -\frac{1}{8}, \frac{1}{16}, \ldots$

▮ Geometric; common ratio $= -\frac{1}{2}$.

12.59 $10, 6, 2, -2, -6, \ldots$

▮ Arithmetic; difference $= -4$.

12.60 $10, 10x, 10x^2, 10x^3, \ldots$

▮ Geometric; ratio $= x$.

For Probs. 12.61 to 12.70, find all indicated quantities, where $a_1, a_2, \ldots, a_n, \ldots$ is an arithmetic sequence; $d = $ common difference; and $S = $ sum of first n terms of the sequence.

12.61 $a_1 = -5$, $d = 4$, $a_2 = ?$, $a_3 = ?$, $a_4 = ?$

▮ $a_2 = a_1 + d = -5 + 4 = -1$; $a_3 = a_2 + d = 3$; $a_4 = a_3 + d = 7$.

12.62 $a_1 = -3$, $d = 5$, $a_{15} = ?$, $S_{11} = ?$

▮ $a_n = a_1 + (n-1)d$ for every $n > 1$. Thus, $a_{15} = -3 + 14(5) = 67$. Also, $S_n = (n/2)(a_1 + a_n) = (n/2)[2a_1 + (n-1)d]$, so $S_{11} = \frac{11}{2}[2(-3) + 10(5)] = \frac{11}{2}(44) = 242$.

12.63 $a_1 = 1$, $a_2 = 5$, $S_{21} = ?$

▮ $S_n = (n/2)[2a_1 + (n-1)d]$. So, $S_{21} = \frac{21}{2}[2(1) + (20)(4)]$ [since $a_2 - a_1 = d = 4$] $= \frac{21}{2}(82) = 861$.

12.64 $a_1 = 7$, $a_2 = 5$, $a_{15} = ?$

▮ $a_n = a_1 + (n-1)d$, for $n \geq 1$; $d = a_2 - a_1 = 5 - 7 = -2$; $a_{15} = 7 + 14(-2) = -21$.

12.65 $a_1 = -12$, $a_{40} = 22$, $S_{40} = ?$

▮ $S_{40} = \frac{40}{2}(a_1 + a_{40}) = 20(-12 + 22) = 20(10) = 200$.

12.66 $a_1 = 24$, $a_{24} = -28$, $S_{24} = ?$

▮ $S_{24} = \frac{24}{2}[24 + (-28)] = 12(-4) = -48$.

12.67 $a_1 = \frac{1}{3}$, $a_2 = \frac{1}{2}$, $a_{11} = ?$, $S_{11} = ?$

▮ $d = a_2 - a_1 = \frac{1}{2} - \frac{1}{3} = \frac{1}{6}$; $a_{11} = a_1 + (n-1)d = \frac{1}{3} + 10(\frac{1}{6}) = \frac{1}{3} + \frac{10}{6} = 2$; $S_{11} = \frac{11}{2}(\frac{1}{3} + 2) = \frac{11}{2} \cdot \frac{7}{3} = \frac{77}{6}$.

12.68 $a_3 = 13$, $a_{10} = 55$, $a_1 = ?$

▮ $a_{10} - a_3 = 42 = 7d$; $d = \frac{42}{7} = 6$; $a_2 = 13 - 6 = 7$; $a_1 = 7 - 6 = 1$.

12.69 $a_9 = -12$, $a_{13} = 4$, $a_1 = ?$

▮ $a_{13} - a_9 = 4 - (-12) = 16 = 4d$; $d = 4$; $a_9 = a_1 + 8d = -12 = a_1 + 8(4)$; $a_1 = -12 - 32 = -44$.

12.70 $a_3 = 13$, $a_{10} = 55$, $a_1 = ?$

▮ $7d = a_{10} - a_3 = 42$; $d = 6$; $a_{10} = a_1 + 9d = a_1 + 54 = 55$; $a_1 = 1$.

For Probs. 12.71 to 12.75, the sequence is geometric, and r stands for the common ratio.

12.71   Find d for the sequence $1, \frac{1}{2}, \frac{1}{4}, \frac{1}{8}, \ldots$

▮ $d = \dfrac{a_{n+1}}{a_n} = \dfrac{\frac{1}{8}}{\frac{1}{4}} = \frac{1}{2}$.   (If the sequence is geometric, any two consecutively numbered terms can be used.)

12.72   Find the sum of the first eight terms of the sequence $1, \frac{1}{3}, \frac{1}{9}, \frac{1}{27}, \ldots$

▮ $S_8 = \dfrac{a_1(1-r^8)}{1-r} = \dfrac{1[1-(\frac{1}{3})^8]}{1-\frac{1}{3}} = \dfrac{1-\frac{1}{6561}}{\frac{2}{3}} = \frac{6560}{6561} \cdot \frac{3}{2} = \frac{19680}{13122}$.

12.73   Find the sum of the first five terms of a geometric sequence where $a_2 = 3$, $r = \frac{1}{2}$.

▮ $S_5 = \dfrac{a_1(1-r^5)}{1-\frac{1}{2}} = \dfrac{6[1-(\frac{1}{2})^5]}{1-\frac{1}{2}} = \dfrac{6(\frac{31}{32})}{\frac{1}{2}} = 12(\frac{31}{32}) = \frac{372}{32}$.

12.74   If $a_1 = 2$, $r = x$ $(x \neq 1)$,   find S_4.

▮ $S_4 = \dfrac{a_1(1-r^4)}{1-r} = \dfrac{2(1-x^4)}{1-x} = \dfrac{2(1+x^2)(1-x)(1+x)}{1-x} = 2(1+x^2)(1+x)$.

12.75   If $r = 1/x$, $S_5 = 10x$,   find a_1.

▮ $S_5 = \dfrac{a_1[1-(1/x)^5]}{1-1/x}$; $10x = \dfrac{a_1(1-1/x^5)}{1-1/x}$; $a_1 = \dfrac{10x(1-1/x)}{1-1/x^5}$.

12.76   The sum of three numbers in arithmetic progression is 24.   If the first is decreased by 1 and the second decreased by 2, the three numbers are in geometric progression.   Find the three numbers.

▮   Let $x = $ 1st number.   Then $x + d = $ 2d   number, and $x + 2d = $ 3d   number (since they are in arithmetic progression).   Furthermore, the sum of these numbers is 24: $x + (x + d) + (x + 2d) = 24$,   or $x + d = 8$.   Also, $x - 1$, $(x + d) - 2$, $x + 2d$ are in geometric progression, so $\dfrac{x+d-2}{x-1} = \dfrac{x+2d}{x+d-2}$, or $(x + d - 2)^2 = (x + 2d)(x - 1)$.   Simplifying: $d^2 - 2d = 3x - 4$.   But, $x + d = 8$,   and solving simultaneously, we find: $d^2 + d - 20 = 0$,   or $d = 4, -5$.   If $d = 4$, $x = 4$.   If $d = -5$, $x = 13$. Therefore, the solutions are   4, 8, 12;   or   13, 8, 3.

12.4 GEOMETRIC SERIES

For Probs. 12.77 to 12.81, tell whether the given series is geometric.

12.77 $1 + \frac{1}{2} + \frac{1}{4} + \frac{1}{8} + \cdots$

▮   We notice that $a_{n+1}/a_n = \frac{1}{2}$ for all successive pairs.   The series is geometric.

12.78 $1 + 2 + 4 + 8 + 16 + \cdots$

▮ $a_{n+1}/a_n = 2$ for all successive pairs. The series is geometric.

12.79 $1 - 2 + 4 - 8 + 16 - \cdots$

▮ $a_{n+1}/a_n = -2$ for all successive pairs. The series is geometric.

12.80 $1 + \frac{1}{4} + \frac{1}{9} + \frac{1}{16} + \cdots$

▮ Notice that $a_2/a_1 \neq a_3/a_2$. Thus, the series is nongeometric.

12.81 $1 + \frac{1}{2} - \frac{1}{4} - \frac{1}{8} + \frac{1}{16} + \frac{1}{25} - \cdots$

▮ $a_2/a_1 \neq a_3/a_2$. The series is nongeometric.

For Probs. 12.82 to 12.86, write the given geometric series using sigma (summation) notation.

12.82 $2 + 1 + \frac{1}{2} + \frac{1}{4} + \frac{1}{8} + \cdots$

▮ These terms are each $\frac{1}{2}$ of the previous term, beginning with 2: $\sum_{i=1}^{\infty} 2(\frac{1}{2})^{i-1}$. When $i = 1$, this is $(\frac{1}{2})^0 = 1$; when $i = 2$, this is $\frac{1}{2}$, etc. See Prob. 12.83.

12.83 $2 + 1 + \frac{1}{2} + \frac{1}{4} + \frac{1}{8} + \cdots$

▮ Each term is a power of $\frac{1}{2}$. The first is $(\frac{1}{2})^{-1}$, the second is $(\frac{1}{2})^0$, the third is $(\frac{1}{2})^1$, etc. $\sum_{i=1}^{\infty} (\frac{1}{2})^{i-2}$ represents this. Note that this is the same series as in Prob. 12.82.

12.84 $x + x^2 + x^3 + \cdots$

▮ $\sum_{i=1}^{\infty} x^i$

12.85 $x - x^2 + x^3 + \cdots$

▮ $\sum_{i=1}^{\infty} x^i(-1)^{i+1}$. This will alternate signs beginning with a positive.

12.86 $\frac{1}{6} - (\frac{1}{6})^2 + (\frac{1}{6})^3 + \cdots$

▮ $\sum_{i=1}^{\infty} (-1)^{i+1}(\frac{1}{6})^i$.

For Probs. 12.87 to 12.99, find the sum, or show that the series has no sum.

12.87 $1 + \frac{1}{2} + \frac{1}{4} + \frac{1}{8} + \cdots$

▮ $S = \dfrac{a}{1-r}$ where $|r| < 1$. In this case $r = \frac{1}{2}$, $a = 1$. $S = \dfrac{1}{1 - \frac{1}{2}} = \dfrac{1}{\frac{1}{2}} = 2$.

12.88 $2 + \frac{2}{3} + \frac{2}{9} + \frac{2}{27} + \cdots$

▮ $r = \frac{1}{3}$, so $|r| < 1$. Thus, $S = \dfrac{a}{1-r} = \dfrac{2}{1 - \frac{1}{3}} = \dfrac{2}{\frac{2}{3}} = 3$.

12.89 $1 + \frac{1}{9} + \frac{1}{81} + \cdots$

▮ $r = \frac{1}{9}$, so $|r| < 1$. Thus, $S = \dfrac{a}{1-r} = \dfrac{1}{1 - \frac{1}{9}} = \dfrac{1}{\frac{8}{9}} = \frac{9}{8}$.

12.90 $3 + 6 + 12 + 24 + \cdots$

▮ Here, $r = 2$; thus, $|r| \geq 1$, and S does not exist.

12.91 $3 - \frac{3}{2} + \frac{3}{4} - \frac{3}{8} + \cdots$

▮ $r = -\frac{1}{2}$, so $|r| < 1$. $S = \dfrac{a}{1-r} = \dfrac{3}{1 - (-\frac{1}{2})} = \dfrac{3}{\frac{3}{2}} = 2$.

12.92 $3 + \frac{3}{2} + \frac{3}{4} + \frac{3}{8} + \cdots$

▮ $r = \frac{1}{2}$, so $|r| < 1$. Thus, $S = \dfrac{a}{1-r} = \dfrac{3}{1-\frac{1}{2}} = \dfrac{3}{\frac{1}{2}} = 6$.   Compare this to Prob. 12.91.

12.93 $\frac{1}{2} + \left(\frac{1}{2}\right)^2 + \left(\frac{1}{2}\right)^3 + \cdots$

▮ $r = \frac{1}{2}$, so $|r| < 1$. Thus, $S = \dfrac{a}{1-r} = \dfrac{\frac{1}{2}}{1-\frac{1}{2}} = \dfrac{\frac{1}{2}}{\frac{1}{2}} = 1$.

12.94 $-\frac{1}{2} + \left(\frac{1}{2}\right)^2 - \left(\frac{1}{2}\right)^3 + \cdots$

▮ This can be rewritten as $-\frac{1}{2} + \frac{1}{4} - \frac{1}{8} + \cdots$, which is geometric, $a = -\frac{1}{2}$, $r = -\frac{1}{2}$. Then $|r| < 1$, and $S = \dfrac{a}{1-r} = \dfrac{-\frac{1}{2}}{1-\left(-\frac{1}{2}\right)} = \dfrac{-\frac{1}{2}}{\frac{3}{2}} = -\frac{1}{3}$.

12.95 $1 - 1 + 1 - 1 + 1 \cdots$

▮ Here, $|r| = 1$, so the geometric series has no sum.

12.96 $x + x^2 + x^3 + \cdots$; $|x| < 1$

▮ $r = x$. Since $|x| < 1$, $S = \dfrac{a}{1-r} = \dfrac{x}{1-x}$.

12.97 $\displaystyle\sum_{i=1}^{\infty} (-1)^{i+1} x^i$, $|x| < \frac{1}{2}$.

▮ Since $|x| < \frac{1}{2}$, this series has a sum. $S = \dfrac{a}{1-r} = \dfrac{x}{1-(-x)} = \dfrac{x}{1+x}$.

12.98 $1 + 0.1 + 0.001 + \cdots$

▮ $r = 0.1$, $a = 1$; $S = \dfrac{a}{1-r} = \dfrac{1}{1-0.1} = \dfrac{1}{0.9} = \dfrac{10}{9}$.

12.99 $-3 + \frac{9}{4} - \frac{27}{16} + \frac{81}{64} + \cdots$

▮ $a = -3$, $r = -\frac{3}{2}$; $S = \dfrac{a}{1-r} = \dfrac{-3}{1-\left(-\frac{3}{2}\right)} = \dfrac{-3}{\frac{5}{2}} = -\dfrac{6}{5}$.

For Probs. 12.100 to 12.104, express the given decimal as a rational number in fraction form.

12.100 $0.\overline{2}$

▮ $0.\overline{2} = 0.222\cdots = 0.2 + 0.02 + 0.002 + \cdots = \dfrac{a}{1-r}$ (where $a = 0.2$; $r = 0.1$) $= \dfrac{0.2}{1-0.1} = \dfrac{0.2}{0.9} = \dfrac{2}{9}$.

12.101 $0.\overline{6}$

▮ $0.\overline{6} = 0.666\cdots = 0.6 + 0.06 + 0.006 + \cdots = \dfrac{a}{1-r}$ (where $a = 0.6$; $r = 0.1$) $= \dfrac{0.6}{1-0.1} = \dfrac{0.6}{0.9} = \dfrac{6}{9}$.

12.102 $0.\overline{12}$

▮ $0.\overline{12} = 0.121212\cdots 0.12 + 0.0012 + 0.000012 + \cdots = \dfrac{a}{1-r}$ (where $a = 0.12$; $r = 0.01$) $= \dfrac{0.12}{1-0.01} = \dfrac{0.12}{0.99} = \dfrac{12}{99}$.

12.103 $0.\overline{121}$

▮ $0.\overline{121} = 0.121 + 0.000121 + 0.000000121 + \cdots = \dfrac{a}{1-r}$ (where $a = 0.121$; $r = 0.001$) $= \dfrac{0.121}{1-0.001} = \dfrac{121}{999}$.

12.104 $0.1\overline{41}$

▮ Here, we rewrite $0.1\overline{41}$ as $0.1 + 0.0\overline{41} = 0.1 + \dfrac{a}{1-r}$ (where $a = 0.041$; $r = 0.01$) $= 0.1 + \dfrac{0.041}{0.99} = 0.1 + \dfrac{41}{990} = \dfrac{130}{990}$.

12.105 A given geometric series has a sum of $\frac{8}{3}$ and its first term is 2. Find the series.

▮ $S = \dfrac{a}{1-r}$; we are given that $a = 2$, $S = \frac{8}{3}$, so $\frac{8}{3} = \dfrac{2}{1-r}$. Then, $2 = \frac{8}{3}(1-r) = \frac{8}{3} - \frac{8}{3}r$, and $\frac{8}{3}r = \frac{2}{3}$, so $r = \dfrac{\frac{2}{3}}{\frac{8}{3}} = \frac{1}{4}$. If $a = 2$, $r = \frac{1}{4}$, then the series is $2 + \frac{1}{2} + \frac{1}{8} + \cdots$

12.106 Evaluate $1 + 1/e + 1/e^2 + 1/e^3 + \cdots$

▮ $r = 1/e$; since $2 < e < 3$, $|r| < 1$ and $S = \dfrac{a}{1-r} = \dfrac{1}{1 - 1/e} = \dfrac{e}{e-1}$.

12.5 THE BINOMIAL THEOREM

For Probs. 12.107 to 12.123, evaluate the given expression.

12.107 $7!$

▮ $7! = 7 \cdot 6 \cdot 5 \cdots 1 = 5040$.

12.108 $\dfrac{7!}{6!}$

▮ $\dfrac{7!}{6!} = \dfrac{7 \cdot 6 \cdots 2 \cdot 1}{6 \cdot 5 \cdots 2 \cdot 1} = 7$.

12.109 $\dfrac{7!}{4!}$

▮ $\dfrac{7!}{4!} = \dfrac{7 \cdot 6 \cdot 5 \cdot 4 \cdots 2 \cdot 1}{4 \cdot 3 \cdot 2 \cdot 1} = 7 \cdot 6 \cdot 5 = 210$.

12.110 $\dfrac{7!}{4!3!}$

▮ $\dfrac{7!}{4!3!} = \dfrac{7 \cdot 6 \cdot 5 \cdot 4 \cdot 3 \cdot 2 \cdot 1}{\underbrace{3 \cdot 2}_{3!} \cdot \underbrace{4 \cdot 3 \cdot 2 \cdot 1}_{4!}} = \dfrac{7 \cdot 6 \cdot 5}{6} = 35$.

12.111 $\dfrac{8!}{6!2!}$

▮ $\dfrac{8!}{6!2!} = \dfrac{8 \cdot 7 \cdot 6 \cdot 5 \cdot 4 \cdot 3 \cdot 2}{2 \cdot 6 \cdot 5 \cdot 4 \cdot 3 \cdot 2} = \dfrac{56}{2} = 28$.

12.112 $\dfrac{14!}{12!2!}$

▮ $\dfrac{14!}{12!2!} = \dfrac{14 \cdot 13 \cdot 12 \cdot 11 \cdot 10 \cdot 9 \cdot 8 \cdot 7 \cdot 6 \cdot 5 \cdot 4 \cdot 3 \cdot 2}{2 \cdot 12 \cdot 11 \cdot 10 \cdot 9 \cdot 8 \cdot 7 \cdot 6 \cdot 5 \cdot 4 \cdot 3 \cdot 2} = \dfrac{182}{2} = 91$

12.113 $\dfrac{102!}{2!100!}$

▮ $\dfrac{102!}{2!100!} = \dfrac{102 \cdot 101 \cdot 100!}{2 \cdot 100!} = 51 \cdot 101 = 5151$.

12.114 $\dfrac{6!}{2!2!2!}$

▮ $\dfrac{6!}{2!2!2!} = \dfrac{6 \cdot 5 \cdot 4 \cdot 3 \cdot 2}{2 \cdot 2 \cdot 2} = 3 \cdot 5 \cdot 2 \cdot 3 = 90$.

12.115 $\dfrac{5!}{2!(5-2)!}$

▮ $\dfrac{5!}{2!(5-2)!} = \dfrac{5!}{2!3!} = \dfrac{5 \cdot 4 \cdot 3 \cdot 2}{2 \cdot 3 \cdot 2} = \dfrac{20}{2} = 10$.

12.116 $\dfrac{15!}{0!(15-0)!}$

▮ $\dfrac{15!}{0!(15-0)!} = \dfrac{15!}{0!15!} = \dfrac{15!}{15!} = 1.$ (Remember: $0! = 1! = 1$.)

12.117 $\dbinom{2}{1}$

▮ $\dbinom{2}{1} = \dfrac{2!}{1!(2-1)!} = \dfrac{2!}{1} = 1,$ since $\dbinom{n}{r} = \dfrac{n!}{r!(n-r)!}$.

12.118 $\dbinom{4}{2}$

▮ $\dbinom{4}{2} = \dfrac{4!}{2!(4-2)!} = \dfrac{4!}{2!2!} = \dfrac{24}{2} = 6.$

12.119 $\dbinom{5}{2}$

▮ $\dbinom{5}{2} = \dfrac{5!}{2!(5-2)!} = \dfrac{5!}{2!3!} = \dfrac{5 \cdot 4 \cdot 3 \cdot 2}{2 \cdot 3 \cdot 2} = \dfrac{20}{2} = 10.$

12.120 $\dbinom{6}{2}$

▮ $\dbinom{6}{2} = \dfrac{6!}{2!4!} = \dfrac{6 \cdot 5 \cdot 4 \cdot 3 \cdot 2}{2 \cdot 4 \cdot 3 \cdot 2} = \dfrac{30}{2} = 15.$

12.121 $\dbinom{7}{0}$

▮ $\dbinom{7}{0} = \dfrac{7!}{0!(7-0)!} = \dfrac{7!}{7!} = 1.$

12.122 $\dbinom{150}{0}$

▮ $\dbinom{150}{0} = \dfrac{150!}{0!150!} = 1.$

12.123 $\dbinom{80}{80}$

▮ $\dbinom{80}{80} = \dfrac{80!}{80!(80-80)!} = \dfrac{80!}{80!} = 1.$

For Probs. 12.124 to 12.131, expand the given expression using the binomial theorem.

12.124 $(x+y)^2$

▮ Remember that $(a+b)^n = \sum\limits_{r=0}^{n} \dbinom{n}{k} a^{n-k}b^k, \quad n \geq 1.$ In this case, $(x+y)^2 = \sum\limits_{k=0}^{2} \dbinom{2}{k} x^{2-k}y^k = \dbinom{2}{0}x^2 + \dbinom{2}{1}xy + \dbinom{2}{2}y^2 \quad (k=0,\ 1,\ \text{and } 2, \quad \text{respectively}) = x^2 + 2xy + y^2.$ All problems in which we use the binomial theorem to expand an exponential form of a binomial are done this way.

12.125 $(x+y)^3$

▮ $(x+y)^3 = \dbinom{3}{0}x^3 + \dbinom{3}{1}x^2y + \dbinom{3}{2}xy^2 + \dbinom{3}{3}y^3 = x^3 + 3x^2y + 3xy^2 + y^3.$ $\left[\text{Here,} \quad (x+y)^3 = \sum\limits_{k=0}^{3} \dbinom{3}{k} x^{3-k}y^k.\right]$

12.126 $(p+q)^4$

▮ $(p+q)^4 = \dbinom{4}{0}p^4 + \dbinom{4}{1}p^3q + \dbinom{4}{2}p^2q^2 + \dbinom{4}{3}pq^3 + \dbinom{4}{4}q^4 = p^4 + 4p^3q + 6p^2q^2 + 4pq^3 + q^4.$

12.127 $(a+3)^3$

▮ $(a+3)^3 = \binom{3}{0}a^3 + \binom{3}{1}a^2 \cdot 3 + \binom{3}{2}a \cdot 3^2 + \binom{3}{3}3^3 = a^3 + 3a^2 \cdot 3 + 3a \cdot 9 + 27 = a^3 + 9a^2 + 27a + 27.$

12.128 $(a-3)^3$

▮ $(a-3)^3 = [a+(-3)]^3 = \binom{3}{0}a^3 + \binom{3}{1}a^2 \cdot (-3) + \binom{3}{2}a \cdot (-3)^2 + \binom{3}{3}(-3)^3 = a^3 + 3a^2(-3) + 3a(9) + (-27) = a^3 - 9a^2 + 27a - 27.$

12.129 $(2a+3)^5$

▮ $(2a+3)^5 = \binom{5}{0}(2a)^5 + \binom{5}{1}(2a)^{4}(3) + \binom{5}{2}(2a)^3 3^2 + \binom{5}{3}(2a)^2 3^3 + \binom{5}{4}(2a)3^4 + \binom{5}{5}3^5 = (2a)^5 + 5(3)(2a)^4 + 10(2a)^3(9) + 10(2a)^2(27) + 5(2a)81 + 243 = 32a^5 + 30a^4 + 180a^3 + 540a^2 + 810a + 243.$

12.130 $(2x-4)^3$

▮ $(2x-4)^3 = \binom{3}{0}(2x)^3 + \binom{3}{1}(2x)^2(-4) + \binom{3}{2}(2x)(-4)^2 + \binom{3}{3}(-4)^3 = 8x^3 + (3)(4x^2)(-4) + 3(2x)(16) - 64 = 8x^3 - 48x^2 + 96x - 64.$

12.131 $(4-2u)^4$

$(4-2u)^4 = \binom{4}{0}4^4 + \binom{4}{1}4^3(-2u) + \binom{4}{2}4^2(-2u)^2 + \binom{4}{3}4(-2a)^3 + \binom{4}{4}(-2u)^4 = 256 + 4 \cdot 4^3(-2u) + 6 \cdot 4^2(-2u)^2 + 4 \cdot 4(-2u)^3 + (-2u)^4 = 256 - 512u + 384u^2 - 128u^3 + 16u^4.$

12.132 Find the coefficient of x^2y^5 in the expansion of $(x-y)^7$.

▮ Recall that $(x-y)^7 = \sum_{k=0}^{7} \binom{7}{k}x^{7-k}(-y)^k$. Thus, the coefficient of x^2y^5 occurs when $7-k=2$, or $k=5$. The coefficient $\binom{7}{k}$, then, is $\binom{7}{5}$ with a negative sign since $(-y)^k < 0$ when k is odd. Then, $-\binom{7}{5} = -\frac{7 \cdot 6 \cdot 5 \cdot 4 \cdot 3 \cdot 2}{2 \cdot 5 \cdot 4 \cdot 3 \cdot 2} = -21.$

12.133 Find the seventh term in the expansion of $(u+v)^{15}$.

▮ The seventh term is $\binom{15}{6}u^9v^6 = 5005u^9v^6.$

For Probs. 12.134 to 12.136, tell whether the statement is true of false.

12.134 $(2 \times 3)! = 2!3!$

▮ False; $2!3! = 4 \cdot 6 = 24$, but $(2 \times 3)! = 6! \neq 24$. Notice that, in general, $(ab)! \neq a!b!$

12.135 If $m = n$, then $m! = n!.$

▮ True; Since $m! = m \cdot (m-1) \cdots 2 \cdot 1$, if $m = n$ then $m!$ must be $n!$.

12.136 If $m! = n!$, then $m = n.$

▮ If $m! = n!$, then $m \cdot (m-1) \cdots 2 \cdot 1 = n(n-1) \cdots 2 \cdot 1$, which means $m = n$ unless $m! = n! = 1$, in which case it is possible that $m = 0$, $n = 1$. False.

12.137 Prove: $\binom{n}{r} = \binom{n}{n-r}$.

▮ $\binom{n}{r} = \frac{n!}{r!(n-r)!}$ and $\binom{n}{n-r} = \frac{n!}{(n-r)![n-(n-r)]!} = \frac{n!}{(n-r)!r!}$. Thus, they are equal.

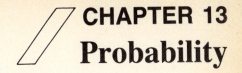

CHAPTER 13
Probability

13.1 PERMUTATIONS AND COMBINATIONS

For Probs. 13.1 to 13.10, calculate the given permutation or combination.

13.1 $P(7, 1)$

▮ (i) $P(n, r) = n(n-1)(n-2) \cdots (n-r+1)$. Thus, $P(7,1) = 7(6)(5) \cdots (7-1+1) = 7$.
(ii) $P(n, r) = \dfrac{n!}{(n-r)!} = \dfrac{7!}{(7-1)!} = \dfrac{7!}{6!} = 7$.

13.2 $P(7, 3)$

▮ $P(7, 3) = \dfrac{7!}{(7-3)!} = \dfrac{7!}{4!} = 7 \cdot 6 \cdot 5 = 210$.

13.3 $P(7, 4)$

▮ $P(7, 4) = \dfrac{7!}{(7-4)!} = \dfrac{7!}{3!} = 7 \cdot 6 \cdot 5 \cdot 4 = 840$.

13.4 $P(7, 7)$

▮ $P(7, 7) = \dfrac{7!}{(7-7)!} = \dfrac{7!}{0!} = 7! = 5040$.

13.5 $P(4, 1) + P(4, 2) + P(4, 3) + P(4, 4)$

▮ $P(4, 1) + P(4, 2) + P(4, 3) + P(4, 4) = \dfrac{4!}{3!} + \dfrac{4!}{2!} + \dfrac{4!}{1!} + \dfrac{4!}{0!} = 4 + 12 + 24 + 24 = 64$.

13.6 $C(31, 2)$

▮ $C(n, r) = \dfrac{n!}{r!(n-r)!}$. Thus, $C(31, 2) = \dfrac{31!}{2!29!} = \dfrac{31 \cdot 30}{2!} = 465$.

13.7 $C(26, 3)$

▮ $C(26, 3) = \dfrac{26!}{3!23!} = \dfrac{26 \cdot 25 \cdot 24}{6} = 2600$.

13.8 $C(24, 4)$

▮ $C(24, 4) = \dfrac{24!}{4!20!} = \dfrac{24 \cdot 23 \cdot 22 \cdot 21}{24} = 10,626$.

13.9 $C(13, 5)$

▮ $C(13, 5) = \dfrac{13!}{5!8!} = \dfrac{13 \cdot 12 \cdots 8 \cdot 7 \cdots 2 \cdot 1}{5!} = \dfrac{13 \cdot 12 \cdot 11 \cdot 10 \cdot 9}{5!} = 1287$.

13.10 $C(16, 4)$

▮ $C(16, 4) = \dfrac{16!}{4!12!} = \dfrac{16 \cdot 15 \cdot 14 \cdot 13 \cdot 12!}{4! \cdot 12!} = \dfrac{16 \cdot 15 \cdot 14 \cdot 13}{4!} = 1820$.

13.11 How many two-digit numbers can be made with the digits 2, 4, 6, 8, if repetitions are allowed?

▮ There are 4 choices (2, 4, 6, 8) for each digit: $4 \cdot 4 = 16$.

13.12 Same as Prob. 13.11, but with no repetitions allowed.

▮ In this case there are 4 choices for the first digit and 3 choices for the second digit: $4 \cdot 3 = 12$.

13.13   How many three-digit numbers can be formed with the digits   1, 2, 3, 4, 5,   if repetitions are allowed?

▮   There are 5 choices for each digit: $5 \cdot 5 \cdot 5 = 125$.

13.14   Same as Prob. 13.13, but with no repetitions allowed.

▮   There are 5 choices for the first digit, 4 for the second, and 3 for the third: $5 \cdot 4 \cdot 3 = 60$.

13.15   How many three-digit numbers (with 0 not the first digit) can be formed with digits   0, 1, 2, 3, 4,   if repetitions are allowed?

▮   There are 4 choices for the first digit   (1, 2, 3, 4)   and 5 choices for the others: $4 \cdot 5 \cdot 5 = 100$.

13.16   Same as Prob. 13.15, but with no repetitions.

▮   There are 4 choices for the first, 4 for the second (since 0 is added in as a possibility), and 3 for the third: $4 \cdot 4 \cdot 3 = 48$.

13.17   How many four-digit numbers can be formed each less than 5000, with the digits   1, 2, 4, 6, 8,   if repetitions are allowed?

▮   The first digit can be   1, 2, or 4.   The second, third, and fourth digits can each be any one of   1, 2, 4, 6, 8: $3 \cdot 5 \cdot 5 \cdot 5 = 375$.

13.18   Same as Prob. 13.17, with no repetitions.

▮   The first digit can be   1, 2, or 4.   The second digit can be   1, 2, 4, 6, or 8   but *not* the digit in the first slot: $3 \cdot 4 \cdot 3 \cdot 2 = 72$.

13.19   How many committees, consisting of 1 freshman, 1 sophomore, and 1 junior can be selected from 40 freshmen, 30 sophomores, and 25 juniors?

▮ $40 \cdot 30 \cdot 25 = 30,000$.

13.20   Five boys are in a room which has 4 doors.   In how many ways can they leave the room?

▮   Each boy has 5 choices. $4^5 = 4 \cdot 4 \cdot 4 \cdot 4 \cdot 4 = 1024$.

For Probs. 13.21 to 13.23, find the number of permutations, each with 7 letters, which can be made with the letters given.

13.21   WYOMING

▮ $P(7, 7)$ [The first 7 indicates that we have 7 letters to choose from; the second 7 indicates that we want to choose 7 letters from those available.] $= \dfrac{7!}{(7-7)!} = \dfrac{7!}{0!} = 7! = 5040$.

13.22   KENTUCKY

▮ $P(8, 7) = \dfrac{8!}{(8-7)!} = \dfrac{8!}{1!} = 8! = 40,320$.

13.23   WASHINGTON

▮ $P(10, 7) = \dfrac{10!}{(10-7)!} = \dfrac{10!}{6} = 604,800$.

13.24   Find how many even numbers of three digits can be made with the digits   1, 2, 3, 4, 5, 6, 7,   if no digit is repeated.

▮   The units digit must be   2, 4, or 6.   The other digits are any one of   1, 2, 3, 4, 5, 6, 7,   with no repetitions: $3 \cdot 6 \cdot 5 = 90$.

13.25   Seven songs are to be given in a program.   In how many different orders could they be rendered?

▮ $7! = 5040$.

For Probs. 13.26 to 13.28, find the number of permutations which can be formed by using all the given letters.

13.26 ALABAMA

▮ $\dfrac{7!}{4!} = 7 \cdot 6 \cdot 5 = 210$ (7 objects, 4 objects are alike).

13.27 ARKANSAS

▮ $\dfrac{8!}{3!2!} = 3360$ (8 letters, 3 A's, 2 Ss).

13.28 MISSISSIPPI

▮ $\dfrac{11!}{4!2!4!} = 34{,}650$ (4 Ss, 2 P's, 4 I's).

13.29 Find the number of ways 2 half dollars, 4 quarters, and 6 dimes can be distributed among 12 boys if each gets a coin.

▮ $\dfrac{12!}{2!4!6!} = 13{,}860.$

13.30 A line is formed by 5 girls and 5 boys, with boys and girls alternating.   Find the number of ways of making the line.

▮ There are 10 spots.   The first is filled by a boy or girl.   There are 10 possibilities.   Then, there are 5 boys or girls left, and 4 left of the other gender, leaving 9 altogether: $10 \cdot 5 \cdot 4 \cdot 4 \cdot 3 \cdot 3 \cdot 2 \cdot 2 \cdot 1 \cdot 1 = 28{,}800.$

13.31 Same as Prob. 13.30, but with a circular arrangement.

▮ $\dfrac{28{,}000}{10} = 2880.$ In a circle, there is no first or last object.

13.32 Show that $P(5,4) = P(5,5)$.

▮ $P(5,4) = \dfrac{5!}{(5-4)!} = \dfrac{5!}{1!} = \dfrac{5!}{0!} = P(5,5)$ since $0! = 1!$.

13.33 Show that $P(n, n-1) = P(n, n)$.

▮ $P(n,n) = \dfrac{n!}{(n-n)!} = \dfrac{n!}{0!}$; $P(n, n-1) = \dfrac{n!}{[n-(n-1)]!} = \dfrac{n!}{1!} = \dfrac{n!}{0!}$ since $0! = 1!$.

13.34 In how many different ways can a tennis team of 4 be chosen from 17 players?

▮ Here we are looking for the number of combinations of 17 objects taken 4 at a time.   The concern here is *not* the order of selection or the arrangement of the players.   Compare this to, for example, Prob. 13.25 which is a permutation: $C(17,4) = \dfrac{17!}{4!13!} = 2380.$

13.35 Nine points, no three of which are on a straight line, are marked on a blackboard.   How many lines, each through two of the points, can be drawn?

▮

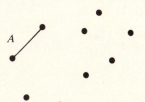

A

Fig. 13.1

See Fig. 13.1.   The point A can be connected to any of the other points. $C(9,2) = \dfrac{9!}{2!7!} = \dfrac{72}{2} = 36.$

13.36 In Prob. 13.35, how many triangles are determined?

▮ $C(9,3) = \dfrac{9!}{3!6!} = 84.$

13.37 Seven different coins are tossed simultaneously. In how many ways can 3 heads and 4 tails come up?

▮ $C(7, 3) \cdot C(4, 4) = 35$. [Choose 3 heads, and then the other 4 must all be tails; $C(4, 4) = 1$.]

13.38 In how many ways can a court of 9 judges make a 5-to-4 decision?

▮ $C(9, 5) \cdot C(4, 4) = \dfrac{9!}{4!5!} \cdot 1 = 126$.

For Probs. 13.39 to 13.41, 4 delegates are to be chosen from 8 members of a club.

13.39 How many choices are possible?

▮ $C(8, 4) = \dfrac{8!}{4!4!} = 70$.

13.40 How many contain member A?

▮ This is the same as choosing 3 from a group of 7, since we are fixing A as a chosen member: $\dfrac{7!}{3!4!} = 35$.

13.41 How many contain A and B?

▮ Here 2 members are fixed. $C(6, 2) = \dfrac{6!}{4!2!} = 15$. The alphabet consists of 21 consonants and 5 vowels. Refer to this information for Probs. 13.42 to 13.45.

13.42 In how many ways can 5 consonants and 3 vowels be selected?

▮ $C(21, 5) \cdot C(5, 3) = 20{,}349 \cdot 40 = 813{,}960$.

13.43 How many words consisting of 5 consonants and 3 vowels can be formed?

▮ 8! words can be found from each combination in Prob. 13.42 by taking all permutations of 8 letters: $C(21, 5) \cdot C(5, 3) \cdot 8!$.

13.44 How many of the words in Prob. 13.43 begin with X?

▮ X's position is fixed. We select 4 other consonants and the 3 vowels. There are 7 choices altogether since X is fixed: $7! \cdot C(20, 4) \cdot C(5, 3)$.

13.45 How many of the words in Prob. 13.44 contain A?

▮ A's position is not fixed. We select 4 other consonants and 2 other vowels: $C(20, 4) \cdot C(4, 2) \cdot 7!$

13.46 Prove: $C(n, r) + C(n, r + 1) = C(n + 1, r + 1)$.

▮ $C(n, r) = \dfrac{n!}{r!(n - r)!}$ and $C(n, r + 1) = \dfrac{n!}{(r + 1)![n - (r + 1)]!}$. Then, $C(n, r) + C(n, r + 1)$

$= \dfrac{n!}{r!(n - r)!} + \dfrac{n!}{(r + 1)!(n - r - 1)!} = \dfrac{n!(r + 1) + n!(n - r)}{(r + 1)!(n - r)!} = \dfrac{(n + 1)1}{(r + 1)!(n - r)!} = C(n + 1, r + 1)$.

13.47 Prove $C(2n, n) = 2C(2n - 1, n - 1)$.

▮ $C(2n, n) = \dfrac{(2n)!}{n!(2n - n)!} = \dfrac{(2n)!}{n!n!} = \dfrac{(2n)!}{(n!)^2}$ and $2C(2n - 1, n - 1) = \dfrac{2(2n - 1)!}{(n - 1)!n!} = \dfrac{(2n)!}{(n!)^2}$.

For Probs. 13.48 to 13.50, solve for n.

13.48 $P(n, 2) = 110$

▮ $P(n, 2) = n(n - 1) = n^2 - n = 110$, or $n = 11$ $(n > 0)$.

13.49 $P(n, 4) = 30P(n, 2)$

▮ $n(n - 1)(n - 2)(n - 3) = 30n(n - 1)$. Thus, $n(n - 1)(n - 8)(n + 3) = 0$ (check the algebra). But, $n \geq 4$, so $n = 8$.

13.50 $C(n, 3) = 84$

▮ $\frac{n!}{3!(n-3)!} = 84$; $\frac{n(n-2)(n-1)}{6} = 84$. $n(n-2)(n-1) = 504$, $n^3 - 3n^2 + 2n - 504 = 0$, $(n-9)(n^2 + 6n + 56) = 0$, or $n = 9$.

13.2 ELEMENTARY PROBABILITY

For Probs. 13.51 to 13.53, a ball is chosen from a bag containing 2 white, 1 black, and 2 red balls.

13.51 Find the probability a white ball is chosen.

▮ $P(W) = \frac{2}{5}$. Since there are 2 whites (we are looking for the probability of drawing white) and there are 5 balls altogether, $P(E) = S/n$ where E is an event, S = number of ways the event can occur, and n = total number of possibilities.

13.52 Find the probability a black ball is chosen.

▮ $P(B) = \frac{1}{5}$. (There is only 1 black; there are 5 balls.)

13.53 Find the probability of choosing a green ball.

▮ $P(G) = \frac{0}{5} = 0$ since there are 0 green balls.

For Probs. 13.54 to 13.57, a single die is rolled.

13.54 Find the probability that it lands on 6.

▮ $P(6) = \frac{1}{6}$ since only one side has a 6, and there are six sides altogether.

13.55 Find the probability it lands on an even number.

▮ $P(\text{even}) = P(2, 4, \text{or } 6) = \frac{3}{6} = \frac{1}{2}$.

13.56 Find the probability it lands on 2 or an odd number.

▮ $P(2 \text{ or odd}) = P(1, 2, 3, 5) = \frac{4}{6} = \frac{2}{3}$.

13.57 Find the probability it lands on a number less than 10.

▮ $P(<10) = \frac{6}{6}$ since all 6 possibilities are less than 10.

For Probs. 13.58 to 13.60, 3 pennies are tossed at the same time.

13.58 Find the probability that all 3 land on heads.

▮ $P(H, H, H) = \frac{1}{8}$ since there are $2^3 = 8$ possibilities. Altogether each coin has 2 possibilities, and only one of these is H, H, H.

13.59 Find the probability that 2 are H and 1 is T.

▮ There are $2^3 = 8$ possibilities altogether. How many are H, H, T? There are $C(3, 1)$ ways of one coin being T. $C(3, 1) = \frac{3!}{1!2!} = 3$. Thus, $P(H, H, T) = \frac{3}{8}$. *Note*: Since $C(3, 1) = C(3, 2)$, we could have used $C(3, 2)$ instead. The number of ways of 1 coin out of 3 being T is the same as 2 coins out of 3 being H.

13.60 Find the probability that 1 is H and 2 are T.

▮ $P(T, T, H) = \frac{C(3, 2)}{8} = \frac{3}{8}$.

For Probs. 13.61 to 13.64, 3 dice are rolled.

13.61 Find the probability that their sum is 18.

▮ $P(\text{sum} = 18) = P(6, 6, 6) = 1/6^3$ (since that is the only way to get 18) $= \frac{1}{216}$.

13.62 Find the probability of the sum being less than 5.

▮ To be less than 5, it must be 3 or 4. To be 3, it must be 1, 1, 1. To be 4, it could be 2, 1, 1 or 1, 2, 1 or 1, 1, 2 [i.e, there are $C(3, 1)$, or 3 ways]. Altogether, there are 3 ways to get 4; altogether, there is 1 way to get 3. $P(<5) = P(3 \text{ or } 4) = \frac{3+1}{6^2} = \frac{4}{36} = \frac{1}{9}$.

13.63 Find the probability that the sum is 6.

▮ $P(6) = P(1, 1, 4) + P(1, 4, 1) + P(4, 1, 1) + P(2, 2, 2) + P(2, 3, 1) + P(2, 1, 3) + P(3, 1, 2) + P(3, 2, 1)$
$+ P(1, 2, 3) + P(1, 3, 2) = \frac{10}{36}$.

13.64 Find the probability that the sum is even.

▮ $P(E) = \frac{8}{16} = \frac{1}{2}$ since there are 16 possibilities $(3 - 18)$ and 8 of those are even $(4, 6, 8, 10, 12, 14, 16, 18)$.

For Probs. 13.65 to 13.67, 3 balls are drawn from a bag containing 6 red and 5 black balls.

13.65 Find the probability that all are red.

▮ $P(\text{all red}) = P(R, R, R)$. There are $C(11, 3)$ possibilities altogether. $C(11, 3) = \frac{11!}{3!8!} = 165$. Since 6 balls are red, there are $C(6, 3)$ ways to pick R, R, R. $C(6, 3) = 20$. $P(R, R, R) = \frac{20}{165} = \frac{4}{33}$.

13.66 Find the probability that all are black.

▮ $P(B, B, B) = \frac{C(5, 3)}{C(11, 3)} = \frac{10}{165} = \frac{2}{33}$.

13.67 Find the probability of 2 being red and 1 being black.

▮ There are $C(6, 2)$ ways to pick 2 red balls, and $C(5, 1)$ ways to pick 1 black ball. Therefore, $P(2R, 1B) = \frac{C(6, 2)C(5, 1)}{C(11, 3)} = \frac{5}{11}$.

For Probs. 13.68 to 13.71, 1 card is picked from an ordinary 52-card deck.

13.68 Find the probability that a red card is chosen.

▮ $P(R) = \frac{26}{52} = \frac{1}{2}$ since 26 cards are red and 26 are black.

13.69 Find the probability that a queen is chosen.

▮ $P(Q) = \frac{4}{52} = \frac{1}{13}$ since there are 4 queens (1 per suit).

13.70 Find the probability that a red 8 is chosen.

▮ $P(\text{red } 8) = \frac{2}{52} = \frac{1}{26}$ since only the 8 of hearts and 8 of diamonds are red.

13.71 Find the probability that the card chosen is less than 5 (counting ace high).

▮ $P(<5) = P(2, 3, \text{or } 4)$. For each of $2, 3, 4$ there are 4 suits. $P(2, 3, \text{ or } 4) = \frac{12}{52} = \frac{3}{13}$.

For Probs. 13.72 to 13.75, 2 cards are chosen simultaneously from an ordinary deck.

13.72 Find the probability both cards chosen are 5s.

▮ There are $C(4, 2)$ ways to choose two 5s from four 5s, and $C(52, 2)$ ways to choose 2 cards from 52 cards. $P(5, 5) = \frac{C(4, 2)}{C(52, 2)} = \frac{6}{1326}$.

13.73 Find the probability both cards chosen are aces.

▮ See Prob. 13.72. This is the exact same situation $P(A, A) = \frac{6}{1326}$.

13.74 Find the probability that 1 ace and 1 king are chosen.

$\blacksquare$ There are $C(4, 1)$ ways to choose 1 ace and 1 king. Therefore, $P(A, K) = \dfrac{C(4, 1)C(4, 1)}{C(52, 2)} = \dfrac{4 \cdot 4}{1275} = \dfrac{16}{1326}$.

13.75 Find the probability of choosing 2 black cards.

$\blacksquare$ $P(B, B) = \dfrac{C(26, 2)}{C(52, 2)} = \dfrac{325}{1326}$.

13.76 One card is drawn from each of 3 decks of 52 cards. What is the probability that all are aces?

$\blacksquare$ $P(A, A, A) = \dfrac{4}{52} \cdot \dfrac{4}{52} \cdot \dfrac{4}{52} = \left(\dfrac{1}{13}\right)^3 = \dfrac{1}{2197}$.

13.77 One card is drawn from each of 3 decks of 52 cards. What is the probability they are all face cards?

$\blacksquare$ $P(\text{face, face, face}) = \left(\dfrac{12}{52}\right)^3 = \dfrac{27}{2197}$.

13.78 Find the probability that 3 spades are chosen when 1 card is chosen from each of three 52-card decks.

$\blacksquare$ $P(S, S, S) = \left(\dfrac{1}{4}\right)^3 = \dfrac{1}{64}$.

For Probs. 13.79 to 13.83, the probability that a brush salesman will make a sale at one house is $\frac{2}{3}$ and at a second house is $\frac{1}{2}$.

13.79 Find the probability he will make both sales.

$\blacksquare$ $P(\text{both sales}) = \dfrac{2}{3} \cdot \dfrac{1}{2} = \dfrac{2}{6} = \dfrac{1}{3}$.

13.80 Find the probability he will make neither sale.

$\blacksquare$ $P(\text{no sales}) = (1 - \frac{2}{3})(1 - \frac{1}{2}) = \frac{1}{3} \cdot \frac{1}{2} = \frac{1}{6}$.

13.81 Find the probability he will make a sale at the first but not the second house.

$\blacksquare$ $P(\text{sale, no sale}) = \frac{2}{3}(1 - \frac{1}{2}) = \frac{2}{3} \cdot \frac{1}{2} = \frac{1}{3}$.

13.82 Find the probability he will make at least 1 sale.

$\blacksquare$ $P(\geq 1) = P(1 \text{ sale}) + P(2 \text{ sales}) = P(\text{sale, no sale}) + P(\text{no sale, sale}) + P(\text{sale, sale}) = \frac{1}{3} + \frac{1}{3} + \frac{1}{6} = \frac{5}{6}$. [Alternately, $P(\geq 1) = 1 - P(\text{no sale}) = 1 - \frac{1}{6} = \frac{5}{6}$.] (See Prob. 13.80.)

13.83 Find the probability he will make *exactly* 1 sale.

$\blacksquare$ $P(1 \text{ sale}) = P(\text{sale, no sale}) + P(\text{no sale, sale}) = \frac{1}{3} + \frac{1}{3} = \frac{2}{3}$.

For Probs. 13.84 to 13.86, the names of 9 freshmen, 8 sophomores, and 7 juniors are written on cards and placed in a box. Find the given probability if 3 cards are drawn from the box, one after the other.

13.84 The probability of drawing 3 freshmen.

$\blacksquare$ After drawing 1 freshman, there will be 8 freshmen and 23 total cards left. Therefore, $P(3F) = \dfrac{9}{24} \cdot \dfrac{8}{23} \cdot \dfrac{7}{22} = \dfrac{21}{506}$.

13.85 The probability of getting 1 card from each class.

$\blacksquare$ This is $\dfrac{9}{24} \cdot \dfrac{8}{23} \cdot \dfrac{7}{22} \cdot 6$ (There are 6 arrangements of the 3 cards.) $= \dfrac{63}{253}$.

13.86 The probability of getting all 3 cards from 1 class.

$\blacksquare$ $P(\text{all 3 from 1 class}) = P(3F) + P(3S) + P(3J)$ (see Prob. 13.84) $= \underbrace{\dfrac{21}{506}}_{\text{Freshmen}} + \underbrace{\dfrac{8}{24} \cdot \dfrac{7}{23} \cdot \dfrac{6}{22}}_{\text{Sophomores}} + \underbrace{\dfrac{7}{24} \cdot \dfrac{6}{23} \cdot \dfrac{5}{22}}_{\text{Juniors}} = \dfrac{175}{2024}$.

For Probs. 13.87 to 13.90, 3 balls are drawn together from a bag containing 8 white and 12 black balls. Find the given probability.

13.87 All are white.

$\blacksquare$ $P(W, W, W) = \dfrac{C(8, 3)}{C(20, 3)} = \dfrac{14}{285}$.

13.88 Just 2 balls are white.

$\blacksquare$ $P(2W) = \dfrac{C(8, 2) \cdot C(12, 1)}{C(20, 3)}$ (We choose 2 whites and 1 black.) $= \dfrac{28}{95}$.

13.89 Just 1 ball is white.

$\blacksquare$ $P(1W) = \dfrac{C(8, 1) \cdot C(12, 2)}{C(20, 3)} = \dfrac{44}{95}$.

13.90 All balls are black.

$\blacksquare$ $P(3B) = \dfrac{C(12, 3)}{C(20, 3)} = \dfrac{11}{27}$.

For Probs. 13.91 to 13.95, 4 cards are drawn from a box containing 10 cards numbered $1, 2, \ldots, 10$. Find the given probability.

13.91 All are even if the cards are all drawn together.

$\blacksquare$ $P(4 \text{ evens}) = \dfrac{C(5, 4)}{C(10, 4)}$ (draw 4 evens from 5 evens) $= \frac{5}{210} = \frac{1}{42}$.

13.92 All are even if each card is replaced before the next is drawn.

$\blacksquare$ $P(\text{1st is even}) = \frac{1}{2}$; $P(\text{2d even given 1st even replaced}) = \frac{1}{2}$,   etc.; $\left(\frac{1}{2}\right)^4 = \frac{1}{16}$.

13.93 Exactly 3 cards are even if the cards are all drawn together.

$\blacksquare$ $P(3 \text{ even}) = P(3 \text{ even}, 1 \text{ odd}) = \dfrac{C(5, 3) \cdot C(5, 1)}{C(10, 4)} = \frac{50}{210} = \frac{5}{21}$.

13.94 Exactly 3 cards are even if the cards are replaced after they are drawn.

$\blacksquare$ $P(\text{1st even}) = \frac{1}{2}$; $P(\text{2d even}) = \frac{1}{2}$; $P(\text{3d even}) = \frac{1}{2}$;   and there are $C(4, 3)$ ways to do this.
$P(3 \text{ even}) = P(3 \text{ even}, 1 \text{ odd}) = \left(\frac{1}{2}\right)^3 \cdot \left(\frac{1}{2}\right)^1 \cdot 4 = \frac{1}{8} \cdot \frac{1}{2} \cdot 4 = \frac{1}{4}$.

13.95 At least 3 cards are even if all cards are drawn at the same time.

$\blacksquare$ $P(\geq 3 \text{ even}) = P(3 \text{ even}) + P(4 \text{ even}) = \frac{5}{21} + \frac{1}{42}$ (See Probs. 13.91 and 13.93.) $= \frac{11}{42}$.

For Probs. 13.96 to 13.98, the probability that X will win a game of checkers is $\frac{2}{5}$. In a five-game match, find the given probability.

13.96 The probability X will win the first, third, and fifth games and lose the others.

$\blacksquare$ The probability of winning 3 games is $\left(\frac{2}{5}\right)^3$. The probability of losing the other 2 is $\left(\frac{3}{5}\right)^2$. Therefore, $P = \left(\frac{2}{5}\right)^3 \cdot \left(\frac{3}{5}\right)^2 = \frac{72}{3125}$. *Note*: $P(\text{a loss}) = 1 - \frac{2}{5} = \frac{3}{5}$.

13.97 The probability that X will win exactly 3 games.

$\blacksquare$ The probability that X will win 3 games is $\left(\frac{2}{5}\right)^3$. The probability that X will lose 2 games is $\left(\frac{3}{5}\right)^2$. There are $C(5, 2)$ ways to choose 2 losses out of 5 games. Therefore, $P = \left(\frac{2}{5}\right)^3 \cdot \left(\frac{3}{5}\right)^2 \cdot C(5, 2) = \frac{72}{3125} \cdot 10 = \frac{720}{3125} = \frac{144}{625}$.

13.98   The probability that X will win at least 3 games.

▐ $P = P(3W) + P(4W) + P(5W)$ [This can also be written as $1 - P(0W) - P(1W) - P(2W)$.]
$= \left(\frac{2}{5}\right)^3 \cdot \left(\frac{3}{5}\right)^2 \cdot C(5,2) + \left(\frac{2}{5}\right)^4 \cdot \left(\frac{3}{5}\right)^1 \cdot C(5,1) + \left(\frac{2}{5}\right)^5 = \frac{992}{3125}$.

For Probs. 13.99 and 13.100, bag 1 contains 2 white and 1 red ball; bag 2 contains 3 white and 1 red ball.   A bag is chosen at random and a ball from it is picked at random.   What is the given probability?

13.99   The ball chosen is red.

▐   The probability of choosing bag 1 is $\frac{1}{2}$.   The probability of choosing a red ball is then $\frac{1}{3}$.   Similarly, the probability of choosing bag 2 is $\frac{1}{2}$, but here the probability of choosing a red ball is $\frac{1}{4}$.   Therefore, $P(\text{red}) = \frac{1}{2} \cdot \frac{1}{3} + \frac{1}{2} \cdot \frac{1}{4} = \frac{1}{6} + \frac{1}{8} = \frac{7}{24}$.

13.100   The ball chosen is white.

▐ $P(\text{white}) = \frac{1}{2} \cdot \frac{2}{3} + \frac{1}{2} \cdot \frac{3}{4} = \frac{17}{24}$.   (Alternately: $P(W) = 1 - P(R) = 1 - \frac{7}{14} - \frac{17}{24}$.)

CHAPTER 14
Conic Sections and Transformations

14.1 THE CIRCLE

For Probs. 14.1 to 14.12, find the center and radius of the given circle.

14.1 $x^2 + y^2 = 14$

▮ If $(x - h)^2 + (y - k)^2 = r^2$, then (h, k) is the center, r is the radius. Here, $C(0, 0)$, $r = 14$.

14.2 $(x - 1)^2 + y^2 = 8$

▮ $C(1, 0)$, $r = \sqrt{8} = 2\sqrt{2}$.

14.3 $(x + 2)^2 + (y + 4)^2 = 10$

▮ $C(-2, -4)$, $r = \sqrt{10}$.

14.4 $(x - 1)^2 + (y - 3)^3 = 0$

▮ This is just the point $(1, 3)$; the only way $(x - 1)^2 + (y - 3)^2 = 0$ is if $(x - 1)^2 = 0$ and $(y - 3)^2 = 0$.

14.5 $x^2 + y^2 = -3$

▮ There are no real numbers satisfying this equation.

14.6 $x^2 + y^2 - 8x + 10y - 12 = 0$

▮ $(x^2 - 8x) + (y^2 + 10y) = 12$, $(x^2 - 8x + 16) + (y^2 + 10y + 25) = 12 + 16 + 25$, and $(x - 4)^2 + (y + 5)^2 = 53$. Here, $C(4, -5)$, $r = \sqrt{53}$.

14.7 $3x^2 + 3y^2 - 4x + 2y + 6 = 0$

▮ $3(x^2 - \frac{4}{3}x + \frac{16}{36}) + 3(y^2 + \frac{2}{3}y + \frac{4}{36}) = -6 + \frac{48}{36} + \frac{12}{36}$, and $3(x - \frac{2}{3})^2 + 3(y + \frac{1}{3})^2 = -6 + \frac{60}{36}$. There are no real x and y satisfying this equation, since $r^2 < 0$.

14.8 $x^2 + y^2 - 8x - 7y = 0$

▮ $x^2 - 8x + 16 + y^2 - 7y + \frac{49}{4} = 16 + \frac{49}{4}$, $(x - 4)^2 + (y - \frac{7}{2})^2 = \frac{113}{4}$, and $C(4, \frac{7}{2})$, $r = \sqrt{113}/2$.

14.9 $2x^2 + 2y^2 - x = 0$

▮ $2(x^2 - x/2 + \frac{1}{16}) + 2y^2 = \frac{2}{16} = \frac{1}{8}$, $(x - \frac{1}{4})^2 + (y - 0)^2 = \frac{1}{16}$, and $C(\frac{1}{4}, 0)$, $r = \frac{1}{4}$.

14.10 $x^2 + y^2 - 6x + 8y - 11 = 0$

▮ $x^2 - 6x + 9 + y^2 + 8y + 16 = 11 + 9 + 16$, $(x - 3)^2 + (y + 4)^2 = 36$, and $C(3, -4)$, $r = 6$.

14.11 $x^2 + y^2 - 4x - 6y - \frac{10}{3} = 0$

▮ $x^2 - 4x + 4 + y^2 - 6y + 9 = \frac{10}{3} + 4 + 9$, $(x - 2)^2 + (y - 3)^2 = 16\frac{1}{3} = \frac{49}{3}$, and $C(2, 3)$, $r = \frac{7}{\sqrt{3}} = \frac{7\sqrt{3}}{3}$.

14.12 $7x^2 + 7y^2 + 14x - 56y - 25 = 0$

▮ $7(x^2 + 2x + 1) + 7(y^2 - 8y + 16) = 25 + 7 + 112 = 144$, $(x + 1)^2 + (y - 4)^2 = \frac{144}{7}$, and $C(-1, 4)$, $r = 12/\sqrt{7}$.

For Probs. 14.13 to 14.24, write an equation of the circle satisfying the given conditions.

14.13 Center $(1, 0)$; radius 2.

▮ $(x - 1)^2 + (y - 0)^2 = 2^2$, or $(x - 1)^2 + y^2 + 4$.

14.14   Center $(0, -6)$; radius 1.

▮ $(x - 0)^2 + [y - (-6)]^2 = 1^2$,   or $x^2 + (y + 6)^2 = 1$.

14.15   Center $(-1, -3)$; radius $\sqrt{3}$.

▮ $(x + 1)^2 + (y + 3)^2 = (\sqrt{3})^2$,   or $(x + 1)^2 + (y + 3)^2 = 3$.

14.16 $C(1, 2)$; passing through $(0,0)$.

▮ The line segment connecting $(1, 2)$ to $(0, 0)$ is a radius: $r = \sqrt{(1 - 0)^2 + (2 - 0)^2} = \sqrt{1 + 4} = \sqrt{5}$.   Then $(x - 1)^2 + (y - 2)^2 = (\sqrt{5})^2 = 5$.

14.17 $C(1, -3)$; passing through $(2, 6)$.

▮ $r = \sqrt{(2 - 1)^2 + [6 - (-3)]^2} = \sqrt{1^2 + 9^2} = \sqrt{82}$,   so the equation is $(x - 1)^2 + (y + 3)^2 = 82$.

14.18   Center at $(1, 2)$;   diameter $= 6$.

▮ If $d = 6$,   then $r = d/2 = 3$,   and the equation is $(x - 1)^2 + (y - 2)^2 = 9$.

14.19   Center $(0, 6)$; radius has $(0, 1)$ as an end point.

▮ $r = \sqrt{(0 - 0)^2 + (1 - 6)^2} = 5$,   so the equation is $x^2 + (y - 6)^2 = 25$.

14.20   Center $(0, 6)$; diameter has $(0, -1)$, $(0, 13)$ as end points.

▮ $r = \dfrac{\sqrt{(0 - 0)^2 + (13 + 1)^2}}{2} = \frac{14}{2} = 7$,   so the equation is $(x - 0)^2 + (y - 6)^2 = 49$.

14.21   Center $(-4, 3)$; tangent to the y axis.

▮ See Fig. 14.1.   If the circle is tangent to the y axis, it must intersect it at P where $\overline{AP} \| x$ axis.   Thus, P has coordinates $(0, 3)$, and $r = \sqrt{(4 - 0)^2 + (3 - 3)^2} = 4$.   Then the equation is $(x + 4)^2 + (y - 3)^2 = 16$.

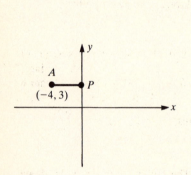

Fig. 14.1

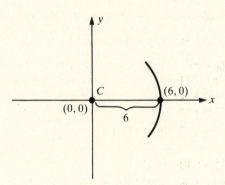

Fig. 14.2

14.22   Center is at the origin; circle crosses the x axis at 6.

▮ See Fig. 14.2.   Then $r = 6$,   and the equation is $(x - 0)^2 + (y - 0)^2 = 36$,   or $x^2 + y^2 = 36$.

14.23   Circle is tangent to both axes; the center is in quadrant I; $r = 8$.

▮ See Fig. 14.3.   The center must be at $(8, 8)$, so the equation is $(x - 8)^2 + (y - 8)^2 = 64$.

14.24   Passes through the origin; $r = 10$;   abscissa of the center is -6.

▮ See Fig. 14.4.   The center is on the line $x = -6$.   We want the distance from $(-6, y)$ to $(0, 0)$ (which is r) to be 10.   Then, $\sqrt{36 + y^2} = 10$, $36 + y^2 = 100$, $y^2 = 64$,   or $y = 8, -8$.   Thus, $C = (-6, 8)$ or $C = (-6, -8)$,   so the equation is either $(x + 6)^2 + (y - 8)^2 = 100$ or $(x + 6)^2 + (y + 8)^2 = 100$.

For Probs. 14.25 to 14.27, find the equation of the circle passing through the three given points.

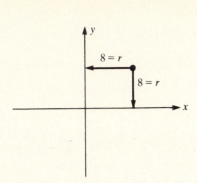

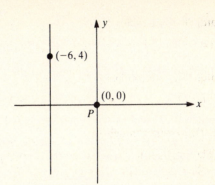

Fig. 14.3   Fig. 14.4

14.25 $(0, 0)$, $(1, 1)$, $(1, 2)$

▮ If $x^2 + y^2 + Ax + By + C = 0$ is the circle, then each of the abscissas and ordinates given must satisfy that equation. Then using $(0, 0)$: $0 + C = 0$, or $C = 0$. Using $(1, 1)$: $1^2 + 1^2 + A + B + C = 0$, but $C = 0$, so $2 + A + B = 0$. Using $(1, 2)$: $1^2 + 2^2 + A + 2B + C = 0$, but again since $C = 0$, $5 + A + 2B = 0$. Then, $5 + (B - 2) + 2B = 0$, $3B + 3 = 0$, and $B = -1$. Using the equation obtained from point $(1, 1)$, $2 + A - 1 = 0$, and $A = -1$. Finally, substituting $A = -1$, $B = -1$, and $C = 0$ into $x^2 + y^2 + Ax + By + C = 0$, we get $x^2 + y^2 - x - y = 0$.

14.26 $(0, 0)$, $(1, -1)$, $(2, 0)$

▮ $x^2 + y^2 + Ax + By + C = 0$. Using $(0, 0)$: $0 + 0 + 0 + 0 + C = 0$, so $C = 0$. Using $(1, -1)$: $1 + 1 + A - B = 0$. Using $(2, 0)$: $4 + 0 + 2A + 0 + 0 = 0$, $2A = -4$, and $A = -2$. Since $2 + A - B = 0$, $2 - 2 - B = 0$, and $B = 0$. Then the equation is $x^2 + y^2 - 2x = 0$ ($B = C = 0$).

14.27 $(0, 1)$, $(1, 0)$, $(0, -1)$

▮ $x^2 + y^2 + Ax + By + C = 0$. Using $(0, 1)$: $0 + 1 + 0 + B + C = 0$. Using $(1, 0)$: $1 + 0 + A + 0 + C = 0$. Using $(0, -1)$: $0 + 1 + 0 - B + C = 0$. Then, $2 + 2C = 0$, so $C = -1$; $1 + A - 1 = 0$, so $A = 0$; and $1 - B - 1 = 0$, so $B = 0$. Thus, $x^2 + y^2 - 1 = 0$, or $x^2 + y^2 = 1$.

For Probs. 14.28 to 14.30, find the center and radius of the circle passing through the given points.

14.28 $(0, 0)$, $(1, 1)$, $(1, 2)$

▮ $(x - h)^2 + (y - k)^2 = r^2$. Using $(0, 0)$: $(-h)^2 + (-k)^2 = r^2$, and $h^2 + k^2 = r^2$. Using $(1, 1)$: $(1 - h)^2 + (1 - k)^2 = r^2$, $1 - 2h + h^2 + 1 - 2k + k^2 = r^2$, and $2 + h^2 - 2h - 2k + k^2 = r^2$. Using $(1, 2)$: $(1 - h)^2 + (2 - k)^2 = r^2$, $1 - 2h + h^2 + 4 - 4k + k^2 = r^2$, and $5 + h^2 - 2h - 4k + k^2 = r^2$. From these equations we have $-2 + 2h + 2k = 0$ and $-3 + 2k = 0$, so $k = \frac{3}{2}$. Thus, $-2 + 2h + \frac{3}{2} = 0$, $2h = \frac{1}{2}$, or $h = \frac{1}{4}$. Then, $\frac{1}{16} + \frac{9}{4} = r^2$, $\dfrac{1 + 36}{16} = r^2$, and $r = \dfrac{\sqrt{37}}{4}$. Thus, $C = (\frac{1}{4}, \frac{3}{2})$, $r = \dfrac{\sqrt{37}}{4}$.

14.29 $(0, 1)$, $(1, 0)$, $(0, 1)$

▮ See Prob. 14.27. If $x^2 + y^2 = 1$, then $C = (0, 0)$, $r = 1$. The technique in Prob. 14.28 will also work.

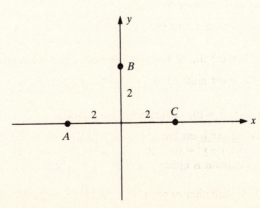

Fig. 14.5

14.30 $(0, 2), (2, 0), (-2, 0)$

▮ See Fig. 14.5. For the circle to pass through A, B, and C, it must have $C = (0, 0)$, $r = 2$.

For Probs. 14.31 to 14.33, find the equation of the circle passing through the given points, with center on the given line.

14.31 Through $(0, 1)$ and $(1, 0)$; center on $x = 0$.

▮ Let $(x - h)^2 + (y = k)^2 = r^2$ be the circle. Then, (h, k) must be the same distance from $(0, 1)$ and $(1, 0)$: $\sqrt{(h - 0)^2 + (k - 1)^2} = \sqrt{(h - 1)^2 + (k - 0)^2}$, $\sqrt{h^2 + (k - 1)^2} = \sqrt{(h - 1)^2 + k^2}$, $h^2 + k^2 - 2k + 1 = h^2 - 2h + 1 + k^2$, so $2k + 2h = 0$. If the center (h, k) lies on $x = 0$, then $h = 0$; since $-k + h = 0$, $k = 0$. Thus, $r = \sqrt{(0 - 0)^2 + (1 + 0)^2} = 1$, and the equation is $x^2 + y^2 = 1$.

14.32 Through $(0, 0)$ and $(1, 0)$; center on $y = 0$.

▮ $\sqrt{(h - 0)^2 + (y - 0)^2} = \sqrt{(h - 1)^2 + y^2}$, $h^2 + y^2 = h^2 - 2h + 1 + y^2$, $-2h = -1$, so $h = \frac{1}{2}$. If the center is on $y = 0$, then $k = 0$. Also, $r = \sqrt{(\frac{1}{2} - 0)^2 + 0} = \sqrt{\frac{1}{4}} = \frac{1}{2}$, so $(x - \frac{1}{2})^2 + y^2 = \frac{1}{4}$.

14.33 Through $(0, 1)$ and $(1, 0)$; center on $y = -1$.

▮ $\sqrt{(h - 0)^2 + (k - 1)^2} = \sqrt{(h - 1)^2 + (k - 0)^2}$, $h^2 + k^2 - 2k + 1 = h^2 - 2h + 1 + k^2$, $-2k + 1 = -2h + 1$, and $k = h$. If the center lies on $y = -1$, then $k = -1$ and $h = -1$. Thus $(h, k) = (-1, -1)$. Therefore, $r = \sqrt{(-1 - 0)^2 + (-1 - 1)^2} = \sqrt{1 + 4} = \sqrt{5}$.

For Probs. 14.34 to 14.38, find the intersection(s) (if any) of the graphs of the given equations. Use algebraic techniques.

14.34 $x^2 + y^2 = 1$ and $x = 1$.

▮ $x^2 + y^2 = 1$ and $x = 1$. Then $1^2 + y^2 = 1$, $y^2 = 0$, and $y = 0$. But $x = 1$, so the intersection point is $(1, 0)$.

14.35 $x^2 + y^2 = 1$ and $x^2 + y^2 = 2$.

▮ $x^2 = 1 - y^2$, $x = \sqrt{1 - y^2}$, or $(1 - y^2) + y^2 = 2$; no solution. These two circles do not intersect.

14.36 $x^2 + y^2 = 1$ and $(x - 1)^2 + y^2 = 1$.

▮ $y^2 = 1 - x^2$, so $(x - 1)^2 + (1 - x^2) = 1$. Then $x^2 - 2x + 1 + 1 - x^2 = 1$, $-2x + 2 = 1$, $-2x = -1$, or $x = \frac{1}{2}$. Since $y^2 = 1 - x^2$, then $y^2 = 1 - (\frac{1}{2})^2 = \frac{3}{4}$. Thus, $y = \pm\sqrt{3}/2$, and $(x, y) = (1/2, \sqrt{3}/2)$ or $(x, y) = (1/2, -\sqrt{3}/2)$.

14.37 $x + y = 2$ and $(x - 1)^2 + (y - 2)^2 = 3$.

▮ $(x - 1)^2 + (2 - x - 2)^2 = 3$, $x^2 - 2x + 1 + x^2 = 3$, $2x^2 - 2x - 2 = 0$, $x^2 - x - 1 = 0$, or $x = \frac{\sqrt{1 \pm (-1)^2 - 4(1)(-1)}}{2} = \frac{1 \pm \sqrt{1 + 4}}{2} = \frac{1 \pm \sqrt{5}}{2}$. If $x = \frac{1 + \sqrt{5}}{2}$, $y = 2 - \frac{1 + \sqrt{5}}{2}$. If $x = \frac{1 - \sqrt{5}}{2}$, $y = 2 - \frac{1 - \sqrt{5}}{2}$.

14.38 $x^2 + y^2 = y$ and $y = 3$

▮ $x^2 + 9 = 4$, or $x^2 = -5$. There is no solution, so there is no intersection.

For Probs. 14.39 to 14.41, tell whether the statement is true or false, and why.

14.39 Two circles either intersect in two places or do not intersect.

▮ False; if they are tangent, they intersect at one point.

14.40 If a circle and line intersect, they intersect twice.

▮ False; if the line is tangent to the circle, it intersects it only once.

14.41 Every triple of points determines a circle.

❚ False; if the three points are collinear, they do not determine a circle.

14.42 Write the equation of the locus of a point, the sum of the squares of whose distances from $(-2, -5)$ and $(3, 4)$ is equal to 70.

❚ Let (x, y) be the point. Then, $d[(x, y), (-2, -5)] = \sqrt{(x + 2)^2 + (y + 5)^2}$ and $d[(x, y), (3, 4)] = \sqrt{(x - 3)^2 + (y - 4)^2}$. We are told that $[(x + 2)^2 + (y + 5)^2] + [(x - 3)^2 + (y - 4)^2] = 70$. Then $x^2 + 4x + 4 + y^2 + 10y + 25 + x^2 - 6x + 9 + y^2 - 8y + 16 = 70$. Thus, $x^2 + y^2 - x + y - 8 = 0$. This is a circle with center at $(1/2, -1/2)$ and radius $\sqrt{17/2}$.

14.43 Find the length of the tangent from $(3, 1)$ to $x^2 + y^2 = 1$.

❚ See Fig. 14.6. Notice that, by the pythagorean theorem, $d^2 = b^2 - a^2$. Then $b = \sqrt{3^2 + 1^2} = \sqrt{10}$ and $a = 1$. Thus, $d^2 = (\sqrt{10})^2 - 1 = 9$ and $d = 3$. Does this result look correct?

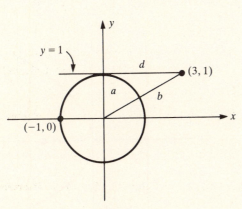

Fig. 14.6

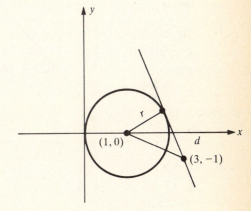

Fig. 14.7

14.44 Find the length of the tangent from $(3, -1)$ to $(x - 1)^2 + y^2 = 1$.

❚ See Fig. 14.7 and Prob. 14.43. Then $d^2 = [\sqrt{(3 - 1)^2 + (-1 - 0)^2}]^2 - r^2 = (\sqrt{4 + 1})^2 - 1 = (\sqrt{5})^2 - 1 = 5 - 1 = 4$, and $d = 2$.

14.2 THE PARABOLA

For Probs. 14.45 to 14.52, find the focus and directrix for the given parabola.

14.45 $y^2 = 9x$

❚ If $y^2 = 4ax$, then the focus is at $(a, 0)$, and the directrix is $x = -a$. In this case, $y^2 = 4 \cdot \frac{9}{4} \cdot x$. Focus: $(\frac{9}{4}, 0)$. Directrix: $x = -\frac{9}{4}$.

14.46 $y^2 = 40x$

❚ $y^2 = 40x$ means $y^2 = 4ax$ where $a = 10$. Focus: $(10, 0)$. Directrix: $x = -10$.

14.47 $y^2 = 2x$

❚ $2x = 4ax$, $2 = 4a$, or $a = \frac{1}{2}$. Focus: $(\frac{1}{2}, 0)$. Directrix: $x = -\frac{1}{2}$.

14.48 $y^2 = -5x$

❚ $-5x = 4ax$, $-5 = 4a$, or $a = -\frac{5}{4}$. Focus: $(-\frac{5}{4}, 0)$. Directrix: $x = \frac{5}{4}$. (Careful! $x = -a$.)

14.49 $y^2 = -x$

❚ $4ax = -x$, $4a = -1$, or $a = -\frac{1}{4}$. Focus: $(-\frac{1}{4}, 0)$. Directrix: $x = \frac{1}{4}$.

14.50 $x^2 = 8y$

▮ If $x^2 = 4ay$, then the focus is $(0, a)$ and the directrix is $y = -a$. In this case, $8 = 4a$, or $a = 2$. Focus: $(0, 2)$. Directrix: $y = -2$.

14.51 $x^2 = y$

▮ Then $4ay = y$, $4a = 1$, or $a = \frac{1}{4}$. Focus: $(0, \frac{1}{4})$. Directrix: $y = -\frac{1}{4}$.

14.52 $x^2 = -16y$

▮ $4a = -16$, or $a = -4$. Focus: $(0, -4)$. Directrix: $y = 4$.

For Probs. 14.53 to 14.62, sketch the graph of the given parabola.

14.53 $y^2 = 4x$

▮ See Fig. 14.8. $4 = 4a$, or $a = 1$. Focus: $(1, 0)$. Directrix: $x = -1$.

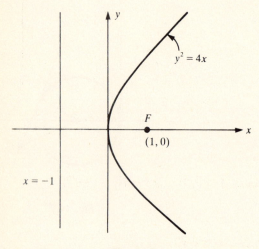

Fig. 14.8

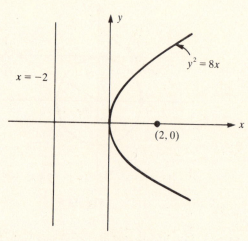

Fig. 14.9

14.54 $y^2 = 8x$

▮ See Fig. 14.9. $8 = 4a$, or $a = 2$.
Focus: $(2, 0)$. Directrix: $x = -2$.

14.55 $y^2 = x$

▮ See Fig. 14.10. Focus: $(\frac{1}{4}, 0)$. Directrix: $x = -\frac{1}{4}$.

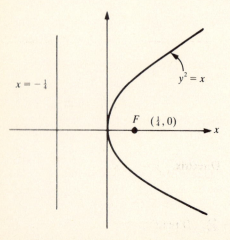

Fig. 14.10

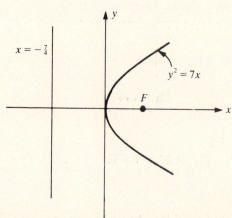

Fig. 14.11

14.56 $y^2 = 7x$

$\blacksquare$ See Fig. 14.11. $4a = 7$, or $a = \frac{7}{4}$. Focus: $(\frac{7}{4}, 0)$. Directrix: $x = -\frac{7}{4}$.

14.57 $y^2 = -4x$

$\blacksquare$ See Fig. 14.12. $4a = -4$, or $a = -1$. Focus: $(-1, 0)$. Directrix: $x = 1$.

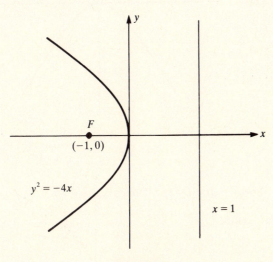

Fig. 14.12

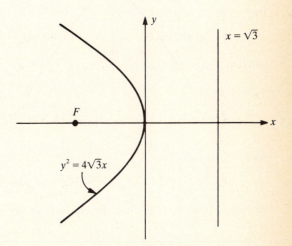

Fig. 14.13

14.58 $y^2 = -4\sqrt{3}x$

$\blacksquare$ See Fig. 14.13. $4a = -4\sqrt{3}$, or $a = -\sqrt{3}$. Focus: $(-\sqrt{3}, 0)$. Directrix: $x = \sqrt{3}$.

14.59 $x^2 = 2y$

$\blacksquare$ See Fig. 14.14. $2 = 4a$, or $a = \frac{1}{2}$. Focus: $(0, \frac{1}{2})$. Directrix: $y = -\frac{1}{2}$.

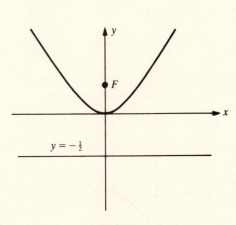

Fig. 14.14

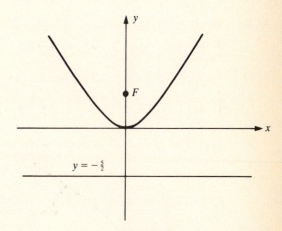

Fig. 14.15

14.60 $x^2 = 10y$

$\blacksquare$ See Fig. 14.15. $10 = 4a$, or $a = \frac{5}{2}$. Focus: $(0, \frac{5}{2})$. Directrix: $y = -\frac{5}{2}$.

14.61 $x^2 = -10y$

$\blacksquare$ See Fig. 14.16. $4a = -10$, or $a = -\frac{5}{2}$. Focus: $(0, -\frac{5}{2})$. Directrix: $y = \frac{5}{2}$.

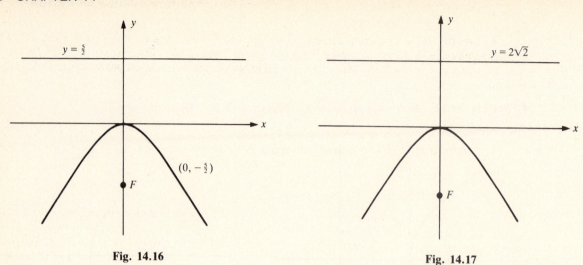

Fig. 14.16   Fig. 14.17

14.62 $x^2 = -8\sqrt{2}\,y$

▮ See Fig. 14.17. $4a = -8\sqrt{2}$, or $a = -2\sqrt{2}$. Focus: $(0, -2\sqrt{2})$. Directrix: $y = 2\sqrt{2}$.

For Probs. 14.63 to 14.67, find the equation of the parabola described.

14.63 Vertex $(0,0)$; focus $(0,1)$; directrix $y = -1$.

▮ Focus $(0, a)$, directrix $y = -a$, and $a > 0$ means the parabola opens up. $x^2 = 4ay$, and $a = 1$; so $x^2 = 4y$.

14.64 Vertex $(0,0)$; focus $(1,0)$; directrix $x = -1$.

▮ Focus $(a, 0)$, directrix $x = -a$, and $a > 0$ means the parabola opens to the right. $y^2 = 4ax$, and $a = 1$; so $y^2 = 4x$.

14.65 Vertex $(0,0)$; focus $(0,2)$; directrix $y = -2$.

▮ $x^2 = 4ay$, and $a = 2$; so $x^2 = 8y$ (opens up).

14.66 Vertex $(0,0)$; focus $(0,-2)$; directrix $y = 2$.

▮ $x^2 = 4ay$, and $a = -2$; $x^2 = -8y$.

14.67 Vertex $(0,0)$; focus $(-3,0)$; directrix $x = 3$.

▮ Focus $(-3, 0)$ and directrix $x = 3$ means $y^2 = 4ax$, $a < 0$. The parabola opens to the left. $y^2 = 4ax$, and $a = -3$; so $y^2 = -12x$.

For Probs. 14.68 to 14.72, find the vertex of the given parabola.

14.68 $x^2 = -10y$

▮ This is of the form $(x - h)^2 = -10(y - k)$, where $(h, k) = (0, 0)$. Vertex: $(0, 0)$.

14.69 $x^2 - 6x + 8y + 25 = 0$

$(x^2 - 6x + 9) + 8y = -25 + 9$, $(x - 3)^2 = -8y - 16$, $(x - 3)^2 = -8(y + 2)$, so $h = 3$, $k = -2$. Vertex: $(3, -2)$.

14.70 $y^2 - 16x + 2y + 4y = 0$

▮ $y^2 + 2y + 1 = 16x - 49 + 1 = 16x - 48$, $(y + 1)^2 = 16(x - 3)$, so $k = -1$, $h = 3$. Vertex: $(3, -1)$.

14.71 $x^2 - 2x - 6y - 53 = 0$

▮ $x^2 - 2x + 1 = 6y + 53 + 1$, $(x - 1)^2 = 6(y + 9)$. Vertex: $(1, -9)$.

14.72 $y^2 + 20x + 4y - 60 = 0$

▮ $y^2 + 4y + 4 = -20x + 60 + 4$, $(y + 2)^2 = -20x + 64$, $(y + 2)^2 = -20(x - \frac{16}{5})$.   Vertex: $(\frac{16}{5}, -2)$.

For Probs. 14.73 to 14.78, find the equation of the parabola satisfying the given conditions, if possible.

14.73   Focus $(3, 0)$;   directrix $x + 3 = 0$.

▮ Focus $(3, 0)$, directrix $x = -3$ means the parabola is of the form $y^2 = 4ax$, $a = 3$.   So the equation is $y^2 = 12x$.

14.74   Vertex $(0, 0)$; axis along the x axis; passing through $(-3, 6)$.

▮ See Fig. 14.18.   The axis passes through the vertex, to the directrix. Then the directrix must be $x = 3$,   and the focus must be $(-3, 0)$.   Then $y^2 = 4ak$, $a < 0$.   When $a = -3$, $y^2 = -12x$.

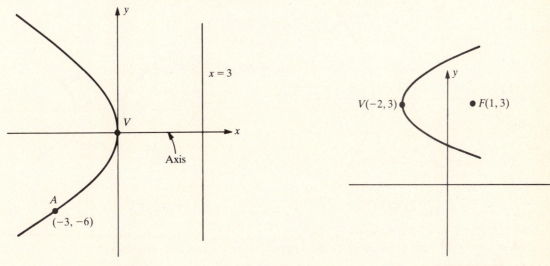

Fig. 14.18   Fig. 14.19

14.75   Vertex $(-2, 3)$; focus $(1, 3)$.

▮ See Fig. 14.19.   Here $a = 3$.   The equation must be of the form $(y - k)^2 = 4a(x - h)$.   Then $(y - 3)^2 = 4 \cdot 3(x + 2) = 12(x + 2)$,   or $y^2 - 6y - 12x - 15 = 0$.

14.76   Vertex $(1, 6)$; focus $(2, 6)$.

▮ Then $a = 6$, $(y - k)^2 = 4a(x - h)$,   and $(y - 6)^2 = 24(x - 1)$.

14.77   Vertex $(2, 5)$; focus $(2, 8)$.

▮ Then $a = 2$,   and $(x - h)^2 = 4a(y - k)$,   and $(x - 2)^2 = 8(y - 5)$.

14.78   Axis parallel to x axis, passing through $(3, 3)$, $(6, 5)$, $(6, -3)$.

▮ If the axis is parallel to the x axis, then $y^2 + Ax + Ey + F = 0$.   Using $(3, 3)$: $9 + 3A + 3E + F = 0$. Using $(6, 5)$: $25 + 6A + 5E + F = 0$.   Using $(6, -3)$: $9 + 6A - 3E + F = 0$.   Then, using these equations, $-3A + 6E = 0$ and $16 + 8E = 0$,   so $E = -2$.   Then $-3A - 12 = 0$,   or $A = -4$; and $9 - 12 - 6 + F = 0$,   or $F = 9$.   Thus, the equation is $y^2 - 2y - 4x + 9 = 0$.

14.79   Find the intersection(s) of $y^2 = x$ and $y = x^2$.

▮ If $y = x^2$,   then $y^2 = x^4$.   From $y^2 = x$ and $y^2 = x^4$ we have $x = x^4$.   Then $x^4 - x = 0$, $x(x^3 - 1) = 0$,   and $x = 0$ or $x^3 - 1 = 0$.   Thus, $x = 0$ or $x = 1$.   If $x = 0$, $y = 0$; if $x = 1$, $y = 1$. Intersection points: $(0, 0)$, $(1, 1)$.

14.80 Find the intersection(s) of $y = (x - 1)^2$ and $y = x + 1$.

▮ Then $x + 1 = (x - 1)^2$ or $x + 1 = x^2 - 2x + 1$. Then $x^2 - 3x = 0$, $x(x - 3) = 0$, and $x = 0$ or $x = 3$. If $x = 0$, $y = 1$; if $x = 3$, $y = 4$. Intersection points: $(0, 1)$ and $(3, 4)$.

14.81 Find the ordinate(s) of the intersection(s) of $y = x^2$ and $x^2 + y^2 = 1$.

▮ $y = x^2$ and $x^2 = 1 - y^2$; $y = 1 - y^2$, or $y^2 + y - 1 = 0$. Then $y = \dfrac{-1 \pm \sqrt{1^2 - 4(1)(-1)}}{2} = \dfrac{-1 \pm \sqrt{1 + 4}}{2}$, so $y = \dfrac{-1 + \sqrt{5}}{2}$, or $y = \dfrac{-1 - \sqrt{5}}{2}$.

14.82 Give an example of a parabola that has no intersection with the circle $x^2 + y^2 = 1$.

▮ See Fig. 14.20. If the parabola's vertex is at V above and it opens to the right, there will be no intersection. Then $(y - k)^2 = 4a(x - h)$. Let $a = 1$ (you could pick others). Then $y^2 = 4(x - 2)$.

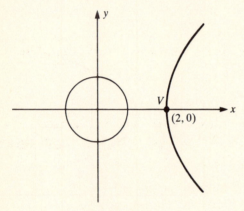

Fig. 14.20

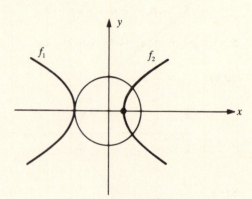

Fig. 14.21

14.83 What are the different possibilities for the intersection of a circle and a parabola?

▮ The possibilities are no intersections (see Fig. 14.20.), one intersection (f_1) or two intersections (f_2) (see Fig. 14.21), three intersections (f_3) (see Fig. 14.22), four intersections (f_4) (see Fig. 14.23).

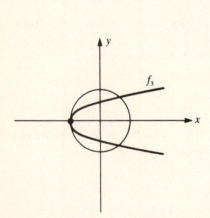

Fig. 14.22

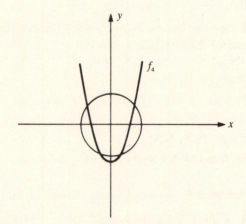

Fig. 14.23

14.3 THE ELLIPSE

For Probs. 14.84 to 14.88, put the given equation in the standard form for an ellipse.

14.84 $x^2 + 8y^2 = 8$

▮ Divide by 8: $\dfrac{x^2}{8} + \dfrac{8y^2}{8} = 1$. Then $\dfrac{x^2}{8} + y^2 = 1$.

14.85 $x^2 + 2y^2 = 3$

▮ Divide by 3: $\dfrac{x^2}{3} + \dfrac{2y^2}{3} = 1,\quad \dfrac{x^2}{3} + \dfrac{y^2}{\frac{3}{2}} = 1.$

14.86 $2x^2 + 5y^2 = 6$

▮ Divide by 6: $\dfrac{2x^2}{6} + \dfrac{5y^2}{6} = 1,\quad \dfrac{x^2}{3} + \dfrac{y^2}{\frac{6}{5}} = 1.$

14.87 $-x^2 - 3y^2 = -1$

▮ Divide by -1: $x^2 + 3y^2 = 1,\quad x^2 + \dfrac{y^2}{\frac{1}{3}} = 1.$

14.88 $(x-1)^2 + 2(y+3)^2 = 4$

▮ Divide by 4: $\dfrac{(x-1)^2}{4} + \dfrac{(y+3)^2}{2} = 1.$

For Probs. 14.89 to 14.98, find the foci and the lengths of the major and minor axes for each given ellipse.

14.89 $\dfrac{x^2}{25} + \dfrac{y^2}{4} = 1$

▮ This is of the form $\dfrac{x^2}{a^2} + \dfrac{y^2}{b^2} = 1,$ where $a > b > 0$. Thus, the foci are at $(\pm c, 0)$, where $c^2 = a^2 - b^2$. Since $c^2 = 25 - 4 = 21,$ $c = \pm\sqrt{21}$ and the foci are at $(\pm\sqrt{21}, 0)$. The length of the major axis $= 2a = 10$ $(a = 5)$; the length of the minor axis $= 2b = 4$ $(b = 2)$.

14.90 $\dfrac{x^2}{9} + \dfrac{y^2}{4} = 1$

▮ This is of the form $\dfrac{x^2}{a^2} + \dfrac{y^2}{b^2} = 1,$ $a > b > 0$ (see Prob. 14.89). Thus, $a = 3$ and $b = 2$. The length of the major axis $= 6$; the length of the minor axis $= 4$. Since $c^2 = a^2 - b^2 = 9 - 4 = 5,$ $c = \pm\sqrt{5}$ and the foci are at $(\pm\sqrt{5}, 0)$.

14.91 $\dfrac{x^2}{4} + \dfrac{y^2}{25} = 1$

▮ Here, $\dfrac{x^2}{b^2} + \dfrac{y^2}{a^2} = 1,$ $a > b > 0$. Then the foci are at $(0, \pm c)$, where $c^2 = a^2 - b^2$. The major axis length $= 2a$; the minor axis length $= 2b$. Since $c^2 = a^2 - b^2 = 25 - 4 = 21,$ the foci are at $(0, \pm\sqrt{21})$. The length of the major axis $= 10$ $(a = 5)$; the length of the minor axis $= 4$ $(b = 2)$.

14.92 $\dfrac{x^2}{4} + \dfrac{y^2}{9} = 1$

▮ See Prob. 14.91. $\dfrac{x^2}{b^2} + \dfrac{y^2}{a^2} = 1,$ $a > b > 0$. Then $a = 3,$ $b = 2$. The length of the major axis $= 6$; the length of the minor axis $= 4$; and since $c^2 = 9 - 4 = 5,$ the foci are at $(0, \pm\sqrt{5})$.

14.93 $x^2 + 9y^2 = 9$

▮ Put this in standard form: $\dfrac{x^2}{9} + \dfrac{y^2}{1} = 1$ or $\dfrac{x^2}{3^2} + \dfrac{y^2}{1^2} = 1.$ The equation is of the form $\dfrac{x^2}{a^2} + \dfrac{y^2}{b^2} = 1$ $(a > b)$. Thus, the length of the major axis $= 6$ $(2 \cdot 3)$; the length of the minor axis $= 2$ $(2 \cdot 1)$; and since $c^2 = 9 - 1 = 8,$ $c = \pm 2\sqrt{2}$ and the foci are at $(\pm 2\sqrt{2}, 0)$.

14.94 $4x^2 + y^2 = 4$

▮ Divide by 4: $\dfrac{x^2}{1^2} + \dfrac{y^2}{2^2} = 1.$ Then $b^2 = 1^2,$ $a^2 = 2^2$; $b = 1,$ $a = 2$. Since $c^2 = 2^2 - 1^2 = 3,$ $c = \pm\sqrt{3}$ and the foci are at $(0, \pm\sqrt{3})$. The length of the major axis $= 4$; the length of the minor axis $= 2$.

14.95 $2x^2 + y^2 = 12$

▮ Divide by 12: $\dfrac{x^2}{6} + \dfrac{y^2}{12} = 1$. Then $b = \sqrt{6}$ and $a = 2\sqrt{3}$. Since $c^2 = 12 - 6 = 6$, $c = \pm\sqrt{6}$ and the foci are at $(0, \pm\sqrt{6})$. The length of the major axis $= 2a = 4\sqrt{3}$; the length of the minor axis $= 2b = 2\sqrt{6}$.

14.96 $4x^2 + 3y^2 = 24$

▮ $\dfrac{x^2}{6} + \dfrac{y^2}{8} = 1$. Then $b^2 = 6$, $a^2 = 8$; thus, c^2 and the foci are at $(0, \pm\sqrt{2})$. The length of the major axis $= 2a = 2\sqrt{8} = 4\sqrt{2}$; the length of the minor axis $= 2b = 2\sqrt{6}$.

14.97 $4x^2 + 7y^2 = 28$

▮ $\dfrac{x^2}{7} + \dfrac{y^2}{4} = 1$. Then $a^2 = 7$, $b^2 = 4$; thus $c^2 = 3$. The foci are at $(\pm\sqrt{3}, 0)$; the length of the major axis $= 2\sqrt{7}$; the length of the minor axis $= 4$.

14.98 $3x^2 + 2y^2 = 24$

▮ $\dfrac{x^2}{8} + \dfrac{y^2}{12} = 1$. Then $b^2 = 8$, $a^2 = 12$; thus $c^2 = 4$. The foci are at $(0, \pm 2)$; the length of the major axis $= 2b = 2\sqrt{8} = 4\sqrt{2}$; the length of the minor axis $= 2a = 2\sqrt{12} = 4\sqrt{3}$.

For Probs. 14.99 to 14.108, sketch the graph of the given ellipse.

14.99 $\dfrac{x^2}{25} + \dfrac{y^2}{4} = 1$

▮ See Fig. 14.24 and Prob. 14.89. $a^2 = 25$ and $b^2 = 4$, where $\dfrac{x^2}{a^2} + \dfrac{y^2}{b^2} = 1$. Thus, the x intercepts $= \pm a$ and the y intercepts $= \pm b$. F_1 and F_2 are the foci.

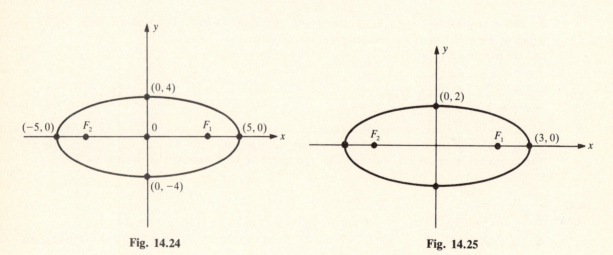

Fig. 14.24   Fig. 14.25

14.100 $\dfrac{x^2}{9} + \dfrac{y^2}{4} = 1$

▮ See Fig. 14.25 and Prob. 14.90. $a^2 = 9$ and $b^2 = 4$. Then the x intercepts $= \pm 3$ and the y intercepts $= \pm 2$.

14.101 $\dfrac{x^2}{4} + \dfrac{y^2}{25} = 1$

▮ See Fig. 14.26 and Prob. 14.91. $a^2 = 25$ and $b^2 = 4$. Then the x intercepts $= \pm 2$ and the y intercepts $= \pm 5$.

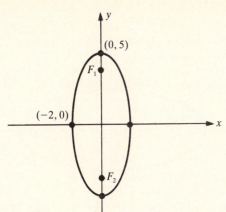

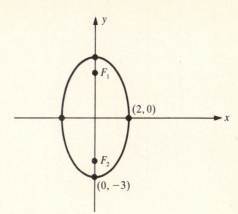

Fig. 14.26

Fig. 14.27

14.102 $\dfrac{x^2}{4} + \dfrac{y^2}{9} = 1$

▮ See Fig. 14.27 and Prob. 14.92. $a^2 = 9$ and $b^2 = 4$. Then the x intercepts $= \pm 2$ and the y intercepts $= \pm 3$.

14.103 $x^2 + 2y^2 = 2$

▮ See Fig. 14.28. Then $x^2/2 + y^2 = 1$. $a^2 = 2$ and $b^2 = 1$. Then the x intercepts $= \pm\sqrt{2}$ and the y intercepts $= \pm 1$.

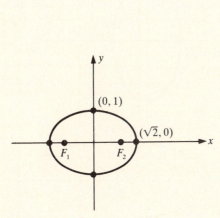

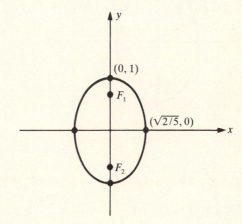

Fig. 14.28

Fig. 14.29

14.104 $5x^2 + 2y^2 = 2$

▮ See Fig. 14.29. Then $5x^2/2 + y^2 = 1$, or $x^2/\frac{2}{5} + y^2 = 1$. $a^2 = 1$ and $b^2 = \frac{2}{5}$. Then the x intercepts $= \pm\sqrt{\frac{2}{5}}$ and the y intercepts $= \pm 1$.

14.105 $10x^2 + 5y^2 = 100$

▮ See Fig. 14.30. Then $\dfrac{x^2}{10} + \dfrac{y^2}{20} = 1$. $a^2 = 20$ and $b^2 = 10$. Then the x intercepts $= \pm\sqrt{10}$ and the y intercepts $= \pm\sqrt{20}$.

14.106 $-x^2 - 2y^2 = -8$

▮ See Fig. 14.31. Then $\dfrac{x^2}{8} + \dfrac{y^2}{4} = 1$. $a^2 = 8$ and $b^2 = 4$. Then the x intercepts $= \pm\sqrt{8}$ and the y intercepts $= \pm 2$.

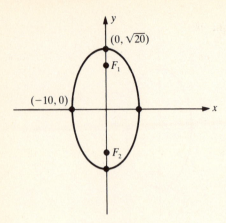

Fig. 14.30

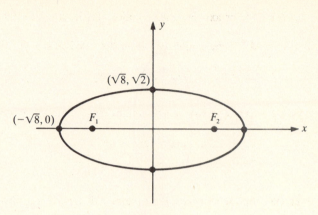

Fig. 14.31

14.107 $\dfrac{x^2}{\sqrt{2}} + \dfrac{y^2}{\sqrt{3}} = 1$

▮ See Fig. 14.32. $\sqrt{3} > \sqrt{2}$. Thus, $a^2 = \sqrt{3}$, $b^2 = \sqrt{2}$, and the x intercepts $= \pm 2^{1/4}$ and the y intercepts $= \pm 3^{1/4}$.

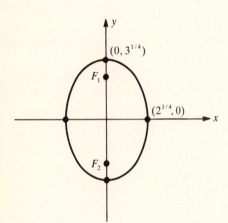

Fig. 14.32

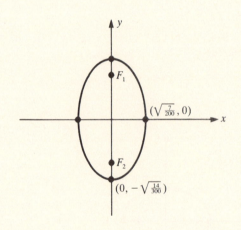

Fig. 14.33

14.108 $\dfrac{2x^2}{0.1} + \dfrac{3y^2}{0.2} = 0.7$

▮ See Fig. 14.33. Then $1 = \dfrac{2x^2}{(0.1)(0.7)} + \dfrac{3y^2}{(0.2)(0.7)} = \dfrac{2x^2}{0.07} + \dfrac{3y^2}{0.14} = \dfrac{2x^2}{\frac{7}{100}} + \dfrac{3y^2}{\frac{14}{100}} = \dfrac{x^2}{\frac{7}{200}} + \dfrac{y^2}{\frac{14}{300}}$. Since $\dfrac{14}{300} > \dfrac{7}{200}$, $a^2 = \dfrac{14}{300}$ and $b^2 = \dfrac{7}{200}$. Then the x intercepts $= \pm\sqrt{\frac{7}{200}}$ and the y intercepts $= \pm\sqrt{\frac{14}{300}}$.

For Probs. 14.109 to 14.116, find an equation of the ellipse with center $(0,0)$ satisfying the given information. Put the equation in standard form for the ellipse.

14.109 Major axis on x axis; major axis length $= 8$; minor axis length $= 6$.

▮ If the major axis is on the x axis, then the ellipse is of the form $\dfrac{x^2}{a^2} + \dfrac{y^2}{b^2} = 1$, $a > b > 0$. Major axis length 8 means $2a = 8$, $a = 4$. Minor axis length 6 means $2b = 6$, $b = 3$. Thus, the equation is $\dfrac{x^2}{16} + \dfrac{y^2}{9} = 1$.

14.110 Major axis on x axis; major axis length 14; minor axis length 10.

▮ If the major axis is on the x axis, then $\dfrac{x^2}{a^2} + \dfrac{y^2}{b^2} = 1$, $a > b > 0$. Then, $2a = 14$, $a = 7$; and $2b = 10$, $b = 5$. Thus, the equation is $\dfrac{x^2}{49} + \dfrac{y^2}{25} = 1$.

14.111 Major axis on y axis; major axis length 22; minor axis length 16.

▮ If the major axis is on the y axis, then $\dfrac{x^2}{b^2} + \dfrac{y^2}{a^2} = 1$, $a > b > 0$. Then $2a = 22$, $a = 11$; and $2b = 16$, $b = 8$. Thus, the equation is $\dfrac{x^2}{64} + \dfrac{y^2}{121} = 1$.

14.112 Major axis on y axis; major axis length 24; minor axis length 18.

▮ Since the major axis is on the y axis, $\dfrac{x^2}{b^2} + \dfrac{y^2}{a^2} = 1$. Then $2a = 24$, $a = 12$; and $2b = 18$, $b = 9$. Thus, the equation is $\dfrac{x^2}{81} + \dfrac{y^2}{144} = 1$.

14.113 Major axis on x axis; major axis length 16; distance of foci from center $= 6$.

▮ See Fig. 14.34. $\dfrac{x^2}{a^2} + \dfrac{y^2}{b^2} = 1$ (since the major axis is on the x axis). Then $2a = 16$, $a = 8$. From the given information $d((0,0), F) = 6$. We know that $b^2 = a^2 - c^2$, and that the distance from the focus to $(0, b)$ is the same as half the length of the major axis (a). Thus, $b^2 = a^2 - c^2 = 64 - 36 = 28$, and $b = \pm\sqrt{28} = \pm 2\sqrt{7}$. Then, $\dfrac{x^2}{64} + \dfrac{y^2}{28} = 1$.

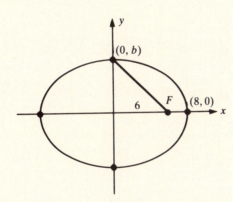

Fig. 14.34

14.114 Major axis on x axis; major axis length 24; distance of foci from center $= 10$.

▮ See Prob. 14.113. $\dfrac{x^2}{a^2} + \dfrac{y^2}{b^2} = 1 (a > b > 0)$. Then, $2a = 24$, $a = 12$. From the given information $d((0,0), F) = 10$. Since $b^2 = a^2 - c^2$, $b^2 = 144 - 100 = 44$. Thus, the equation is $\dfrac{x^2}{100} + \dfrac{y^2}{44} = 1$.

14.115 Major axis on y axis; minor axis length 20; distance of foci from center $= \sqrt{70}$.

▮ Then, $\dfrac{x^2}{b^2} + \dfrac{y^2}{a^2} = 1$, and $2b = 20$, or $b = 10$. Since $a^2 = b^2 + c^2$, $a^2 = 100 + 70 = 170$. Thus, the equation is $\dfrac{x^2}{100} + \dfrac{y^2}{170} = 1$.

14.116 Major axis on y axis; minor axis length 14; distance of foci from center $= \sqrt{200}$.

▮ Then, $\dfrac{x^2}{b^2} + \dfrac{y^2}{a^2} = 1$, and $2b = 14$, or $b = 7$. Since $a^2 = b^2 + c^2$, $a^2 = 49 + 200 = 249$. Thus, the equation is $\dfrac{x^2}{49} + \dfrac{y^2}{249} = 1$.

For Probs. 14.117 to 14.120, find the coordinates of the center of the ellipse and put the equation in standard form for an ellipse.

14.117 $4x^2 + 4x + y^2 = 10$

▐ $4(x^2 + x + \frac{1}{4}) + y^2 = 10 + 1$, $4(x + \frac{1}{2})^2 + y^2 = 11$, and $\dfrac{(x + \frac{1}{2})^2}{\frac{11}{4}} + \dfrac{4^2}{11} = 1$. Center: $(-\frac{1}{2}, 0)$, since this
is of the form $\dfrac{(x - h)^2}{b^2} + \dfrac{(y - k)^2}{a^2} = 1$.

14.118 $2x^2 + y^2 + 2y = 15$

▐ $2x^2 + (y^2 + 2y + 1) = 15 + 1$, $2x^2 + (y + 1)^2 = 16$, $\dfrac{2x^2}{16} + \dfrac{(y + 1)^2}{16} = 1$, and $\dfrac{x^2}{8} + \dfrac{(y + 1)^2}{16} = 1$.
Center: $(0, -1)$.

14.119 $x^2 + 4y^2 - 6x + 32y + 69 = 0$

▐ $(x^2 - 6x + 9) + 4(y^2 + 8y + 16) = -69 + 9 + 64$, $(x - 3)^2 + 4(y + 4)^2 = 4$, and $\dfrac{(x - 3)^2}{4} + (y + 4)^2 = 1$.
Center: $(3, -4)$.

14.120 $16x^2 + 9y^2 + 32x - 36y - 92 = 0$

▐ $16(x^2 + 2x + 1) + 9(y^2 - 4y + 4) = 92 + 16 + 36$, $16(x + 1)^2 + 9(y - 2)^2 = 144$, and
$\dfrac{(x + 1)^2}{9} + \dfrac{(y - 2)^2}{16} = 1$. Center: $(-1, 2)$.

For Probs. 14.121 and 14.122, find the vertices and foci for the given ellipse.

14.121 $x^2 + 4y^2 - 6x + 32y + 69 = 0$

▐ See Fig. 14.35 and Prob. 14.119. $\dfrac{(x - 3)^2}{4} + \dfrac{(y + 4)^2}{1} = 1$. Then $a^2 = 4$, $a = 2$; and $b^2 = 1$, $b = 1$.
Thus, the vertices must be at a distance of a $(=2)$ from the center. The vertices are also along the
major axis (here, the major axis is $y = -4$ since a^2 is the denominator of the x term). Vertices:
$V_1(5, -4)$ and $V_2(1, -4)$. Also, $c = \sqrt{a^2 - b^2} = \sqrt{4 - 1} = \sqrt{3}$. The foci are along the major axis at a
distance c (here, $c = 3$) from the center. Foci: $(3 \pm \sqrt{3}, 4)$

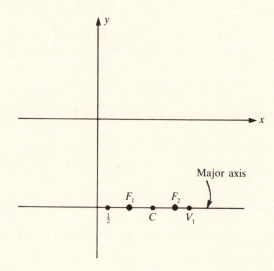

Fig. 14.35

14.122 $16x^2 + 9y^2 + 32x - 36y - 92 = 0$

▐ See Prob. 14.120. $\dfrac{(x + 1)^2}{9} + \dfrac{(y - 2)^2}{16} = 1$. The center is at $(-1, 2)$. Then, $a^2 = 16$, $b^2 = 9$ and
$c^2 = \sqrt{a^2 - b^2} = \sqrt{16 - 9} = \sqrt{7}$. The vertices are along the major axis. Vertices: $V(-1, 6)$ and $(-1, -2)$.
Foci: $(-1, 2 \pm \sqrt{7})$.

For Probs. 14.123 and 14.124, find the equation of the ellipse satisfying the given conditions.

14.123 One vertex at $(0, 10)$; one focus at $(0, -2)$; center at $(0, 0)$.

▮ The major axis must be along the y axis. Since the center is $(0,0)$, $a = 10$, $c = 2$, and $b^2 = a^2 - c^2 = 100 - 4 = 96$. Thus, $\dfrac{x^2}{96} + \dfrac{y^2}{100} = 1$.

14.124 Vertices $(7, 2)$ and $(-3, 2)$; one focus at $(6, 2)$.

▮ The center is the midpoint of the segment joining the two vertices. Center: $\left(\dfrac{7-3}{2}, 2\right) = (2, 2)$. Then $a = d(C, V) = 7 - 2 = 5$, $c = d(C, F) = 6 - 2 = 4$, and $b^2 = a^2 - c^2 = 25 - 16 = 9$. The major axis is parallel to the x axis, so $\dfrac{(x-2)^2}{25} + \dfrac{(y-2)^2}{9} = 1$.

For Probs. 14.125 to 14.127, find the intersection(s) of the graphs of the given relations.

14.125 $\dfrac{x^2}{1} + \dfrac{y^2}{4} = 1$ and $y = x$.

▮ If $y = x$, then $x^2 + x^2/4 = 1$ where they meet. Then $5x^2/4 = 1$, $x^2 = \frac{4}{5}$, or $x = \frac{\pm 2}{\sqrt{5}}$. If $x = 2/\sqrt{5}$, then $y = 2/\sqrt{5}$; if $x = -2/\sqrt{5}$, then $y = -2/\sqrt{5}$. Intersection points: $(2/\sqrt{5}, 2/\sqrt{5})$ and $(-2/\sqrt{5}, -2/\sqrt{5})$.

14.126 $\dfrac{x^2}{1} + \dfrac{y^2}{4} = 1$ and $x^2 + y^2 = 1$.

▮ $x^2 = 1 - y^2/4$. Thus, where they meet, $1 - y^2/4 + y^2 = 1$. Then $3y^2/4 = 0$, $y = 0$. If $y = 0$, then $x^2 + 0 = 1$, or $x = \pm 1$. Intersection points: $(1, 0)$ and $(-1, 0)$.

14.127 $\dfrac{x^2}{1} + \dfrac{y^2}{4} = 1$ and $y = x^2$.

▮ If $y = x^2$, then, where they meet, $y + y^2/4 = 1$. So $4y + y^2 = 4$, $y^2 + 4y - 4 = 0$, and $y = \dfrac{-4 \pm \sqrt{4^2 - 4(1)(-4)}}{2(1)} = \dfrac{-4 \pm \sqrt{32}}{2} = \dfrac{-4 \pm 4\sqrt{2}}{2} = -2 \pm 2\sqrt{2}$. Then, if $y = x^2$, $x = \pm\sqrt{y}$. If $y = -2 + 2\sqrt{2}$, $x = \sqrt{-2 + 2\sqrt{2}}$ or $x = -\sqrt{-2 + 2\sqrt{2}}$. If $y = -2 - 2\sqrt{2}$, then there is no x corresponding since $-2 - 2\sqrt{2} < 0$. Intersection points: $(\pm\sqrt{-2 + 2\sqrt{2}}, -2 + 2\sqrt{2})$.

14.4 THE HYPERBOLA

For Probs. 14.128 to 14.132, put the equation in standard form for the hyperbola.

14.128 $x^2 - 5y^2 = 10$

▮ The standard form for the hyperbola with center at $(0,0)$ is $\dfrac{x^2}{a^2} - \dfrac{y^2}{b^2} = 1$, or $\dfrac{y^2}{a^2} - \dfrac{x^2}{b^2} = 1$. Dividing by 10 above, we get: $\dfrac{x^2}{10} - \dfrac{5y^2}{10} = 1$, or $\dfrac{x^2}{10} - \dfrac{y^2}{2} = 1$.

14.129 $-6y^2 + 7x^2 = 15$

▮ Divide by 15: $\dfrac{-6y^2}{15} + \dfrac{7x^2}{15} = 1$, or $\dfrac{x^2}{\frac{15}{7}} + \dfrac{y^2}{\frac{15}{6}} = 1$.

14.130 $x^2 + 7y^2 = 14$

▮ This is an ellipse, not a hyperbola. It cannot be made to fit either form in Prob. 14.128 above.

14.131 $-x^2 + \sqrt{2}y^2 = 17$

▮ $1 = \dfrac{-x^2}{17} + \dfrac{\sqrt{2}}{17}y^2 = \dfrac{\sqrt{2}}{17}y^2 - \dfrac{x^2}{17} = \dfrac{y^2}{17/\sqrt{2}} = \dfrac{x^2}{17}$.

14.132 $x^2 - 2y^2 = \dfrac{1}{\sqrt{3}}$

▮ $1 = \sqrt{3}x^2 - 2\sqrt{3}y^2 = \dfrac{x^2}{1/\sqrt{3}} - \dfrac{y^2}{1/2\sqrt{3}}$.

For Probs. 14.133 to 14.144, find the foci and the lengths of the transverse and conjugate axes.

14.133 $\dfrac{x^2}{9} - \dfrac{y^2}{4} = 1$

▮   If $\dfrac{x^2}{a^2} - \dfrac{y^2}{b^2} = 1$, then the foci are $(\pm c, 0)$ where $c^2 = a^2 + b^2$, and the lengths of the transverse and conjugate axes are $2a$ and $2b$, respectively. Here, $a^2 = 9$, $b^2 = 4$, and $c^2 = 9 + 4 = 13$. Thus, the foci are $(\pm\sqrt{3}, 0)$. The length of the transverse axis $= 6$ $(=2a)$; and the length of the conjugate axis $= 4$ $(= 2b)$.

14.134 $\dfrac{x^2}{9} - \dfrac{y^2}{25} = 1$

▮   See Prob. 14.133. $\dfrac{x^2}{a^2} - \dfrac{y^2}{b^2} = 1$; $a^2 = 9$, $b^2 = 25$; thus, $c^2 = 9 + 25 = 34$. The foci are at $(\pm\sqrt{34}, 0)$; the length of the transverse axis $= 6$; the length of the conjugate axis $= 10$.

14.135 $\dfrac{y^2}{4} - \dfrac{x^2}{9} = 1$

▮   Here, $\dfrac{y^2}{a^2} - \dfrac{x^2}{b^2} = 1$. If $c^2 = a^2 + b^2$, then the foci are at $(0, \pm c)$; the transverse axis length is $2a$, the conjugate axis length is $2b$. Then, $c^2 = 4 + 9 = 13$. The foci are at $(0, \pm\sqrt{13})$; the length of the transverse axis $= 4$; and the length of the conjugate axis $= 6$.

14.136 $\dfrac{y^2}{25} - \dfrac{x^2}{9} = 1$

▮   See Prob. 14.135. $c^2 = a^2 + b^2 = 25 + 9 = 34$. The foci are at $(0, \pm\sqrt{34})$; the length of the transverse axis $= 10$; the length of the conjugate axis $= 6$.

14.137 $4x^2 - y^2 = 16$

▮   Divide by 16: $\dfrac{x^2}{4} - \dfrac{y^2}{16} = 1$. Then $c^2 = a^2 + b^2 = 20$, and $c = \pm\sqrt{20} = \pm2\sqrt{5}$. The foci are at $(\pm2\sqrt{5}, 0)$; the length of the transverse axis $= 4$; and the length of the conjugate axis $= 8$.

14.138 $x^2 - 9y^2 = 9$

▮ $x^2/9 - y^2 = 1$. Then $c^2 = a^2 + b^2 = 9 + 10 = 19$. The foci at at $(\pm\sqrt{10}, 0)$; the length of the transverse axis $= 6$; and the length of the conjugate axis $= 2$.

14.139 $9y^2 - 16x^2 = 144$

▮ $\dfrac{y^2}{16} - \dfrac{x^2}{9} = 1$. Then $c^2 = a^2 + b^2 = 16 + 9 = 25$, and $c = \pm5$. The foci are at $(0, \pm5)$; the length of the transverse axis $= 8$; and the length of the conjugate axis $= 6$

14.140 $4y^2 - 25x^2 = 100$

▮ $\dfrac{y^2}{25} - \dfrac{x^2}{4} = 1$. Then $c^2 = a^2 + b^2 = 25 + 4 = 29$. The foci are at $(0, \pm\sqrt{29})$; the length of the transverse axis $= 10$; and the length of the conjugate axis $= 4$.

14.141 $3x^2 - 2y^2 = 12$

▮ $\dfrac{3x^2}{12} - \dfrac{2y^2}{12} = 1$; $\dfrac{x^2}{4} - \dfrac{y^2}{6} = 1$. Then $c^2 = 4 + 6 = 10$. The foci are at $(\pm\sqrt{10}, 0)$; the length of the transverse axis $= 4$; and the length of the conjugate axis $= 2\sqrt{6}$.

14.142 $3x^2 - 4y^2 = 24$

  ▮ $\dfrac{3x^2}{24} - \dfrac{4y^2}{24} = 1$; $\dfrac{x^2}{8} - \dfrac{y^2}{6} = 1$.   Then $c^2 = a^2 + b^2 = 8 + 6 = 14$.   The foci are at $(\pm\sqrt{14}, 0)$;   the length of the transverse axis $= 2a = 2(2\sqrt{2}) = 4\sqrt{2}$;   and   the length of the conjugate axis $= 2b = 2\sqrt{6}$.

14.143 $7y^2 - 4x^2 = 28$

  ▮ $\dfrac{y^2}{4} - \dfrac{x^2}{7} = 1$.   Then $c^2 = a^2 + b^2 = 4 + 7 = 11$.   The foci are at $(0, \pm\sqrt{11})$;   the length of the transverse axis $= 4$;   and   the length of the conjugate axis $= 2\sqrt{7}$.

14.144 $3y^2 - 2x^2 = 24$

  ▮ $\dfrac{y^2}{8} - \dfrac{x^2}{12} = 1$.   Then $c^2 = 8 + 12 = 20 = 2\sqrt{5}$.   The foci are at $(0, \pm 2\sqrt{5})$;   the length of the transverse axis $= 2(2\sqrt{2}) = 4\sqrt{2}$;   and   the length of the conjugate axis $= 2(2\sqrt{3}) = 4\sqrt{3}$.

For Probs. 14.145 to 14.156, sketch the given hyperbola. *Sketching hint*: Sketch the rectangle formed by the two axes first; then the diagonals to get the asymptotes; then the vertices and the hyperbola.

14.145 $\dfrac{x^2}{9} - \dfrac{y^2}{4} = 1$

  ▮ See Fig. 14.36 and Prob. 14.133. Vertices are $V_1, V_2 (\pm 3, 0)$ (when $y = 0$). The center is at $C(0, 0)$; the length of the transverse axis $= 6$;   and   the length of the conjugate axis $= 10$.

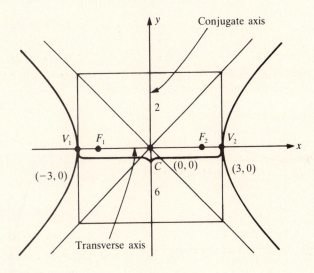

Fig. 14.36

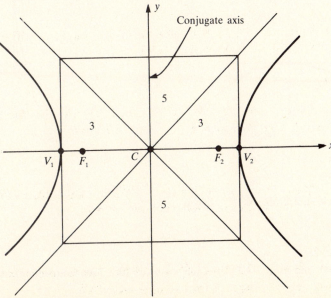

Fig. 14.37

14.146 $\dfrac{x^2}{9} - \dfrac{y^2}{25} = 1$

▮ See Fig. 14.37 and Prob. 14.134.

14.147 $\dfrac{y^2}{4} - \dfrac{x^2}{9} = 1$

▮ See Fig. 14.38 and Prob. 14.135. Vertices at $(0, \pm 2)$ (when $x = 0$).

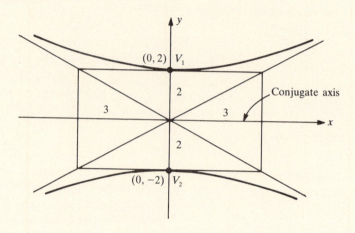

Fig. 14.38

14.148 $\dfrac{y^2}{25} - \dfrac{x^2}{9} = 1$

▮ See Fig. 14.39 and Prob. 14.136.

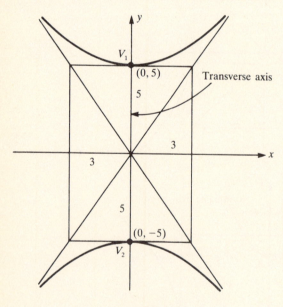

Fig. 14.39

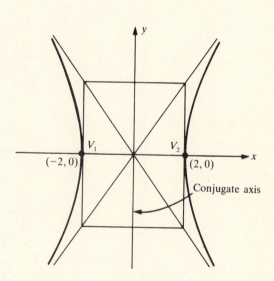

Fig. 14.40

14.149 $4x^2 - y^2 = 16$

▮ See Fig. 14.40 and Prob. 14.137. Then $\dfrac{x^2}{4} - \dfrac{y^2}{16} = 1$.

14.150 $x^2 - 9y^2 = 9$

▮ See Fig. 14.41 and Prob. 14.138. Then $\dfrac{x^2}{9} - y^2 = 1$. Ask yourself, what significance does the y axis have?

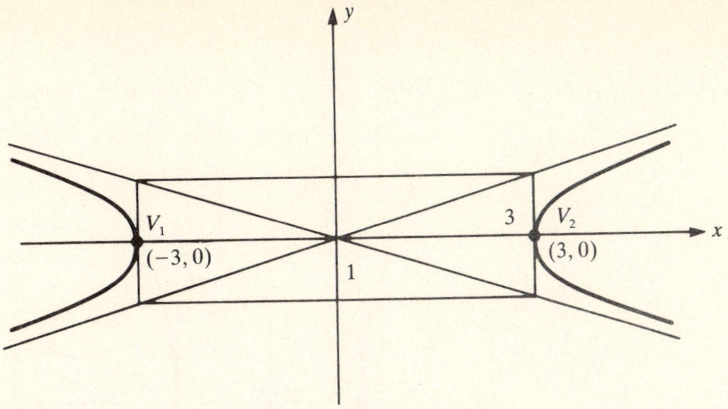

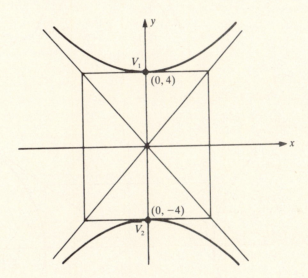

Fig. 14.41

14.151 $9y^2 - 16x^2 = 144$

▮ See Fig. 14.42 and Prob. 14.139. Then $\dfrac{y^2}{16} - \dfrac{x^2}{9} = 1$. $\overline{V_1V_2}$ is the transverse axis.

Fig. 14.42

14.152 $4y^2 - 25x^2 = 100$

▮ See Fig. 14.43 and Prob. 14.140. Then $\dfrac{y^2}{25} - \dfrac{x^2}{4} = 1$.

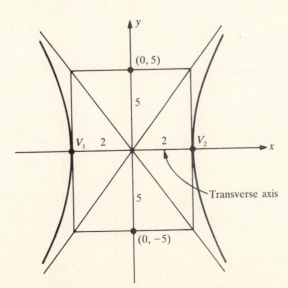

Fig. 14.43

14.153 $3x^2 - 2y^2 = 12$

⬛ See Fig. 14.44 and Prob. 14.141.   Then $\dfrac{x^2}{4} - \dfrac{y^2}{6} = 1$.   What is the significance of the x axis?

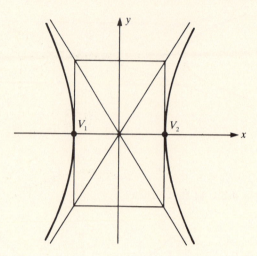

Fig. 14.44

14.154 $3x^2 - 4y^2 = 24$

⬛ See Fig. 14.45 and Prob. 14.142.   Then $\dfrac{x^2}{8} - \dfrac{y^2}{6} = 1$.

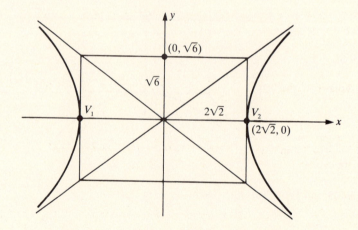

Fig. 14.45

14.155 $7y^2 - 4x^2 = 28$

⬛ See Fig. 14.46 and Prob. 14.143.   Then $\dfrac{y^2}{4} - \dfrac{x^2}{7} = 1$.

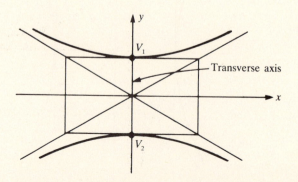

Fig. 14.46

14.156 $3y^2 - 2x^2 = 24$

$\quad\blacksquare$ See Fig. 14.47 and Prob. 14.144. Then $\dfrac{y^2}{8} - \dfrac{x^2}{12} = 1$.

For Probs. 14.157 to 14.162, find the equations of the asymptotes for the given hyperbola.

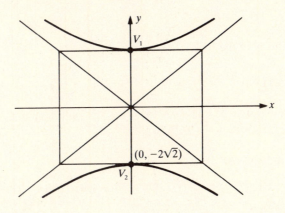

Fig. 14.47

14.157 $x^2 - \dfrac{y^2}{4} = 1$

$\quad\blacksquare$ The asymptotes for an equation of the form $\dfrac{x^2}{a^2} - \dfrac{y^2}{b^2} = 1$ are $y = \pm \dfrac{b}{a} x$; here $y = \pm \dfrac{2}{1} x$ or $y = 2x$ and $y = -2x$.

14.158 $\dfrac{x^2}{4} - \dfrac{y^2}{9} = 1$

$\quad\blacksquare$ Here, $a = 2$, $b = 3$; $y = \pm \dfrac{3}{2} x$, or $y = \dfrac{3}{2} x$ and $y = -\dfrac{3}{2} x$.

14.159 $\dfrac{y^2}{3} - \dfrac{x^2}{2} = 1$

$\quad\blacksquare$ This hyperbola is of the form $\dfrac{y^2}{a^2} - \dfrac{x^2}{b^2} = 1$. Here, $y = \pm \dfrac{\sqrt{3}}{\sqrt{2}} x$, or $y = \pm \sqrt{\dfrac{3}{2}} x$.

14.160 $y^2 - 4x^2 = 1$

$\quad\blacksquare$ This is $y^2 - \dfrac{x^2}{\frac{1}{4}} = 1$ in standard form. Then $a = 1$, $b = \dfrac{1}{2}$, and $y = \pm \dfrac{a}{b} x = \pm \dfrac{1}{\frac{1}{2}} x = \pm 2x$.

14.161 $4x^2 - y^2 = 2$

$\quad\blacksquare$ $2x^2 - \dfrac{y^2}{2} = 1$, $\dfrac{x^2}{\frac{1}{2}} - \dfrac{y^2}{2} = 1$. Then $a = \sqrt{\dfrac{1}{2}}$, $b = \sqrt{2}$, and $y = \pm \dfrac{b}{a} x = \pm \dfrac{\sqrt{2}}{1/\sqrt{2}} x = \pm 2x$ (compare with Prob. 14.160).

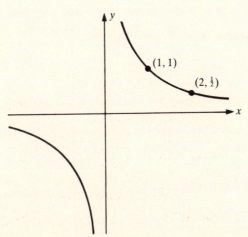

Fig. 14.48

14.162 $xy = 1$

▮ See Fig. 14.48 $xy = a$ is a hyperbola whose asymptotes are the x and y axes.

For Probs. 14.163 to 14.165, find the center of the given hyperbola and put the equation in standard hyperbola form.

14.163 $16y^2 - 9x^2 = 144$

▮ $\dfrac{y^2}{9} - \dfrac{x^2}{16} = 1$. This is of the form $\dfrac{(y-k)^2}{a^2} - \dfrac{(x-h)^2}{b^2} = 1$, where $h = k = 0$. The center is at $(0,0)$.

14.164 $x^2 - 4y^2 + 6x + 16y - 11 = 0$

▮ $(x^2 + 6x + 9) - 4(y^2 - 4y + 4) = 11 + 9 - 16$, $(x + 3)^2 - 4(y - 2)^2 = 4$, and $\dfrac{(x+3)^2}{4} - \dfrac{(y-2)^2}{1} = 1$.
The center is at $(-3, 2)$ since $h = -3$, $k = 2$.

14.165 $144x^2 - 25y^2 - 576x + 200y + 3776 = 0$

▮ $144(x^2 - 4x + 4) - 25(y^2 - 8y + 16) = -3776 + 576 - 400$, $144(x - 2)^2 - 25(y - 4)^2 = -3600$,
$\dfrac{25(y-4)^2}{3600} - \dfrac{144(x-2)^2}{3600} = 1$, $\dfrac{(y-4)^2}{\frac{3600}{25}} - \dfrac{(x-2)^2}{\frac{3600}{144}} = 1$, $\dfrac{(y-4)^2}{144} - \dfrac{(x-2)^2}{25} = 1$. The center is at $(2, 4)$.

For Probs. 14.166 and 14.167, find the vertices, the foci, and the asymptotes.

14.166 $x^2 - 4y^2 + 6x + 16y - 11 = 0$

▮ See Prob. 14.164. Then $\dfrac{(x+3)^2}{4} - \dfrac{(y-2)^2}{1} = 1$. Thus, $a = 2$, $b = 1$, and $c = \sqrt{a^2 + b^2} = \sqrt{5}$.
The vertices are on the transverse axis at a distance a from the center $(-3, 2)$. Thus, the vertices are
$(-1, 2)$ and $(-5, 2)$. The foci are at a distance c from the center $(-3, 2)$. Thus, the foci are $(-3 \pm \sqrt{5}, 2)$.

14.167 $144x^2 - 25y^2 - 576x + 200y + 3776 = 0$

▮ See Prob. 14.165. Then $\dfrac{(y-4)^2}{144} - \dfrac{(x-2)^2}{25} = 1$. Thus, $a = 12$, $b = 5$, and $c = \sqrt{a^2 + b^2} = \sqrt{144 + 25} = 13$. Then the vertices are $V_1 = (2, 4 + 12) = (2, 16)$ and $V_2 = (2, 4 - 12) = (2, -8)$. The foci
are $F_1 = (2, 4 + 13) = (2, 17)$ and $F_2 = (2, 4 - 13) = (2, -9)$.

For Probs. 14.168 to 14.173, find the equation of the hyperbola with center $(0, 0)$ satisfying the given conditions.

14.168 Transverse axis is on the x axis; transverse axis length is 14; conjugate axis length is 10.

▮ Since the transverse axis is on the x axis, $\dfrac{x^2}{a^2} - \dfrac{y^2}{b^2} = 1$. Then $2a = 14$, $a = 7$; and $2b = 10$, $b = 5$.
Thus, the equation is $\dfrac{x^2}{49} - \dfrac{y^2}{25} = 1$.

14.169 Transverse axis is on the x axis; transverse axis length is 8; conjugate axis length is 6.

▮ See Prob. 14.168. Again, $\dfrac{x^2}{a^2} - \dfrac{y^2}{b^2} = 1$. Then $2a = 8$, $a = 4$; and $2b = 6$, $b = 3$. Thus, the
equation is $\dfrac{x^2}{16} - \dfrac{y^2}{9} = 1$.

14.170 Transverse axis is on the y axis; transverse axis length is 24; conjugate axis length is 18.

▮ Here, $\dfrac{y^2}{a^2} - \dfrac{x^2}{b^2} = 1$, since the transverse axis is on the y axis. Then $2a = 24$, $a = 12$; and $2b = 18$,
$b = 9$. Thus, the equation is $\dfrac{y^2}{144} - \dfrac{x^2}{81} = 1$.

14.171 Transverse axis is on the y axis; transverse axis length is 16; conjugate axis length is 22.

▮ $\dfrac{y^2}{a^2} - \dfrac{x^2}{b^2} = 1$. Then $2a = 16$, $a = 8$; and $2b = 22$, $b = 11$. Thus, the equation is $\dfrac{y^2}{64} - \dfrac{x^2}{121} = 1$.

14.172 Transverse axis on the x axis; transverse axis length is 18; distance from foci to center is 11.

▮ $\dfrac{x^2}{a^2} - \dfrac{y^2}{b^2} = 1$. Then $2a = 18$, $a = 9$; and $c = 11 =$ distance from foci to center. Since $a^2 + b^2 = c^2$, $b^2 = c^2 - a^2 = 121 - 81 = 40$. Thus, the equation is $\dfrac{x^2}{81} - \dfrac{y^2}{40} = 1$.

14.173 Conjugate axis is on the x axis; conjugate axis length is 14; distance from foci to center is $\sqrt{200}$.

▮ The transverse axis is on the y axis, so $\dfrac{y^2}{a^2} - \dfrac{x^2}{b^2} = 1$. Then $2b = 14$, $b = 7$; and $a^2 = c^2 - b^2 = 200 - 49 = 151$. Thus, $\dfrac{y^2}{151} - \dfrac{x^2}{49} = 1$.

For Probs. 14.174 to 14.176, find the equation of the hyperbola fitting the given conditions.

14.174 Vertices $(\pm 5, 0)$; focus $(13, 0)$.

▮ Here, the transverse axis is on the x axis since $(5, 0)$ and $(-5, 0)$ are vertices. Also, $c =$ distance from center to focus $= 13$, since the center is at $(0, 0)$ [midway between $(5, 0)$ and $(-5, 0)$]. Thus, $a = 5$, $c = 13$, and $b^2 = c^2 - a^2 = 169 - 25 = 144$. Thus, $\dfrac{x^2}{25} - \dfrac{y^2}{144} = 1$.

14.175 Center $(0, 0)$; $a = 5$; focus $(0, 6)$.

▮ If the focus is at $(0, 6)$, then the transverse axis is on the y axis and $c = 6$. Then $b^2 = c^2 - a^2 = 36 - 25 = 11$. Thus, $\dfrac{y^2}{25} - \dfrac{x^2}{11} = 1$.

14.176 Center $(2, -3)$; vertex $(7, -3)$; asymptote $3x - 5y - 21 = 0$.

▮ If the center is at $(2, -3)$ and one vertex is $(7, -3)$, the other must be at $(-3, -3)$ so that c is the midpoint of $\overline{V_1 V_2}$. If $3x - 5y - 21 = 0$, then $5y = 3x - 21$ and $y = \frac{3}{5}x - \frac{21}{5}$. The slope, $\frac{3}{5}$, of the asymptotes is b/a. But, $CV = a = 5$; thus $\frac{3}{5} = b/5$, $b = 3$. $\dfrac{(x - 2)^2}{25} - \dfrac{(y + 3)^2}{9} = 1$.

For Probs. 14.177 to 14.180, find the intersection(s) of the given curves.

14.177 $x^2 - y^2 = 1$ and $y = x$.

▮ Then $x^2 - x^2 = 1$, $0 = 1$; no intersection.

14.178 $x^2 - 2y^2 = 1$ and $x^2 + y^2 = 4$.

▮ $x^2 = 4 - y^2$. Thus, $4 - y^2 - 2y^2 = 1$, $-3y^2 = -3$, $y^2 = 1$, or $y = \pm 1$. If $y = 1$, $x^2 = 4 - 1$, or $x = \pm\sqrt{3}$. If $y = -1$, $x^2 = 4 - 1$, or $x = \pm\sqrt{3}$. Intersections: $(\sqrt{3}, 1)$, $(-\sqrt{3}, 1)$, $(\sqrt{3}, -1)$, $(-\sqrt{3}, -1)$.

14.179 $x^2 - 2y^2 = 1$ and $x^2 + 2y^2 = 1$.

▮ Adding, we get $2x^2 = 2$, $x^2 = 1$, or $x = \pm 1$. If $x = 1$, $1 + 2y^2 = 1$, or $y = 0$. If $x = -1$, $1 + 2y^2 = 1$, or $y = 0$. Intersections: $(1, 0)$, $(-1, 0)$.

14.180 $x^2 - 2y^2 = 1$ and $y = 2x^2$.

▮ $x^2 - 2(2x^2)^2 = 1$, $x^2 - 8x^4 = 1$, and $8x^4 - x^2 + 1 = 0$. Let $u = x^2$. Then $8u^2 - u + 1 = 0$ and $u = \dfrac{1 \pm \sqrt{1 - 4(8)(1)}}{16}$. There is no solution (negative discriminant), and, therefore, there are no intersections.

14.5 TRANSFORMATIONS IN THE PLANE

For Probs. 14.181 to 14.185, find the coordinates asked for given equations of translation $x = x' + 2$ and $y = y' - 5$.

14.181 The coordinates for $O(0, 0)$ in the translated system.

▮ If $x = x' + 2$ and $y = y' - 5$, then $x' = x - 2$, $y' = y + 5$, and $O' = (0 - 2, 0 + 5) = (-2, +5)$.

14.182 The coordinates for $P(-2, 4)$ in the translated system.

❙ If $x' = x - 2$ and $y' = y + 5$, then $x' = -2 - 2 = -4$, $y' = 4 + 5 = 9$, and $P' = (-4, 9)$.

14.183 The coordinates of $P'(-2, 4)$ in the original (untranslated) system.

❙ $x = x' + 2 = -2 + 2 = 0$ and $y = y' - 5 = 4 - 5 = -1$. So $P = (0, -1)$.

14.184 The coordinates of the x intercept for $5x + 2y = 0$ in the translated system.

❙ Let $y = 0$; then $5x = 0$, $x = 0$; and if $0 + 2y = 0$, $y = 0$. The x intercept is $O(0, 0)$. In the translated system $x' = x - 2 = 0 - 2 = -2$ and $y' = y + 5 = 0 + 5 = 5$. So $O' = (-2, 5)$.

14.185 The coordinates for the y intercept for $x - 2y + 4 = 0$ in the translated system.

❙ If $x = 0$, $2y = 4$, $y = 2$; then $P(0, 2)$ is the y intercept. $P' = (0 - 2, 2 + 5) = (-2, 7)$.

For Probs. 14.186 to 14.192, use the translation equations $x = x' - 4$ and $y = y' + 6$ to find the equation of the given curve (or line) in the translated system.

14.186 $2x - 3y + 18 = 0$

❙ Since $x = x' - 4$, $y = y' + 6$.

Then, $2(x' - 4) - 3(y' + 6) + 18 = 0$, $2x' - 8 - 3y' - 18 + 18 = 0$, and $2x' - 3y' - 8 = 0$.

14.187 $-3y + 4x = 10$

❙ $-3(y' + 6) + 4(x' - 4) + 18 = 0$, $-3y' - 18 + 4x' - 16 + 18 = 0$, and $4x' - 3y' - 16 = 0$.

14.188 $y = mx + b$

❙ $y' + 6 = m(x' - 4) + b$, $y' = mx' - 4m + b - 6$, and $y' = mx' + (6 - 4m + b)$. Do you notice something interesting? This equation tells you that the image of a nonvertical line (under a translation) is a nonvertical line with the same slope!

14.189 $x = a$

❙ Since $x = x' - 4$, $x' - 4 = a$, or $x' = a + 4$. Thus, the image of a vertical line under a translation is a vertical line.

14.190 $x^2 + y^2 = 4$

❙ $(x' - 4)^2 + (y' + 6)^2 = 4$.
Notice that this is a circle in the translated system. Compare what happened here to what takes place in the next example.

14.191 $(x + 4)^2 + (y - 6)^2 = 10$

❙ Since $x = x' - 4$ and $y = y' + 6$, $x + 4 = x' - 4 + 4 = x'$ and $y - 6 = y' + 6 - 6 = y'$. Thus, $(x')^2 + (y')^2 = 10$ or $x'^2 + y'^2 = 10$. In this case, the translation made the equation simpler.

14.192 $(y - 6)^2 = x$

❙ To translate this parabola, $y = y' + 6$, so $y - 6 = y' + 6 - 6 = y'$. Thus, since $x = x' - 4$, $y'^2 = x' - 4$. Notice that the vertex of the parabola translated is not at $(0, 0)$ in the translated system.

For Probs. 14.193 to 14.208, find translation equations that change each given equation to a standard form. Write the equation of the curve for the translated system, and identify the curve.

14.193 $(x + 1)^2 = 4(y - 3)$

❙ Let $x' = x + 1$ and $y' = y - 3$. Then $x'^2 = 4y'$, which is a parabola.

14.194 $(y - 1)^2 = 8(x - 2)$

❙ Let $y' = y - 1$ and $x' = x - 2$. Then $y'^2 = 8x'$, which is a parabola.

14.195 $(x-1)^2 + (y-5)^2 = 100$

▮ If we let $x' = x - 1$ and $y' = y - 5$, then $(x')^2 + (y')^2 = 100$, which is in the standard form for a circle.

14.196 $\dfrac{(x+3)^2}{10} + \dfrac{(y-2)^2}{8} = 1$

▮ Let $x' = x + 3$ and $y' = y - 2$. Then $\dfrac{x'^2}{10} + \dfrac{y'^2}{8} = 1$.

14.197 $16(x-3)^2 - 9(y+2)^2 = 144$

▮ Let $x' = x - 3$ and $y' = y + 2$. Then $16x'^2 - 9y'^2 = 144$ and $\dfrac{x'^2}{9} - \dfrac{y'^2}{16} = 1$ which is a hyperbola.

14.198 $(y+2)^2 - 12(x-3) = 0$

▮ Let $y' = y + 2$ and $x' = x - 3$. Then $(y')^2 - 12(x') = 0$, and $12(x') = (y')^2$, which is a parabola.

14.199 $6(x+5)^2 + 5(y+7)^2 = 30$

▮ Let $x' = x + 5$, and $y' = y + 7$. Then $6x'^2 + 5y'^2 = 30$, and $\dfrac{x'^2}{5} + \dfrac{y'^2}{6} = 1$, which is an ellipse.

14.200 $12(y-5)^2 - 8(x-3)^2 = 24$

▮ Let $y' = y - 5$ and $x' = x - 3$. Then $12y'^2 - 8x'^2 = 24$, and $\dfrac{y'^2}{2} - \dfrac{x'^2}{3} = 1$, which is a hyperbola.

14.201 $4x^2 + 9y^2 - 16x - 36y + 16 = 0$

▮ $4(x^2 - 4x) + 9(y^2 - 4y) = -16$. Then $4(x^2 - 4x + 4) + 9(y^2 - 4y + 4) = -16 + 16 + 36$, $4(x-2)^2 + 9(y-2)^2 = 36$, and $\dfrac{(x-2)^2}{9} + \dfrac{(y-2)^2}{4} = 1$. Let $x' = x - 2$, and $y' = y - 2$. Then $\dfrac{x'^2}{9} + \dfrac{y'^2}{4} = 1$, which is an ellipse. Notice that the coordinates of the origin of the translated system are $(2, 2)$, since when $x' = x - h$ and $y' = y - k$, (h, k) are the coordinates of the origin.

14.202 $16x^2 + 9y^2 + 64x + 54y + 1 = 0$

▮ $16(x^2 + 4x + 4) + 9(y^2 + 6y + 9) = -1 + 64 + 81 = 144$. Then $16(x+2)^2 + 9(y+3)^2 = 144$, and $\dfrac{(x+2)^2}{9} + \dfrac{(y+3)^2}{16} = 1$. Let $x' = x + 2$ and $y' = y + 3$. Then $\dfrac{x'^2}{9} + \dfrac{y'^2}{16} = 1$. This is an ellipse with center $(0', 0')$ in the translated system, where $h = -2$, $k = -3$.

14.203 $x^2 + 8x + 8y = 0$

▮ $x^2 + 8x + 16 = -8y + 16$, and $(x+4)^2 = -8(y-2)$. Let $x' = x + 4$ and $y' = y - 2$. Then $x'^2 = -8y'$, which is a parabola. Here, $h = -4$, $k = 2$.

14.204 $y^2 + 12x + 4y - 32 = 0$

$y^2 + 4y + 4 = -12x + 32 + 4$, $(y+2)^2 = -12x + 36$, and $(y+2)^2 = -12(x-3)$. Let $y' = y + 2$ and $x' = x - 3$. Then $y'^2 = -12x'$, which is a parabola, where $h = 3$, $k = -2$.

14.205 $x^2 + y^2 + 12x + 10y + 45 = 0$

▮ $x^2 + 12x + 36 + y^2 + 10y + 25 = -45 + 36 + 25$, and $(x+6)^2 + (y+5)^2 = 16$. Let $x + 6 = x'$ and $y + 5 = y'$. Then $x'^2 + y'^2 = 16$, which is a circle with center at the origin of the new system.

14.206 $x^2 + y^2 - 8x - 6y = 0$

▮ $x^2 - 8x + 16 + y^2 - 6y + 9 = 16 + 9$, and $(x-4)^2 + (y-3)^2 = 25$. Let $x' = x - 4$ and $y' = y - 3$. Then $x'^2 + y'^2 = 25$, which is a circle.

14.207 $-9x^2 + 16y^2 - 72x - 96y - 144 = 0$

▮ $-9(x^2 + 8x + 16) + 16(y^2 - 6y + 9) = 144 - 144 + 144$, $-9(x+4)^2 + 16(y-3)^2 = 144$, and $\dfrac{-(x+4)^2}{16} + \dfrac{(y-3)^2}{9} = 1$. Let $x' = x + 4$ and $y' = y - 3$. Then $\dfrac{y'^2}{9} - \dfrac{x'^2}{16} = 1$, which is a hyperbola.

14.208 $16x^2 - 25y^2 - 160x = 0$

▮ $16(x^2 - 10x + 25) - 25y^2 = 400$, $16(x-5)^2 - 25y^2 = 400$, and $\dfrac{(x-5)^2}{25} - \dfrac{y^2}{16} = 1$. Let $x' = x - 5$ and $y' = y$. Be careful here! Then $\dfrac{x'^2}{25} - \dfrac{y'^2}{16} = 1$, which is an ellipse.

For Probs. 14.209 to 14.211, simplify the given equation using a translation, and sketch the graph showing both sets of axes.

14.209 $4x^2 + y^2 - 16x + 6y - 11 = 0$

▮ See Fig. 14.49. $4(x^2 - 4x + 4) + y^2 + 6y + 9 = 11 + 9 + 16$, $4(x-2)^2 + (y+3)^2 = 36$, and $\dfrac{(x-2)^2}{9} + \dfrac{(y+3)^2}{36} = 1$. Let $x' = x - 2$ and $y' = y + 3$. Then $\dfrac{x'^2}{9} + \dfrac{y'^2}{36} - 1$. Then center of the (x', y') system is $(2, -3)$ since $x' = x - h$, and $y' = y - k$.

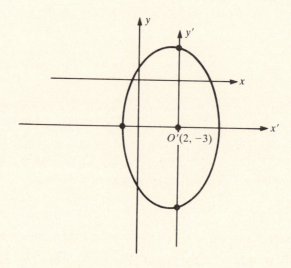

Fig. 14.49

14.210 $x^2 - 12x - 8y - 4 = 0$

▮ See Fig. 14.50. $x^2 - 12x + 36 = 8y + 4 + 36$, and $(x-6)^2 = 8(y+5)$. Let $x' = x - 6$ and $y' = y + 5$. Then $x'^2 = 8y'$, where the new center is $(6, -5)$.

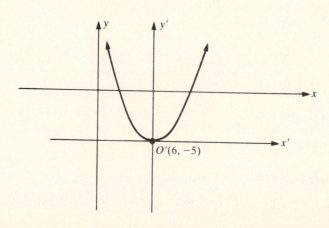

Fig. 14.50

14.211 $16y^2 + 5x + 32y + 6 = 0$

▮ See Fig. 14.51. $16(y^2 + 2y + 1) = -5x - 6 + 16$, $16(y + 1)^2 = -5x + 10$ and $16(y + 1)^2 = -5(x - 2)$. Let $y' = y + 1$ and $x' = x - 2$. Then $y'^2 = -\frac{5}{16}x'$.

For Probs. 14.212 to 14.215, transform the given equation by rotating the axes through the indicated angle.

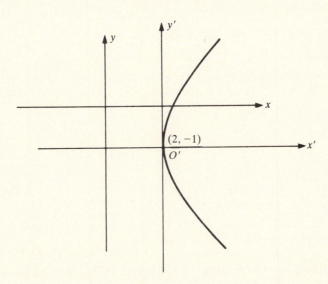

Fig. 14.51

14.212 $x^2 + y^2 = 16$; $\theta = 30°$.

▮ We use the rotation equations $x = x' \cos \theta - y' \sin \theta$ and $y = x' \sin \theta + y' \cos \theta$. Then $x = x' \cos 30° - y' \sin 30° = x' \frac{\sqrt{3}}{2} - \frac{y'}{2}$ and $y = x' \sin 30° + y' \cos 30° = \frac{x'}{2} + y' \frac{\sqrt{3}}{2}$. Then, if $x^2 + y^2 = 16$, $\left(\frac{x'\sqrt{3}}{2} - \frac{y'}{2}\right)^2 + \left(\frac{x'}{2} + y' \frac{\sqrt{3}}{2}\right)^2 = 16$. Then $x'^2 + y'^2 = 16$. What do you notice? What is the rotated image of a circle centered at the origin?

14.213 $x^2 + y^2 = 4$; $\theta = 45°$.

▮ Then $x = x' \cos 45° - y' \sin 45°$ and $y = x' \sin 45° + y' \cos 45°$. Thus, $x = \frac{\sqrt{2}}{2} x' - \frac{\sqrt{2}}{2} y'$ and $y = x' \frac{\sqrt{2}}{2} + y' \frac{\sqrt{2}}{2}$. Then $\left(\frac{\sqrt{2}}{2} x' - \frac{\sqrt{2}}{2} y'\right)^2 + \left(x' \frac{\sqrt{2}}{2} + y' \frac{\sqrt{2}}{2}\right)^2 = 4$, and $x'^2 + y'^2 = 4$. (There it is again. Think about it? If you have a circle centered at the origin and you rotate it, what is the image?)

14.214 $(x - 1)^2 + (y - 2)^2 = 5$; $\theta = 60°$.

▮ Using the same reasoning, the image of a circle at $(1, 2)$ rotated is the same circle! $(x' - 1)^2 + (y' - 2)^2 = 5$.

14.215 $2x^2 + \sqrt{3}xy + y^2 - 10 = 0$; $\theta = 30°$.

▮ $x = x' \frac{\sqrt{3}}{2} - y' \frac{1}{2}$ and $y = x' \frac{1}{2} + y' \frac{\sqrt{3}}{2}$. Then $2\left(\frac{\sqrt{3}}{2} x' - \frac{y'}{2}\right)^2 + \sqrt{3}\left(x' \frac{\sqrt{3}}{2} - \frac{y'}{2}\right)\left(\frac{x'}{2} + y' \frac{\sqrt{3}}{2}\right) + \left(\frac{x'}{2} + y' \frac{\sqrt{3}}{2}\right)^2 - 10 = 0$, and $\frac{x'^2}{4} + \frac{y'^2}{20} = 1$.

For Probs. 14.216 to 14.221, find an angle of rotation so that the transformed equation will have no $x'y'$ term.

14.216 $x^2 - 4xy + y^2 = 12$

▮ We are looking for $\theta \in (0, 90°)$ such that $\cot 2\theta = \frac{A - C}{B}$ where $Ax^2 + Bxy + Cy^2 + Dx + Ey + F = 0$. Here, $A = 1$, $C = 1$, $B = -4$. Thus, $\cot 2\theta = \frac{1 - 1}{-4} = 0$. Then $2\theta = 90°$, and $\theta = 45°$.

14.217 $x^2 + xy + y^2 = 6$

▮ See Prob. 14.217. $A = 1$, $C = 1$, $B = 1$; $\cot 2\theta = \dfrac{1-1}{1} = 0$. Then $2\theta = 90°$, and $\theta = 45°$.

14.218 $8x^2 - 4xy + 5y^2 = 36$

▮ $A = 8$, $C = 5$, $B = -4$; $\cot 2\theta = \dfrac{8-5}{-4} = \dfrac{-3}{4}$. Then $2\theta = \operatorname{Cot}^{-1}(-\tfrac{3}{4})$, and $\theta = \dfrac{\operatorname{Cot}^{-1}(-\frac{3}{4})}{2}$.

14.219 $5x^2 - 4xy + 8y^2 = 36$

▮ $A = 5$, $B = 8$, $C = -4$; $\cot 2\theta = \dfrac{5-8}{-4} = \dfrac{3}{4}$. Then $\theta = \dfrac{\operatorname{Cot}^{-1}\frac{3}{4}}{2}$.

14.220 $x^2 - 2\sqrt{3}xy + 3y^2 - 16\sqrt{3}x - 16y = 0$

▮ $A = 1$, $C = 3$, $B = -2\sqrt{3}$; $\cot 2\theta = \dfrac{A-C}{B} = \dfrac{1-3}{-2\sqrt{3}} = \dfrac{-2}{-2\sqrt{3}} \cdot \dfrac{\sqrt{3}}{\sqrt{3}} = \dfrac{\sqrt{3}}{3}$. Then $2\theta = 60°$, and $\theta = 30°$.

14.221 $x^2 + 2\sqrt{3}xy + 3y^2 + 8\sqrt{3}x - 8y = 0$

▮ See Fig. 14.52. $A = 1$, $C = 3$, $B = 2\sqrt{3}$. $\cot 2\theta = \dfrac{A-C}{B} = \dfrac{1-3}{2\sqrt{3}} = \dfrac{-2}{2\sqrt{3}} = \dfrac{-\sqrt{3}}{3}$. Then 2θ is a quadrantal angle, and $2\theta = 120°$, so $\theta = 60°$.

For Probs. 14.222 to 14.226, rotate the coordinate axes through 45° and find the given item.

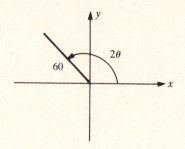

Fig. 14.52

14.222 $P(\sqrt{2}, 3\sqrt{2})'$

▮ $x = x' \cos 45° - y' \sin 45°$ and $y = x' \sin 45° + y' \cos 45°$. Then $x = \dfrac{1}{\sqrt{2}}(x' - y')$ and $y = \dfrac{1}{\sqrt{2}}(x' + y')$. Thus, $x = \dfrac{1}{\sqrt{2}}(2 - 3\sqrt{2}) = -2$ and $y = \dfrac{1}{\sqrt{2}}(\sqrt{2} + 3\sqrt{2}) = 4$.

14.223 $\theta(0, 0)$ where referred to the new system.

▮ $x' = \dfrac{1}{\sqrt{2}}(x + y)$ and $y' = \dfrac{-1}{\sqrt{2}}(x - y)$. Let $x = y = 0$. Then $x' = 0$, and $y' = 0$. Does this surprise you? It shouldn't!

14.224 $P(\sqrt{2}, 3\sqrt{2})$

▮ Here, $x = \sqrt{2}$, and $y = 3\sqrt{2}$. Then $x' = \dfrac{1}{\sqrt{2}}(\sqrt{2} + 3\sqrt{2}) = 4$ and $y' = \dfrac{-1}{\sqrt{2}}(\sqrt{2} - 3\sqrt{2}) = 2$. $P(\sqrt{2}, 3\sqrt{2})$ is $(4, 2)'$ in the rotated system.

14.225 $x + y + 3\sqrt{2} = 0$ in the new (rotated) system.

▮ $x = \dfrac{1}{\sqrt{2}}(x' + y')$ $y = \dfrac{1}{\sqrt{2}}(x' + y')$. Then $x + y + 3\sqrt{2} = 0$ becomes $\dfrac{1}{\sqrt{2}}(x' - y') + \dfrac{1}{\sqrt{2}}(x' + y') + 3\sqrt{2} = 0$, and $\dfrac{2}{\sqrt{2}}x' + 3\sqrt{2} = 0$; multiply by $\sqrt{2}$. $2x' + 6 = 0$ or $x' = -3$. This is a vertical line relative to the new system. It is *not* perpendicular to the x axis. It *is* perpendicular to the x' axis.

14.226 $3x - 3y + 4 = 0$ in the rotated system.

▮ $x = \dfrac{1}{\sqrt{2}}(x' - y')$ and $y = \dfrac{1}{\sqrt{2}}(x' + y')$. Then $3\left[\dfrac{1}{\sqrt{2}}(x' - y')\right] - 3\left[\dfrac{1}{\sqrt{2}}(x' + y')\right] + 4 = 0$; the x' terms drop, and $3y' = 2\sqrt{2}$. This is horizontal in the rotated system. What axis is it parallel to?

For Probs. 14.227 to 14.229, answer true or false, and why?

14.227 If two lines intersect in the (x, y) system, they must intersect in the translated (x', y') system.

▮ See Fig. 14.53. True. If $P \in l_1$ and $P \in l_2$, then $l'_1 \cap l'_2 = P'$ where P' is the image of P under the translation.

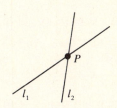

Fig. 14.53

14.228 The image of a circle with center at $(0, 0)$ under a rotation is a circle with center at $(0, 0)'$.

▮ True. See Prob. 14.212. Also, if $x^2 + y^2 = r^2$, then $(x' \cos \theta - y' \sin \theta)^2 + (x' \sin \theta + y' \cos \theta)^2 = r^2$. Thus, $x'^2 \cos^2 \theta - 2x'y' \sin \theta \cos \theta - y'^2 \sin^2 \theta + x'^2 \sin^2 \theta + 2x'y' \sin \theta \cos \theta + y'^2 \cos^2 \theta = r^2$. Then $x'^2(\sin^2 \theta + \cos^2 \theta) + y'^2(\sin^2 \theta + \cos^2 \theta) = 1$ and $x'^2 + y'^2 = r^2$.

14.229 The image of $y = x^2$ under a rotation is a parabola.

▮ See Fig. 14.54. True. If this system is rotated $\theta°$, the parabola shifts position, but *not* shape.

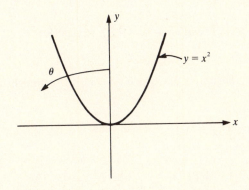

Fig. 14.54

14.230 Transform $x^2 + 2\sqrt{3}xy + 3y^2 - 8\sqrt{3}x - 84 - 4 = 0$ into a standard form using a rotation followed by a translation.

▮ $\cot 2\theta = \dfrac{A - C}{B} = \dfrac{1 - 3}{2\sqrt{3}} = \dfrac{-2}{2\sqrt{3}} = \dfrac{-\sqrt{3}}{3}$; then $2\theta = 120°$, and $\theta = 60°$. Then

$$x = x' \cos 60° - y' \sin 60° = \frac{x'}{2} - \frac{\sqrt{3}}{2}y'$$

$$y = x' \sin 60° - y' \cos 60° = \frac{\sqrt{3}x'}{2} - y'$$

Thus,

$$\left(\frac{x'}{2} - \frac{\sqrt{3}y'}{2}\right)^2 + 2\sqrt{3}\left(\frac{x'}{2} - \frac{\sqrt{3}y'}{2}\right)\left(\frac{\sqrt{3}x'}{2} - \frac{y'}{2}\right) +$$

$$3\left(\frac{\sqrt{3}x'}{2} - \frac{y'}{2}\right)^2 - 8\sqrt{3}\left(\frac{x'}{2} - \frac{\sqrt{3}x'}{2} - \frac{y'}{2}\right) = 4$$

and

$$x'^2 - 2\sqrt{3}x' + 2y' = 1.$$

Then $x'^2 - 2\sqrt{3}x' + 3 = 1 - 2y' + 3$, and $(x' - \sqrt{3})^2 = -2(y' - 2)$. Thus, translating, let $x'' = x' - \sqrt{3}$ and $y'' = y' - 2$. Then $x''^2 = -2y''$.

14.6 MISCELLANEOUS PROBLEMS

For Probs. 14.231 to 14.234, find the equation of the locus of points satisfying the given condition(s).

14.231 The distance of the points from $(3, 0)$ is equal to their distance from the line $x + 3 = 0$.

∎ $\sqrt{(x-3)^2 + (y-0)^2} = \sqrt{(x+3)^2}$. Thus, $x^2 - 6x + 9 + y^2 = x^2 + 6x + 9$, and $y^2 - 12x = 0$. (This is a parabola.)

14.232 The distance of the points from $(-3, 4)$ is twice their distance from the line $y + 4 = 0$.

∎ $\sqrt{(x+3)^2 + (y-4)^2} = 2\sqrt{(y+4)^2}$, $x^2 + 6x + 9 + y^2 - 8y + 16 = 4y^2 + 32y + 64$, and $x^2 - 3y^2 + 6x - 40y - 39 = 0$. What curve is this?

14.233 The sum of the distances of the points from $(0, 4)$ and $(0, -4)$ is 10.

∎ $\sqrt{x^2 + (y-4)^2} + \sqrt{x^2 + (y+4)^2} = 10$. Then $\sqrt{x^2 + (y-4)^2} = 10 - \sqrt{x^2 + (y+4)^2}$. Squaring, we get $x^2 + (y-4)^2 = 100 = 20\sqrt{x^2 + (y+4)^2} + x^2 + (y+4)^2$. Then $x^2 + (y-4)^2 - 100 - x^2 - (y+4)^2 = -20\sqrt{x^2 + (y+4)^2}$. Squaring, $25x^2 + 9y^2 - 225 = 0$. This is an ellipse.

14.234 The difference of the distances of the points from $(-4, 0)$ and $(4, 0)$ is 6.

∎ $\sqrt{(x+4)^2 + y^2} - \sqrt{(x-4)^2 + y^2} = 6$. Then $\sqrt{(x+4)^2 + y^2} = 6 + \sqrt{(x-4)^2 + y^2}$, $(x+4)^2 + y^2 = 36 + 12\sqrt{(x-4)^2 + y^2} + (x-4)^2 + y^2$, and $(x+4)^2 - 36 - (x-4)^2 = 12\sqrt{(x-4)^2 + y^2}$. Squaring, we get $7x^2 - 9y^2 = 63$. What curve is this?

For Probs. 14.235 to 14.247, eliminate the parameter to obtain an equation in x and y.

14.235 $x = -t$, $y = 2t - 2$

∎ Then $t = -x$, and $y = 2(-x) - 2 = -2x - 2$, a straight line.

14.236 $x = t$, $y = 3t - 5$

∎ Then, since $x = t$, $t = x$, and $y = 3x - 5$, a straight line.

14.237 $x = 3t$, $y = 5t - 6$

∎ If $x = 3t$, $t = x/3$. Then $y = 5(x/3) - 6 = 5x/3 - 6$, a straight line.

14.238 $x = t^2$, $y = 3t^2 + 1$

∎ If $x = t^2$, $t^2 = x$, and $y = 3x + 1$, $x \geq 0$; this is a ray.

14.239 $x = \frac{1}{4}t^4$, $y = t^2$

∎ Then, $x = \frac{1}{4}(t^2)^2 = \frac{1}{4}y^2$ (since $t^2 = y$) where $y \geq 0$. This is the upper half of a parabola.

14.240 $x = \sin \theta$, $y = \cos \theta$

∎ Then $x^2 = \sin^2 \theta$ and $y^2 = \cos^2 \theta$. Adding, we get $x^2 + y^2 = \sin^2 \theta + \cos^2 \theta$, or $x^2 + y^2 = 1$.

14.241 $x = \sin \theta$, $y = 2 \cos \theta$

∎ $x^2 = \sin^2 \theta$ and $y^2 = 4 \cos^2 \theta$. But $\sin^2 \theta + \cos^2 \theta = 1$; thus, we consider $y^2/4$. Since $y^2/4 = \cos^2 \theta$, $x^2 + y^2/4 = \sin^2 \theta + \cos^2 \theta = 1$, and $x^2 + y^2/4 = 1$.

14.242 $x = 2 + 2 \sin \theta$ and $y = 3 + 2 \cos \theta$.

∎ Then $x - 2 = 2 \sin \theta$ and $y - 3 = 2 \cos \theta$. Thus, $(x - 2)^2 + (y - 3)^2 = 4(\sin^2 \theta + \cos^2 \theta)$, and $(x - 2)^2 + (y - 3)^2 = 4$; this is a circle.

14.243 $x = t - 2$ and $y = \dfrac{2}{2 - t}$; $t \neq 2$.

∎ Note that $2 - t = -(t - 2)$. Then $y = 2/-x$ and $xy = -2$; this is a hyperbola.

14.244 $x = t - 1$ and $y = \sqrt{t}$; $t \geq 0$.

∎ Then $x = y^2 - 1$ where $y \geq 0$ ($y = \sqrt{t}$) and $x \geq -1$ ($x = t - 1$). This is the upper half of a parabola.

14.245 $x = t^2$ and $y = t^{-2}$, $t \neq 0$.

▮ $t^2 = 1/t^{-2}$. Thus, $x = 1/y$, or $xy = 1$, $x > 0$ (since $x = t^2$, and $t \neq 0$). This is a single branch of a hyperbola.

14.246 $x = e^t$ and $y = e^{-t}$.

▮ $e^t = 1/e^{-t}$. Thus, $x = 1/y$, or $xy = 1$, $x > 0$ (since $x = e^t$). Compare this to Prob. 14.245 above.

14.247 $x = \dfrac{8}{t^2 + 4}$ and $y = \dfrac{4t}{t^2 + 4}$.

▮ $4t = \frac{8}{2} \cdot t$. Thus, $y = (x/2)t$, and $y^2 = (x^2/4)t^2$. But $(t^2 + 4)x = 8$, $t^2 + 4 = 8/x$, and $t^2 = 8/x - 4$. Then $y^2 = (x^2/4)(8/x - 4) = 2x - x^2$, and $x^2 + y^2 - 2x = 0$, $x \neq 0$ ($t^2 = 8/x - 4$). This is a circle with a hole in it when $x = 0$.

For Probs. 14.248 to 14.250, use the "$B^2 - AC$ method" to identify the locus of points satisfying the given equation.

14.248 $3x^2 - 10xy + 3y^2 + x - 32 = 0$

▮ Notice that this is in semireduced form. $Ax^2 + 2Bxy + Cy^2 + 2Dx + 2Ey + F = 0$. Then $B^2 - AC = 25 - (3)(3) > 0$. This is a hyperbola or a pair of intersecting lines.

14.249 $41x^2 - 84xy + 76y^2 = 168$

▮ $B^2 = AC = (-41)^2 - (41)(76) < 0$. This is an ellipse, a point, or imaginary.

14.250 $16x^2 + 23xy + 9y^2 - 30x + 40y = 0$

▮ $B^2 - AC = (12)^2 - (16)(9) = 0$. This is a parabola or a pair of parallel lines.

For Probs. 14.251 to 14.253, an object follows a path given by $x = 5\sin(6\pi t)$ and $y = 5\cos(6\pi t)$ with $t \geq 0$ (t in seconds; x, y in feet).

14.251 What are the coordinates of the object when $t = 0.1$ second?

▮ See Fig. 14.55. When $t = 0.1$, $x = 5\sin[6\pi(0.1)] = 5\sin(0.6\pi)$ and $y = 5\cos[6\pi(0.1)] = 5\cos(0.6\pi)$. Using 3.14 as an approximation for π, we have $0.4\pi \approx 1.256$ in quadrant II. Then $5\sin(1.256) \approx 4.8$ (to the nearest tenth) and $5\cos(1.256) \approx -1.5$ (to the nearest tenth). (Do not forget you are in quadrant II!)

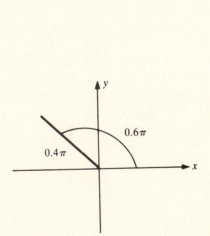

Fig. 14.55

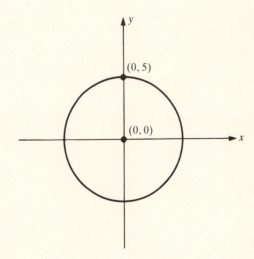

Fig. 14.56

14.252 Eliminate the parameter t.

▮ $x^2 = 25\sin^2(6\pi t)$ and $y^2 = 25\cos^2(6\pi t)$. $x^2 + y^2 = 25[\sin^2(6\pi t) + \cos^2(6\pi t)] = 25(1) = 25$.

14.253 Sketch the path of the moving object.

▮ See Fig. 14.56. The path is a circle with center $(0, 0)$ and $r = 5$.

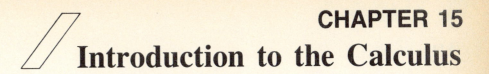

CHAPTER 15
Introduction to the Calculus

15.1 LIMITS AND THE TANGENT TO A GRAPH

For Probs. 15.1 to 15.30, find the indicated limit or show that it does not exist.

15.1 $\lim\limits_{x \to 2} (2x + 3)$

$\blacksquare$ As x gets arbitrarily close to 2, $2x + 3$ gets arbitrarily close to $2(2) + 3 = 7$. Alternately, recall that if $P(x)$ is a polynomial, then $\lim\limits_{x \to a} P(x) = P(a)$. $\lim\limits_{x \to 2} (2x + 3) = 7$.

15.2 $\lim\limits_{x \to -3} 4x$

$\blacksquare$ $\lim\limits_{x \to -3} 4x = 4(-3) = -12$ since $4x$ is a polynomial.

15.3 $\lim\limits_{x \to 5} (2x^2 - 6)$

$\blacksquare$ $\lim\limits_{x \to 5} (2x^2 - 6) = 2(5)^2 - 6$ since $2x^2 - 6$ is a polynomial. See Prob. 15.4.

15.4 $\lim\limits_{x \to 5} (2x^2 - 6)$

$\blacksquare$ Another way to approach this problem is to use the following limit theorems, which assume that all limits exist and no denominators are zero:

$$\lim\limits_{x \to a} x = a \quad \lim\limits_{x \to b} [f(x) \pm g(x)] = f(b) \pm g(b) \quad \lim\limits_{x \to e} k(fx) = k \lim\limits_{x \to e} f(x) \quad \lim\limits_{x \to d} f(x)g(x) = \lim\limits_{x \to d} f(x) \lim\limits_{x \to d} g(x)$$

$$\lim\limits_{x \to e} \frac{f(x)}{g(x)} = \frac{\lim\limits_{x \to e} f(x)}{\lim\limits_{x \to e} g(x)}$$

Then, $\lim\limits_{x \to 5} (2x^2 - 6) = \lim\limits_{x \to 5} 2x^2 - \lim\limits_{x \to 5} 6 = 2 \lim\limits_{x \to 5} x \lim\limits_{x \to 5} x - \lim\limits_{x \to 5} 6 = 2 \cdot 5 \cdot 5 - 6 = 44$. We will use both these theorems and the method used in Prob. 15.3 in subsequent problems.

15.5 $\lim\limits_{x \to 1} (4x^3 - 6x + 1)$

$\blacksquare$ Since $4x^3 - 6x + 1$ is a polynomial, $\lim\limits_{x \to 1} (4x^3 - 6x + 1) = 4(1)^3 - 6(1) + 1 = 4 - 6 + 1 = -1$.

15.6 $\lim\limits_{x \to 2} (-10)$

$\blacksquare$ $\lim\limits_{x \to 2} (-10) = -10$ since, in this case, the function is *constantly* -10, and thus -10 will always be its limit.

15.7 $\lim\limits_{x \to d} (4x - 6)$

$\blacksquare$ $\lim\limits_{x \to d} (4x - 6) = 4d - 6$.

15.8 $\lim\limits_{x \to 2} (3x^2 + ax)$

$\blacksquare$ $\lim\limits_{x \to 2} (3x^2 + ax) = 3 \lim\limits_{x \to 2} x \lim\limits_{x \to 2} x + a \lim\limits_{x \to 2} x = 12 + 2a$.

15.9 $\lim\limits_{x \to 4} (2x^3 - 2x^2 + 2x - 2)$

$\blacksquare$ $\lim\limits_{x \to 4} (2x^3 - 2x^2 + 2x - 2) = 2 \cdot 4^3 - 2 \cdot 4^2 + 2(4) - 2 = 128 - 32 + 8 - 2 = 102$.

15.10 $\lim\limits_{x \to \pi} (\sin 2x)$

▮ As x gets close to π, $2x$ gets close to 2π, and $\sin 2x$ gets close to $\sin 2\pi$ which is 0.   Thus, $\lim\limits_{x \to \pi} (\sin 2x) = 0$.

15.11 $\lim\limits_{x \to \pi/2} (\cos x)$

▮ $\lim\limits_{x \to \pi/2} (\cos x) = 0$ since, as x gets near $\pi/2$, $\cos x$ gets near $\cos \pi/2$ which is 0.

15.12 $\lim\limits_{x \to \pi/2} [3 \cos (2x + \pi)]$

▮ As x gets near $\pi/2$, $2x + \pi$ gets near 2π.   Then, since $\cos 2\pi = \cos 0 = 1$,   and $3 \cdot 1 = 3$,   this limit $= 3$.

15.13 $\lim\limits_{x \to \pi/2} (\tan x)$

▮ $\tan \pi/2$ is undefined; thus, since as $x \to \pi/2$, $\tan x \to \tan \pi/2$, $\lim\limits_{x \to \pi/2} (\tan x)$ does not exist.

15.14 $\lim\limits_{x \to \pi} (2 \tan x/2)$

$\lim\limits_{x \to \pi} (2 \tan x/2) = 2 \lim\limits_{x \to \pi} (\tan x/2)$,   which does not exist since $x/2 \to \pi/2$ as $x \to \pi$.

15.15 $\lim\limits_{x \to 1} (\text{Tan}^{-1} x)$

▮ As x gets close to 1, $\text{Tan}^{-1} x$ gets close to $\text{Tan}^{-1} 1$, which is $\pi/4$.   Thus, $\lim\limits_{x \to 1} (\text{Tan}^{-1} x) = 1$.

15.16 $\lim\limits_{x \to 0} [3 \text{Sin}^{-1} (x + 0.5)]$

▮ $3 \text{Sin}^{-1} (x + 0.5) \to 3 \text{Sin}^{-1} 0.5$ as $x \to 0$. $\text{Sin}^{-1} 0.5 = \pi/6$,   so $\lim\limits_{x \to 0} [3 \text{Sin}^{-1} (x + 0.5)] = 3(\pi/6) = \pi/2$.

15.17 $\lim\limits_{x \to 1} (\text{Sin}^{-1} x \, \text{Cos}^{-1} x)$

▮ $\lim\limits_{x \to 1} (\text{Sin}^{-1} x \, \text{Cos}^{-1} x) = \lim\limits_{x \to 1} (\text{Sin}^{-1} x) \lim\limits_{x \to 1} (\text{Cos}^{-1} x) = \pi/2 \cdot 0 = 0$.

15.18 $\lim\limits_{x \to 1} (\text{Sin}^{-1})^2$

▮ $\lim\limits_{x \to 1} (\text{Sin}^{-1} x)^2 = (\lim\limits_{x \to 1} \text{Sin}^{-1} x)^2 = (\pi/2)^2 = \pi^2/4$.

15.19 $\lim\limits_{x \to e} (\ln x)$

▮ As $x \to e$, $\ln x \to \ln e = 1$.   Thus, $\lim\limits_{x \to e} (\ln x) = 1$.

15.20 $\lim\limits_{x \to 10} (\ln x)$

▮ As $x \to 10$, $\ln x \to \ln 10$.   Thus, $\lim\limits_{x \to 10} (\ln x) = \ln 10$.

15.21 $\lim\limits_{x \to 100} (\log x)$

▮ As $x \to 100$, $\log x \to \log 100$ and $\log 100 = 2$.   Thus, $\lim\limits_{x \to 100} (\log x) = 2$.

15.22 $\lim\limits_{x \to 0} \dfrac{2x^3 - 6}{x + 1}$

▮ Since $\lim\limits_{x \to 0} (2x^3 - 6) = -6$ and $\lim\limits_{x \to 0} (x + 1) = 1$, $\lim\limits_{x \to 0} \dfrac{2x^3 - 6}{x + 1} = \dfrac{-6}{1} = -6$.

15.23 $\lim\limits_{x \to 1} \dfrac{2x^4 - 3x + 9}{x^3 - 11}$

▮ $\lim\limits_{x \to 1} (2x^4 - 3x + 9) = 8$, and $\lim\limits_{x \to 1} (x^3 - 11) = -10$; thus, $\lim\limits_{x \to 1} \dfrac{2x^4 - 3x + 9}{x^3 - 11} = \dfrac{8}{-10} = \dfrac{-4}{5}$.

15.24 $\lim\limits_{x \to 2} \dfrac{x + 2}{x - 2}$

As $x \to 2$, $x - 2 \to 0$; but $x + 2 \to 4$, so the limit does not exist. Actually, since the quotient is growing without bound as $x \to 2$, we say $\lim\limits_{x \to 2} \dfrac{x + 2}{x - 2} = \infty$. That does *not* mean, however, that the limit exists! It does not!

15.25 $\lim\limits_{x \to 0} \dfrac{3x^3 - 6}{x}$

▮ Since $3x^3 - 6 \to 6$ and $x \to 0$ (as $x \to 0$), this limit does not exist.

15.26 $\lim\limits_{x \to 1} \dfrac{x^2 - 3x + 2}{x - 1}$

▮ Notice that $x^2 - 3x + 2 = (x - 1)(x - 2)$. However, since $x \to 1$, x is *not* 1; therefore, $\dfrac{x^2 - 3x + 2}{x - 1} =$
$x - 2$, and $\lim\limits_{x \to 1} \dfrac{x^2 - 3x + 2}{x - 1} = \lim\limits_{x \to 1} (x - 2) = -1$.

15.27 $\lim\limits_{x \to 0} \dfrac{x^3 - 2x}{x}$

▮ Since $\dfrac{x^3 - 2x}{x} = \dfrac{x(x^2 - 2)}{x} = x^2 - 2$ (when $x \neq 0$), $\lim\limits_{x \to 0} \dfrac{x^3 - 2x}{x} = \lim\limits_{x \to 0} (x^2 - 2) = -2$.

15.28 $\lim\limits_{x \to 3} \dfrac{x^2 + x - 12}{x - 3}$

▮ $\lim\limits_{x \to 3} \dfrac{x^2 + x - 12}{x - 3} = \lim\limits_{x \to 3} \dfrac{(x - 3)(x + 4)}{x - 3} = \lim\limits_{x \to 3} (x + 4)$(since $x \neq 3$) $= 7$.

15.29 $\lim\limits_{x \to \infty} 3/x$

▮ As x grows without bound, $3/x$ gets arbitrarily close to 0. Thus, $\lim\limits_{x \to \infty} 3/x = 0$.

15.30 $\lim\limits_{x \to \infty} (2x^2 - 9)$

▮ As x gets arbitrarily large, so does $2x^2 - 9$. Thus, $\lim\limits_{x \to \infty} (2x^2 - 9)$ does not exist.

For Probs. 15.31 to 15.40, answer true or false.

15.31 If $\lim\limits_{x \to a} f(x) = \lim\limits_{x \to a} g(x)$, then $f(x) = g(x)$.

▮ False. $\lim\limits_{x \to 0} x^2 = 0 = \lim\limits_{x \to 0} x$.

15.32 If $\lim\limits_{x \to 1} f(x) = L$, then $\lim\limits_{x \to 1} [f(x) + 1] = L + 1$.

▮ True. $\lim\limits_{x \to 1} [f(x) + 1] = \lim\limits_{x \to 1} f(x) + \lim\limits_{x \to 1} 1 = L + 1$.

15.33 If $f(x) = g(x)$, then $\lim\limits_{x \to a} f(x) = \lim\limits_{x \to a} g(x)$.

▮ True. They are the same function, so their limits must be the same.

15.34 If $\lim_{x \to 3} f(x) = \infty$, and $g(x) > f(x)$ for all x, then $\lim_{x \to 3} g(x) = \infty$.

I True. If $f(x)$ grows without bound, then $g(x)$ must also if $g(x) > f(x)$.

15.35 $\lim_{x \to a} f(x) + \lim_{x \to a} g(x) = \lim_{x \to a} [f(x) + g(x)]$.

I False. We must first know that the individual limits exist.

15.36 $\lim_{x \to 4} \dfrac{x^2 - 4x}{x - 4}$ does not exist.

I False. $\lim_{x \to 4} \dfrac{x^2 - 4x}{x - 4} = \lim_{x \to 4} \dfrac{x(x - 4)}{x - 4} = 4$.

15.37 If $p(x)$ is a polynomial, then $\lim_{x \to a} p(x) = p(a)$ for all x and a.

I True. This is a fundamental limit theorem.

15.38 $\lim_{x \to e} [\operatorname{Sin}^{-1} (\ln x)]$ is undefined.

I False. $\ln x \to 1$, and $\operatorname{Sin}^{-1} 1 = \pi/2$. This limit is $\pi/2$.

15.39 $\lim_{x \to 3\pi/2} (\tan x/3) = \lim_{x \to 3\pi/2} (\tan x/2)$

I False. LHS does not exist since $\tan \pi/2$ is undefined.

15.40 $\lim_{x \to 1} (\operatorname{Csc}^{-1} x) = \lim_{x \to 0} (\sin x + \pi/2)$

I True. They are both $\pi/2$. $\sin 0 = 0$ and $\operatorname{Csc}^{-1} 1 = \operatorname{Sin}^{-1} \frac{1}{1} = \pi/2$.

15.2 THE DERIVATIVE

For Probs. 15.41 to 15.47, find the derivative of the given function using the definition of the limit:

15.41 $f(x) = 2x$

I $f'(x) = \lim_{x \to h} \dfrac{f(x + h) - f(x)}{h} = \lim_{h \to 0} \dfrac{2(x + h) - 2x}{h} = \lim_{x \to 0} \dfrac{2h}{h} = \lim_{h \to 0} 2 = 2$.

15.42 $f(x) = 2x + 3$

I $f'(x) = \lim_{h \to 0} \dfrac{2(x + h) + 3 - (2x + 3)}{h} = \lim_{h \to 0} \dfrac{2h}{h} = 2$.

15.43 $g(x) = 7x - 8$

I $g'(x) = \lim_{h \to 0} \dfrac{7(x + h) - 8 - (7x - 8)}{h} = 7$.

15.44 $h(x) = x^2$

I $h' = \lim_{h \to 0} \dfrac{(x + h)^2 - x^2}{h} = \lim_{h \to 0} \dfrac{x^2 + 2xh + h^2 - x^2}{h} = \lim_{h \to 0} \dfrac{2xh + h^2}{h} = \lim (2x + h) = 2x$.

15.45 $f(x) = 3x^2$

I $f'(x) = \lim_{h \to 0} \dfrac{3(x + h)^2 - 3x^2}{h} = \lim_{h \to 0} \dfrac{2xh + h^2}{h} = 2x$.

15.46 $g(x) = x^3$

I $f'(x) = \lim_{h \to 0} \dfrac{(x + h)^3 - x^3}{h} = \lim_{h \to 0} \dfrac{x^3 + 3x^2h + 3xh^2 + h^3 - x^3}{h} = \lim_{h \to 0} (3x^2 + 3xh + h^2) = 3x^2$.

15.47 $h(x) = 7$

▮ $h'(x) = \lim\limits_{h \to 0} \dfrac{h(x + h) - h(x)}{h} = \lim\limits_{h \to 0} \dfrac{7 - 7}{h} = \lim\limits_{h \to 0} 0 = 0.$

15.48 Find $f'(x)$ if $f(x) = 3x + 8$.

▮ $f'(x) = 3.$ $\left[\text{Recall that } \dfrac{d}{dx} x^n = nx^{n-1}, \quad \dfrac{d}{dx}(f + g) = f' + g', \quad \text{and} \quad \dfrac{d}{dx} af(x) = af'(x)\right].$

15.49 Find $f'(x)$ if $f(x) = x^2 + 8x$.

▮ Since $\dfrac{d}{dx} x^2 = 2x$ and $\dfrac{d}{dx} 8x = 8$ and we want the sum of the two derivatives, $f'(x) = 2x + 8$.

15.50 Find $g'(x)$ if $g(x) = 2x^3 + 8x^2$.

▮ $g'(x) = 6x^2 + 16x,$ since $\dfrac{d}{dx} 2x^3 = 6x^2$ and $\dfrac{d}{dx} 8x^2 = 16x.$

15.51 Find $h'(x)$ if $h(x) = 4x^2 + 6x - 11$.

▮ $h'(x) = 8x + 6.$ $\left(\text{Remember: } \dfrac{d}{dx} c = 0 \quad \text{where} \quad c \in \mathcal{R}.\right)$

15.52 Find $f'(x)$ if $f(x) = 7t$.

▮ $f'(x) = 0.$ Do you see why? t is not the variable.

15.53 Find $f'(x)$ if $f(x) = -6x^3 + 4x^2 + 8x$.

▮ $f'(x) = -18x^2 + 8x + 8.$

For Probs. 15.54 to 15.67, find the indicated derivative for the given function.

15.54 $g'(x); g(x) = (x + 1)(x - 7)$

▮ $g'(x) = (x + 1)(1) + (x - 7)(1) = x + 1 + x - 7 = 2x - 6.$

$$\left(\text{Remember: } \dfrac{d}{dx} fg = fg' + gx' \quad \text{and} \quad \dfrac{d}{dx} \dfrac{f}{g} = \dfrac{gf' - fg'}{g^2}\right).$$

15.55 $f'(x); f(x) = (x^2 - 6)(x^3 + 4)$

▮ $f'(x) = (x^2 - 6)(3x^2) + (x^3 + 4)(2x) = 3x^4 - 18x^2 + 2x^4 + 8x = 5x^4 - 18x^2 + 8x.$

15.56 $g'(x); g(x) = \dfrac{x}{x + 1}$

▮ $g'(x) = \dfrac{(x + 1)(1) - x(1)}{(x + 1)^2} = \dfrac{1}{(x + 1)^2} = (x + 1)^{-2}.$

15.57 $h'(x); h(x) = \dfrac{x^2}{x - 3}$

▮ $h'(x) = \dfrac{(x - 3)(2x) - x^2(1)}{(x - 3)^2} = \dfrac{2x^2 - 6x - x^2}{(x - 3)^2} = \dfrac{x^2 - 6x}{(x - 3)^2}.$

15.58 $y'; y = (x - 2)^4$

▮ $y' = 4(x - 2)^3.$ $\left[\text{Recall: If } u = u(k), \text{ then } \dfrac{d}{dx} u^m = mu^{m-1} \cdot u'(x). \text{ This is a version of the chain rule.}\right]$

15.59 $y'; y = (x^2 + 2)^3$

▮ $y' = 3(x^2 + 2)^2 \cdot 2x = 6x(x^2 + 2)^2.$

15.60 $h'(x); h(x) = (4 - x^2)^{10}$

▮ $h'(x) = 10(4 - x^2)^9 \cdot (-2x) = -20x(4 - x^2)^9.$

15.61 $f'(x); f(x) = (2x^2 + 4x - 5)^6$

▮ $f'(x) = 6(2x^2 + 4x - 5)^5(4x + 4) = 24(x + 1)(2x^2 + 4x - 5)^5.$

15.62 $h'(x); h(x) = (2x^{1/3} - 6)^4$

▮ $h'(x) = 4(2x^{1/3} - 6)^3 \cdot \frac{2}{3}x^{-2/3} = \frac{8}{3}x^{-2/3}(2x^{1/3} - 6)^3.$

15.63 $y'; y = (1 - x^2)^{1/2}$

▮ $y' = \frac{1}{2}(1 - x^2)^{-1/2} \cdot (-2x) = \frac{-x}{(1 - x^2)^{1/2}}.$

15.64 Find $h'(1)$ if $h(x) = (4 - x^2)^{10}$. (See Prob. 15.60.)

▮ $h'(1) = -20(1)(4 - 1^2)^9 = -20(3^9).$

15.65 $y'; y = \sqrt{x + 1}$

▮ $y = (x + 1)^{1/2}.$
$y' = \frac{1}{2}(x + 1)^{-1/2} = \frac{1}{2(x + 1)^{1/2}} = \frac{1}{2\sqrt{x + 1}}.$

15.66 $f'(x); f(x) = \sqrt{2x + 1} = (2x + 1)^{1/2}$

▮ $f'(x) = \frac{1}{2}(2x + 1)^{-1/2} \cdot 2.$ (The derivative of $2x + 1$ is 2.) $= \frac{1}{\sqrt{2x + 1}}.$

15.67 $f'(1); f(x) = (2x + 1)^{1/2}$. (See Prob. 15.66.)

▮ $f'(1) = \frac{1}{\sqrt{2(1) + 1}} = \frac{1}{\sqrt{3}}.$

For Probs. 15.68 to 15.70, $f(x) = 3x^4 - 8x^3 + 12x^2 + 5.$

15.68 Find $f'(x)$.

▮ $f'(x) = 12x^3 - 24x^2 + 24x.$

15.69 Find $f''(x)$.

▮ $f''(x) = 36x^2 - 48x + 24.$

15.70 Find $f'''(x)$.

▮ $f'''(x) = 72x - 48.$

For Probs. 15.71 to 15.73, $f(x) = \frac{1}{4 - x}.$

15.71 Find $f'(x)$.

▮ $f'(x) = -(4 - x)^{-2}(-1) \left[\text{since } \frac{1}{4 - x} = (4 - x)^{-1} \right] = (4 - x)^{-2}.$

15.72   Find $f''(x)$.

$$f''(x) = -2(4-x)^{-3}(-1) = 2(4-x)^{-3}.$$

15.73   Find $f'''(x)$.

▮ $f'''(x) = -6(4-x)^{-4}(-1) = 6(4-x)^{-4}.$

For Probs. 15.74 to 15.79, find the indicated derivative for the given function.

15.74 $f''(x)$; $f(x) = 6x$

▮ $f'(x) = 6$, $f''(x) = 0$.

15.75 $f'''(x)$; $f(x) = 167x^2 - 18x$

▮ $f'''(x) = 0$,   since x^2 is the highest order of the variable appearing.

15.76 $f^{(100)}(x)$; $f(x) = x^{91} - x^{84}$

▮ $f^{(100)}(x) = 0$.

15.77 $f''(x)$; $f(x) = 1/x$

▮ $f'(x) = -x^{-2}$; $f''(x) = 2x^{-3}$.

15.78 $f^{(4)}(x)$; $f(x) = 1/x$

▮ $f'''(x) = -6x^{-4}$; $f^{(4)}(x) = 24x^{-5}$.   (Notice the pattern.)

15.79 $f^{(7)}(x)$; $f(x) = 1/x$

▮ $f^{(7)}(x) = 7!x^{-8}(-1) = -7!x^{-8}.$

15.3.   CONTINUITY

For Probs. 15.80 to 15.90, find $\lim_{x \to 1} f(x)$, and find $f(1)$.   What do you conclude concerning continuity of $f(x)$ at 1?

15.80 $f(x) = 2x + 3$

▮ $f(1) = 5$; $\lim_{x \to 1} 2x + 3 = 5$. $f(1) = \lim_{x \to 1} f(x)$ implies that f is continuous at 1.   Recall that $f(x)$ is continuous at a if $f(a)$ and $\lim_{x \to a} f(x)$ both exist and are equal.

15.81 $f(x) = 3$

▮ $f(1) = 3$; $\lim_{x \to 1} f(x) = 3$. f is continuous at 1.

15.82 $f(x) = 3x^2$

▮ $f(1) = 3$; $\lim_{x \to 1} f(x) = 3$. f is continuous at 1.

15.83 $f(x) = 2(x - 1)^3$

▮ $f(1) = 0$; $\lim_{x \to 1} 2(x - 1)^3 = 0$. f is continuous at 1.

15.84 $f(x) = \sqrt{x^3 + 1}$

▮ $f(1) = \sqrt{2}$; $\lim_{x \to 1} f(x) = \sqrt{2}$. f is continuous at 1.

15.85 $f(x) = \sqrt{x^3 - 2}$

▮ $f(1)$ does not exist. f is not continuous at 1.

15.86 $f(x) = [x]$

▮ $f(1) = [1] = 1$; $\lim_{x \to 1} [x]$ does not exist, since $\lim_{x \to 1^+} [x] = 1$ but $\lim_{x \to 1^-} [x] = 0$. f is not continuous at 1.

15.87 $f(x) = |x - 1|$

▮ $f(1) = 0$; $\lim_{x \to 1} |x - 1| = 0$. f is continuous at 1.

15.88 $f(x) = [x/2]$

▮ $f(1) = [1/2] = 0$; $\lim_{x \to 1} [x/2] = 0$. f is continuous at 1.

15.89 $f(x) = x^g - 11$

▮ $f(1) = -10$; $\lim_{x \to 1} f(x) = -10$. f is continuous at 1.

15.90 $f(x) = \dfrac{1}{2 - |x|}$

▮ $f(1) = 1$; $\lim_{x \to 1} \dfrac{1}{2 - |x|} = 1$. f is continuous at 1.

For Probs. 15.91 to 15.101, indicate any points of discontinuity.

15.91 $f(x) = 2x - 3$

▮   No points of discontinuity.   Polynomial functions are continuous everywhere.

15.92 $f(x) = -3x^3 - 14x^2 - 118x + 11$

▮   No points of discontinuity.   This is a polynomial function.

15.93 $h(x) = 7$

▮   None; constant functions are continuous everywhere.

15.94 $f(x) = 1/x$

▮   Discontinuous when $x = 0$.   If $x = 0$, $1/x$ is undefined.

15.95 $g(x) = 1/|x|$

▮   Discontinuous when $x = 0$.   If $x = 0$, $1/|x|$ is undefined.

15.96 $f(x) = |x|/x$

▮   Discontinuous when $x = 0$. $0/0$ is indeterminate. $\lim_{x \to 0^+} |x|/x = 1 \neq \lim_{x \to 0^-} |x|/x = -1$.

15.97 $f(x) = \begin{cases} 1, & x < 0 \\ x + 1, & 0 \le x < 2 \\ 2, & x \ge 2 \end{cases}$

▮ $\lim_{x \to 2^+} f(x) = \lim_{x \to 2} 2 = 2$; $\lim_{x \to 2^-} f(x) = \lim_{x \to 2} x + 1 = 3$.   Since $2 \neq 3$, $f(x)$ is discontinuous at 2.   Note that it is continuous everywhere else.

15.98 $f(x) = \begin{cases} 0, & \text{if } x < 0 \\ 4 - x^2, & \text{if } 0 \le x < 2 \\ 0, & \text{if } x > 2 \end{cases}$

▮ $\lim_{x \to 0^-} f(x) = 0$; $\lim_{x \to 0^+} f(x) = \lim_{x \to 0} 4 - x^2 = 4$. f is discontinuous at 0.

15.99 $f(x) = [x]$

▮ Discontinuous at all integers. For example, $\lim_{x \to 1^+} [x] = 1$, but $\lim_{x \to 1^-} [x] = 0$.

15.100 $f(x) = -[x - 1]$

▮ Discontinuous at all integers. See Prob. 15.99! The reasoning is the same.

15.101 $f(x) = x - [x]$

▮ Discontinuous at all integers. For example, $\lim_{x \to 3^+} x - [x] = 3 - 3 = 0$, but $\lim_{x \to 3^-} x - [x] = 3 - 2 = 1$.

For Probs. 15.102 to 15.110, tell whether the given statement is true or false, and why.

15.102 $f(x) = ax^2 + bx + c$ is continuous everywhere.

▮ True; $f(x)$ is a polynomial.

15.103 If $g(x)$ is never zero, then $\dfrac{ax^2 + bx}{g(x)}$ is continuous everywhere.

▮ False; what if $g(x) = [x]$? It depends on what $g(x)$ is.

15.104 If $f(x)$ and $g(x)$ are polynomials, and $g(5) \neq 0$, then $f(x)/g(x)$ is continuous at 5.

▮ True; the only possible discontinuity would be when $g(x) = 0$.

15.105 If $f(x)$ and $g(x)$ are continuous at b, then $f(x) + g(x)$ is continuous at b.

▮ True; this is a basic theorem.

15.106 If $f(x)$ and $g(x)$ are continuous everywhere, then so is $f(x)g(x)$.

▮ True; this is a basic theorem.

15.107 $[x]$ is continuous at all nonintegers.

▮ True; $\lim_{x \to a} [x] = [a]$ unless $a \in \mathscr{L}$.

15.108 $[x/3]$ is continuous at $x = 1$.

▮ True; $\lim_{x \to 1} [x/3] = [1/3] = 0$.

15.109 If $g(x)$ is never zero, and $f(x)$ and $g(x)$ are continuous everywhere, then f/g is continuous everywhere.

▮ True; this is a basic theorem.

15.110 $f(x) = \dfrac{|x|}{1 - |x|}$ is discontinuous only when $x = 1$.

▮ False; $1 - |x| = 0$ when $x = 1$ or $x = -1$.

15.111 Suppose that $f(x)$ is continuous everywhere. What is the maximum number of discontinuities for $g(x) = \dfrac{f(x)}{ax^2 + bx + c}$?

▮ Two; $ax^2 + bx + c = 0$ has at most two real solutions.

For Probs. 15.112 to 15.118, find where the given function is discontinuous and describe the discontinuity in terms of limits or in terms of the domain of the function.

15.112 See Fig. 15.1.

▮ $f(x)$ is continuous everywhere.

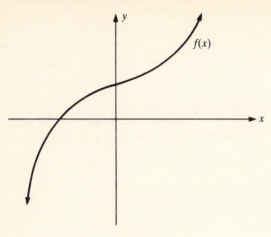

Fig. 15.1

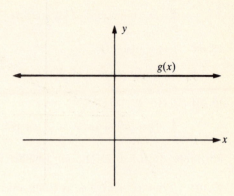

Fig. 15.2

15.113 See Fig. 15.2.

▮ $g(x)$ is continuous everywhere.

15.114 See Fig. 15.3.

▮ Discontinuous at $x = -1$ since $\lim\limits_{x \to -1^-} f(x) \neq \lim\limits_{x \to -1^+} f(x)$.

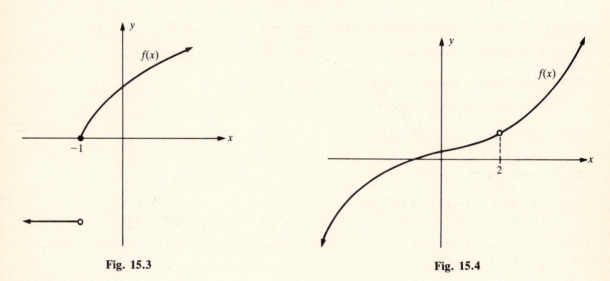

Fig. 15.3

Fig. 15.4

15.115 See Fig. 15.4.

▮ Discontinuous at $x = 2$. $f(2)$ is not defined. Note that $\lim\limits_{x \to 2} f(x)$ *does* exist.

15.116 See Fig. 15.5.

▮ Discontinuous at $x = 3$. $\lim\limits_{x \to 3} f(x) \neq f(3)$; Note that they do both exist however.

15.117 See Fig. 15.6.

▮ f is discontinuous at -1 and 0. $\lim\limits_{x \to -1} f(x) \neq f(-1)$. $\lim\limits_{x \to 0} f(x)$ does not exist.

15.118 See Fig. 15.7.

▮ Discontinuous at $0, 1$, and 2. $\lim\limits_{x \to a} f(x)$ does not exist for $a = 0, 1, 2$. Also discontinuous at -1; $f(-1) \neq \lim\limits_{x \to -1} f(x)$.

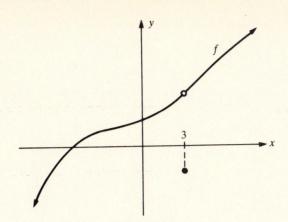

Fig. 15.5

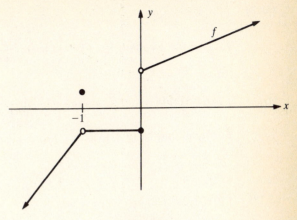

Fig. 15.6

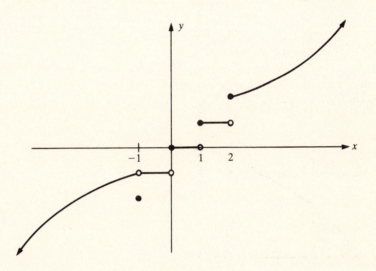

Fig. 15.7

15.119 Find $\lim_{x \to 0} |x|/x$ where $a \neq 0$.

▮ Since $|x| = x$ if $x > 0$ and $|x| = -x$ if $x < 0$, we need to look at the one-sided limits: $\lim_{x \to 0^+} |x|/x$ $= x/x = 1$ and $\lim_{x \to 0^-} |x|/x = -x/x = -1$. Thus, since they are unequal, $\lim_{x \to 0} |x|/x$ does not exist.

15.120 Find $\lim_{x \to 1.5^+} [x - 0.5]$.

▮ $\lim_{x \to 1.5^+} [x - 0.5] = [1.5 - 0.5] = [1] = 1$.

15.121 Find $\lim_{x \to 1.5^-} [x - 0.5]$.

▮ $\lim_{x \to 1.5^-} [x - 0.5] = 0$. Look at the graph!

15.4 THE INTEGRAL AND THE FUNDAMENTAL THEOREM

For Probs. 15.122 to 15.141, evaluate the given indefinite integral (antiderivative).

15.122 $\int 2x \, dx$

▮ Recall that $\int kf(x) \, dx = kf(k) \, dx$, and $\int x^m \, dx = \frac{x^{m+1}}{m+1} + c$ $(m \neq -1)$. In this case, $\int 2x \, dx =$ $2 \int x \, dx = \frac{2x^2}{2} + c = x^2 + c$. $\left[\text{Check: } \frac{d}{dx} (x^2 + c) = 2x . \right]$

15.123 $\int 3t^4\,dt$

▮ $\int 3t^4\,dt = 3\int t^4\,dt = \dfrac{3t^5}{5} + c.$ $\left[\text{Check: } \dfrac{d}{dt}\left(\dfrac{3t^5}{5}\right) = 3t^4.\right]$

15.124 $\int (3x + 4x^2)\,dx$

▮ Recall that $\int (f + g)\,dx = \int f\,dx \pm \int g\,dx.$ Then, $\int (3x + 4x^2)\,dx = 3\int x\,dx + 4\int x^2\,dx = \dfrac{3x^2}{2} + \dfrac{4x^3}{3} + c.$

15.125 $\int (3x^4 - 2x^3 + 5)\,dx$

▮ $\int (3x^4 - 2x^3 + 5)\,dx = \int 3x^4\,dx - \int 2x^3\,dx + \int 5\,dx = \dfrac{3x^5}{5} - \dfrac{2x^4}{4} + \dfrac{5x}{1} + c = \dfrac{3x^5}{5} - \dfrac{x^4}{2} + 5x + c.$

15.126 $\int \sqrt{x}\,dx$

▮ $\int \sqrt{x}\,dx = \int x^{1/2}\,dx = \dfrac{x^{3/2}}{\frac{3}{2}} + c = \tfrac{2}{3}x^{3/2} + c.$

15.127 $\int 2x^{2/3}\,dx$

▮ $\int 2x^{2/3}\,dx = \dfrac{2x^{5/3}}{\frac{5}{3}} + c = \tfrac{6}{5}x^{3/2} + c.$

15.128 $\int (-3\sqrt{x} + 4\sqrt[3]{x})\,dx$

▮ $\int (-3\sqrt{x} + 4\sqrt[3]{x})\,dx = \dfrac{-3x^{3/2}}{\frac{3}{2}} + \dfrac{4x^{4/3}}{\frac{4}{3}} + c = -2x^{3/2} + 3x^{4/3} + c.$

15.129 $\int \dfrac{1}{x^2}\,dx$

▮ $\int \dfrac{1}{x^2}\,dx = \int x^{-2}\,dx = \dfrac{x^{-1}}{-1} + c = \dfrac{-1}{x} + c.$

15.130 $\int \left(2 - \dfrac{1}{\sqrt[4]{x}}\right)dx$

▮ $\int \left(2 - \dfrac{1}{\sqrt[4]{x}}\right)dx = \int 2\,dx - \int \dfrac{1}{\sqrt[4]{x}}\,dx = 2x - \dfrac{x^{3/4}}{\frac{3}{4}} + c = 2x - \tfrac{4}{3}x^{3/4} + c.$

15.131 $\int (2x^{-4} - 7)\,dx$

▮ $\int (2x^{-4} - 7)\,dx = \dfrac{2x^{-3}}{-3} - 7x + c = -\tfrac{2}{3}x^{-3} - 7x + c.$

15.132 $\int 3x^2\,dt$

▮ Be careful here! t is the variable, so $\int 3x^2\,dt = 3x^2 t + c.$

15.133 $4\int (x^2 + 3x^{-2})\,dx$

$\blacksquare$ $4\int (x^2 + 3x^{-2})\,dx = 4\left(\dfrac{x^3}{3} + \dfrac{3x^{-1}}{-1}\right) + c = \frac{4}{3}x^3 - 12x^{-1} + c.$

15.134 $\int x(x^2 + 1)\,dx$

$\blacksquare$ $\int x(x^2 + 1)\,dx = \int (x^3 + x)\,dx = \dfrac{x^4}{4} + \dfrac{x^2}{2} + c.$

15.135 $\int (x + 1)^2\,dx$

$\blacksquare$ $\int (x + 1)^2\,dx = \int (x^2 + 2x + 1)\,dx = \dfrac{x^3}{3} + x^2 + x + c.$

15.136 $\int (2x - 1)^2\,dx$

$\blacksquare$ $\int (2x - 1)^2\,dx = \int (4x^2 - 4x + 1)\,dx = \dfrac{4x^3}{3} - 2x^2 + x + c.$

15.137 $\int x(x - 1)\,dx$

$\blacksquare$ $\int \sqrt{x}(x - 1)\,dx = \int x^{1/2}(x - 1)\,dx = \int (x^{3/2} - x^{1/2})\,dx = \frac{2}{5}x^{5/2} - \frac{2}{3}x^{3/2} + c.$

15.138 $\int (x^2 - 1)(x^2 + 1)\,dx$

$\blacksquare$ $\int (x^2 - 1)(x^2 + 1)\,dx = \int (x^4 - 1)\,dx = \dfrac{x^5}{5} - x + c.$

15.139 $\int \dfrac{x^2 - 1}{x^4}\,dx$

$\blacksquare$ $\dfrac{x^2 - 1}{x^4} = \dfrac{x^2}{x^4} - \dfrac{1}{x^4} = x^{-2} - x^{-4}.$ Thus, $\int \dfrac{x^2 - 1}{x^4}\,dx = \int (x^{-2} - x^{-4})\,dx = -x^{-1} + \frac{1}{3}x^{-3} + c.$

15.140 $\int \dfrac{1 - x}{\sqrt{x}}\,dx$

$\blacksquare$ $\int \dfrac{1 - x}{\sqrt{x}}\,dx = \int \dfrac{1}{\sqrt{x}}\,dx - \int \dfrac{x}{\sqrt{x}}\,dx = \int x^{-1/2}\,dx - \int x^{1/2}\,dx = 2x^{1/2} - \frac{2}{3}x^{3/2} + c.$

15.141 $\int \dfrac{(1 - x)^2}{\sqrt{x}}\,dx$

$\blacksquare$ $\int \dfrac{(1 - x)^2}{\sqrt{x}}\,dx = \int \dfrac{1 - 2x + x^2}{x^{1/2}}\,dx = \int (x^{-1/2} - 2x^{1/2} + x^{3/2})\,dx = 2x^{1/2} - \frac{4}{3}x^{3/2} + \frac{2}{5}x^{5/2} + c.$

For Probs. 15.142 to 15.151, use the following formulas

$$\frac{d}{dx}\sin x = \cos x \qquad \frac{d}{dx}\cos x = -\sin x \qquad \frac{d}{dx}\tan x = \sec^2 x$$

and find each antiderivative.

15.142 $\int \sin x\,dx$

$\blacksquare$ $\int \sin x\,dx = -\cos x + c$ since $\dfrac{d}{dx}(-\cos x) = \dfrac{-d}{dx}\cos x = \sin x.$

15.143 $\displaystyle\int -2 \sin x \, dx$

▮ $\displaystyle\int -2 \sin x \, dx = -2 \int \sin x \, dx = -2(-\cos x) + c = 2 \cos x + c.$

15.144 $\displaystyle\int \cos x \, dx$

▮ $\displaystyle\int \cos x \, dx = \sin x + c$ since $\dfrac{d}{dx} \sin x = \cos x.$

15.145 $\displaystyle\int (\sin x + \cos x) \, dx$

▮ $\displaystyle\int (\sin x + \cos x) \, dx = \int \sin x \, dx + \int \cos x \, dx = -\cos x + \sin x + c.$

15.146 $\displaystyle\int \sec^2 x \, dx$

▮ $\displaystyle\int \sec^2 x \, dx = \tan x + c$ since $\dfrac{d}{dx} \tan x = \sec^2 x.$

15.147 $\displaystyle\int (2 \sec^2 x - \sin x) \, dx$

▮ $\displaystyle\int (2 \sec^2 x - \sin x) \, dx = \int 2 \sec^2 x \, dx - \int \sin x \, dx = 2 \tan x + \cos x + c.$

15.148 $\displaystyle\int \dfrac{\sin x}{\tan x} \, dx$

▮ $\dfrac{\sin x}{\tan x} = \dfrac{\sin x}{\sin x / \cos x} = \cos x.$ $\displaystyle\int \dfrac{\sin x}{\tan x} \, dx = \int \cos x \, dx = \sin x + c.$

15.149 $\displaystyle\int x\left(1 - \dfrac{\sin x}{x}\right) dx$

▮ $\displaystyle\int x\left(1 - \dfrac{\sin x}{x}\right) dx = \int (x - \sin x) \, dx = \dfrac{x^2}{2} + \cos x + c.$

15.150 $\displaystyle\int (2 \sin x - 4 \cos x + x^2) \, dx$

▮ $\displaystyle\int (2 \sin x - 4 \cos x + x^2) \, dx = -2 \cos x = 4 \sin x + \dfrac{x^3}{3} + c.$

15.151 $\displaystyle\int (3 \sin \theta + \cos \theta) \, dx$

▮ Since θ is constant here, the integral is $x(3 \sin \theta + \cos \theta) + c.$

For Probs. 15.152 to 15.161, evaluate the definite integral using the fundamental theorem of calculus.

15.152 $\displaystyle\int_1^2 x \, dx$

▮ Recall the basic technique: (a) Find $\displaystyle\int x \, dx$ (the antiderivative), (b) "plug in" the endpoints (here they are 2 and 1), and (c) subtract to find the definite integral. $\displaystyle\int_1^2 x \, dx = \dfrac{x^2}{2}\Big|_1^2 = \dfrac{2^2}{2} - \dfrac{1^2}{2} = 2 - \tfrac{1}{2} = \tfrac{3}{2}.$

15.153 $\displaystyle\int_0^1 3\,dx$

$\blacksquare$ $\displaystyle\int_0^1 3\,dx = 3x\Big|_0^1 = 3 - 0 = 3.$

15.154 $\displaystyle\int_{-1}^1 x^2\,dx$

$\blacksquare$ $\displaystyle\int_{-1}^1 x^2\,dx = \frac{x^3}{3}\Big|_{-1}^1 = \tfrac13 - (-\tfrac13) = \tfrac23.$

15.155 $\displaystyle\int_{-1}^1 x^4\,dx$

$\blacksquare$ $\displaystyle\int_{-1}^1 x^4\,dx = \frac{x^5}{5}\Big|_{-1}^1 = \tfrac15 - (-\tfrac15) = \tfrac25.$

15.156 $\displaystyle\int_{-1}^1 x^3\,dx$

$\blacksquare$ $\displaystyle\int_{-1}^1 x^3\,dx = \frac{x^4}{4}\Big|_{-1}^1 = \tfrac14 - \tfrac14 = 0.$

15.157 $\displaystyle\int_{-1}^1 x^7\,dx$

$\blacksquare$ $\displaystyle\int_{-1}^1 x^7\,dx = \frac{x^8}{8}\Big|_{-1}^1 = 0.$ $\left(\text{Do you see a pattern developing? If so, what is } \int_{-1}^1 x^{101}\,dx?\right)$

15.158 $\displaystyle\int_0^2 (2x - x^2)\,dx$

$\blacksquare$ $\displaystyle\int (2x - x^2)\,dx = \left(x^2 - \frac{x^3}{3}\right)\Big|_0^2 = (4 - \tfrac83) - 0 = \tfrac43.$

15.159 $\displaystyle\int_1^2 (x-1)^2\,dx$

$\blacksquare$ $(x-1)^2 = x^2 - 2x + 1;\ \displaystyle\int_1^2 (x^2 - 2x + 1)\,dx = \left(\frac{x^3}{3} - x^2 + x\right)\Big|_1^2 = \tfrac83 - 4 + 2 - (\tfrac13 - 1 + 1) = \tfrac23 - \tfrac13 = \tfrac13.$

15.160 $\displaystyle\int_0^\pi \sin x\,dx$

$\blacksquare$ $\displaystyle\int_0^\pi \sin x\,dx = -\cos x\Big|_0^\pi = -\cos\pi - (-\cos 0) = -(-1) - (-1) = 1 + 1 = 2.$

15.161 $\displaystyle\int_{-\pi}^\pi \cos x\,dx$

$\blacksquare$ $\displaystyle\int_{-\pi}^\pi \cos x\,dx = \sin x\Big|_{-\pi}^\pi = \sin\pi - \sin(-\pi) = 0 - 0 = 0.$

15.5 APPLICATIONS OF THE DERIVATIVE

For Probs. 15.162 to 15.168, find the slope of the tangent line to $y = f(x)$ when $x = 1$.

15.162 $y = x^2$

$\blacksquare$ $y'(x) = 2x$ and $y'(1) = 2(1) = 2 =$ slope of tangent line when $x = 1$.

15.163 $y = 3x^4$

$\blacksquare$ $y'(x) = 12x^3$ and $y'(1) = 12(1^3) = 12 =$ slope of tangent line when $x = 1$.

15.164 $y = 17$

$\blacksquare$ $y'(x) = 0 = y'(1) =$ slope of tangent line when $x = 1$.

15.165 $y = \sin x$

$\blacksquare$ $y'(x) = \cos x$ and $y'(1) = \cos 1 =$ slope of tangent line when $x = 1$.

15.166 $y + x^2 + x^3$

$\blacksquare$ $y'(x) = 2x + 3x^2$ and $y'(1) = 2 + 3 = 5 =$ slope of tangent line when $x = 1$.

15.167 $y = x(x^2 + 3)$

$\blacksquare$ Since $y = x^3 + 3x$, $y'(x) = 3x^2 + 3$ and $y'(1) = 3 + 3 = 6 =$ slope of tangent line when $x = 1$.

15.168 $y = \dfrac{1}{\sqrt{x}}$

$\blacksquare$ $y'(x) = -\dfrac{1}{2x^{3/2}}$ and $y'(1) = -\frac{1}{2} =$ slope of tangent line when $x = 1$.

For Probs. 15.169 to 15.176, find the slope of the normal line to $y = f(x)$ when $x = 1$.

15.169 $y = x^3 + 2$

$\blacksquare$ $y'(x) = 3x^2$ and $y'(1) = 3 =$ slope of tangent. Thus, the slope of the normal line $= -\frac{1}{3}$.

15.170 $y = 1/x$

$\blacksquare$ $y'(x) = -1/x^2$ and $y'(1) = -1$. Thus the slope of the normal line $= 1$.

15.171 $y = \sqrt{x} - 3$

$\blacksquare$ $y'(x) = \dfrac{1}{2\sqrt{x}}$ and $y'(1) = \frac{1}{2}$. Thus, the slope of the normal line $= -2$.

15.172 $y = 2 \cos x$

$\blacksquare$ $y'(x) = -2 \sin x$ and $y'(1) = -2 \sin 1$. Thus, the slope of the normal line $= \dfrac{1}{2 \sin 1}$.

15.173 $y = \dfrac{1}{x + 1}$

$\blacksquare$ $y'(x) = -\dfrac{1}{(x + 1)^2}$ $y'(1) = -\dfrac{1}{2^2} = -\dfrac{1}{4}$. Thus, the slope of the normal line $= 4$.

15.174 $y = \dfrac{2x}{x^2 - 8}$

$\blacksquare$ $y'(x) = \dfrac{(x^2 - 8)(2) - 2x(2x)}{(x^2 - 8)^2} = -\dfrac{2x^2 - 16}{(x^2 - 8)^2}$ and $y'(1) = -\frac{18}{49}$. Thus, the slope of the normal line $= \frac{49}{18}$.

15.175 $y = x\sqrt{x+1}$

▮ $y'(x) = x\dfrac{1}{2\sqrt{x+1}} + \sqrt{x+1}$ and $y'(1) = \dfrac{1}{2\sqrt{2}} + \sqrt{2} = \dfrac{5}{2\sqrt{2}}$. Thus, the slope of the normal line $= \dfrac{-2\sqrt{2}}{5}$.

15.176 $y = \tan x$

▮ $y'(x) = \sec^2 x$ and $y'(1) = \sec^2 1$. Thus, the slope of the normal line $= -\dfrac{1}{\sec^2 1} = -\cos^2 1$.

For Probs. 15.177 to 15.180, find the equation of the tangent to the given curve at the given point.

15.177 $y = x^2 + 2$; $P(1,3)$

▮ $y' = 2x$; $y'(1) = 2$. Then, using the point-slope form, $y - 3 = 2(x-1)$, so, $2x - y + 1 = 0$ is the equation of the tangent line.

15.178 $y = 2x^2 - 3x$; $P(1,-1)$

▮ $y' = 4x - 3$; $y'(1) = 1$. Then, $y + 1 = 1(x-1)$, so $x - y - 2 = 0$ is the equation of the tangent line.

15.179 $y = x^2 - 4x + 5$; $P(1,2)$

▮ $y' = 2x - 4$; $y'(1) = 2 - 4 = -2$. Then, $y - 2 = -2(x-1)$, so $2x + y - 4 = 0$ is the equation of the tangent line.

15.180 $y = x^2 + 3x - 10$; $P(2,0)$

▮ $y' = 2x + 3$; $y'(2) = 7$. Then, $y - 0 = 7(x-2)$, so $7x - y - 14 = 0$ is the equation of the tangent line.

For Probs. 15.181 to 15.184, find the equation of the normal line to the given curve at the given point.

15.181 $y = x^2 + 2$; $P(1,3)$

▮ $y' = 2x$; $y'(1) = 2$. Then, $y - 3 = -\frac{1}{2}(x-1)$, where $-\frac{1}{2}$ is the slope, so $x + 2y - 7 = 0$ is the equation of the normal line.

15.182 $y = 2x^2 - 3x$; $P(1,-1)$

▮ $y' = 4x - 3$; $y'(1) = 1$. Then, $y + 1 = -1(x-1)$, so $x + y = 0$ is the equation of the normal line.

15.183 $y = x^2 - 4x + 5$; $P(1,2)$

▮ $y' = 2x - 4$; $y'(1) = -2$. Then, $y - 2 = \frac{1}{2}(x-1)$, so $x - 2y + 3 = 0$ is the equation of the normal line.

15.184 $y = x^2 + 3x - 10$; $P(2,0)$

▮ $y' = 2x + 3$; $y'(2) = 7$. Then, $y - 0 = -\frac{1}{7}(x-2)$, so $x + 7y - 2 = 0$ is the equation of the normal line.

For Probs. 15.185 to 15.188, find the intervals on which f is increasing.

15.185 $f(x) = x^2$

▮ $f'(x) = 2x$. When $2x > 0$, $x > 0$; f is increasing when $x > 0$.

15.186 $f(x) = 4 - x^2$

▮ $f'(x) = -2x$. When $-2x > 0$, $2x < 0$ and $x < 0$; f is increasing when $x < 0$.

15.187 $f(x) = x^2 + 6x - 5$

▮ $f'(x) = 2x + 6$. When $2x + 6 > 0$, $2x > -6$ and $x > -3$; f is increasing when $x > -3$.

15.188 $f(x) = 3x^2 + 6x + 18$

▮ $f'(x) = 6x + 6$.   When $6x + 6 > 0$, $x > -1$; f is increasing when $x > -1$.

For Probs. 15.189 to 15.191, find those intervals where f is decreasing.

15.189 $f(x) = x^2$

▮ $f'(x) = 2x$.   When $2x < 0$, $x < 0$; f is decreasing when $x < 0$.

15.190 $f(x) = x^3$

▮ $f'(x) = 3x^2$.   When $3x^2 < 0$, $x^2 < 0$,   which is impossible; f is never decreasing.

15.191 $f(x) = x^5$

▮ $f'(x) = 5x^4$.   When $5x^4 < 0$, $x^4 < 0$,   which is impossible; thus, f is never decreasing.   Do you see a pattern?   Look at the graphs of $f(x) = x^3$ and $f(x) = x^5$.

For Probs. 15.192 to 15.195, find all relative minimum and maximum values $f(x)$.

15.192 $f(x) = x^2$

▮ $f'(x) = 2x$.   When $2x = 0$, $x = 0$ and $(0, 0)$ is a critical point.   But x^2 is increasing for $x > 0$,   so 0 is the minimum value.   There is no maximum.

15.193 $f(x) = x^3$

▮ $f'(x) = 3x^2$.   When $3x^2 = 0$, $x^2 = 0$ and $x = 0$,   so $(0, 0)$ is a critical point.   But x^3 is always increasing, so there are no relative maxima or minima.   [$(0, 0)$ is an inflection point.]

15.194 $f(x) = 4 - x^2$

▮ $f'(x) = -2x$.   When $-2x = 0$, $x = 0$ and $(0, 4)$ is a critical point.   Since f is increasing for $x < 0$,   4 is a maximum.

15.195 $f(x) = x^2 + 6x - 5$

▮ $f'(x) = 2x + 6$.   When $2x + 6 = 0$, $x = -3$ and $(-3, -14)$ is a critical point. -14 is a minimum since f is decreasing for $x < -3$.

For Probs. 15.196 to 15.201, tell where f is increasing, decreasing, and has relative maxima and minima.

15.196   See Fig. 15.8.

▮ f is increasing on $(-\infty, 0)$ and decreasing on $(0, \infty)$; there is a maximum at $(0, 3)$.

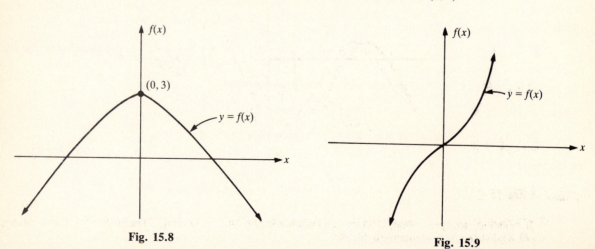

Fig. 15.8   **Fig. 15.9**

15.197 See Fig. 15.9.

▮ f is always increasing; there are no maxima or minima.

15.198 See Fig. 15.10.

▮ f is increasing on $(-\infty, a) \cup (b, \infty)$; f is constant on (a, b); there are no maxima or minima.

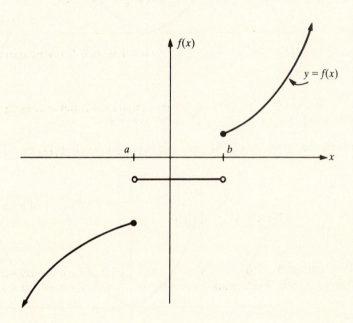

Fig. 15.10

15.199 See Fig. 15.11.

▮ f is increasing on $(-\infty, f) \cup (g, h) \cup (j, \infty)$ and decreasing on $(f, g) \cup (h, j)$. There are maxima when $x = f, h$ and minima when $x = g, j$.

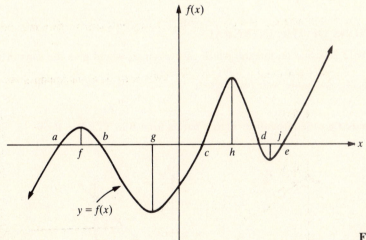

Fig. 15.11

15.200 See Fig. 15.12.

▮ f is increasing on $(-\infty, a) \cup (0, c)$ and decreasing on $(a, 0) \cup (c, \infty)$. There are maxima at (a, b), (c, d), and there is a minimum at $(0, 0)$.

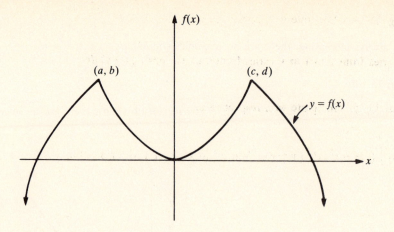

Fig. 15.12

15.201 See Fig. 15.13.

▮ f is decreasing on $(-\infty, a) \cup (a, 0)$ and increasing on $(0, b) \cup (b, \infty)$. There is a minimum at $(0, 0)$.

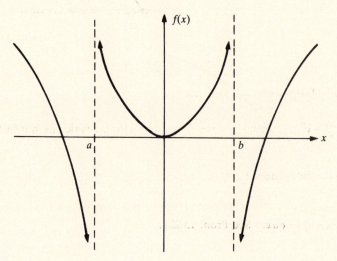

Fig. 15.13

15.6 APPLICATIONS OF THE INTEGRAL

For Probs. 15.202 to 15.209, find an integral which, if evaluated, would give the area of the shaded region.

15.202 See Fig. 15.14. Integrate with respect to x.

▮ $A = \int_0^1 x \, dx$ (y varies from 0 to x, as x varies from 0 to 1). See Prob. 15.203.

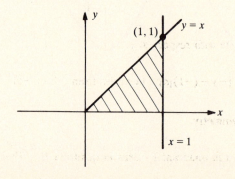

Fig. 15.14

15.203 See Fig. 15.14. Integrate with respect to y.

$\blacksquare$ x varies from y to 1 as y varies from 0 to 1. $a = \int_0^1 (1-y)\,dy$.

15.204 See Fig. 15.15. Integrate with respect to x.

$\blacksquare$ $A = \int_0^1 (\sqrt{x} - x^2)\,dx$ (y varies from x^2 to $\sqrt{x}$ as x varies from 0 to 1).

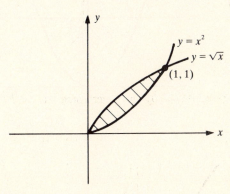

$y = x^2$
$y = \sqrt{x}$
$(1,1)$

Fig. 15.15

15.205 See Fig. 15.15 Integrate with respect to y.

$\blacksquare$ $A = \int_0^1 (\sqrt{y} - y^2)\,dy$, since x varies from y^2 to $\sqrt{y}$ as y varies from 0 to 1.

15.206 See Fig. 15.16. Integrate with respect to x.

$\blacksquare$ $A = \int_0^1 x\,dx + \int_{-1}^0 -x\,dx$. See Prob. 15.207.

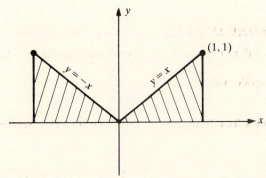

$y = -x$
$y = x$
$(1,1)$

Fig. 15.16

15.207 See Fig. 15.16. Integrate with respect to y.

$\blacksquare$ $A = \int_0^1 (1-y)\,dy + \int_0^1 [-y-(-1)]\,dy$ (x varies from -1 to $-y$). See Prob. 15.208.

15.208 See Fig. 15.16. Use symmetry.

$\blacksquare$ Notice that the area in quadrant I = area in quadrant II. Thus $A = 2\int_0^1 x\,dx$.

15.209 See Fig. 15.17.

▮ $A = 2 \int_0^1 (1 - x^2)\, dx$, since y varies from x^2 to 1 while x varies from 0 to 1. We double the area of quadrant I because of the symmetry of $y = x^2$.

For Probs. 15.210 to 15.213, find the equation of the family of curves whose slope is the given function of x.

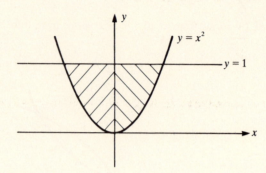

Fig. 15.17

15.210 $m = 1$

▮ $y = \int 1\, dx = x + c.$

15.211 $m = -6x$

▮ $y = \int -6x\, dx = \dfrac{-6x^2}{2} + c = -3x^2 + c.$

15.212 $m = 3x^2 + 2x$

▮ $y = \int (3x^2 + 2x)\, dx = x^3 + x^2 + c.$

15.213 $m = 6x^2$

▮ $y = \int 6x^2\, dx = 2x^3 + c.$

For Probs. 15.214 to 15.217, find the equation of that curve (of the family of curves with given slope) passing through the given point.

15.214 $m = 1,\ (1, -2)$

▮ See Prob. 15.210. $y = x + c;\ x = 1,\ y = -2.$ Then, $-2 = 1 + c,$ and $c = -3.$ $y = x + (-3)$ or $y = x - 3.$

15.215 $m = -6x,\ (0, 0)$

▮ See Prob. 15.211. $y = -3x^2 + c,\ y = 0,\ x = 0.$ Then, $0 = 0 + c,\ c = 0.$ $y = -3x^2 + 0$ or $y = -3x^2.$

15.216 $m = 3x^2 + 2x,\ (1, -3)$

▮ See Prob. 15.212. $y = x^3 + x^2 + c,\ x = 1,\ y = -3.$ Then, $-3 = 1 + 1 + c,\ c = -5.$ $y = x^3 + x^2 - 5.$

15.217 $m = 6x^2,\ (0, 1)$

▮ See Prob. 15.213. $y = 2x^3 + c,\ x = 0,\ y = 1.$ Then, $1 = 0 + c,$ or $c = 1.$ $y = 2x^3 + 1.$

For Probs. 15.218 to 15.225, draw the graph of the region bounded as follows:

15.218 $y = x^2$, $y = 0$, $x = 3$

▌ See Fig. 15.18.

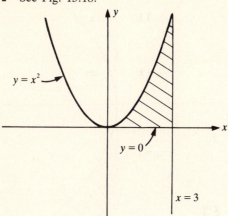

Fig. 15.18

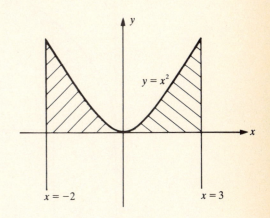

Fig. 15.19

15.219 $y = x^2$, $y = 0$, $x = -2$, $x = 3$

▌ See Fig. 15.19.

15.220 $y = x^2$, $y = 9$

▌ See Fig. 15.20.

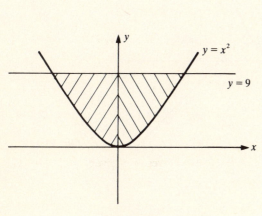

Fig. 15.20

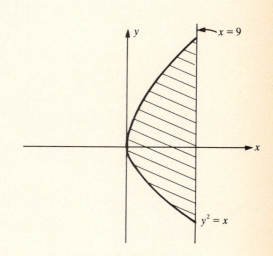

Fig. 15.21

15.221 $y^2 = x$, $x = 9$

▌ See Fig. 15.21.

15.222 $y^2 = x$, $x = 4$

▌ See Fig. 15.22.

15.223 $y^2 = x$, $x = 0$, $y = -2$

▌ See Fig. 15.23.

15.224 $y = x^2$, $x + y = 2$

▌ See Fig. 15.24.

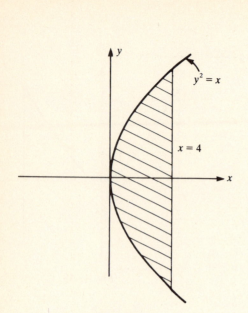

Fig. 15.22

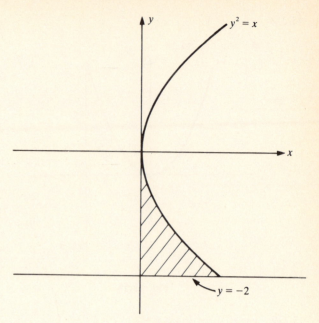

Fig. 15.23

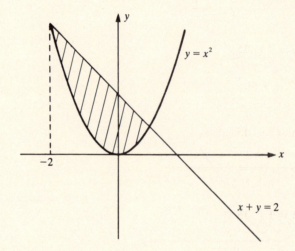

Fig. 15.24

15.225 $y = x^3, x = 0, \quad y = 8$

▮ See Fig. 15.25.

For Probs. 15.226 to 15.233, find the area of the shaded region in the given figure.

15.226 The shaded region in Fig. 15.18.

▮ $A = \int_0^3 x^2 \, dx = x^3/3 \Big|_0^3 = \frac{27}{3} - 0 = 9.$

15.227 The shaded region in Fig. 15.19.

▮ $A = \int_{-2}^0 x^2 \, dx + \int_0^3 x^2 \, dx = x^3/3 \Big|_{-2}^0 + x^3/3 \Big|_0^3 = 0 - (-\frac{8}{3}) + (\frac{27}{3} - 0) = \frac{8}{3} + \frac{27}{3} = \frac{35}{3}.$

15.228 The shaded region in Fig. 15.20.

▮ $A = 2 \int_0^3 (9 - x^2) \, dx = 2(9x - x^3/3) \Big|_0^3 = 2(27 - 9) = 2(18) = 36.$

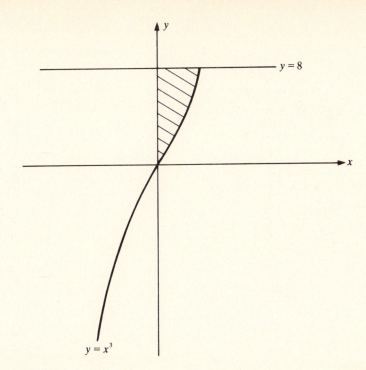

$y = 8$

$y = x^3$

Fig. 15.25

15.229 The shaded region in Fig. 15.21.

$$A = \int_0^9 \sqrt{x}\, dx = \tfrac{2}{3}x^{3/2}\Big|_0^9 = 18.$$

15.230 The shaded region in Fig. 15.22.

▮ $$A = \int_0^4 [\sqrt{x} - (-\sqrt{x})]\,dx = \int_0^4 2\sqrt{x}\, dx = 2\int_0^4 \sqrt{x}\, dx = \frac{2x^{3/2}}{\tfrac{3}{2}} = \tfrac{4}{3}x^{3/2}\Big|_0^4 = \tfrac{32}{3}.$$

15.231 The shaded region in Fig. 15.23.

▮ $$A = \int_{-2}^0 y^2\, dy = y^3/3\Big|_{-2}^0 = 0 - (-\tfrac{8}{3}) = \tfrac{8}{3}.$$

15.232 The shaded region in Fig. 15.24.

▮ $$A = \int_{-2}^1 (2 - x - x^2)\,dx = (2x - x^2/2 - x^3/3)\Big|_{-2}^1 = \tfrac{9}{2}.$$

15.233 The shaded region in Fig. 15.25.

▮ $$A = \int_0^2 (8 - x^3)\,dx = (8x - x^4/4)\Big|_0^2 = 16 - \tfrac{16}{4} - 0 = 12.$$

TABLE A.1 The Exponential Function

x	e^x	e^{-x}	x	e^x	e^{-x}
.00	1.00000	1.00000	.60	1.82212	.54881
.01	1.01005	.99005	.61	1.84043	.54335
.02	1.02020	.98020	.62	1.85893	.53794
.03	1.03045	.97045	.63	1.87761	.53259
.04	1.04081	.96079	.64	1.89648	.52729
.05	1.05127	.95123	.65	1.91554	.52205
.06	1.06184	.94176	.66	1.93479	.51685
.07	1.07251	.93239	.67	1.95424	.51171
.08	1.08329	.92312	.68	1.97388	.50662
.09	1.09417	.91393	.69	1.99372	.50158
.10	1.10517	.90484	.70	2.01375	.49659
.11	1.11628	.89583	.71	2.03399	.49164
.12	1.12750	.88692	.72	2.05443	.48675
.13	1.13883	.87810	.73	2.07508	.48191
.14	1.15027	.86936	.74	2.09594	.47711
.15	1.16183	.86071	.75	2.11700	.47237
.16	1.17351	.85214	.76	2.13828	.46767
.17	1.18530	.84366	.77	2.15977	.46301
.18	1.19722	.83527	.78	2.18147	.45841
.19	1.20925	.82696	.79	2.20340	.45384
.20	1.22140	.81873	.80	2.22554	.44933
.21	1.23368	.81058	.81	2.24791	.44486
.22	1.24608	.80252	.82	2.27050	.44043
.23	1.25860	.79453	.83	2.29332	.43605
.24	1.27125	.78663	.84	2.31637	.43171
.25	1.28403	.77880	.85	2.33965	.42741
.26	1.29693	.77105	.86	2.36316	.42316
.27	1.30996	.76338	.87	2.38691	.41895
.28	1.32313	.75578	.88	2.41090	.41478
.29	1.33643	.74826	.89	2.43513	.41066
.30	1.34986	.74082	.90	2.45960	.40657
.31	1.36343	.73345	.91	2.48432	.40252
.32	1.37713	.72615	.92	2.50929	.39852
.33	1.39097	.71892	.93	2.53451	.39455
.34	1.40495	.71177	.94	2.55998	.39063
.35	1.41907	.70469	.95	2.58571	.38674
.36	1.43333	.69768	.96	2.61170	.38289
.37	1.44773	.69073	.97	2.63794	.37908
.38	1.46228	.68386	.98	2.66446	.37531
.39	1.47698	.67706	.99	2.69123	.37158
.40	1.49182	.67032	1.00	2.71828	.36788
.41	1.50682	.66365	1.01	2.74560	.36422
.42	1.52196	.65705	1.02	2.77319	.36059
.43	1.53726	.65051	1.03	2.80107	.35701
.44	1.55271	.64404	1.04	2.82922	.35345
.45	1.56831	.63763	1.05	2.85765	.34994
.46	1.58407	.63128	1.06	2.88637	.34646
.47	1.59999	.62500	1.07	2.91538	.34301
.48	1.61607	.61878	1.08	2.94468	.33960
.49	1.63232	.61263	1.09	2.97427	.33622
.50	1.64872	.60653	1.10	3.00417	.33287
.51	1.66529	.60050	1.11	3.03436	.32956
.52	1.68203	.59452	1.12	3.06485	.32628
.53	1.69893	.58860	1.13	3.09566	.32303
.54	1.71601	.58275	1.14	3.12677	.31982
.55	1.73325	.57695	1.15	3.15819	.31664
.56	1.75067	.57121	1.16	3.18993	.31349
.57	1.76827	.56553	1.17	3.22199	.31037
.58	1.78604	.55990	1.18	3.25437	.30728
.59	1.80399	.55433	1.19	3.28708	.30422

From Koshy: *College Algebra and Trigonometry with Applications*, 1986. McGraw-Hill. Reprinted with permission.

TABLE A.1 The Exponential Function (*Continued*)

x	e^x	e^{-x}	x	e^x	e^{-x}
1.20	3.32012	.30119	1.80	6.04965	.16530
1.21	3.35348	.29820	1.81	6.11045	.16365
1.22	3.38719	.29523	1.82	6.17186	.16203
1.23	3.42123	.29229	1.83	6.23389	.16041
1.24	3.45561	.28938	1.84	6.29654	.15882
1.25	3.49034	.28650	1.85	6.35982	.15724
1.26	3.52542	.28365	1.86	6.42374	.15567
1.27	3.56085	.28083	1.87	6.48830	.15412
1.28	3.59664	.27804	1.88	6.55350	.15259
1.29	3.63279	.27527	1.89	6.61937	.15107
1.30	3.66930	.27253	1.90	6.68589	.14957
1.31	3.70617	.26982	1.91	6.75309	.14808
1.32	3.74342	.26714	1.92	6.82096	.14661
1.33	3.78104	.26448	1.93	6.88951	.14515
1.34	3.81904	.26185	1.94	6.95875	.14370
1.35	3.85743	.25924	1.95	7.02869	.14227
1.36	3.89619	.25666	1.96	7.09933	.14086
1.37	3.93535	.25411	1.97	7.17068	.13946
1.38	3.97490	.25158	1.98	7.24274	.13807
1.39	4.01485	.24908	1.99	7.31553	.13670
1.40	4.05520	.24660	2.00	7.38906	.13534
1.41	4.09596	.24414	2.01	7.46332	.13399
1.42	4.13712	.24171	2.02	7.53832	.13266
1.43	4.17870	.23931	2.03	7.61409	.13134
1.44	4.22070	.23693	2.04	7.69061	.13003
1.45	4.26311	.23457	2.05	7.76790	.12873
1.46	4.30596	.23224	2.06	7.84597	.12745
1.47	4.34924	.22993	2.07	7.92482	.12619
1.48	4.39295	.22764	2.08	8.00447	.12493
1.49	4.43710	.22537	2.09	8.08491	.12369
1.50	4.48169	.22313	2.10	8.16617	.12246
1.51	4.52673	.22091	2.11	8.24824	.12124
1.52	4.57223	.21871	2.12	8.33114	.12003
1.53	4.61818	.21654	2.13	8.41487	.11884
1.54	4.66459	.21438	2.14	8.49944	.11765
1.55	4.71147	.21225	2.15	8.58486	.11648
1.56	4.75882	.21014	2.16	8.67114	.11533
1.57	4.80665	.20805	2.17	8.75828	.11418
1.58	4.85496	.20598	2.18	8.84631	.11304
1.59	4.90375	.20393	2.19	8.93521	.11192
1.60	4.95303	.20190	2.20	9.02501	.11080
1.61	5.00281	.19989	2.21	9.11572	.10970
1.62	5.05309	.19790	2.22	9.20733	.10861
1.63	5.10387	.19593	2.23	9.29987	.10753
1.64	5.15517	.19398	2.24	9.39333	.10646
1.65	5.20698	.19205	2.25	9.48774	.10540
1.66	5.25931	.19014	2.26	9.58309	.10435
1.67	5.31217	.18825	2.27	9.67940	.10331
1.68	5.36556	.18637	2.28	9.77668	.10228
1.69	5.41948	.18452	2.29	9.87494	.10127
1.70	5.47395	.18268	2.30	9.97418	.10026
1.71	5.52896	.18087	2.31	10.07442	.09926
1.72	5.58453	.17907	2.32	10.17567	.09827
1.73	5.64065	.17728	2.33	10.27794	.09730
1.74	5.69734	.17552	2.34	10.38124	.09633
1.75	5.75460	.17377	2.35	10.48557	.09537
1.76	5.81244	.17204	2.36	10.59095	.09442
1.77	5.87085	.17033	2.37	10.69739	.09348
1.78	5.92986	.16864	2.38	10.80490	.09255
1.79	5.98945	.16696	2.39	10.91349	.09163

TABLE A.1 The Exponential Function (Continued)

x	e^x	e^{-x}	x	e^x	e^{-x}
2.40	11.02318	.09072	3.00	20.08554	.04979
2.41	11.13396	.08982	3.01	20.28740	.04929
2.42	11.24586	.08892	3.02	20.49129	.04880
2.43	11.35888	.08804	3.03	20.69723	.04832
2.44	11.47304	.08716	3.04	20.90524	.04783
2.45	11.58835	.08629	3.05	21.11534	.04736
2.46	11.70481	.08543	3.06	21.32756	.04689
2.47	11.82245	.08458	3.07	21.54190	.04642
2.48	11.94126	.08374	3.08	21.75840	.04596
2.49	12.06128	.08291	3.09	21.97708	.04550
2.50	12.18249	.08208	3.10	22.19795	.04505
2.51	12.30493	.08127	3.11	22.42104	.04460
2.52	12.42860	.08046	3.12	22.64638	.04416
2.53	12.55351	.07966	3.13	22.87398	.04372
2.54	12.67967	.07887	3.14	23.10387	.04328
2.55	12.80710	.07808	3.15	23.33606	.04285
2.56	12.93582	.07730	3.16	23.57060	.04243
2.57	13.06582	.07654	3.17	23.80748	.04200
2.58	13.19714	.07577	3.18	24.04675	.04159
2.59	13.32977	.07502	3.19	24.28843	.04117
2.60	13.46374	.07427	3.20	24.53253	.04076
2.61	13.59905	.07353	3.21	24.77909	.04036
2.62	13.73572	.07280	3.22	25.02812	.03996
2.63	13.87377	.07208	3.23	25.27966	.03956
2.64	14.01320	.07136	3.24	25.53372	.03916
2.65	14.15404	.07065	3.25	25.79034	.03877
2.66	14.29629	.06995	3.26	26.04954	.03839
2.67	14.43997	.06925	3.27	26.31134	.03801
2.68	14.58509	.06856	3.28	26.57577	.03763
2.69	14.73168	.06788	3.29	26.84286	.03725
2.70	14.87973	.06721	3.30	27.11264	.03688
2.71	15.02928	.06654	3.31	27.38512	.03652
2.72	15.18032	.06587	3.32	27.66035	.03615
2.73	15.33289	.06522	3.33	27.93834	.03579
2.74	15.48698	.06457	3.34	28.21913	.03544
2.75	15.64263	.06393	3.35	28.50273	.03508
2.76	15.79984	.06329	3.36	28.78919	.03474
2.77	15.95863	.06266	3.37	29.07853	.03439
2.78	16.11902	.06204	3.38	29.37077	.03405
2.79	16.28102	.06142	3.39	29.66595	.03371
2.80	16.44465	.06081	3.40	29.96410	.03337
2.81	16.60992	.06020	3.41	30.26524	.03304
2.82	16.77685	.05961	3.42	30.56941	.03271
2.83	16.94546	.05901	3.43	30.87664	.03239
2.84	17.11577	.05843	3.44	31.18696	.03206
2.85	17.28778	.05784	3.45	31.50039	.03175
2.86	17.46153	.05727	3.46	31.81698	.03143
2.87	17.63702	.05670	3.47	32.13674	.03112
2.88	17.81427	.05613	3.48	32.45972	.03081
2.89	17.99331	.05558	3.49	32.78595	.03050
2.90	18.17414	.05502	3.50	33.11545	.03020
2.91	18.35680	.05448	3.51	33.44827	.02990
2.92	18.54129	.05393	3.52	33.78443	.02960
2.93	18.72763	.05340	3.53	34.12397	.02930
2.94	18.91585	.05287	3.54	34.46692	.02901
2.95	19.10595	.05234	3.55	34.81332	.02872
2.96	19.29797	.05182	3.56	35.16320	.02844
2.97	19.49192	.05130	3.57	35.51659	.02816
2.98	19.68782	.05079	3.58	35.87354	.02788
2.99	19.88568	.05029	3.59	36.23408	.02760

TABLE A.1 The Exponential Function (*Continued*)

x	e^x	e^{-x}	x	e^x	e^{-x}
3.60	36.59823	.02732	4.20	66.68633	.01500
3.61	36.96605	.02705	4.21	67.35654	.01485
3.62	37.33757	.02678	4.22	68.03348	.01470
3.63	37.71282	.02652	4.23	68.71723	.01455
3.64	38.09184	.02625	4.24	69.40785	.01441
3.65	38.47467	.02599	4.25	70.10541	.01426
3.66	38.86134	.02573	4.26	70.80998	.01412
3.67	39.25191	.02548	4.27	71.52163	.01398
3.68	39.64639	.02522	4.28	72.24044	.01384
3.69	40.04485	.02497	4.29	72.96647	.01370
3.70	40.44730	.02472	4.30	73.69979	.01357
3.71	40.85381	.02448	4.31	74.44049	.01343
3.72	41.26439	.02423	4.32	75.18863	.01330
3.73	41.67911	.02399	4.33	75.94429	.01317
3.74	42.09799	.02375	4.34	76.70754	.01304
3.75	42.52108	.02352	4.35	77.47846	.01291
3.76	42.94843	.02328	4.36	78.25713	.01278
3.77	43.38006	.02305	4.37	79.04363	.01265
3.78	43.81604	.02282	4.38	79.83803	.01253
3.79	44.25640	.02260	4.39	80.64042	.01240
3.80	44.70118	.02237	4.40	81.45087	.01228
3.81	45.15044	.02215	4.41	82.26946	.01216
3.82	45.60421	.02193	4.42	83.09628	.01203
3.83	46.06254	.02171	4.43	83.93141	.01191
3.84	46.52547	.02149	4.44	84.77494	.01180
3.85	46.99306	.02128	4.45	85.62694	.01168
3.86	47.46535	.02107	4.46	86.48751	.01156
3.87	47.94238	.02086	4.47	87.35672	.01145
3.88	48.42421	.02065	4.48	88.23467	.01133
3.89	48.91089	.02045	4.49	89.12144	.01122
3.90	49.40245	.02024	4.50	90.01713	.01111
3.91	49.89895	.02004	4.51	90.92182	.01100
3.92	50.40044	.01984	4.52	91.83560	.01089
3.93	50.90698	.01964	4.53	92.75856	.01078
3.94	51.41860	.01945	4.54	93.69080	.01067
3.95	51.93537	.01925	4.55	94.63240	.01057
3.96	52.45732	.01906	4.56	95.58347	.01046
3.97	52.98453	.01887	4.57	96.54411	.01036
3.98	53.51703	.01869	4.58	97.51439	.01025
3.99	54.05489	.01850	4.59	98.49443	.01015
4.00	54.59815	.01832	4.60	99.48431	.01005
4.01	55.14687	.01813	4.61	100.48415	.00995
4.02	55.70110	.01795	4.62	101.49403	.00985
4.03	56.26091	.01777	4.63	102.51406	.00975
4.04	56.82634	.01760	4.64	103.54435	.00966
4.05	57.39745	.01742	4.65	104.58498	.00956
4.06	57.97431	.01725	4.66	105.63608	.00947
4.07	58.55696	.01708	4.67	106.69774	.00937
4.08	59.14547	.01691	4.68	107.77007	.00928
4.09	59.73989	.01674	4.69	108.85318	.00919
4.10	60.34029	.01657	4.70	109.94717	.00910
4.11	60.94671	.01641	4.71	111.05216	.00900
4.12	61.55924	.01624	4.72	112.16825	.00892
4.13	62.17792	.01608	4.73	113.29556	.00883
4.14	62.80282	.01592	4.74	114.43420	.00874
4.15	63.43400	.01576	4.75	115.58428	.00865
4.16	64.07152	.01561	4.76	116.74592	.00857
4.17	64.71545	.01545	4.77	117.91924	.00848
4.18	65.36585	.01530	4.78	119.10435	.00840
4.19	66.02279	.01515	4.79	120.30136	.00831

TABLE A.1 The Exponential Function (*Continued*)

x	e^x	e^{-x}	x	e^x	e^{-x}
4.80	121.51041	.00823	**8.00**	2980.95779	.00034
4.81	122.73161	.00815	8.10	3294.46777	.00030
4.82	123.96509	.00807	8.20	3640.95004	.00027
4.83	125.21096	.00799	8.30	4023.87219	.00025
4.84	126.46935	.00791	8.40	4447.06665	.00022
4.85	127.74039	.00783	8.50	4914.76886	.00020
4.86	129.02420	.00775	8.60	5431.65906	.00018
4.87	130.32091	.00767	8.70	6002.91180	.00017
4.88	131.63066	.00760	8.80	6634.24371	.00015
4.89	132.95357	.00752	8.90	7331.97339	.00014
4.90	134.28978	.00745	**9.00**	8103.08295	.00012
4.91	135.63941	.00737	9.10	8955.29187	.00011
4.92	137.00261	.00730	9.20	9897.12830	.00010
4.93	138.37951	.00723	9.30	10938.01868	.00009
4.94	139.77024	.00715	9.40	12088.38049	.00008
4.95	141.17496	.00708	9.50	13359.72522	.00007
4.96	142.59379	.00701	9.60	14764.78015	.00007
4.97	144.02688	.00694	9.70	16317.60608	.00006
4.98	145.47438	.00687	9.80	18033.74414	.00006
4.99	146.93642	.00681	9.90	19930.36987	.00005
5.00	148.41316	.00674	**10.00**	22026.46313	.00005
5.10	164.02190	.00610	10.10	24343.00708	.00004
5.20	181.27224	.00552	10.20	26903.18408	.00004
5.30	200.33680	.00499	10.30	29732.61743	.00003
5.40	221.40641	.00452	10.40	32859.62500	.00003
5.50	244.69192	.00409	10.50	36315.49854	.00003
5.60	270.42640	.00370	10.60	40134.83350	.00002
5.70	298.86740	.00335	10.70	44355.85205	.00002
5.80	330.29955	.00303	10.80	49020.79883	.00002
5.90	365.03746	.00274	10.90	54176.36230	.00002
6.00	403.42877	.00248	**11.00**	59874.13477	.00002
6.10	445.85775	.00224	11.10	66171.15430	.00002
6.20	492.74903	.00203	11.20	73130.43652	.00001
6.30	544.57188	.00184	11.30	80821.63379	.00001
6.40	601.84502	.00166	11.40	89321.72168	.00001
6.50	665.14159	.00150	11.50	98715.75879	.00001
6.60	735.09516	.00136	11.60	109097.78906	.00001
6.70	812.40582	.00123	11.70	120571.70605	.00001
6.80	897.84725	.00111	11.80	133252.34570	.00001
6.90	992.27469	.00101	11.90	147266.62109	.00001
7.00	1096.63309	.00091	—	—	—
7.10	1211.96703	.00083	—	—	—
7.20	1339.43076	.00075	—	—	—
7.30	1480.29985	.00068	—	—	—
7.40	1635.98439	.00061	—	—	—
7.50	1808.04231	.00055	—	—	—
7.60	1998.19582	.00050	—	—	—
7.70	2208.34796	.00045	—	—	—
7.80	2440.60187	.00041	—	—	—
7.90	2697.28226	.00037	—	—	—

TABLE A.2 Common Logarithms

N	0	1	2	3	4	5	6	7	8	9
1.0	.0000	.0043	.0086	.0128	.0170	.0212	.0253	.0294	.0334	.0374
1.1	.0414	.0453	.0492	.0531	.0569	.0607	.0645	.0682	.0719	.0755
1.2	.0792	.0828	.0864	.0899	.0934	.0969	.1004	.1038	.1072	.1106
1.3	.1139	.1173	.1206	.1239	.1271	.1303	.1335	.1367	.1399	.1430
1.4	.1461	.1492	.1523	.1553	.1584	.1614	.1644	.1673	.1703	.1732
1.5	.1761	.1790	.1818	.1847	.1875	.1903	.1931	.1959	.1987	.2014
1.6	.2041	.2068	.2095	.2122	.2148	.2175	.2201	.2227	.2253	.2279
1.7	.2304	.2330	.2355	.2380	.2405	.2430	.2455	.2480	.2504	.2529
1.8	.2553	.2577	.2601	.2625	.2648	.2672	.2695	.2718	.2742	.2765
1.9	.2788	.2810	.2833	.2856	.2878	.2900	.2923	.2945	.2967	.2989
2.0	.3010	.3032	.3054	.3075	.3096	.3118	.3139	.3160	.3181	.3201
2.1	.3222	.3243	.3263	.3284	.3304	.3324	.3345	.3365	.3385	.3404
2.2	.3424	.3444	.3464	.3483	.3502	.3522	.3541	.3560	.3579	.3598
2.3	.3617	.3636	.3655	.3674	.3692	.3711	.3729	.3747	.3766	.3784
2.4	.3802	.3820	.3838	.3856	.3874	.3892	.3909	.3927	.3945	.3962
2.5	.3979	.3997	.4014	.4031	.4048	.4065	.4082	.4099	.4116	.4133
2.6	.4150	.4166	.4183	.4200	.4216	.4232	.4249	.4265	.4281	.4298
2.7	.4314	.4330	.4346	.4362	.4378	.4393	.4409	.4425	.4440	.4456
2.8	.4472	.4487	.4502	.4518	.4533	.4548	.4564	.4579	.4594	.4609
2.9	.4624	.4639	.4654	.4669	.4683	.4698	.4713	.4728	.4742	.4757
3.0	.4771	.4786	.4800	.4814	.4829	.4843	.4857	.4871	.4886	.4900
3.1	.4914	.4928	.4942	.4955	.4969	.4983	.4997	.5011	.5024	.5038
3.2	.5051	.5065	.5079	.5092	.5105	.5119	.5132	.5145	.5159	.5172
3.3	.5185	.5198	.5211	.5224	.5237	.5250	.5263	.5276	.5289	.5302
3.4	.5315	.5328	.5340	.5353	.5366	.5378	.5391	.5403	.5416	.5428
3.5	.5441	.5453	.5465	.5478	.5490	.5502	.5514	.5527	.5539	.5551
3.6	.5563	.5575	.5587	.5599	.5611	.5623	.5635	.5647	.5658	.5670
3.7	.5682	.5694	.5705	.5717	.5729	.5740	.5752	.5763	.5775	.5786
3.8	.5798	.5809	.5821	.5832	.5843	.5855	.5866	.5877	.5888	.5899
3.9	.5911	.5922	.5933	.5944	.5955	.5966	.5977	.5988	.5999	.6010
4.0	.6021	.6031	.6042	.6053	.6064	.6075	.6085	.6096	.6107	.6117
4.1	.6128	.6138	.6149	.6160	.6170	.6180	.6191	.6201	.6212	.6222
4.2	.6232	.6243	.6253	.6263	.6274	.6284	.6294	.6304	.6314	.6325
4.3	.6335	.6345	.6355	.6365	.6375	.6385	.6395	.6405	.6415	.6425
4.4	.6435	.6444	.6454	.6464	.6474	.6484	.6493	.6503	.6513	.6522
4.5	.6532	.6542	.6551	.6561	.6571	.6580	.6590	.6599	.6609	.6618
4.6	.6628	.6637	.6646	.6656	.6665	.6675	.6684	.6693	.6702	.6712
4.7	.6721	.6730	.6739	.6749	.6758	.6767	.6776	.6785	.6794	.6803
4.8	.6812	.6821	.6830	.6839	.6848	.6857	.6866	.6875	.6884	.6893
4.9	.6902	.6911	.6920	.6928	.6937	.6946	.6955	.6964	.6972	.6981
5.0	.6990	.6998	.7007	.7016	.7024	.7033	.7042	.7050	.7059	.7067
5.1	.7076	.7084	.7093	.7101	.7110	.7118	.7126	.7135	.7143	.7152
5.2	.7160	.7168	.7177	.7185	.7193	.7202	.7210	.7218	.7226	.7235
5.3	.7243	.7251	.7259	.7267	.7275	.7284	.7292	.7300	.7308	.7316
5.4	.7324	.7332	.7340	.7348	.7356	.7364	.7372	.7380	.7388	.7396

From Koshy: *College Algebra and Trigonometry with Applications*, 1986. McGraw-Hill. Reprinted with permission.

TABLE A.2 Common Logarithms (*Continued*)

N	0	1	2	3	4	5	6	7	8	9
5.5	.7404	.7412	.7419	.7427	.7435	.7443	.7451	.7459	.7466	.7474
5.6	.7482	.7490	.7497	.7505	.7513	.7520	.7528	.7536	.7543	.7551
5.7	.7559	.7566	.7574	.7582	.7589	.7597	.7604	.7612	.7619	.7627
5.8	.7634	.7642	.7649	.7657	.7664	.7672	.7679	.7686	.7694	.7701
5.9	.7709	.7716	.7723	.7731	.7738	.7745	.7752	.7760	.7767	.7774
6.0	.7782	.7789	.7796	.7803	.7810	.7818	.7825	.7832	.7839	.7846
6.1	.7853	.7860	.7868	.7875	.7882	.7889	.7896	.7903	.7910	.7917
6.2	.7924	.7931	.7938	.7945	.7952	.7959	.7966	.7973	.7980	.7987
6.3	.7993	.8000	.8007	.8014	.8021	.8028	.8035	.8041	.8048	.8055
6.4	.8062	.8069	.8075	.8082	.8089	.8096	.8102	.8109	.8116	.8122
6.5	.8129	.8136	.8142	.8149	.8156	.8162	.8169	.8176	.8182	.8189
6.6	.8195	.8202	.8209	.8215	.8222	.8228	.8235	.8241	.8248	.8254
6.7	.8261	.8267	.8274	.8280	.8287	.8293	.8299	.8306	.8312	.8319
6.8	.8325	.8331	.8338	.8344	.8351	.8357	.8363	.8370	.8376	.8382
6.9	.8388	.8395	.8401	.8407	.8414	.8420	.8426	.8432	.8439	.8445
7.0	.8451	.8457	.8463	.8470	.8476	.8482	.8488	.8494	.8500	.8506
7.1	.8513	.8519	.8525	.8531	.8537	.8543	.8549	.8555	.8561	.8567
7.2	.8573	.8579	.8585	.8591	.8597	.8603	.8609	.8615	.8621	.8627
7.3	.8633	.8639	.8645	.8651	.8657	.8663	.8669	.8675	.8681	.8686
7.4	.8692	.8698	.8704	.8710	.8716	.8722	.8727	.8733	.8739	.8745
7.5	.8751	.8756	.8762	.8768	.8774	.8779	.8785	.8791	.8797	.8802
7.6	.8808	.8814	.8820	.8825	.8831	.8837	.8842	.8848	.8854	.8859
7.7	.8865	.8871	.8876	.8882	.8887	.8893	.8899	.8904	.8910	.8915
7.8	.8921	.8927	.8932	.8938	.8943	.8949	.8954	.8960	.8965	.8971
7.9	.8976	.8982	.8987	.8993	.8998	.9004	.9009	.9015	.9020	.9025
8.0	.9031	.9036	.9042	.9047	.9053	.9058	.9063	.9069	.9074	.9079
8.1	.9085	.9090	.9096	.9101	.9106	.9112	.9117	.9122	.9128	.9133
8.2	.9138	.9143	.9149	.9154	.9159	.9165	.9170	.9175	.9180	.9186
8.3	.9191	.9196	.9201	.9206	.9212	.9217	.9222	.9227	.9232	.9238
8.4	.9243	.9248	.9253	.9258	.9263	.9269	.9274	.9279	.9284	.9289
8.5	.9294	.9299	.9304	.9309	.9315	.9320	.9325	.9330	.9335	.9340
8.6	.9345	.9350	.9355	.9360	.9365	.9370	.9375	.9380	.9385	.9390
8.7	.9395	.9400	.9405	.9410	.9415	.9420	.9425	.9430	.9435	.9440
8.8	.9445	.9450	.9455	.9460	.9465	.9469	.9474	.9479	.9484	.9489
8.9	.9494	.9499	.9504	.9509	.9513	.9518	.9523	.9528	.9533	.9538
9.0	.9542	.9547	.9552	.9557	.9562	.9566	.9571	.9576	.9581	.9586
9.1	.9590	.9595	.9600	.9605	.9609	.9614	.9619	.9624	.9628	.9633
9.2	.9638	.9643	.9647	.9652	.9657	.9661	.9666	.9671	.9675	.9680
9.3	.9685	.9689	.9694	.9699	.9703	.9708	.9713	.9717	.9722	.9727
9.4	.9731	.9736	.9741	.9745	.9750	.9754	.9759	.9763	.9768	.9773
9.5	.9777	.9782	.9786	.9791	.9795	.9800	.9805	.9809	.9814	.9818
9.6	.9823	.9827	.9832	.9836	.9841	.9845	.9850	.9854	.9859	.9863
9.7	.9868	.9872	.9877	.9881	.9886	.9890	.9894	.9899	.9903	.9908
9.8	.9912	.9917	.9921	.9926	.9930	.9934	.9939	.9943	.9948	.9952
9.9	.9956	.9961	.9965	.9969	.9974	.9978	.9983	.9987	.9991	.9996

TABLE A.3 Trigonometric Functions of an Angle

t	t degrees	sin t	cos t	tan t	cot t	sec t	csc t		
.0000	0° 00′	.0000	1.0000	.0000		1.000		90° 00′	1.5708
.0029	10	.0029	1.0000	.0029	343.8	1.000	343.8	50	1.5679
.0058	20	.0058	1.0000	.0058	171.9	1.000	171.9	40	1.5650
.0087	30	.0087	1.0000	.0087	114.6	1.000	114.6	30	1.5621
.0116	40	.0116	.9999	.0116	85.94	1.000	85.95	20	1.5592
.0145	50	.0145	.9999	.0145	68.75	1.000	68.76	10	1.5563
.0175	1° 00′	.0175	.9998	.0175	57.29	1.000	57.30	89° 00′	1.5533
.0204	10	.0204	.9998	.0204	49.10	1.000	49.11	50	1.5504
.0233	20	.0233	.9997	.0233	42.96	1.000	42.98	40	1.5475
.0262	30	.0262	.9997	.0262	38.19	1.000	38.20	30	1.5446
.0291	40	.0291	.9996	.0291	34.37	1.000	34.38	20	1.5417
.0320	50	.0320	.9995	.0320	31.24	1.001	31.26	10	1.5388
.0349	2° 00′	.0349	.9994	.0349	28.64	1.001	28.65	88° 00′	1.5359
.0378	10	.0378	.9993	.0378	26.43	1.001	26.45	50	1.5330
.0407	20	.0407	.9992	.0407	24.54	1.001	24.56	40	1.5301
.0436	30	.0436	.9990	.0437	22.90	1.001	22.93	30	1.5272
.0465	40	.0465	.9989	.0466	21.47	1.001	21.49	20	1.5243
.0495	50	.0494	.9988	.0495	20.21	1.001	20.23	10	1.5213
.0524	3° 00′	.0523	.9986	.0524	19.08	1.001	19.11	87° 00′	1.5184
.0553	10	.0552	.9985	.0553	18.07	1.002	18.10	50	1.5155
.0582	20	.0581	.9983	.0582	17.17	1.002	17.20	40	1.5126
.0611	30	.0610	.9981	.0612	16.35	1.002	16.38	30	1.5097
.0640	40	.0640	.9980	.0641	15.60	1.002	15.64	20	1.5068
.0669	50	.0669	.9978	.0670	14.92	1.002	14.96	10	1.5039
.0698	4° 00′	.0698	.9976	.0699	14.30	1.002	14.34	86° 00′	1.5010
.0727	10	.0727	.9974	.0729	13.73	1.003	13.76	50	1.4981
.0756	20	.0756	.9971	.0758	13.20	1.003	13.23	40	1.4952
.0785	30	.0785	.9969	.0787	12.71	1.003	12.75	30	1.4923
.0814	40	.0814	.9967	.0816	12.25	1.003	12.29	20	1.4893
.0844	50	.0843	.9964	.0846	11.83	1.004	11.87	10	1.4864
.0873	5° 00′	.0872	.9962	.0875	11.43	1.004	11.47	85° 00′	1.4835
.0902	10	.0901	.9959	.0904	11.06	1.004	11.10	50	1.4806
.0931	20	.0929	.9957	.0934	10.71	1.004	10.76	40	1.4777
.0960	30	.0958	.9954	.0963	10.39	1.005	10.43	30	1.4748
.0989	40	.0987	.9951	.0992	10.08	1.005	10.13	20	1.4719
.1018	50	.1016	.9948	.1022	9.788	1.005	9.839	10	1.4690
.1047	6° 00′	.1045	.9945	.1051	9.514	1.006	9.567	84° 00′	1.4661
.1076	10	.1074	.9942	.1080	9.255	1.006	9.309	50	1.4632
.1105	20	.1103	.9939	.1110	9.010	1.006	9.065	40	1.4603
.1134	30	.1132	.9936	.1139	8.777	1.006	8.834	30	1.4573
.1164	40	.1161	.9932	.1169	8.556	1.007	8.614	20	1.4544
.1193	50	.1190	.9929	.1198	8.345	1.007	8.405	10	1.4515
.1222	7° 00′	.1219	.9925	.1228	8.144	1.008	8.206	83° 00′	1.4486
		cos t	sin t	cot t	tan t	csc t	sec t	t degrees	t

From Koshy: *College Algebra and Trigonometry with Applications*, 1986. McGraw-Hill. Reprinted with permission.

TABLE A.3 Trigonometric Functions of an Angle (*Continued*)

t	t degrees	sin t	cos t	tan t	cot t	sec t	csc t		
.1222	**7° 00'**	.1219	.9925	.1228	8.144	1.008	8.206	**83° 00'**	1.4486
.1251	10	.1248	.9922	.1257	7.953	1.008	8.016	50	1.4457
.1280	20	.1276	.9918	.1287	7.770	1.008	7.834	40	1.4428
.1309	30	.1305	.9914	.1317	7.596	1.009	7.661	30	1.4399
.1338	40	.1334	.9911	.1346	7.429	1.009	7.496	20	1.4370
.1367	50	.1363	.9907	.1376	7.269	1.009	7.337	10	1.4341
.1396	**8° 00'**	.1392	.9903	.1405	7.115	1.010	7.185	**82° 00'**	1.4312
.1425	10	.1421	.9899	.1435	6.968	1.010	7.040	50	1.4283
.1454	20	.1449	.9894	.1465	6.827	1.011	6.900	40	1.4254
.1484	30	.1478	.9890	.1495	6.691	1.011	6.765	30	1.4224
.1513	40	.1507	.9886	.1524	6.561	1.012	6.636	20	1.4195
.1542	50	.1536	.9881	.1554	6.435	1.012	6.512	10	1.4166
.1571	**9° 00'**	.1564	.9877	.1584	6.314	1.012	6.392	**81° 00'**	1.4137
.1600	10	.1593	.9872	.1614	6.197	1.013	6.277	50	1.4108
.1629	20	.1622	.9868	.1644	6.084	1.013	6.166	40	1.4079
.1658	30	.1650	.9863	.1673	5.976	1.014	6.059	30	1.4050
.1687	40	.1679	.9858	.1703	5.871	1.014	5.955	20	1.4021
.1716	50	.1708	.9853	.1733	5.769	1.015	5.855	10	1.3992
.1745	**10° 00'**	.1736	.9848	.1763	5.671	1.015	5.759	**80° 00'**	1.3963
.1774	10	.1765	.9843	.1793	5.576	1.016	5.665	50	1.3934
.1804	20	.1794	.9838	.1823	5.485	1.016	5.575	40	1.3904
.1833	30	.1822	.9833	.1853	5.396	1.017	5.487	30	1.3875
.1862	40	.1851	.9827	.1883	5.309	1.018	5.403	20	1.3846
.1891	50	.1880	.9822	.1914	5.226	1.018	5.320	10	1.3817
.1920	**11° 00'**	.1908	.9816	.1944	5.145	1.019	5.241	**79° 00'**	1.3788
.1949	10	.1937	.9811	.1974	5.066	1.019	5.164	50	1.3759
.1978	20	.1965	.9805	.2004	4.989	1.020	5.089	40	1.3730
.2007	30	.1994	.9799	.2035	4.915	1.020	5.016	30	1.3701
.2036	40	.2022	.9793	.2065	4.843	1.021	4.945	20	1.3672
.2065	50	.2051	.9787	.2095	4.773	1.022	4.876	10	1.3643
.2094	**12° 00'**	.2079	.9781	.2126	4.705	1.022	4.810	**78° 00'**	1.3614
.2123	10	.2108	.9775	.2156	4.638	1.023	4.745	50	1.3584
.2153	20	.2136	.9769	.2186	4.574	1.024	4.682	40	1.3555
.2182	30	.2164	.9763	.2217	4.511	1.024	4.620	30	1.3526
.2211	40	.2193	.9757	.2247	4.449	1.025	4.560	20	1.3497
.2240	50	.2221	.9750	.2278	4.390	1.026	4.502	10	1.3468
.2269	**13° 00'**	.2250	.9744	.2309	4.331	1.026	4.445	**77° 00'**	1.3439
.2298	10	.2278	.9737	.2339	4.275	1.027	4.390	50	1.3410
.2327	20	.2306	.9730	.2370	4.219	1.028	4.336	40	1.3381
.2356	30	.2334	.9724	.2401	4.165	1.028	4.284	30	1.3352
.2385	40	.2363	.9717	.2432	4.113	1.029	4.232	20	1.3323
.2414	50	.2391	.9710	.2462	4.061	1.030	4.182	10	1.3294
.2443	**14° 00'**	.2419	.9703	.2493	4.011	1.031	4.134	**76° 00'**	1.3265
		cos t	sin t	cot t	tan t	csc t	sec t	t degrees	t

TABLE A.3 Trigonometric Functions of an Angle (*Continued*)

t	t degrees	$\sin t$	$\cos t$	$\tan t$	$\cot t$	$\sec t$	$\csc t$		
.2443	**14° 00′**	.2419	.9703	.2493	4.011	1.031	4.134	**76° 00′**	1.3265
.2473	10	.2447	.9696	.2524	3.962	1.031	4.086	50	1.3235
.2502	20	.2476	.9689	.2555	3.914	1.032	4.039	40	1.3206
.2531	30	.2504	.9681	.2586	3.867	1.033	3.994	30	1.3177
.2560	40	.2532	.9674	.2617	3.821	1.034	3.950	20	1.3148
.2589	50	.2560	.9667	.2648	3.776	1.034	3.906	10	1.3119
.2618	**15° 00′**	.2588	.9659	.2679	3.732	1.035	3.864	**75° 00′**	1.3090
.2647	10	.2616	.9652	.2711	3.689	1.036	3.822	50	1.3061
.2676	20	.2644	.9644	.2742	3.647	1.037	3.782	40	1.3032
.2705	30	.2672	.9636	.2773	3.606	1.038	3.742	30	1.3003
.2734	40	.2700	.9628	.2805	3.566	1.039	3.703	20	1.2974
.2763	50	.2728	.9621	.2836	3.526	1.039	3.665	10	1.2945
.2793	**16° 00′**	.2756	.9613	.2867	3.487	1.040	3.628	**74° 00′**	1.2915
.2822	10	.2784	.9605	.2899	3.450	1.041	3.592	50	1.2886
.2851	20	.2812	.9596	.2931	3.412	1.042	3.556	40	1.2857
.2880	30	.2840	.9588	.2962	3.376	1.043	3.521	30	1.2828
.2909	40	.2868	.9580	.2994	3.340	1.044	3.487	20	1.2799
.2938	50	.2896	.9572	.3026	3.305	1.045	3.453	10	1.2770
.2967	**17° 00′**	.2924	.9563	.3057	3.271	1.046	3.420	**73° 00′**	1.2741
.2996	10	.2952	.9555	.3089	3.237	1.047	3.388	50	1.2712
.3025	20	.2979	.9546	.3121	3.204	1.048	3.356	40	1.2683
.3054	30	.3007	.9537	.3153	3.172	1.049	3.326	30	1.2654
.3083	40	.3035	.9528	.3185	3.140	1.049	3.295	20	1.2625
.3113	50	.3062	.9520	.3217	3.108	1.050	3.265	10	1.2595
.3142	**18° 00′**	.3090	.9511	.3249	3.078	1.051	3.236	**72° 00′**	1.2566
.3171	10	.3118	.9502	.3281	3.047	1.052	3.207	50	1.2537
.3200	20	.3145	.9492	.3314	3.018	1.053	3.179	40	1.2508
.3229	30	.3173	.9483	.3346	2.989	1.054	3.152	30	1.2479
.3258	40	.3201	.9474	.3378	2.960	1.056	3.124	20	1.2450
.3287	50	.3228	.9465	.3411	2.932	1.057	3.098	10	1.2421
.3316	**19° 00′**	.3256	.9455	.3443	2.904	1.058	3.072	**71° 00′**	1.2392
.3345	10	.3283	.9446	.3476	2.877	1.059	3.046	50	1.2363
.3374	20	.3311	.9436	.3508	2.850	1.060	3.021	40	1.2334
.3403	30	.3338	.9426	.3541	2.824	1.061	2.996	30	1.2305
.3432	40	.3365	.9417	.3574	2.798	1.062	2.971	20	1.2275
.3462	50	.3393	.9407	.3607	2.773	1.063	2.947	10	1.2246
.3491	**20° 00′**	.3420	.9397	.3640	2.747	1.064	2.924	**70° 00′**	1.2217
.3520	10	.3448	.9387	.3673	2.723	1.065	2.901	50	1.2188
.3549	20	.3475	.9377	.3706	2.699	1.066	2.878	40	1.2159
.3578	30	.3502	.9367	.3739	2.675	1.068	2.855	30	1.2130
.3607	40	.3529	.9356	.3772	2.651	1.069	2.833	20	1.2101
.3636	50	.3557	.9346	.3805	2.628	1.070	2.812	10	1.2072
.3665	**21° 00′**	.3584	.9336	.3839	2.605	1.071	2.790	**69° 00′**	1.2043
		$\cos t$	$\sin t$	$\cot t$	$\tan t$	$\csc t$	$\sec t$	t degrees	t

TABLE A.3 Trigonometric Functions of an Angle (*Continued*)

t	t degrees	$\sin t$	$\cos t$	$\tan t$	$\cot t$	$\sec t$	$\csc t$		
.3665	**21° 00′**	.3584	.9336	.3839	2.605	1.071	2.790	**69° 00′**	1.2043
.3694	10	.3611	.9325	.3872	2.583	1.072	2.769	50	1.2014
.3723	20	.3638	.9315	.3906	2.560	1.074	2.749	40	1.1985
.3752	30	.3665	.9304	.3939	2.539	1.075	2.729	30	1.1956
.3782	40	.3692	.9293	.3973	2.517	1.076	2.709	20	1.1926
.3811	50	.3719	.9283	.4006	2.496	1.077	2.689	10	1.1897
.3840	**22° 00′**	.3746	.9272	.4040	2.475	1.079	2.669	**68° 00′**	1.1868
.3869	10	.3773	.9261	.4074	2.455	1.080	2.650	50	1.1839
.3898	20	.3800	.9250	.4108	2.434	1.081	2.632	40	1.1810
.3927	30	.3827	.9239	.4142	2.414	1.082	2.613	30	1.1781
.3956	40	.3854	.9228	.4176	2.394	1.084	2.595	20	1.1752
.3985	50	.3881	.9216	.4210	2.375	1.085	2.577	10	1.1723
.4014	**23° 00′**	.3907	.9205	.4245	2.356	1.086	2.559	**67° 00′**	1.1694
.4043	10	.3934	.9194	.4279	2.337	1.088	2.542	50	1.1665
.4072	20	.3961	.9182	.4314	2.318	1.089	2.525	40	1.1636
.4102	30	.3987	.9171	.4348	2.300	1.090	2.508	30	1.1606
.4131	40	.4014	.9159	.4383	2.282	1.092	2.491	20	1.1577
.4160	50	.4041	.9147	.4417	2.264	1.093	2.475	10	1.1548
.4189	**24° 00′**	.4067	.9135	.4452	2.246	1.095	2.459	**66° 00′**	1.1519
.4218	10	.4094	.9124	.4487	2.229	1.096	2.443	50	1.1490
.4247	20	.4120	.9112	.4522	2.211	1.097	2.427	40	1.1461
.4276	30	.4147	.9100	.4557	2.194	1.099	2.411	30	1.1432
.4305	40	.4173	.9088	.4592	2.177	1.100	2.396	20	1.1403
.4334	50	.4200	.9075	.4628	2.161	1.102	2.381	10	1.1374
.4363	**25° 00′**	.4226	.9063	.4663	2.145	1.103	2.366	**65° 00′**	1.1345
.4392	10	.4253	.9051	.4699	2.128	1.105	2.352	50	1.1316
.4422	20	.4279	.9038	.4734	2.112	1.106	2.337	40	1.1286
.4451	30	.4305	.9026	.4770	2.097	1.108	2.323	30	1.1257
.4480	40	.4331	.9013	.4806	2.081	1.109	2.309	20	1.1228
.4509	50	.4358	.9001	.4841	2.066	1.111	2.295	10	1.1199
.4538	**26° 00′**	.4384	.8988	.4877	2.050	1.113	2.281	**64° 00′**	1.1170
.4567	10	.4410	.8975	.4913	2.035	1.114	2.268	50	1.1141
.4596	20	.4436	.8962	.4950	2.020	1.116	2.254	40	1.1112
.4625	30	.4462	.8949	.4986	2.006	1.117	2.241	30	1.1083
.4654	40	.4488	.8936	.5022	1.991	1.119	2.228	20	1.1054
.4683	50	.4514	.8923	.5059	1.977	1.121	2.215	10	1.1025
.4712	**27° 00′**	.4540	.8910	.5095	1.963	1.122	2.203	**63° 00′**	1.0996
.4741	10	.4566	.8897	.5132	1.949	1.124	2.190	50	1.0966
.4771	20	.4592	.8884	.5169	1.935	1.126	2.178	40	1.0937
.4800	30	.4617	.8870	.5206	1.921	1.127	2.166	30	1.0908
.4829	40	.4643	.8857	.5243	1.907	1.129	2.154	20	1.0879
.4858	50	.4669	.8843	.5280	1.894	1.131	2.142	10	1.0850
.4887	**28° 00′**	.4695	.8829	.5317	1.881	1.133	2.130	**62° 00′**	1.0821
		$\cos t$	$\sin t$	$\cot t$	$\tan t$	$\csc t$	$\sec t$	t degrees	t

TABLE A.3 Trigonometric Functions of an Angle (*Continued*)

t	t degrees	sin t	cos t	tan t	cot t	sec t	csc t		
.4887	28° 00′	.4695	.8829	.5317	1.881	1.133	2.130	62° 00′	1.0821
.4916	10	.4720	.8816	.5354	1.868	1.134	2.118	50	1.0792
.4945	20	.4746	.8802	.5392	1.855	1.136	2.107	40	1.0763
.4974	30	.4772	.8788	.5430	1.842	1.138	2.096	30	1.0734
.5003	40	.4797	.8774	.5467	1.829	1.140	2.085	20	1.0705
.5032	50	.4823	.8760	.5505	1.816	1.142	2.074	10	1.0676
.5061	29° 00′	.4848	.8746	.5543	1.804	1.143	2.063	61° 00′	1.0647
.5091	10	.4874	.8732	.5581	1.792	1.145	2.052	50	1.0617
.5120	20	.4899	.8718	.5619	1.780	1.147	2.041	40	1.0588
.5149	30	.4924	.8704	.5658	1.767	1.149	2.031	30	1.0559
.5178	40	.4950	.8689	.5696	1.756	1.151	2.020	20	1.0530
.5207	50	.4975	.8675	.5735	1.744	1.153	2.010	10	1.0501
.5236	30° 00′	.5000	.8660	.5774	1.732	1.155	2.000	60° 00′	1.0472
.5265	10	.5025	.8646	.5812	1.720	1.157	1.990	50	1.0443
.5294	20	.5050	.8631	.5851	1.709	1.159	1.980	40	1.0414
.5323	30	.5075	.8616	.5890	1.698	1.161	1.970	30	1.0385
.5352	40	.5100	.8601	.5930	1.686	1.163	1.961	20	1.0356
.5381	50	.5125	.8587	.5969	1.675	1.165	1.951	10	1.0327
.5411	31° 00′	.5150	.8572	.6009	1.664	1.167	1.942	59° 00′	1.0297
.5440	10	.5175	.8557	.6048	1.653	1.169	1.932	50	1.0268
.5469	20	.5200	.8542	.6088	1.643	1.171	1.923	40	1.0239
.5498	30	.5225	.8526	.6128	1.632	1.173	1.914	30	1.0210
.5527	40	.5250	.8511	.6168	1.621	1.175	1.905	20	1.0181
.5556	50	.5275	.8496	.6208	1.611	1.177	1.896	10	1.0152
.5585	32° 00′	.5299	.8480	.6249	1.600	1.179	1.887	58° 00′	1.0123
.5614	10	.5324	.8465	.6289	1.590	1.181	1.878	50	1.0094
.5643	20	.5348	.8450	.6330	1.580	1.184	1.870	40	1.0065
.5672	30	.5373	.8434	.6371	1.570	1.186	1.861	30	1.0036
.5701	40	.5398	.8418	.6412	1.560	1.188	1.853	20	1.0007
.5730	50	.5422	.8403	.6453	1.550	1.190	1.844	10	.9977
.5760	33° 00′	.5446	.8387	.6494	1.540	1.192	1.836	57° 00′	.9948
.5789	10	.5471	.8371	.6536	1.530	1.195	1.828	50	.9919
.5818	20	.5495	.8355	.6577	1.520	1.197	1.820	40	.9890
.5847	30	.5519	.8339	.6619	1.511	1.199	1.812	30	.9861
.5876	40	.5544	.8323	.6661	1.501	1.202	1.804	20	.9832
.5905	50	.5568	.8307	.6703	1.492	1.204	1.796	10	.9803
.5934	34° 00′	.5592	.8290	.6745	1.483	1.206	1.788	56° 00′	.9774
.5963	10	.5616	.8274	.6787	1.473	1.209	1.781	50	.9745
.5992	20	.5640	.8258	.6830	1.464	1.211	1.773	40	.9716
.6021	30	.5664	.8241	.6873	1.455	1.213	1.766	30	.9687
.6050	40	.5688	.8225	.6916	1.446	1.216	1.758	20	.9657
.6080	50	.5712	.8208	.6959	1.437	1.218	1.751	10	.9628
.6109	35° 00′	.5736	.8192	.7002	1.428	1.221	1.743	55° 00′	.9599
		cos t	sin t	cot t	tan t	csc t	sec t	t degrees	t

TABLE A.3 Trigonometric Functions of an Angle (*Continued*)

t	t degrees	$\sin t$	$\cos t$	$\tan t$	$\cot t$	$\sec t$	$\csc t$		
.6109	**35° 00′**	.5736	.8192	.7002	1.428	1.221	1.743	**55° 00′**	.9599
.6138	10	.5760	.8175	.7046	1.419	1.223	1.736	50	.9570
.6167	20	.5783	.8158	.7089	1.411	1.226	1.729	40	.9541
.6196	30	.5807	.8141	.7133	1.402	1.228	1.722	30	.9512
.6225	40	.5831	.8124	.7177	1.393	1.231	1.715	20	.9483
.6254	50	.5854	.8107	.7221	1.385	1.233	1.708	10	.9454
.6283	**36° 00′**	.5878	.8090	.7265	1.376	1.236	1.701	**54° 00′**	.9425
.6312	10	.5901	.8073	.7310	1.368	1.239	1.695	50	.9396
.6341	20	.5925	.8056	.7355	1.360	1.241	1.688	40	.9367
.6370	30	.5948	.8039	.7400	1.351	1.244	1.681	30	.9338
.6400	40	.5972	.8021	.7445	1.343	1.247	1.675	20	.9308
.6429	50	.5995	.8004	.7490	1.335	1.249	1.668	10	.9279
.6458	**37° 00′**	.6018	.7986	.7536	1.327	1.252	1.662	**53° 00′**	.9250
.6487	10	.6041	.7969	.7581	1.319	1.255	1.655	50	.9221
.6516	20	.6065	.7951	.7627	1.311	1.258	1.649	40	.9192
.6545	30	.6088	.7934	.7673	1.303	1.260	1.643	30	.9163
.6574	40	.6111	.7916	.7720	1.295	1.263	1.636	20	.9134
.6603	50	.6134	.7898	.7766	1.288	1.266	1.630	10	.9105
.6632	**38° 00′**	.6157	.7880	.7813	1.280	1.269	1.624	**52° 00′**	.9076
.6661	10	.6180	.7862	.7860	1.272	1.272	1.618	50	.9047
.6690	20	.6202	.7844	.7907	1.265	1.275	1.612	40	.0918
.6720	30	.6225	.7826	.7954	1.257	1.278	1.606	30	.8988
.6749	40	.6248	.7808	.8002	1.250	1.281	1.601	20	.8959
.6778	50	.6271	.7790	.8050	1.242	1.284	1.595	10	.8930
.6807	**39° 00′**	.6293	.7771	.8098	1.235	1.287	1.589	**51° 00′**	.8901
.6836	10	.6316	.7753	.8146	1.228	1.290	1.583	50	.8872
.6865	20	.6338	.7735	.8195	1.220	1.293	1.578	40	.8843
.6894	30	.6361	.7716	.8243	1.213	1.296	1.572	30	.8814
.6923	40	.6383	.7698	.8292	1.206	1.299	1.567	20	8785
.6952	50	.6406	.7679	.8342	1.199	1.302	1.561	10	.8756
.6981	**40° 00′**	.6428	.7660	.8391	1.192	1.305	1.556	**50° 00′**	.8727
.7010	10	.6450	.7642	.8441	1.185	1.309	1.550	50	.8698
.7039	20	.6472	.7623	.8491	1.178	1.312	1.545	40	.8668
.7069	30	.6494	.7604	.8541	1.171	1.315	1.540	30	.8639
.7098	40	.6517	.7585	.8591	1.164	1.318	1.535	20	.8610
.7127	50	.6539	.7566	.8642	1.157	1.322	1.529	10	.8581
.7156	**41° 00′**	.6561	.7547	.8693	1.150	1.325	1.524	**49° 00′**	.8552
.7185	10	.6583	.7528	.8744	1.144	1.328	1.519	50	.8523
.7214	20	.6604	.7509	.8796	1.137	1.332	1.514	40	.8494
.7243	30	.6626	.7490	.8847	1.130	1.335	1.509	30	.8465
.7272	40	.6648	.7470	.8899	1.124	1.339	1.504	20	.8436
.7301	50	.6670	.7451	.8952	1.117	1.342	1.499	10	.8407
.7330	**42° 00′**	.6691	.7431	.9004	1.111	1.346	1.494	**48° 00′**	.8378
		$\cos t$	$\sin t$	$\cot t$	$\tan t$	$\csc t$	$\sec t$	t degrees	t

TABLE A.3 Trigonometric Functions of an Angle (*Continued*)

t	t degrees	$\sin t$	$\cos t$	$\tan t$	$\cot t$	$\sec t$	$\csc t$		
.7330	**42° 00′**	.6691	.7431	.9004	1.111	1.346	1.494	**48° 00′**	.8378
.7359	10	.6713	.7412	.9057	1.104	1.349	1.490	50	.8348
.7389	20	.6734	.7392	.9110	1.098	1.353	1.485	40	.8319
.7418	30	.6756	.7373	.9163	1.091	1.356	1.480	30	.8290
.7447	40	.6777	.7353	.9217	1.085	1.360	1.476	20	.8261
.7476	50	.6799	.7333	.9271	1.079	1.364	1.471	10	.8232
.7505	**43° 00′**	.6820	.7314	.9325	1.072	1.367	1.466	**47° 00′**	.8203
.7534	10	.6841	.7294	.9380	1.066	1.371	1.462	50	.8174
.7563	20	.6862	.7274	.9435	1.060	1.375	1.457	40	.8145
.7592	30	.6884	.7254	.9490	1.054	1.379	1.453	30	.8116
.7621	40	.6905	.7234	.9545	1.048	1.382	1.448	20	.8087
.7650	50	.6926	.7214	.9601	1.042	1.386	1.444	10	.8058
.7679	**44° 00′**	.6947	.7193	.9657	1.036	1.390	1.440	**46° 00′**	.8029
.7709	10	.6967	.7173	.9713	1.030	1.394	1.435	50	.7999
.7738	20	.6988	.7153	.9770	1.024	1.398	1.431	40	.7970
.7767	30	.7009	.7133	.9827	1.018	1.402	1.427	30	.7941
.7796	40	.7030	.7112	.9884	1.012	1.406	1.423	20	.7912
.7825	50	.7050	.7092	.9942	1.006	1.410	1.418	10	.7883
.7854	**45° 00′**	.7071	.7071	1.0000	1.0000	1.414	1.414	**45° 00′**	.7854
		$\cos t$	$\sin t$	$\cot t$	$\tan t$	$\csc t$	$\sec t$	t degrees	t

TABLE A.4 Trigonometric Functions of a Real Number

t	sin t	cos t	tan t	cot t	sec t	csc t
.00	.0000	1.0000	.0000		1.000	
.01	.0100	1.0000	.0100	99.997	1.000	100.00
.02	.0200	.9998	.0200	49.993	1.000	50.00
.03	.0300	.9996	.0300	33.323	1.000	33.34
.04	.0400	.9992	.0400	24.987	1.001	25.01
.05	.0500	.9988	.0500	19.983	1.001	20.01
.06	.0600	.9982	.0601	16.647	1.002	16.68
.07	.0699	.9976	.0701	14.262	1.002	14.30
.08	.0799	.9968	.0802	12.473	1.003	12.51
.09	.0899	.9960	.0902	11.081	1.004	11.13
.10	.0998	.9950	.1003	9.967	1.005	10.02
.11	.1098	.9940	.1104	9.054	1.006	9.109
.12	.1197	.9928	.1206	8.293	1.007	8.353
.13	.1296	.9916	.1307	7.649	1.009	7.714
.14	.1395	.9902	.1409	7.096	1.010	7.166
.15	.1494	.9888	.1511	6.617	1.011	6.692
.16	.1593	.9872	.1614	6.197	1.013	6.277
.17	.1692	.9856	.1717	5.826	1.015	5.911
.18	.1790	.9838	.1820	5.495	1.016	5.586
.19	.1889	.9820	.1923	5.200	1.018	5.295
.20	.1987	.9801	.2027	4.933	1.020	5.033
.21	.2085	.9780	.2131	4.692	1.022	4.797
.22	.2182	.9759	.2236	4.472	1.025	4.582
.23	.2280	.9737	.2341	4.271	1.027	4.386
.24	.2377	.9713	.2447	4.086	1.030	4.207
.25	.2474	.9689	.2553	3.916	1.032	4.042
.26	.2571	.9664	.2660	3.759	1.035	3.890
.27	.2667	.9638	.2768	3.613	1.038	3.749
.28	.2764	.9611	.2876	3.478	1.041	3.619
.29	.2860	.9582	.2984	3.351	1.044	3.497
.30	.2955	.9553	.3093	3.233	1.047	3.384
.31	.3051	.9523	.3203	3.122	1.050	3.278
.32	.3146	.9492	.3314	3.018	1.053	3.179
.33	.3240	.9460	.3425	2.920	1.057	3.086
.34	.3335	.9428	.3537	2.827	1.061	2.999
.35	.3429	.9394	.3650	2.740	1.065	2.916
.36	.3523	.9359	.3764	2.657	1.068	2.839
.37	.3616	.9323	.3879	2.578	1.073	2.765
.38	.3709	.9287	.3994	2.504	1.077	2.696
.39	.3802	.9249	.4111	2.433	1.081	2.630

From Koshy: *College Algebra and Trigonometry with Applications*, 1986. McGraw-Hill. Reprinted with permission.

TABLE A.4 Trigonometric Functions of a Real Number (*Continued*)

t	$\sin t$	$\cos t$	$\tan t$	$\cot t$	$\sec t$	$\csc t$
.40	.3894	.9211	.4228	2.365	1.086	2.568
.41	.3986	.9171	.4346	2.301	1.090	2.509
.42	.4078	.9131	.4466	2.239	1.095	2.452
.43	.4169	.9090	.4586	2.180	1.100	2.399
.44	.4259	.9048	.4708	2.124	1.105	2.348
.45	.4350	.9004	.4831	2.070	1.111	2.299
.46	.4439	.8961	.4954	2.018	1.116	2.253
.47	.4529	.8916	.5080	1.969	1.122	2.208
.48	.4618	.8870	.5206	1.921	1.127	2.166
.49	.4706	.8823	.5334	1.875	1.133	2.125
.50	.4794	.8776	.5463	1.830	1.139	2.086
.51	.4882	.8727	.5594	1.788	1.146	2.048
.52	.4969	.8678	.5726	1.747	1.152	2.013
.53	.5055	.8628	.5859	1.707	1.159	1.978
.54	.5141	.8577	.5994	1.668	1.166	1.945
.55	.5227	.8525	.6131	1.631	1.173	1.913
.56	.5312	.8473	.6269	1.595	1.180	1.883
.57	.5396	.8419	.6410	1.560	1.188	1.853
.58	.5480	.8365	.6552	1.526	1.196	1.825
.59	.5564	.8309	.6696	1.494	1.203	1.797
.60	.5646	.8253	.6841	1.462	1.212	1.771
.61	.5729	.8196	.6989	1.431	1.220	1.746
.62	.5810	.8139	.7139	1.401	1.229	1.721
.63	.5891	.8080	.7291	1.372	1.238	1.697
.64	.5972	.8021	.7445	1.343	1.247	1.674
.65	.6052	.7961	.7602	1.315	1.256	1.652
.66	.6131	.7900	.7761	1.288	1.266	1.631
.67	.6210	.7838	.7923	1.262	1.276	1.610
.68	.6288	.7776	.8087	1.237	1.286	1.590
.69	.6365	.7712	.8253	1.212	1.297	1.571
.70	.6442	.7648	.8423	1.187	1.307	1.552
.71	.6518	.7584	.8595	1.163	1.319	1.534
.72	.6594	.7518	.8771	1.140	1.330	1.517
.73	.6669	.7452	.8949	1.117	1.342	1.500
.74	.6743	.7385	.9131	1.095	1.354	1.483
.75	.6816	.7317	.9316	1.073	1.367	1.467
.76	.6889	.7248	.9505	1.052	1.380	1.452
.77	.6961	.7179	.9697	1.031	1.393	1.437
.78	.7033	.7109	.9893	1.011	1.407	1.422
.79	.7104	.7038	1.009	.9908	1.421	1.408

TABLE A.4	Trigonometric Functions of a Real Number (*Continued*)					
t	sin t	cos t	tan t	cot t	sec t	csc t
.80	.7174	.6967	1.030	.9712	1.435	1.394
.81	.7243	.6895	1.050	.9520	1.450	1.381
.82	.7311	.6822	1.072	.9331	1.466	1.368
.83	.7379	.6749	1.093	.9146	1.482	1.355
.84	.7446	.6675	1.116	.8964	1.498	1.343
.85	.7513	.6600	1.138	.8785	1.515	1.331
.86	.7578	.6524	1.162	.8609	1.533	1.320
.87	.7643	.6448	1.185	.8437	1.551	1.308
.88	.7707	.6372	1.210	.8267	1.569	1.297
.89	.7771	.6294	1.235	.8100	1.589	1.287
.90	.7833	.6216	1.260	.7936	1.609	1.277
.91	.7895	.6137	1.286	.7774	1.629	1.267
.92	.7956	.6058	1.313	.7615	1.651	1.257
.93	.8016	.5978	1.341	.7458	1.673	1.247
.94	.8076	.5898	1.369	.7303	1.696	1.238
.95	.8134	.5817	1.398	.7151	1.719	1.229
.96	.8192	.5735	1.428	.7001	1.744	1.221
.97	.8249	.5653	1.459	.6853	1.769	1.212
.98	.8305	.5570	1.491	.6707	1.795	1.204
.99	.8360	.5487	1.524	.6563	1.823	1.196
1.00	.8415	.5403	1.557	.6421	1.851	1.188
1.01	.8468	.5319	1.592	.6281	1.880	1.181
1.02	.8521	.5234	1.628	.6142	1.911	1.174
1.03	.8573	.5148	1.665	.6005	1.942	1.166
1.04	.8624	.5062	1.704	.5870	1.975	1.160
1.05	.8674	.4976	1.743	.5736	2.010	1.153
1.06	.8724	.4889	1.784	.5604	2.046	1.146
1.07	.8772	.4801	1.827	.5473	2.083	1.140
1.08	.8820	.4713	1.871	.5344	2.122	1.134
1.09	.8866	.4625	1.917	.5216	2.162	1.128
1.10	.8912	.4536	1.965	.5090	2.205	1.122
1.11	.8957	.4447	2.014	.4964	2.249	1.116
1.12	.9001	.4357	2.066	.4840	2.295	1.111
1.13	.9044	.4267	2.120	.4718	2.344	1.106
1.14	.9086	.4176	2.176	.4596	2.395	1.101
1.15	.9128	.4085	2.234	.4475	2.448	1.096
1.16	.9168	.3993	2.296	.4356	2.504	1.091
1.17	.9208	.3902	2.360	.4237	2.563	1.086
1.18	.9246	.3809	2.427	.4120	2.625	1.082
1.19	.9284	.3717	2.498	.4003	2.691	1.077

TABLE A.4 Trigonometric Functions of a Real Number (*Continued*)

t	sin t	cos t	tan t	cot t	sec t	csc t
1.20	.9320	.3624	2.572	.3888	2.760	1.073
1.21	.9356	.3530	2.650	.3773	2.833	1.069
1.22	.9391	.3436	2.733	.3659	2.910	1.065
1.23	.9425	.3342	2.820	.3546	2.992	1.061
1.24	.9458	.3248	2.912	.3434	3.079	1.057
1.25	.9490	.3153	3.010	.3323	3.171	1.054
1.26	.9521	.3058	3.113	.3212	3.270	1.050
1.27	.9551	.2963	3.224	.3102	3.375	1.047
1.28	.9580	.2867	3.341	.2993	3.488	1.044
1.29	.9608	.2771	3.467	.2884	3.609	1.041
1.30	.9636	.2675	3.602	.2776	3.738	1.038
1.31	.9662	.2579	3.747	.2669	3.878	1.035
1.32	.9687	.2482	3.903	.2562	4.029	1.032
1.33	.9711	.2385	4.072	.2456	4.193	1.030
1.34	.9735	.2288	4.256	.2350	4.372	1.027
1.35	.9757	.2190	4.455	.2245	4.566	1.025
1.36	.9779	.2092	4.673	.2140	4.779	1.023
1.37	.9799	.1994	4.913	.2035	5.014	1.021
1.38	.9819	.1896	5.177	.1931	5.273	1.018
1.39	.9837	.1798	5.471	.1828	5.561	1.017
1.40	.9854	.1700	5.798	.1725	5.883	1.015
1.41	.9871	.1601	6.165	.1622	6.246	1.013
1.42	.9887	.1502	6.581	.1519	6.657	1.011
1.43	.9901	.1403	7.055	.1417	7.126	1.010
1.44	.9915	.1304	7.602	.1315	7.667	1.009
1.45	.9927	.1205	8.238	.1214	8.299	1.007
1.46	.9939	.1106	8.989	.1113	9.044	1.006
1.47	.9949	.1006	9.887	.1011	9.938	1.005
1.48	.9959	.0907	10.983	.0910	11.029	1.004
1.49	.9967	.0807	12.350	.0810	12.390	1.003
1.50	.9975	.0707	14.101	.0709	14.137	1.003
1.51	.9982	.0608	16.428	.0609	16.458	1.002
1.52	.9987	.0508	19.670	.0508	19.695	1.001
1.53	.9992	.0408	24.498	.0408	24.519	1.001
1.54	.9995	.0308	32.461	.0308	32.476	1.000
1.55	.9998	.0208	48.078	.0208	48.089	1.000
1.56	.9999	.0108	92.620	.0108	92.626	1.000
1.57	1.0000	.0008	1255.8	.0008	1255.8	1.000